CW00705934

Microwave Technology

Microwave Technology

Dennis Roddy

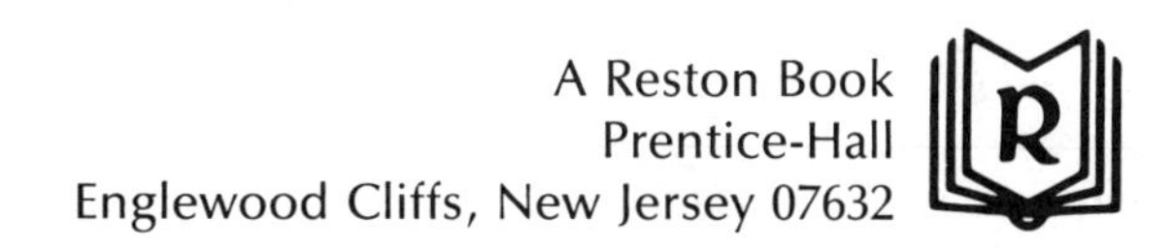
A Reston Book
Prentice-Hall
Englewood Cliffs, New Jersey 07632

Library of Congress Cataloging-in-Publication Data

Roddy, D.
Microwave technology.

Includes bibliographies and index.
1. Microwave devices. I. Title.
TK7874.R62 1986 621.381'3 85-14367
ISBN 0-8359-4390-9

A Reston Book
Published by Prentice-Hall
A Division of Simon & Schuster, Inc.
Englewood Cliffs, NJ 07632

10 9 8 7 6 5 4 3 2 1

PRINTED IN THE UNITED STATES OF AMERICA

Contents

five: Substrate Lines and Components, 141

six: Microwave Ferrite Components, 177

seven: Microwave Diodes and Diode Circuits, 219

Preface

This book is intended for use in the final year of technology programs which offer a course in microwaves. It is also hoped that it will prove useful to technical personnel working in the field of microwaves.

In order to provide a comprehensive text at the technology level, emphasis is given to applications rather than theory. Sufficient theoretical background is included where this appears to be helpful, but the mathematics required are limited to algebra and trigonometry. The book also covers the principles of operation, and constructional features of a wide range of microwave hardware.

I am much indebted to the many companies and organizations who provided the detailed information necessary in a book of this kind. Although the sources are acknowledged in the text, I would like to thank in particular the following companies for providing a great deal of additional technical information, photographs, and figures for use in the book: The Hewlett Packard Company; N.V. Philips' Gloeilampenfabrieken; Marconi Instruments. Also the following professional institutions which kindly gave permission for use of material from their publications: The Institute of Electrical and Electronic Engineers (IEEE); The Institution of Electrical Engineers (IEE); and The Institution of Electronic & Radio Engineers (IERE).

Much of the writing of the book was carried out while I was on sabbatical leave at the Communications Research Centre, Department of Communications, Ottawa. I am very happy to acknowledge here the generous assistance provided by the Department, both financially and by way of access to resources. I would like to thank in particular the library staff for all their help. My thanks also go to Lakehead University for providing me with sabbatical leave.

The book originated under the auspices of the Reston Publishing Company. It has been a pleasure working with the company, and in particular I would like to thank Ben Wentzell, Greg Michael, Norma Karlin, and Alice Cave for their courteous help and cooperation.

D. Roddy.
Lakehead University,
Thunder Bay.

one
Transmission Lines

1.1 INTRODUCTION

Voltage and current *waves* are widely encountered in electrical and electronics technology, and for example the oscilloscope provides a familiar means of displaying voltage–time waveforms. Although these are referred to as waveforms, they show one aspect only of a wave, the amplitude–time variation. Equally important is the fact that a wave quantity has an amplitude–distance variation. The *wavelength* is a characteristic of the amplitude–distance curve (this will be defined shortly) and when the wavelength is very much larger than the physical dimensions of the circuit, the amplitude–distance variation can be ignored. When, however, the wavelength is small compared to circuit dimensions, the amplitude–distance variation must be taken into account. In fact, much of microwave technology utilizes properties resulting from the amplitude–distance variation.

An illustration of the effect of the amplitude–distance variation may be seen by considering the simple circuit of Fig. 1.1, in which a wire is shown connecting a voltage generator to a load through two ammeters. Assuming that the ammeters are accurate, then at low frequencies or long wavelengths, much longer than the wire length l, the reading I_1 on the first ammeter should be the same as the reading I_2 on the second ammeter, as expected from Kirchhoff's current law. At microwave wavelengths the situation changes drastically, and in general the reading I_1 will not be equal to I_2. This is not because Kirchhoff's law is invalidated but because other factors, negligible at low frequencies or long wavelengths, must now be taken into account.

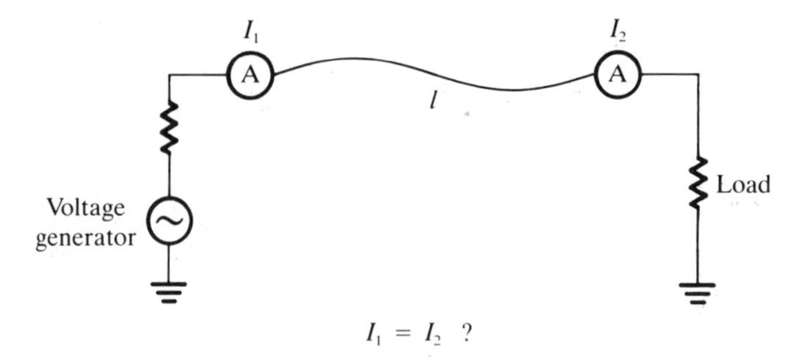

FIGURE 1.1. Current flow in a wire connecting a voltage generator to a load.

Another good illustration of the special character of microwave circuits is the rectangular waveguide sketched in Fig. 1.2. The rectangular waveguide consists of a rectangular metal tube capable of carrying microwave energy. Unlike the more familiar coaxial cable, there is no center conductor associated with the waveguide, and the normal concepts of "go" and "return" wires cannot be used. It would be difficult, for example, to find a current in the waveguide which gave a clear measure of the power flow (although as will be shown later, currents do flow in the waveguide walls in rather complicated patterns). Thus the usual circuit variables of voltage and current are of limited use with microwaves and it becomes necessary to introduce new variables. An important principle underlying the introduction of any variable in a technological context is that it should be relatively easy to measure the variable, or obtain it from measurements, and this will be seen to be the case with the new variables as they are introduced later.

Much of microwave theory is based on the assumption of sinusoidal signals (the term *sinusoidal* is meant to include cosinusoidal), as is the case

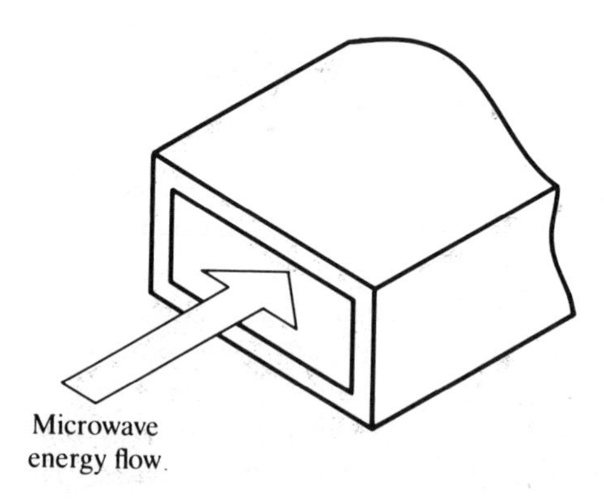

FIGURE 1.2. Rectangular waveguide.

with much of alternating-current (ac) circuit theory. Real signals are nonsinusoidal, some more so than others, and care must be taken to recognize this fact. It has an important bearing on results derived from the amplitude–distance and amplitude–time variations based on sine waves, and careful interpretation of these results is required when extending them to nonsinusoidal situations. At this point, however, it should be noted that sinusoidal variations will be assumed unless otherwise stated, so that the familiar ideas of peak values, root-mean-square (rms) values, impedance, and admittance may be used.

Microwave theory and applications of transmission lines are described in this chapter. A more general treatment of transmission lines will usually be found in most books on communications; see e.g., Glazier & Lamont (1958); Roddy & Coolen (1984).

1.2 THE UNIFORM TRANSMISSION LINE

A uniform transmission line is one for which the *primary line constants* do not change with distance along the line. The primary line constants are the *series resistance* R, *series inductance* L, *shunt conductance* G, and *shunt capacitance* C, all determined in "per unit length" values. The series resistance is the resistance of the wire conductors, and the series inductance the self-inductance of the wires. These elements are classed as series elements because the line current flows through them. The shunt capacitance is the capacitance of the dielectric between the conductors, and the shunt conductance the conductance of this dielectric. Where the dielectric is mostly air, the shunt conductance will be very small and will be that of the conductor supports. These elements are classed as shunt elements because the line voltage appears across them.

Because these line constants are uniformly distributed over the line, it is not possible to show them exactly by means of a circuit diagram. An approximate circuit diagram representation is shown in Fig. 1.3. One section represents a small length δl of line, and the difficulty becomes immediately apparent since one is forced to show the series and shunt ele-

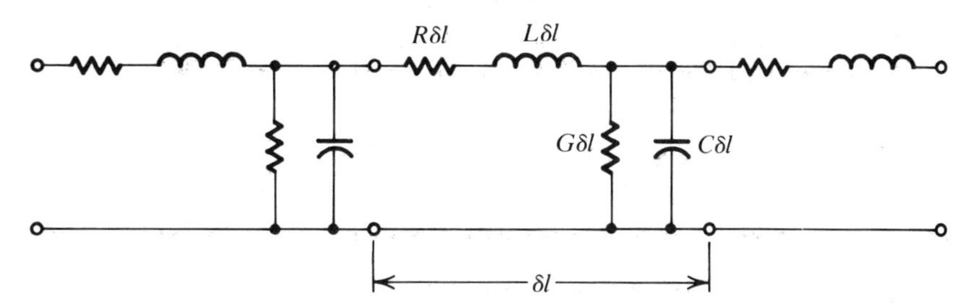

FIGURE 1.3. Circuit representation of a uniform line.

ments as being separate. Also note that the series elements for the return wire are lumped together with those for the go wire. By making the length infinitesimally small, the section becomes, in the limit, an accurate representation of the real line. In terms of basic units the primary line constants are:

R	ohms/meter (Ω/m)
L	henrys/meter (H/m)
G	siemens/meter (S/m)
C	farads/meter (F/m)

For a sinusoidal signal of angular frequency ω radians per second (rad/s), the series impedance and shunt admittance, both per unit length, become

$$Z = R + j\omega L \qquad \Omega/\text{m} \tag{1.1}$$

$$Y = G + j\omega C \qquad \text{S/m} \tag{1.2}$$

The primary line constants are constant in that they do not vary with voltage or current. They are, however, frequency dependent to some extent. The most important frequency dependence is that of the resistive components. The series resistance increases with frequency as a result of skin effect, in which the high-frequency magnetic fields internal to the wire conductors force the currents to crowd toward the outer surface (hence the name *skin effect*). This reduces the effective cross-sectional area available for conduction and therefore increases the resistance. The power absorbed by the dielectric and dissipated as heat also increases with frequency, and this can be represented by a decrease in shunt resistance, or equivalently an increase in shunt conductance. Recall that the line voltage appears across the conductance, and the power loss is given by V^2G, where V is the rms voltage. It is this increase with frequency in both the power loss components R and G that makes transmission lines unsuitable for carrying large amounts of power at microwaves and forces the use of alternatives such as the rectangular waveguide shown in Fig. 1.2.

The two basic types of line are the *balanced two-wire line*, shown in Fig. 1.4(a), and the *unbalanced coaxial line or cable*, shown in Fig. 1.4(b). The two-wire line is balanced in the sense that both conductors are symmetrically arranged with respect to their surroundings; for example, the capacitance to ground would be the same for each. The disadvantage of the two-wire line is that radiation takes place from the wires, and this increases with frequency. The radiation loss can be reduced to negligible levels by the use of coaxial cable since the outer conductor confines the electromagnetic energy to the space between the conductors. The coaxial

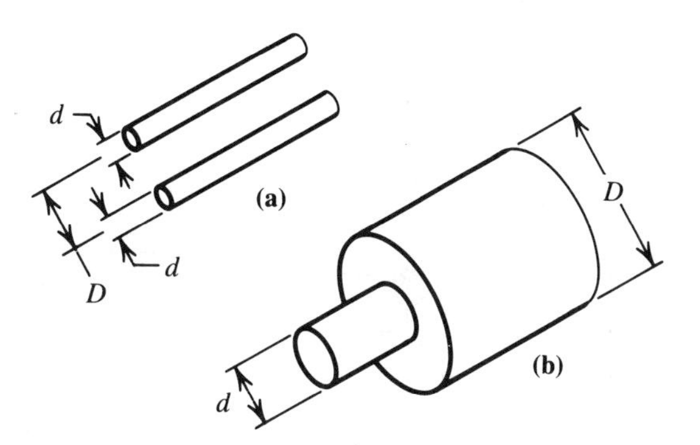

FIGURE 1.4. (a) Important dimensions for a two-wire transmission line; (b) corresponding dimensions for a coaxial cable.

cable is an unbalanced arrangement, as the capacitance to ground is that for the outer conductor alone. There are certain advantages to operating in a balanced mode (some of these are discussed in Chapter 10) and special transformers are required to connect coaxial cables to balanced loads.

The inductance per unit length and the capacitance per unit length are determined respectively by the magnetic permeability μ and the electric permittivity ε of the dielectric, and by the line geometry. For the moment it may be assumed that permeability and permittivity are constants that are known (these are discussed in detail later). The geometric factor is a logarithmic function of the line dimensions d and D, these dimensions being shown for the two basic line arrangements in Fig. 1.4. The equations for L and C for both cases are:

Two-wire line:

$$L = \frac{\mu}{\pi} \ln \frac{2D}{d} \qquad \text{H/m} \tag{1.3}$$

$$C = \frac{\pi\varepsilon}{\ln (2D/d)} \qquad \text{F/m} \tag{1.4}$$

Coaxial line:

$$L = \frac{\mu}{2\pi} \ln \frac{D}{d} \qquad \text{H/m} \tag{1.5}$$

$$C = \frac{2\pi\varepsilon}{\ln (D/d)} \qquad \text{F/m} \tag{1.6}$$

Although these equations involve certain approximations, they can be considered highly accurate for all practical purposes.

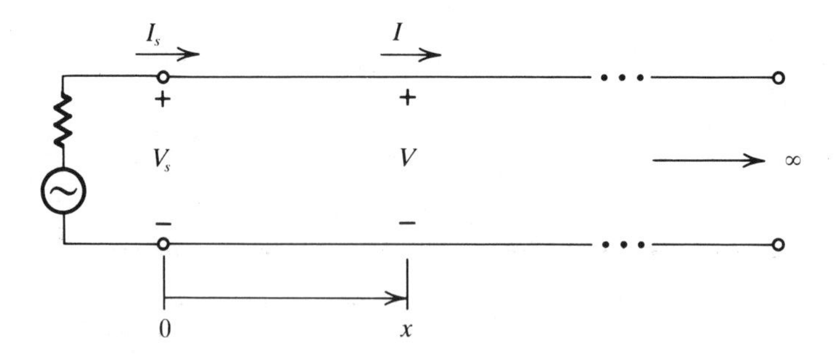

FIGURE 1.5. Section of an infinitely long line.

1.3 THE INFINITELY LONG LINE

It may seem strange to introduce the nonrealizable dimension of infinite length into the discussion. The reason for doing so is that it allows the effects of the line parameters on the signal to be studied independently of the load parameters; with an infinite line, the signal never reaches the load! It is very important to note that this is not the same thing as removing the load. A line of finite length, open-circuited at the end, behaves quite differently from an infinitely long line. There are two important line characteristics which can be defined for the infinite line and which can also be used to characterize a finite line, these being the *propagation coefficient* and the *characteristic impedance*.

Consider a section of an infinitely long, uniform line carrying a sinusoidal current. This is shown in Fig. 1.5. The input voltage and current are denoted by V_s and I_s, and the corresponding values at distance x from the source are indicated by V and I. These are all rms values. The voltage and current are found to vary exponentially with distance x according to the relationships

$$V = V_s e^{-\gamma x} \tag{1.7}$$

$$I = I_s e^{-\gamma x} \tag{1.8}$$

The constant γ is known as the *propagation coefficient.*

1.4 THE PROPAGATION COEFFICIENT

As shown in Section 1.3, the propagation coefficient determines the exponential rate of change with distance for a sinusoidal signal being propagated along a uniform, infinite line. The propagation coefficient is in general a complex quantity which may be written as

$$\gamma = \alpha + j\beta \tag{1.9}$$

where α is known as the *attenuation coefficient*, and β as the *phase-shift coefficient*. The current equation (1.8) may be rewritten with γ in its complex form as

$$\begin{aligned} I &= I_s e^{-(\alpha + j\beta)x} \\ &= I_s e^{-\alpha x} e^{-j\beta x} \\ &= |I| \angle{-\beta x} \end{aligned} \tag{1.10}$$

In the final form, Eq. (1.10), the current at position x is written in terms of its magnitude $|I|$ and phase angle $-\beta x$, the *reference current* being I_s; that is, the input current I_s has zero phase angle.

From Eq. (1.10) the magnitude is seen to be given by

$$|I| = I_s e^{-\alpha x} \tag{1.11}$$

The attenuation of the current in *nepers* is defined as

$$N = -\ln \frac{|I|}{I_s} \tag{1.12}$$

Hence

$$N = \alpha x \qquad \text{nepers} \tag{1.13}$$

The word *neper* is derived from the Naperian logarithm. It will be seen that by including a minus sign in the defining equation, N comes out as a positive quantity (α is always positive). Equation (1.13) also shows why α is referred to as the attenuation coefficient; it is the constant of proportionality between attenuation in nepers and distance x in meters.

The neper is a useful unit in theoretical work, but in practical work the decibel (dB) is more commonly used. The relationship between the decibel value D and the neper value N of a given current ratio is

$$D = 8.686N \tag{1.14}$$

Hence, for the transmission line, where $N = \alpha x$,

$$D = 8.686\alpha x \qquad \text{dB} \tag{1.15}$$

Therefore, if the attenuation coefficient is *defined* in decibels per unit length, say [α], then

$$[\alpha] = 8.686\alpha \qquad \text{dB/m} \tag{1.16}$$

The phase difference between the current at position x and the reference current I_s is, from Eq. (1.10),

$$\theta = -\beta x \qquad \text{radians} \tag{1.17}$$

This shows that the phase difference is proportional to the distance x, the constant of proportionality being $-\beta$; hence the name "phase-shift coefficient" for β. The variation of current amplitude and phase as functions of distance are sketched in Fig. 1.6.

On the matter of units, it should be noted that the factor $e^{-\gamma x}$ has to be dimensionless and therefore γ must have reciprocal units of distance, or m^{-1}. This also applies to both α and β. It is also true that α can be expressed in both nepers per meter and decibels per meter, but because both nepers and decibels are logarithmic units they are dimensionless in terms of fundamental units. It therefore follows that m^{-1} is the basic unit for α, but in practice it is very useful to express α in terms of either nepers/m or dB/m, depending on which is being used. Similarly, from Eq. (1.17) β can be expressed in terms of radians per meter. Radians are dimensionless in terms of fundamental units but again, it is more useful practically to use the units of radians per meter.

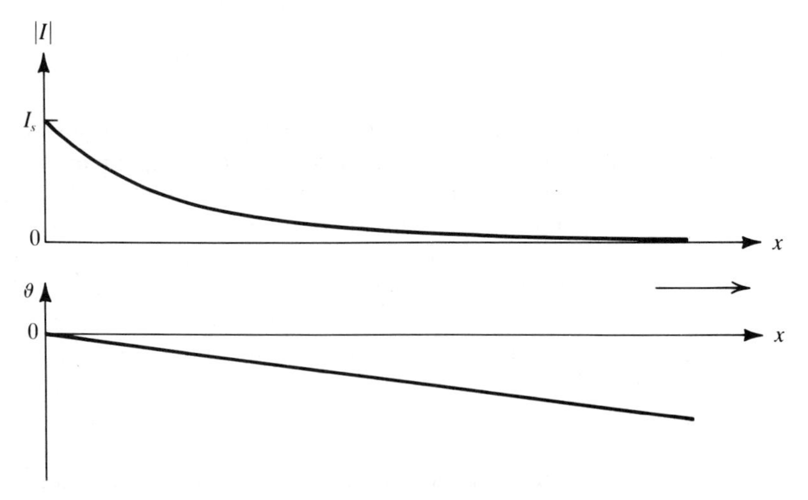

FIGURE 1.6. Variations of current amplitude and phase with distance as given by Eqs. (1.11) and (1.17).

EXAMPLE 1.1

(a) For a very long transmission line, which may be assumed infinite, the propagation coefficient is known to be $0.0006 + j\pi/10\ \text{m}^{-1}$. For an input current of $1.0\ \angle 0$ mA, determine the current at a distance of 1 km along the line.

(b) Calculate the attenuation in decibels per foot.

SOLUTION

(a) $\alpha x = 0.0006 \times 10^3 = 0.6$
From Eq. (1.11),

$$|I| = (1\ \text{mA}) \times e^{-0.6} = \mathbf{0.549\ mA}$$

Also,

$$\beta x = \frac{\pi}{10} \times 10^3 = \mathbf{100\pi\ rad}$$

From Eq. (1.10),

$$I = |I|\ \angle -\beta x = 0.549\ \angle -100\pi\ \text{rad} \qquad \text{mA}$$

(b) From Eq. (1.16),

$$[\alpha] = 8.686 \times 0.0006$$
$$= 0.0052\ \text{dB/m}$$

Since 1 ft = 0.305 m, the attenuation in dB/ft is

$$0.0052 \times 0.305 = \mathbf{0.00159\ dB/ft}$$

1.5 THE CHARACTERISTIC IMPEDANCE

The characteristic impedance is defined for an infinitely long line as the ratio of voltage V to current I at any position x:

$$Z_0 = \frac{V}{I} \tag{1.18}$$

For the uniform line, the current and voltage equations (1.7) and (1.8) can be used, the exponential factor will cancel out, and therefore the characteristic impedance is seen to be independent of position x. This means that

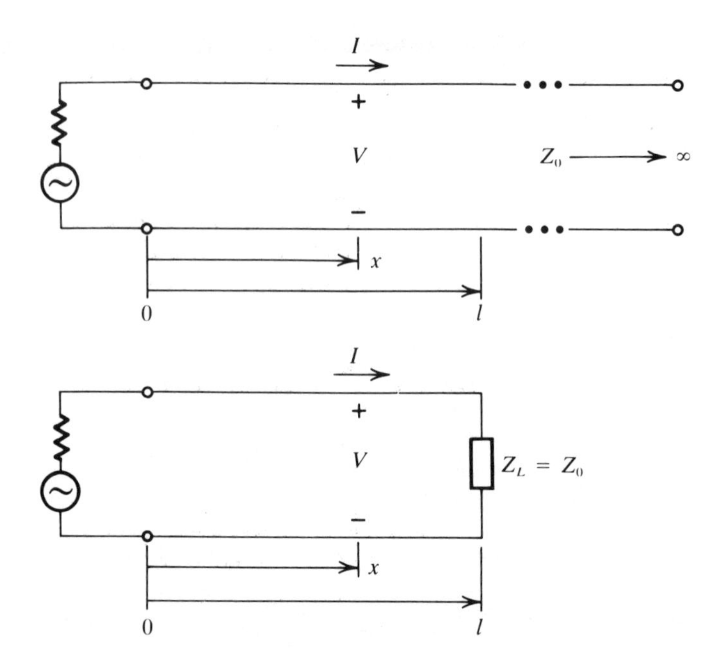

FIGURE 1.7. Showing how an infinitely long line may be replaced by a finite line correctly terminated. $Z_0 = V/I$ and is independent of position x.

at any position on the infinite line, the ratio of V to I is a constant equal to Z_0. This being so, the hypothetical infinite line may be cut at some position l, as shown in Fig. 1.7, and an impedance Z_L, having exactly the value Z_0 substituted for the infinite line remaining to the right of the cut. This substitution will not affect voltage and current values to the left of the cut, so that the propagation coefficient and the characteristic impedance remain the same as they were for the infinite line. In other words, a finite length of line terminated correctly has the same characteristics as the infinite line. Terminating a line in this manner is known as *reflectionless matching*, or sometimes simply as *matching*. Matching is of great practical significance and will be discussed in considerable detail later.

1.6 WAVE PROPAGATION ALONG A TRANSMISSION LINE

Consider a finite section of line, matched correctly to simulate an infinite line. Let a sinusoidal current be applied to the input, the instantaneous value being given by

$$i_s = I_{s\max} \sin \omega t \tag{1.19}$$

Equation (1.11) shows that the amplitude is altered by a factor $e^{-\alpha x}$, and Eq. (1.17), that the phase is altered by amount $-\beta x$. The equation for instantaneous current at any position x measured along the line from the source is therefore given by

$$i = I_{s\max} e^{-\alpha x} \sin(\omega t - \beta x) \tag{1.20}$$

This is the equation for an exponentially damped wave. It will be seen that the variables time and distance are both involved. The amplitude–time curve (x a constant) is shown in Fig. 1.8(a), and the amplitude–distance curve (t a constant) in Fig. 1.8(b).

The *periodic time* T is the time interval covered by one complete cycle, as shown in Fig. 1.8(a). For convenience the periodic time is shown between two zero crossings, but any two periodic points could have been chosen. Since one complete cycle represents a phase change of 2π radians, the angular velocity, which is radians per unit time, is

$$\omega = \frac{2\pi}{T} \tag{1.21}$$

Also, for a frequency of f hertz (or f cycles per second), the periodic time, which is the time for one cycle, is therefore

$$T = \frac{1}{f} \quad \text{seconds} \tag{1.22}$$

This means that an alternative expression for ω is

$$\omega = 2\pi f \tag{1.23}$$

The *wavelength* λ is the distance interval between alternate zero crossings, as shown in Fig. 1.8(b). Because of the exponential damping factor the wave does not repeat in the sense of identical cycles. However, the amplitude–distance curve can be thought of as the product of two functions, and since time is considered to be constant, $t = T/2$ may be chosen to illustrate this. With $t = T/2$, the amplitude–distance function is

$$e^{-\alpha x} \sin \beta x \tag{1.24}$$

The sin βx term is periodic, and the wavelength λ is seen to be the distance covered by one cycle, and therefore it performs the same function for the distance variable as the periodic time T does for the time variable. In particular, since one complete cycle represents a phase change of 2π ra-

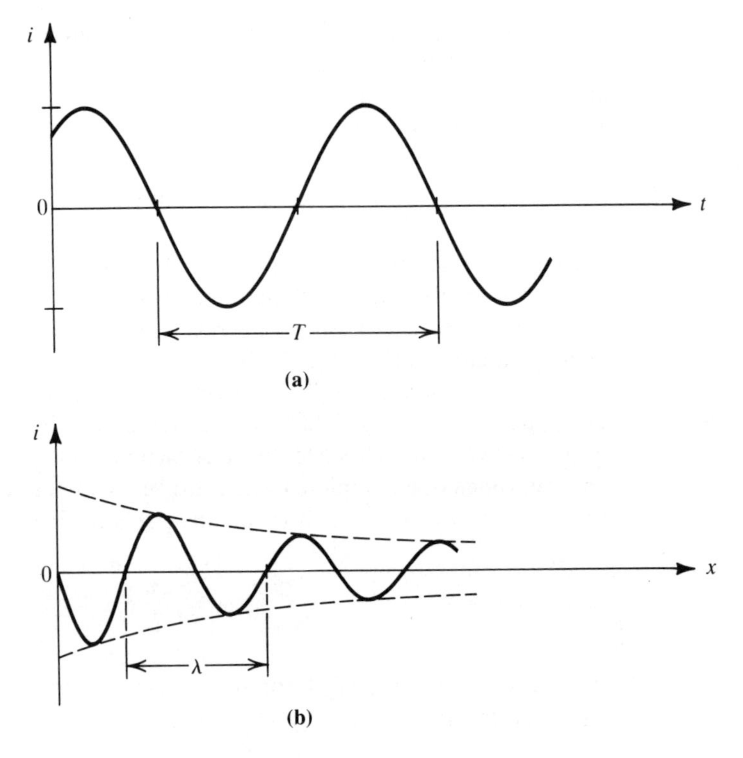

FIGURE 1.8. (a) Amplitude–time curve and (b) amplitude–distance curve for a wave on a transmission line.

dians, the phase change per unit length, which is the *phase-change coefficient*, is

$$\beta = \frac{2\pi}{\lambda} \tag{1.25}$$

Because the current exhibits a phase lag, given by $-\beta x$, a current amplitude observed at some one position appears at a later time to have moved down the line, and of course the amplitude will have been diminished by the factor $e^{-\alpha x}$. The speed at which any chosen amplitude value appears to travel is termed the *phase velocity*, denoted here by v_p. For the amplitude to be affected only by the exponential term $e^{-\alpha x}$, the sinusoidal term in Eq. (1.20) must remain constant, which means in turn that the angle $(\omega t - \beta x)$ must remain constant. This total angle is termed the *phase*. The phase velocity can be determined from the condition that

$$\omega t - \beta x = \text{constant} \tag{1.26}$$

Differentiating through with respect to time gives

$$\omega - \beta \frac{dx}{dt} = 0 \qquad \text{or} \qquad \frac{dx}{dt} = \frac{\omega}{\beta} \tag{1.27}$$

But dx/dt is the phase velocity for the conditions imposed, and therefore

$$v_p = \frac{\omega}{\beta} \tag{1.28}$$

It is also useful to obtain v_p in terms of frequency f and wavelength λ. Substituting $\omega = 2\pi f$ and $\beta = 2\pi/\lambda$ gives

$$f\lambda = v_p \tag{1.29}$$

EXAMPLE 1.2

A transmission line has a solid dielectric which reduces the phase velocity to 2×10^8 m/s from its air value. Calculate the wavelength of a 500-MHz sine wave on this line.

SOLUTION

$$\lambda = \frac{v_p}{f}$$

$$= \frac{2 \times 10^8}{5 \times 10^8}\ \text{m}$$

$$= \mathbf{0.4\ m} \quad \text{or} \quad \mathbf{40\ cm}$$

1.7 RELATIONSHIP BETWEEN PRIMARY AND SECONDARY LINE CONSTANTS

The primary line constants are the series resistance and inductance per unit length, R and L, and the shunt conductance and capacitance per unit length G and C, as described in Section 1.2. The secondary line constants are the propagation coefficient γ and the characteristic impedance Z_0, as described in Sections 1.4 and 1.5. The secondary constants depend on the primary constants, and expressions are available showing the relationships. The derivation of these results is too lengthy to include here, but the results are important and will be used.

The characteristic impedance is given by

$$Z_0 = \sqrt{\frac{R + j\omega L}{G + j\omega C}} \tag{1.30}$$

Now at high frequencies such that $\omega L >> R$ and $\omega C >> G$, the expression for Z_0 is seen to reduce to

$$Z_0 \simeq \sqrt{\frac{L}{C}} \tag{1.31}$$

This expression is sufficiently accurate at microwaves for the "approximately equal to" sign to be replaced by "equal to," which shall be used from now on. A remarkable feature of Eq. (1.30) is that it shows that the characteristic impedance is *purely resistive*, although it is determined solely by the reactive elements L and C.

The characteristic impedance being purely resistive does *not* mean that the line is a resistance, and this distinction must be carefully observed. Suppose, for example, that the characteristic impedance of a certain line is 50 Ω, purely resistive, and the line is correctly matched to simulate an infinite line. This means that the load impedance on the line must be a 50-Ω resistance. A sinusoidal current of rms value 1 ampere (A) flowing in the line will dissipate zero power in Z_0 but will dissipate a power I^2R_L or 50 watts (W) in the load.

In Section 1.2 it is shown that both L and C are functions of the line dimensions D and d, as well as permittivity and permeability. The characteristic impedance for the two-wire line may be found in terms of the primary line constants by substituting Eqs. (1.3) and (1.4) for L and C in Eq. (1.31) to get

$$Z_0 = \frac{1}{\pi}\sqrt{\frac{\mu}{\varepsilon}}\ln\frac{2D}{d} \tag{1.32}$$

A similar result may be obtained for the coaxial cable by substituting from Eqs. (1.5) and (1.6):

$$Z_0 = \frac{1}{2\pi}\sqrt{\frac{\mu}{\varepsilon}}\ln\frac{D}{d} \tag{1.33}$$

For practical dielectrics the permeability is equal to the free-space value, which is $\mu = \mu_0 = 4\pi \times 10^{-7}$ H/m. The permittivity is given by $\varepsilon = \varepsilon_r\varepsilon_0$, where $\varepsilon_0 = 8.854 \times 10^{-12}$ F/m is the free-space value of permittivity and ε_r is the *relative permittivity* or *dielectric constant*. Substituting these values into Eqs. (1.32) and (1.33) gives, for the characteristic impedance in each case:

Two-wire line:

$$Z_0 = \frac{120}{\sqrt{\varepsilon_r}}\ln\frac{2D}{d} \quad \Omega \tag{1.34}$$

Coaxial:

$$Z_0 = \frac{60}{\sqrt{\varepsilon_r}} \ln \frac{D}{d} \quad \Omega \tag{1.35}$$

For a given dielectric constant the characteristic impedance is seen to depend on the D/d ratio (see Fig. 1.4). Practical values for dielectric constant range between 1 and 5, and practical limitations on the ratio of D/d limit Z_0 to the range of about 40 to 150 Ω for coaxial cables and 150 to 600 Ω for the two-wire line.

In terms of the primary line constants, the propagation coefficient is given by

$$\gamma = \sqrt{(R + j\omega L)(G + j\omega C)} \tag{1.36}$$

To be useful this expression must be expanded into its real and imaginary parts to give α, the attenuation coefficent, and β, the phase-shift coefficient. Again, in the interests of clarity the mathematical manipulations will be omitted, but the results that will be used are

$$\alpha \simeq \frac{R}{2Z_0} + \frac{GZ_0}{2} \tag{1.37}$$

$$\beta \simeq \omega\sqrt{LC} \tag{1.38}$$

In Eq. (1.37) the characteristic impedance has the value $\sqrt{L/C}$. Again, in using these expressions the "approximately equal to" signs will be replaced by "equal to."

The attenuation coefficient is seen to depend on both loss components, R and G. As mentioned previously, both of these increase with frequency, and therefore the attenuation coefficient also increases with frequency. In practice, values of attenuation coefficient will be specified for a number of different frequencies, and this is illustrated in Fig. 1.9.

From Eq. (1.38) the phase-shift coefficient is seen to be proportional to frequency, and combining Eqs. (1.28) and (1.38) gives for the phase velocity v_p:

$$v_p = \frac{1}{\sqrt{LC}} \tag{1.39}$$

This shows that the phase velocity decreases as either L or C is increased, which is important to know. Also, from the results of Section 1.2, $LC = \mu\varepsilon$ and therefore

$$v_p = \frac{1}{\sqrt{\mu\varepsilon}} \tag{1.40}$$

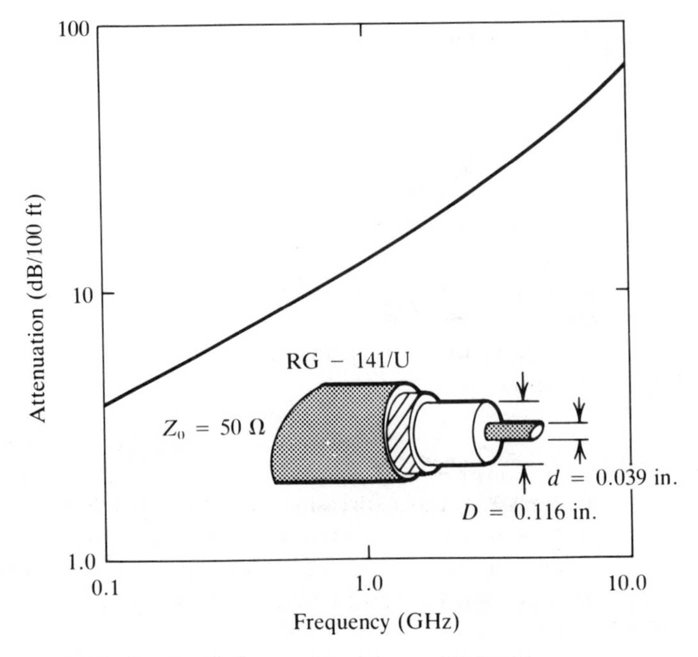

FIGURE 1.9. Details for coaxial cable type RG-141/U.

For an air dielectric, the phase velocity is the same as that for light, normally denoted by the symbol c. For all practical purposes, $c = 3 \times 10^8$ m/s, and

$$c = \frac{1}{\sqrt{\mu_0 \varepsilon_0}} = 3 \times 10^8 \text{ m/s} \tag{1.41}$$

For other dielectrics it may be assumed that $\mu = \mu_0$ and $\varepsilon = \varepsilon_r \varepsilon_0$, as described for characteristic impedance. Combining these values with Eqs. (1.40) and (1.41) gives

$$v_p = \frac{c}{\sqrt{\varepsilon_r}} \tag{1.42}$$

This result has great practical significance, and the factor $1/\sqrt{\varepsilon_r}$ is sometimes referred to as the *velocity factor*.

EXAMPLE 1.3

The primary constants for a coaxial cable, at a frequency of 50 MHz, are $L = 234$ nH/m, $C = 93.5$ pF/m, $R = 0.6$ Ω/m, and $G \approx 0$. Determine (a) the attenuation coefficient, (b) the phase-shift coefficient, (c) the phase velocity, and (d) the relative permittivity.

SOLUTION

$$\omega L = 2\pi \times 50 \times 10^6 \times 234 \times 10^{-9} = 73.5\ \Omega/\text{m}.$$

Hence, since $\omega L >> R$ and $\omega C >> G$, the approximation for Z_0, Eq. (1.31), can be used, giving

$$Z_0 = \sqrt{\frac{234 \times 10^{-9}}{93.5 \times 10^{-12}}} = 50\ \Omega$$

(a) From Eq. (1.37),

$$\alpha \simeq \frac{0.6}{2 \times 50} = \mathbf{0.006\ N/m}$$

(b) From Eq. (1.38),

$$\begin{aligned}\beta &= 2\pi \times 50 \times 10^6 \sqrt{234 \times 10^{-9} \times 93.5 \times 10^{-12}} \\ &= \mathbf{0.468\pi\ rad/m}\end{aligned}$$

(c) From Eq. (1.39),

$$\begin{aligned}v_p &= \frac{1}{\sqrt{234 \times 10^{-9} \times 93.5 \times 10^{-12}}} \\ &= \mathbf{2.14 \times 10^8\ m/s}\end{aligned}$$

(d) From Eq. (1.42),

$$\varepsilon_r = \left(\frac{3 \times 10^8}{2.14 \times 10^8}\right)^2 = \mathbf{1.97}$$

1.8 COAXIAL LINES FOR MICROWAVES

Open wire lines are unsuitable for microwave transmission because of the high radiation losses that occur at high frequencies. Coaxial lines are used over the lower part of the microwave spectrum. For power transmission, they may be used up to frequencies of about 3 GHz, and for small-signal applications, up to about 18 GHz. The ratio D/d is important in determining a number of operating limits for coaxial cable. For cables in which the conductance G of the dielectric is negligible, it can be shown that the ratio $D/d = 3.6$ gives minimum attenuation coefficient. When this value is substituted in Eq. (1.35) for the characteristic impedance, then

$$Z_0 \text{ for minimum } \alpha = \frac{60}{\sqrt{\varepsilon_r}} \ln 3.6$$

$$= \frac{76.86}{\sqrt{\varepsilon_r}} \tag{1.43}$$

Thus, with an air dielectric, $\varepsilon_r = 1$ and $Z_0 = 76.86\ \Omega$. With a solid polyethylene dielectric, $\varepsilon_r = 2.3$ and $Z_0 = 50.67\ \Omega$. The attenuation coefficient does not vary sharply with changes in D/d, so in practice, 50-Ω lines are very common.

Dielectric breakdown is another factor that must be taken into account. The maximum field strength E_{max} in volts per meter will be determined by the cable dielectric breakdown. This will be a fixed value, and therefore assuming that it is known, the peak voltage that the line can handle can be shown to be given by

$$V_p = E_{max} d \ln \frac{D}{d} \tag{1.44}$$

But from Eq. (1.35), $\ln (D/d) = Z_0\sqrt{\varepsilon_r}/60$ and therefore

$$V_p = \frac{E_{max} d Z_0 \sqrt{\varepsilon_r}}{60} \tag{1.45}$$

Let a sinusoidal voltage of peak value V_p be applied to the input of a *matched* coaxial line. Since the line is matched, the input impedance is Z_0 (purely resistive at microwaves), and therefore the average input power is

$$P = \left(\frac{V_p}{\sqrt{2}}\right)^2 \frac{1}{Z_0}$$

Substituting for V_p from Eq. (1.45) and simplifying gives

$$P = \frac{E_{max}^2 d^2 Z_0^2 \varepsilon_r}{7200} \quad \text{watts} \tag{1.46}$$

The factor d^2 in this equation is also dependent on Z_0 and from Eq. (1.35) is found to be $d^2 = D^2 e^{-aZ_0}$, where $a = \sqrt{\varepsilon_r}/30$. Substituting this in Eq. (1.46) gives

$$P = Ke^{-aZ_0} Z_0 \tag{1.47}$$

where K is a constant. The value of Z_0 that maximizes P may be found by the usual calculus method and is

$$Z_0 \text{ for maximum } P = \frac{30}{\sqrt{\varepsilon_r}} \quad \text{ohms} \tag{1.48}$$

This, in fact, corresponds to a D/d ratio of 1.65. Thus it will be seen that the value of Z_0 for minimum attenuation is not the same as that for maximum power input capacity. Fortunately, these parameters do not vary sharply with D/d and compromise between conflicting requirements is possible.

Another constraint on the value of Z_0 is that it should equal the load impedance for matching. A common load is an antenna, and in Chapter 10 it is shown that the radiation resistance of a half-wave dipole is 73 Ω. When the dipole is loaded with parasitic elements, its impedance drops. However, the range of values encountered is similar to the range determined by minimum attenuation and maximum power handling, which is fortunate.

The power dissipated in a cable is an important practical consideration. Again considering a matched cable with a sinusoidal input, let V_s and I_s be the rms input values of voltage and current. Since Z_0 is purely resistive, the load must also be purely resistive, and the load voltage and current will be in phase, and are given by

$$V_L = V_s e^{-\alpha l} \tag{1.49}$$

$$I_L = I_s e^{-\alpha l} \tag{1.50}$$

The load power is therefore

$$\begin{aligned} P_L = V_L I_L &= V_s I_s e^{-2\alpha l} \\ &= P_s e^{-2\alpha l} \end{aligned} \tag{1.51}$$

Here P_s is the input power and l is the length of line between source and load. The power loss in the line is

$$\begin{aligned} P_{\text{loss}} &= P_s - P_L \\ &= P_s(1 - e^{-2\alpha l}) \end{aligned} \tag{1.52}$$

Normally, the attenuation is very small, such that $\alpha l \ll 1$, and therefore the factor $(1 - e^{-2\alpha l})$ can be approximated by $2\alpha l$. Thus the power loss becomes

$$P_{\text{loss}} = P_s \cdot 2\alpha l \tag{1.53}$$

Recalling that α increases with increase in frequency, then for a specified P_{loss}, the power-handling capacity P_s must be reduced as frequency is increased. Manufacturers usually supply a derating curve to take this into account, as shown in the specifications sheets (Fig. 1.10). P_{loss} will be fixed by the permissible temperature rise in the cable, since P_{loss} is converted

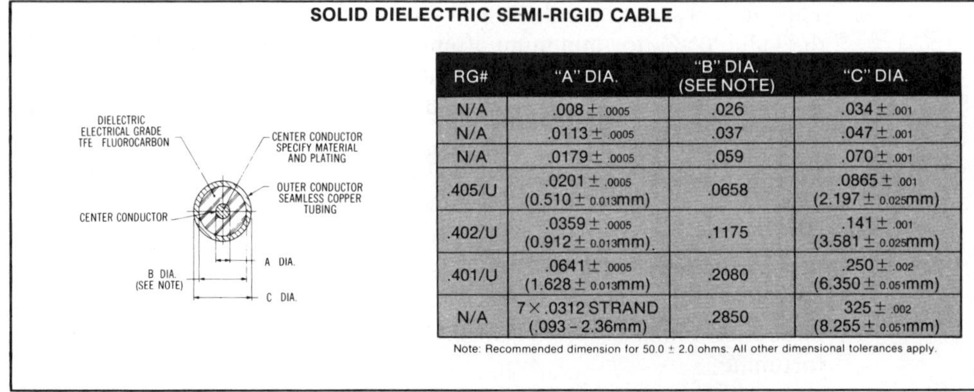

RG#	"A" DIA.	"B" DIA. (SEE NOTE)	"C" DIA.
N/A	.008 ± .0005	.026	.034 ± .001
N/A	.0113 ± .0005	.037	.047 ± .001
N/A	.0179 ± .0005	.059	.070 ± .001
.405/U	.0201 ± .0005 (0.510 ± 0.013mm)	.0658	.0865 ± .001 (2.197 ± 0.025mm)
.402/U	.0359 ± .0005 (0.912 ± 0.013mm).	.1175	.141 ± .001 (3.581 ± 0.025mm)
.401/U	.0641 ± .0005 (1.628 ± 0.013mm)	.2080	.250 ± .002 (6.350 ± 0.051mm)
N/A	7 × .0312 STRAND (.093 - 2.36mm)	.2850	325 ± .002 (8.255 ± 0.051mm)

Note: Recommended dimension for 50.0 ± 2.0 ohms. All other dimensional tolerances apply.

MICRO-POROUS DIELECTRIC SEMI-RIGID CABLE

MICRO-POROUS TEFLON DIELECTRIC
CENTER CONDUCTOR
D MATERIAL
A DIA.
"B" DIA. (SEE NOTE)
C DIA.

RG#	"A" DIA.	"B" DIA. (SEE NOTE)	"C" DIA.	"D" MATERIAL
N/A	.043 ± .001	.116	.141 ± .001	COPPER
N/A	.074 ± .001 (1.88 ± .025mm)	.210	.250 ± .002 (6.35 ± .178mm)	ALUMINUM
N/A	.114 ± .001 (2.90 ± .025mm)	.327	.375 ± .007 (9.53 ± .18 mm)	ALUMINUM
385/U	.153 ± .001 (3.89 ± .025mm)	.430	.495 ± .007 (12.57 ± .18 mm)	ALUMINUM

Note: Recommended dimension for 50.0 ± 2.0 ohms. All other dimensional tolerances apply.

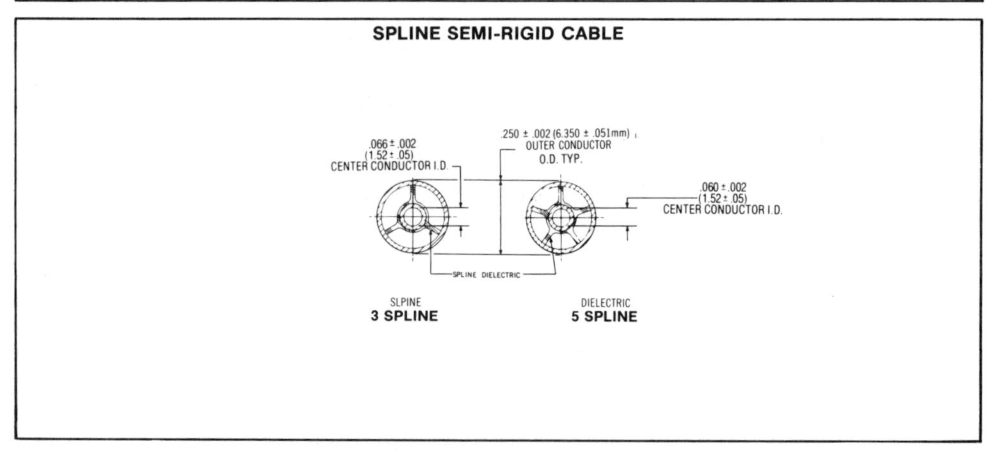

FIGURE 1.10. (a) Details of semirigid coaxial cables for microwaves; (b) attenuation–frequency and power rating–frequency curves for the cables of part (a); (c) details of flexible coaxial cables for microwaves. (*Courtesy of Microwave Connector Division, Omni-Spectra, Inc., Waltham, Mass.*)

TYPICAL ATTENUATION CHARACTERISTICS OF SEMI-RIGID COAXIAL CABLES

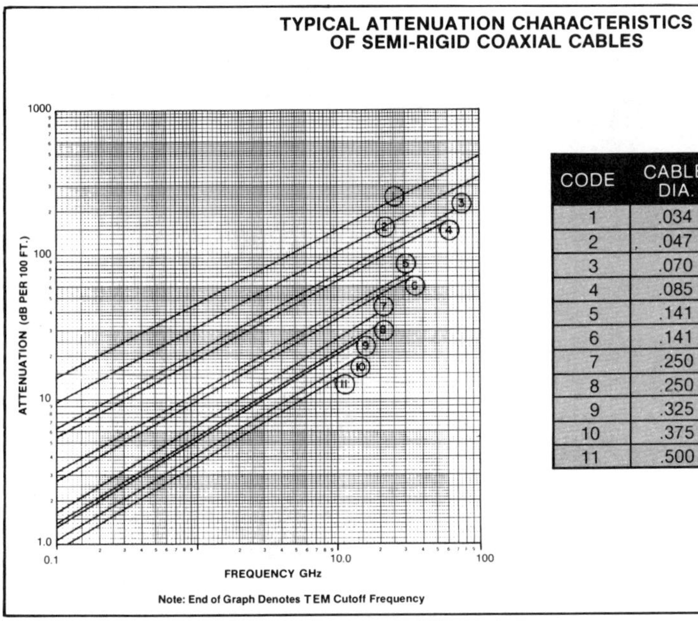

CODE	CABLE DIA.	TYPE
1	.034	SEMI-RIGID
2	.047	SEMI-RIGID
3	.070	SEMI-RIGID
4	.085	SEMI-RIGID
5	.141	SEMI-RIGID
6	.141	MICRO-POROUS
7	.250	SEMI-RIGID
8	.250	MICRO-POROUS
9	.325	SEMI-RIGID
10	.375	MICRO-POROUS
11	.500	MICRO-POROUS

TYPICAL POWER RATING CHARACTERISTICS OF SEMI-RIGID COAXIAL CABLES

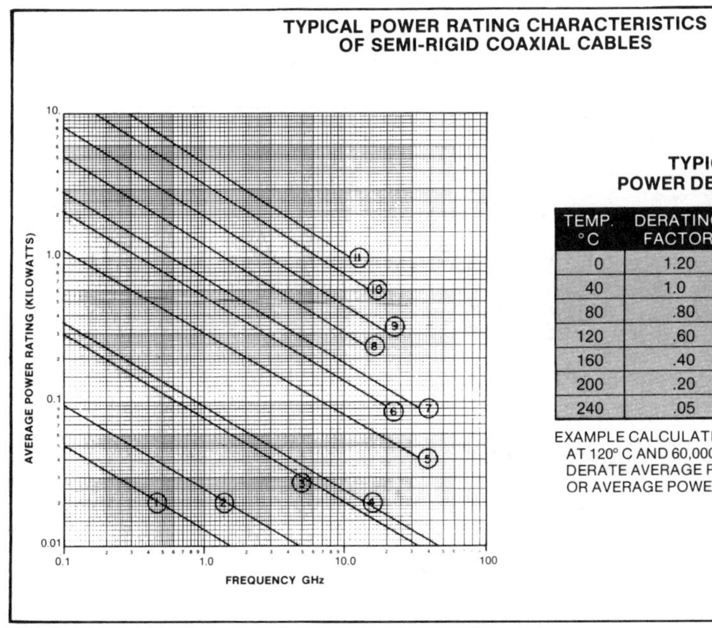

TYPICAL AVERAGE POWER DERATING FACTORS

TEMP. °C	DERATING FACTOR	ALTITUDE × 1000 FT.	DERATING FACTOR
0	1.20	0	1.0
40	1.0	20	.80
80	.80	30	.70
120	.60	40	.60
160	.40	50	.50
200	.20	60	.40
240	.05	70	.30

EXAMPLE CALCULATION:
AT 120° C AND 60,000 FT:
DERATE AVERAGE POWER BY .60 x .40
OR AVERAGE POWER x .24

FIGURE 1.10. Continued.

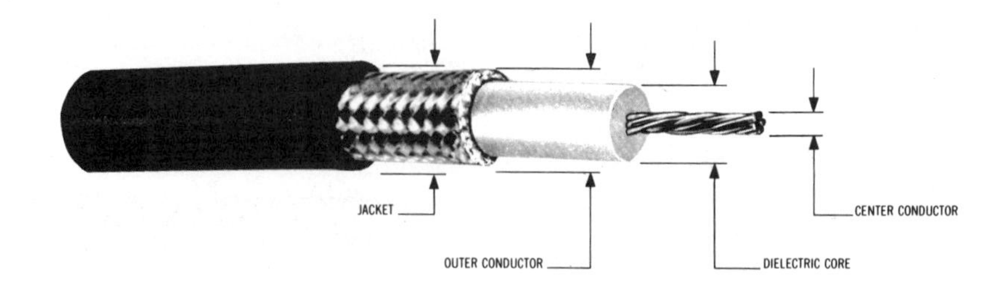

FLEXIBLE CABLES

	CABLE TYPE	RG-55/U	RG-58/U	RG-141/U	RG-142/U	RG-174/U	RG-178/U	RG-179/U	RG-180/U
	IMPEDANCE (OHMS)	53.5	50	50	50	50	50	75	95
DIA	JACKET	.216 MAX	.195± $_{004}$	.190± $_{005}$	.195± $_{005}$	.100± $_{005}$	.075 MAX	.100± $_{005}$	.145 MAX
DIA	OUTER CONDUCTOR	.176 MAX	.150 MAX	.146 MAX	.171 MAX	.088 MAX	.054 MAX	.084 MAX	.124 MAX
DIA	DIELECTRIC CORE	.116± $_{005}$	.116± $_{004}$	.116± $_{005}$	.116± $_{005}$	.060± $_{003}$	.034± $_{002}$	.063± $_{003}$	.102± $_{003}$
DIA	CENTER CONDUCTOR	.032 NOM	.0375 MAX	.039± $_{001}$	.039± $_{001}$	.020 NOM	.012 NOM	.012 NOM	.012 NOM

	CABLE TYPE	RG-187/U	RG-188/U	RG-195/U	RG-196/U	RG-214/U	RG-223/U	RG-303/U	RG-316/U
	IMPEDANCE (OHMS)	75	50	95	50	50	50	50	50
DIA	JACKET	.110 MAX	.110 MAX	.155 MAX	.080 MAX	.425± $_{007}$	.216 MAX	.170± $_{005}$	.102 MAX
DIA	OUTER CONDUCTOR	.084 MAX	.081 MAX	.124 MAX	.054 MAX	.360 MAX	.176 MAX	.146 MAX	.081 MAX
DIA	DIELECTRIC CORE	.060± $^{006}_{000}$	.060± $_{003}$	.102± $_{003}$	.034± $_{002}$	.285± $_{007}$	.116± $_{004}$	.116± $_{005}$	.060± $_{003}$
DIA	CENTER CONDUCTOR	.012 NOM	.020 NOM	.012 NOM	.012 NOM	.089± $_{001}$	.035± $_{001}$	.039± $_{001}$	.020 NOM

RG CABLE	ATTENUATION RATING TYP dB/100 FT. AT FREQUENCY, GHz							POWER RATING MAXIMUM WATTS AT FREQUENCY, GHz							
	.1	.2	.4	1	3	5	10	.1	.2	.4	1	3	5	10
55	4.8	7.0	10.0	16.5	30.5	46.0	>100.0	480	320	215	120	60	40	—
58	4.6	6.9	10.5	17.5	37.5	60.0	>100.0	300	200	135	80	40	20	—
141	3.9	5.6	8.0	13.5	27.0	39.0	70.0	1,700	1,200	830	450	220	140	65
142	3.9	5.6	8.0	13.5	27.0	39.0	70.0	1,800	1,300	800	530	265	175	100
174	8.9	12.0	17.5	30.0	64.0	99.0	>100.0	110	80	60	35	15	10	—
178	14.0	19.0	28.0	46.0	85.0	>100.0	>100.0	240	180	120	75	40	—	—
179	10.0	12.5	16.0	24.0	44.0	65.0	>100.0	480	420	320	190	100	73	—
180	5.7	7.6	10.8	17.0	35.0	50.0	88.0	800	570	400	240	130	90	50
187	10.0	12.5	16.0	24.0	44.0	69.0	>100.0	480	420	320	190	100	73	—
188	11.4	14.2	16.7	31.0	60.0	82.0	>100.0	400	325	275	150	80	55	—
195	5.7	7.6	10.8	17.0	35.0	50.0	88.0	800	570	400	240	130	90	50
196	14.0	19.0	28.0	46.0	85.0	>100.0	>100.0	240	180	120	75	40	—	—
214	2.3	3.3	5.0	8.8	18.0	27.0	45.0	780	550	360	200	100	65	40
223	4.8	7.0	10.0	16.5	30.5	46.0	>100.0	480	320	215	120	60	40	—
303	3.9	5.6	8.0	13.5	27.0	39.0	70.0	1,800	1,300	900	530	265	175	100
316	11.4	14.2	16.7	31.0	60.0	82.0	>100.0	400	325	275	150	80	55	—

* This information listed is typical. Consult your cable source for further details.

FIGURE 1.10. Continued.

into heat. Note that the power rating for a cable is the maximum permissible input P_s.

It is also found that voltage breakdown of dielectrics decreases with decrease in atmospheric pressure, for a wide range of pressures, and Fig. 1.10 shows the derating factor to be used for the cables listed. This is of obvious importance in airborne and spacecraft applications.

Although the power loss may be calculated using Eq. (1.53), a more direct method is often possible using the catalog specifications for the cable, as the following example illustrates.

EXAMPLE 1.4

For the RG-214 cable listed in Fig. 1.10(c) determine the power loss at 0.1 GHz, when the input power is the maximum rated, for a length of 5 m.

SOLUTION

Method 1: From Fig. 1.10(c) the attenuation is specified as 2.3 dB/100 ft at 0.1 GHz. Hence

$$\alpha = \frac{2.3}{100} \times \frac{1}{0.305} \times \frac{1}{8.686}$$

$$= 0.00868 \text{ N/m}$$

From Eq. (1.53),

$$P_{\text{loss}} = 780 \times 0.00868 \times 2 \times 5$$

$$= \mathbf{67.72\ W}$$

Method 2: 5 m in ft:

$$5 \times \frac{1}{0.305} = 16.39 \text{ ft}$$

Therefore, attenuation is (2.3)/100 × 16.39 = 0.377 dB. But $-10 \log (P_{\text{out}}/P_s) = 0.377$ and solving for P_{out} gives $P_{\text{out}} = 715$ W. Thus

$$P_{\text{loss}} = 780 - 715 = \mathbf{64.85\ W}$$

The first solution differs from this because of the approximation used in arriving at Eq. (1.53) from Eq. (1.52).

Care must also be taken when using Eq. (1.53) to keep in mind that it is valid only for $\alpha l \ll 1$. For example, from Fig. 1.10(c), RG-174 has a loss of 99 dB/100 ft at 5 GHz. This is equivalent to 0.374 N/m, and therefore for lengths greater than about 1 m the approximation is not valid.

1.9 ELECTRIC AND MAGNETIC FIELD DISTRIBUTIONS

The electric and magnetic field distributions for a two-wire line are shown in Fig. 1.11(a), and for the coaxial line in Fig. 1.11(b). The cross indicates current "into page," and the dot, current "out of page." At any point of

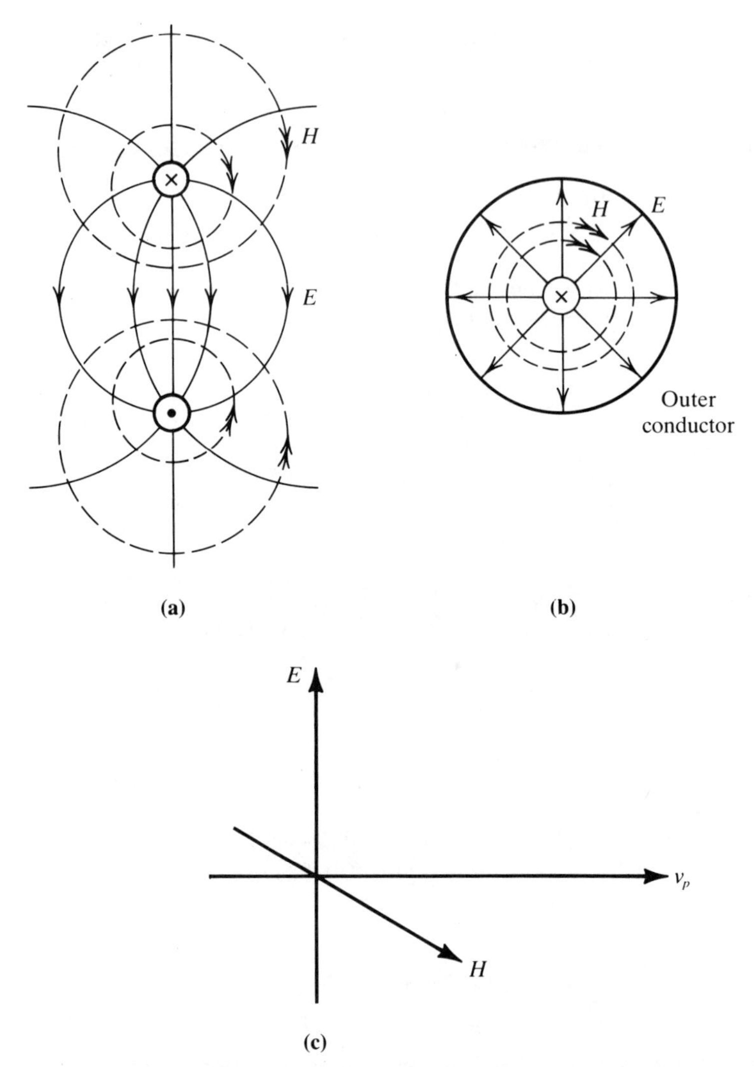

FIGURE 1.11. Electromagnetic field distribution: (a) two-wire line; (b) coaxial line. (c) Vector directions for a TEM wave.

intersection of the fields, the tangent lines to the field lines are at right angles. Furthermore, both fields are at right angles to the direction of propagation. This is illustrated by the vector diagram of Fig. 1.11(c). The direction of each field can also be said to be transverse to the direction of propagation, and for this reason this type of wave is referred to as a *Transverse ElectroMagnetic* (TEM) wave. Later it will be shown that more complex type of waves, such as exist in waveguides, can be made up from two or more TEM waves.

It is also important to observe the relative directions of the three vectors shown in Fig. 1.11(c). They form what is termed a right-hand set; that is, rotation in the manner of a right-hand screw is required to go from E to H while looking along the direction of v_p. An easy way of remembering the rotation rule is that it follows the alphabetical sequence of the letters—E to H looking along v_p. The significance of the right-hand set is that when the direction of the wave is reversed, *either E or H* (but not both) must be reversed to maintain right-hand-set conditions. The direction of propagation can be reversed through reflection, and this is discussed further in the next section.

1.10 REFLECTED WAVES

When the load impedance does not match the characteristic impedance of the line, the voltage and current distribution along the line can differ considerably from the matched (or infinite line) situation. Part of the energy of the incident wave is absorbed in the load, and rest is reflected back along the line with change in both amplitude and phase. This reflected energy is carried in a reflected wave and at any point on the line the voltage and current is given by the phasor sum of the incident and reflected waves at that point.

Consider first the situation at the load itself. The load voltage V_L will consist of two components, the incident voltage V_{IL} and the reflected voltage V_{RL}. Thus

$$V_L = V_{IL} + V_{RL} \tag{1.54}$$

The reflected voltage originates from the incident voltage and can be written as

$$V_{RL} = \Gamma_L V_{IL} \tag{1.55}$$

Γ_L is termed the *voltage reflection coefficient* at the load, and in fact Eq. (1.55) can be taken as the defining equation for Γ_L. Equations (1.54) and (1.55) can be combined to give

$$V_L = V_{IL}(1 + \Gamma_L) \tag{1.56}$$

Reflection also occurs for the current wave and from what was expressed in Section 1.9, if the reflected voltage is related to incident voltage as $V_{RL} = \Gamma_L V_{IL}$, then reflected current will be related to incident current as

$$I_{RL} = -\Gamma_L I_{IL} \tag{1.57}$$

The load current is therefore given by

$$I_L = I_{IL}(1 - \Gamma_L) \tag{1.58}$$

A very useful expression can now be obtained for the voltage reflection coefficient. Taking the ratio of Eqs. (1.56) and (1.58) gives

$$\frac{V_L}{I_L} = \frac{V_{IL}(1 + \Gamma_L)}{I_{IL}(1 - \Gamma_L)} \tag{1.59}$$

But V_L/I_L is equal to Z_L, the load impedance, and V_{IL}/I_{IL} is equal to Z_0, the characteristic impedance of the line. Therefore, from Eq. (1.59) it follows that

$$Z_L = Z_0\frac{1 + \Gamma_L}{1 - \Gamma_L} \tag{1.60}$$

Equation (1.60) can be arranged to give

$$\Gamma_L = \frac{Z_L - Z_0}{Z_L + Z_0} \tag{1.61}$$

Equation (1.60) has considerable practical value. The voltage reflection coefficient can be obtained by measurement, and from this and a knowledge of Z_0, the load impedance can be determined. Also, because Γ_L can be obtained by measurement it plays a very important part in practical applications of transmission lines.

EXAMPLE 1.5

A 200-Ω load is connected to a 50-Ω line. Calculate (a) the voltage reflection coefficient at the load, and (b) the ratio of reflected current to incident current at the load.

SOLUTION

(a) From Eq. (1.61),

$$\Gamma_L = \frac{200 - 50}{200 + 50} = \mathbf{0.6}$$

(b) From Eq. (1.57),

$$\frac{I_{RL}}{I_{IL}} = -\Gamma_L = \mathbf{-0.6}$$

EXAMPLE 1.6

A 50-Ω line is terminated in a 20 − j30-Ω load. Calculate the voltage reflection coefficient at the load.

SOLUTION From Eq. (1.61),

$$\Gamma_L = \frac{20 - j30 - 50}{20 - j30 + 50}$$

$$= \frac{-30 - j30}{70 - j30}$$

$$= -\sqrt{\frac{1800}{5800}}\,\frac{\angle 45^\circ}{\angle -23.2^\circ}$$

$$= -0.557 \angle 68.2^\circ$$

$$= \mathbf{0.557 \angle 248.2^\circ}$$

A reflected voltage wave, starting out as V_{RL} at the load, will be propagated back along the line from load to source. In the general case in which neither source impedance nor load impedance match the characteristic impedance of the line, multiple reflections will occur at each end. Now the resultant of any number of sine waves of a given frequency, whatever the relative phase relationships, is a single sine wave of the same frequency. With a sinusoidal source, therefore, *all* of the waves traveling from source to load may be represented by a single sinusoidal incident wave, having a value V_{IS} at the source terminals and V_{IL} at the load terminals. Similarly, all of the waves traveling from load to source may be lumped into a single sinusoidal reflected wave, having a value V_{RL} at the load and V_{RS} at the source. The situation is illustrated in Fig. 1.12.

The voltage reflection coefficient for the source is defined as

$$\Gamma_S = \frac{V_{RS}}{V_{IS}} \tag{1.62}$$

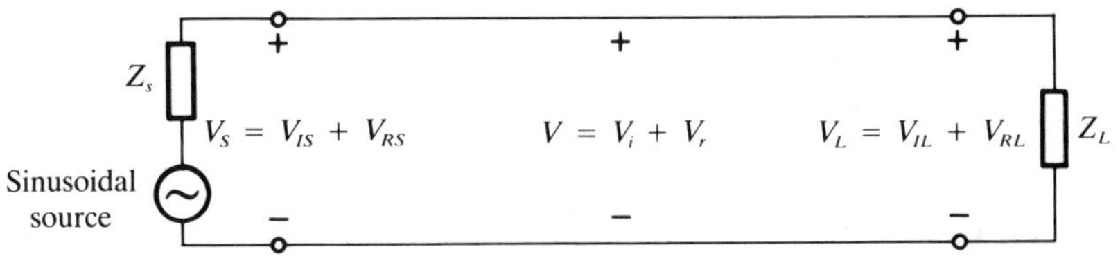

FIGURE 1.12. Incident and reflected voltages at specific points on a line.

In terms of impedances, the analogous expression to Eq. (1.61) is

$$\Gamma_S = \frac{Z_S - Z_0}{Z_S + Z_0} \tag{1.63}$$

Now, to find the effective voltage at any point on the line, Eq. (1.7) can be applied to both the incident and reflected waves, but care must be taken to use the parameters and variables associated with each component. Let l represent the distance *measured from load in the direction of the source;* then Eq. (1.7) modified for the reflected voltage wave is

$$V_r = V_{RL}e^{-\gamma l} \tag{1.64}$$

Similarly, Eq. (1.7), modified for the incident voltage wave, is

$$V_i = V_{IS}e^{-\gamma x} \tag{1.65}$$

It is inconvenient to have both l and x as measures of distance. In practice, it proves more convenient to use l. Let X represent the total length of line; then $x = X - l$ and Eq. (1.65) becomes

$$\begin{aligned} V_i &= V_{IS}e^{-\gamma(X-l)} \\ &= V_{IS}e^{-\gamma X}e^{\gamma l} \\ &= V_{IL}e^{\gamma l} \end{aligned} \tag{1.66}$$

where

$$V_{IL} = V_{IS}e^{-\gamma X} \tag{1.67}$$

Up to this point, reflection coefficients have been defined with reference to the load impedance and the source impedance only. A more general definition that applies for any position on the line *referred to load* is

$$\Gamma \triangleq \frac{V_r}{V_i} \tag{1.68}$$

Equations (1.64) and (1.66) may be used to give

$$\begin{aligned} \Gamma &= \frac{V_{RL}e^{-\gamma l}}{V_{IL}e^{\gamma l}} \\ &= \Gamma_L e^{-2\gamma l} \end{aligned} \tag{1.69}$$

Equation (1.65) shows that the voltage reflection coefficient at any point on the line is simply related to the voltage reflection coefficient at the load.

It will be seen that the reflection coefficient is a complex quantity in general, and measurements of voltage reflection coefficient usually yield its magnitude, denoted here by ρ, and its phase angle, ϕ. Thus,

$$\begin{aligned}\Gamma &= \rho \angle\phi \\ &= \rho e^{j\phi}\end{aligned} \tag{1.70}$$

Appropriate subscripts can be appended to both ρ and ϕ where necessary. In all cases, the current reflection coefficient is the negative of the voltage reflection coefficient.

The main points that may be summarized at this juncture are:

1. The "independent quantities" in the system are Z_S, Z_L, Z_0, γ, X, and E_S, the EMF of the sinusoidal source.
2. The voltage reflection coefficients Γ_S and Γ_L may be calculated knowing the impedance values Z_S, Z_L, and Z_0. In practice, the voltage reflection coefficients can be measured (the method of measurement is described in Chapter 15).
3. The voltage existing on the line at any point is the phasor sum of the incident and reflected voltages, and a voltage measurement (which may be very difficult to make) would yield the total. It would not be possible to separate out the incident and reflected components from this one measurement. On the other hand, the voltage reflection coefficient can be measured in a relatively straightforward manner at any point on the line, and related to the load impedance.
4. The current reflection coefficient is the negative of the voltage reflection coefficient.

EXAMPLE 1.7

A 75-Ω source is connected through a 50-Ω line to a 30-Ω load. Calculate (a) the source voltage reflection coefficient, and (b) the load voltage reflection coefficient.

SOLUTION (a)

$$\Gamma_S = \frac{75 - 50}{75 + 50} = \mathbf{0.2}$$

(b)

$$\Gamma_L = \frac{30 - 50}{30 + 50} = \mathbf{-0.25}$$

EXAMPLE 1.8

The voltage reflection coefficient at a load is $0.5 \underline{/45^\circ}$. Determine the voltage reflection coefficient at a distance of 0.2λ toward the generator. Neglect line attenuation.

SOLUTION

$$\gamma l \simeq \left(0 + j\frac{2\pi}{\lambda}\right)l = j2\pi \times 0.2 \text{ rad} = j72^\circ$$

From Eq. (1.69),

$$\Gamma = 0.5e^{j45}e^{-j144}$$
$$= \mathbf{0.5} \underline{/-99^\circ}$$

1.11 PHASOR REPRESENTATION OF VOLTAGES AND CURRENTS

The incident voltage is seen to be given by Eq. (1.66) as

$$V_i = V_{IL}e^{\gamma l}$$

Now, we know from Eq. (1.9) that $\gamma = \alpha + j\beta$, and therefore the incident voltage equation may be written as

$$V_i = V_{IL}e^{\alpha l}e^{j\beta l} \tag{1.71}$$

Using the voltage V_{IL} as reference, Eq. (1.71) may be represented on a phasor diagram by means of a rotating phasor, the angle of rotation being proportional to l as shown by βl. The length of the phasor is also dependent on l, as shown in Eq. (1.71). The path traced out by the phasor is therefore a spiral, as shown in Fig. 1.13(a).

A similar sort of phasor diagram can be drawn for the reflected voltage component, the difference here being that the load reflection coefficient $\rho_L \underline{/\phi_L}$ must be taken into account. Since $V_{RL} = \rho_L V_{IL} \underline{/\phi_L}$, Eq. (1.64) can be written as

$$V_r = \rho_L V_{IL}e^{-\alpha l}e^{-j(\beta l - \phi_L)} \tag{1.72}$$

The phasor diagram for this is shown in Fig. 1.13(b).

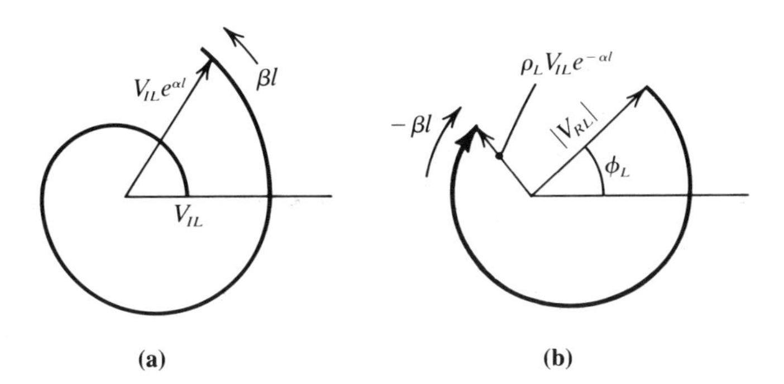

FIGURE 1.13. Rotating phasor representation of voltage: (a) incident voltage wave; (b) reflected voltage wave.

Current phasors may be shown in the same way except that it will be recalled that a phase reversal occurs for the reflected current relative to reflected voltage. The current equations are

$$I_i = \frac{V_i}{Z_0} \tag{1.73}$$

$$I_r = -\frac{V_r}{Z_0} \tag{1.74}$$

Since Z_0 is purely resistive, the phasor diagrams of Fig. 1.13 may be scaled by the factor Z_0^{-1}, and the reflected phasor reversed in direction, to give the current phasors.

1.12 STANDING WAVES

As shown by the phasor diagram of Fig. 1.13, the phase shifts for incident and reflected waves are in opposite directions. At certain points, therefore, the phasors will be in phase, as shown in Fig. 1.14(a), while at other points they will be in antiphase, as shown in Fig. 1.14(b). The phasor sum, which gives the resultant voltage at any point on the line, $V = V_i + V_r$, will therefore go through a series of maxima and minima, as shown in Fig. 1.14(c). Note that rms voltage is indicated on Fig. 1.14(c), and this is to emphasize the fact that the voltage pattern shown is stationary in time. For this reason it is referred to as a *voltage standing wave* (VSW). However, the voltage itself (as distinct from the standing-wave pattern) is of course time varying at the frequency of the input sinusoidal signal. For example,

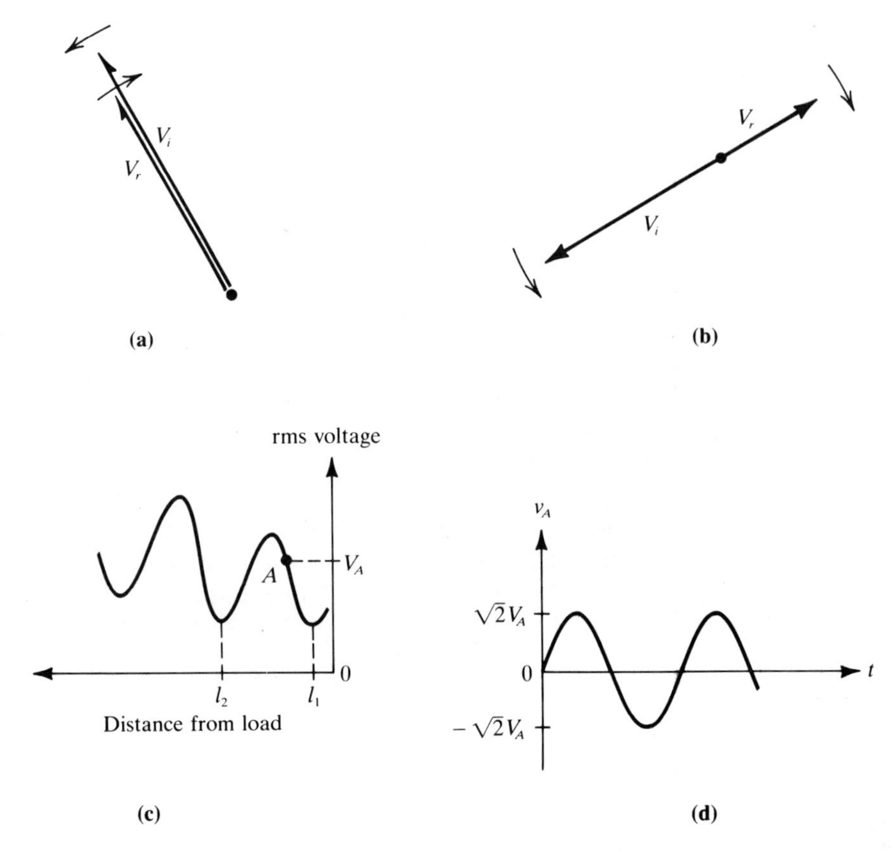

FIGURE 1.14. Incident and reflected voltages: (a) in phase; (b) antiphase. (c) The voltage standing-wave pattern. (d) The time variations at point A.

at point A, where the rms voltage is V_A, the time equation would be

$$v_A = \sqrt{2}\, V_A \sin \omega t \tag{1.75}$$

This is sketched in Fig. 1.14(d).

A knowledge of the distance between any two successive minima is of importance in measurements involving standing waves. This distance is exactly one-half wavelength, which may be shown as follows. Starting from position l_1, at which the phasors are antiphase [Fig. 1.14(b)], each phasor rotates by amount $\beta(l_2 - l_1)$ to arrive at the next antiphase position. But this corresponds to a rotation of π radians for each phasor, and therefore

$$\beta(l_2 - l_1) = \pi$$

or

$$\frac{2\pi}{\lambda}(l_2 - l_1) = \pi$$

or

$$l_2 - l_1 = \frac{\lambda}{2} \tag{1.76}$$

In a similar manner it can be shown that any two successive maxima are separated by $\lambda/2$, and the separation between a minimum and neighboring maxima is $\lambda/4$.

1.13 THE VOLTAGE STANDING-WAVE RATIO ON LOSSLESS LINES

There are many situations for which the attenuation coefficient α may be assumed equal to zero; in effect, the line is assumed to be lossless. This would apply, for example, for a short section of good-quality line used as a circuit element, an application to be described later.

With $\alpha = 0$, the magnitude of the incident voltage is, from Eq. (1.66),

$$|V_i| = V_{IL} \tag{1.77}$$

Also, from Eqs. (1.55) and (1.70), the magnitude of the reflected voltage is

$$|V_r| = \rho_L V_{IL} \tag{1.78}$$

At a voltage maximum the voltage is

$$\begin{aligned} V_{\max} &= |V_i| + |V_r| \\ &= V_{IL}(1 + \rho_L) \end{aligned} \tag{1.79}$$

(Note that this is *not* the same as $|V_i + V_r|$.) At a voltage minimum the voltage is

$$\begin{aligned} V_{\min} &= |V_i| - |V_r| \\ &= V_{IL}(1 - \rho_L) \end{aligned} \tag{1.80}$$

The voltage standing-wave ratio (VSWR) is defined as the ratio of maximum to minimum voltage:

$$\text{VSWR} \triangleq \frac{V_{\max}}{V_{\min}} \tag{1.81}$$

Substituting from Eqs. (1.79) and (1.80) gives the important relationship

$$\text{VSWR} = \frac{1 + \rho_L}{1 - \rho_L} \tag{1.82}$$

Also, Eq. (1.82) is easily rearranged to give

$$\rho_L = \frac{\text{VSWR} - 1}{\text{VSWR} + 1} \tag{1.83}$$

The importance of this relationship between ρ_L and VSWR is that the VSWR is a measurable quantity, and thus ρ_L can also be determined. Special test equipment known as *slotted line equipment* enables VSWR to be measured, as described in Section 14.5.

1.14 IMPEDANCE TRANSFORMATION

A length of transmission line may be used to transform an impedance from one value to another. Referring to Fig. 1.15, the length of transmission line between source and load is X, and the electrical length is

$$\theta = \beta l = 360 \frac{X}{\lambda} \tag{1.84}$$

The electrical length is stated in degrees.

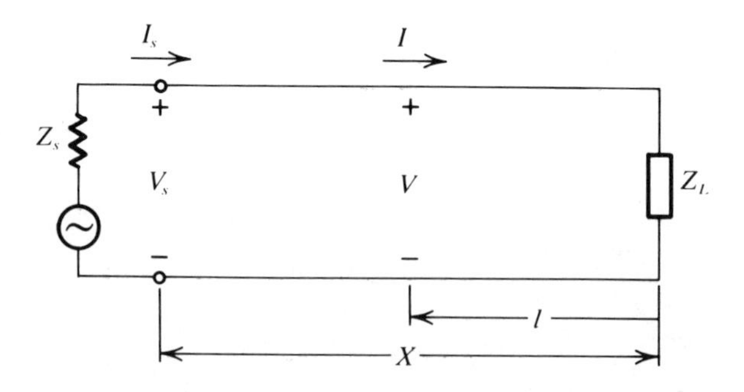

FIGURE 1.15. Section of transmission line used as a transformer.

The sending-end voltage V_S is the sum of the incident and reflected voltages, and is

$$\begin{aligned} V_S &= V_{IS} + V_{RS} \\ &= V_{IL}e^{j\theta} + V_{RL}e^{-j\theta} \\ &= (V_{IL} + V_{RL}) \cos \theta + j(V_{IL} - V_{RL}) \sin \theta \end{aligned} \tag{1.85}$$

where, as before, V_{IL} is the incident voltage at the load and V_{RL} is the reflected voltage at the load. Now from Eqs. (1.73) and (1.74),

$$V_{IL} = I_{IL}Z_0 \tag{1.86}$$

$$V_{RL} = -I_{RL}Z_0 \tag{1.87}$$

From Eqs. (1.86) and (1.87), $(V_{IL} - V_{RL}) = (I_{IL} + I_{RL})Z_0 = I_LZ_0$. Also, since $V_L = (V_{IL} + V_{RL})$, Eq. (1.85) can be rewritten as

$$V_S = V_L \cos \theta + jI_LZ_0 \sin \theta \tag{1.88}$$

The sending current I_S is

$$\begin{aligned} I_S &= I_{IS} + I_{RS} \\ &= I_{IL}e^{j\theta} + I_{RL}e^{-j\theta} \\ &= (I_{IL} + I_{RL}) \cos \theta + j(I_{IL} - I_{RL}) \sin \theta \end{aligned} \tag{1.89}$$

Again, from Eqs. (1.86) and (1.87), $(I_{IL} - I_{RL}) = (V_{IL} + V_{RL})/Z_0 = V_L/Z_0$. Also, since $I_L = (I_{IL} + I_{RL})$, Eq. (1.89) can be rewritten as

$$I_S = I_L \cos \theta + j\frac{V_L}{Z_0} \sin \theta \tag{1.90}$$

Equations (1.88) and (1.90) relate the sending-end voltage and current to the load-end voltage and current for a given electrical length θ. The input impedance at the sending end is

$$\begin{aligned} Z_{\text{in}} &= \frac{V_S}{I_S} \\ &= \frac{V_L \cos \theta + jI_LZ_0 \sin \theta}{I_L \cos \theta + j\dfrac{V_L}{Z_0} \sin \theta} \end{aligned} \tag{1.91}$$

On dividing through by I_L and multiplying up by Z_0, this becomes

$$Z_{\text{in}} = Z_0 \frac{Z_L \cos\theta + jZ_0 \sin\theta}{Z_0 \cos\theta + jZ_L \sin\theta} \tag{1.92}$$

EXAMPLE 1.9

A 50-Ω line has an electrical length of 36° and a relative permittivity of 2.32. It is terminated in a 100 $\angle 0°$-Ω load. For a signal frequency of 1 GHz, determine (a) the physical length of the line, and (b) the input impedance.

SOLUTION

(a) $\lambda_0 = \dfrac{3 \times 10^8}{1 \times 10^9} = 30 \text{ cm}$

$$\lambda = \frac{\lambda_0}{\sqrt{2.32}} = \mathbf{19.7\ cm}$$

From Eq. (1.84),

$$\begin{aligned} X &= \frac{\theta}{360}\lambda \\ &= \frac{36}{360} \times 19.7 \\ &= \mathbf{1.97\ cm} \end{aligned}$$

(b) From Eq. (1.92),

$$\begin{aligned} Z &= 50\left(\frac{100 \cos 36 + j50 \sin 36}{50 \cos 36 + j100 \sin 36}\right) \\ &= 50\left(\frac{80.9 + j29.39}{40.45 + j58.78}\right) \\ &= \frac{86.07 \angle 19.97°}{71.35 \angle 55.46°} \\ &= \mathbf{1.21 \angle -35.49°\ \Omega} \end{aligned}$$

Bearing in mind that the load impedance Z_L will in general be complex, Eq. (1.92) is usually solved by means of the Smith chart, to be described later. However, a specific result of practical importance is the $\frac{1}{4}\lambda$ transformer, which occurs when $X = \lambda/4$. The angle βX is then equal to 90°, and Eq. (1.92) simplifies to

$$Z_{in} = \frac{Z_0^2}{Z_L} \tag{1.93}$$

The λ/4 transformer is widely used as an impedance transformer for matching antennas to transmission lines. This is illustrated in Fig. 1.16, which shows a 73-Ω antenna being matched to a 300-Ω main feeder. One must be careful to distinguish between the characteristic impedance of the main line or feeder and that of the λ/4 matching section, and the latter is denoted here by Z_{01}. The requirement is that the input impedance to the λ/4 section equal the characteristic impedance of the main feeder, and therefore application of Eq. (1.93) gives

$$300 = \frac{Z_{01}^2}{73}$$

or

$$Z_{01} = \mathbf{148\ \Omega}$$

Because the main transmission line is matched, the VSWR on it will be unity, which is the ideal condition. In effect, no standing wave exists on the main line; however, a standing wave is present on the λ/4 matching section.

It will also be obvious that the λ/4 transformer can be fully effective at only one frequency, that which makes the length equal to one-quarter wavelength. With narrowband signals the variation in impedance transformation may be tolerable but with wideband signals, some form of "broadbanding" must be employed. One method is to use two or more matching sections in tandem so that the transfer of impedance values from load value to the final required value takes place in stages rather than abruptly. A useful practical design, which gives maximal flat response of VSWR against frequency, is the binomial transformer, so named because the impedance values of the λ/4 sections are related to Z_0 and Z_L in a

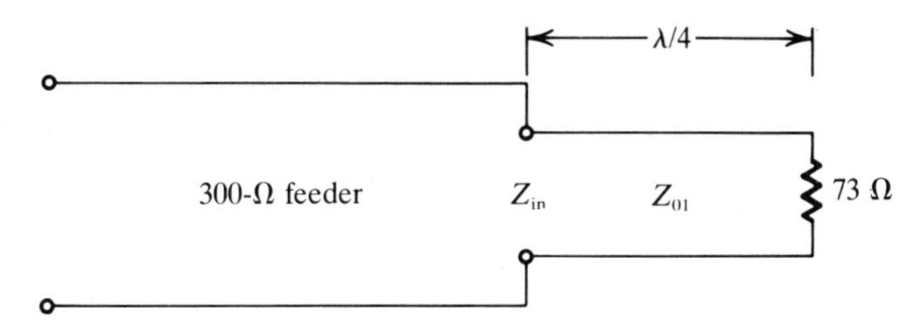

FIGURE 1.16. Matching a 73-Ω load to a 300-Ω line. $Z_{01} = 148\ \Omega$.

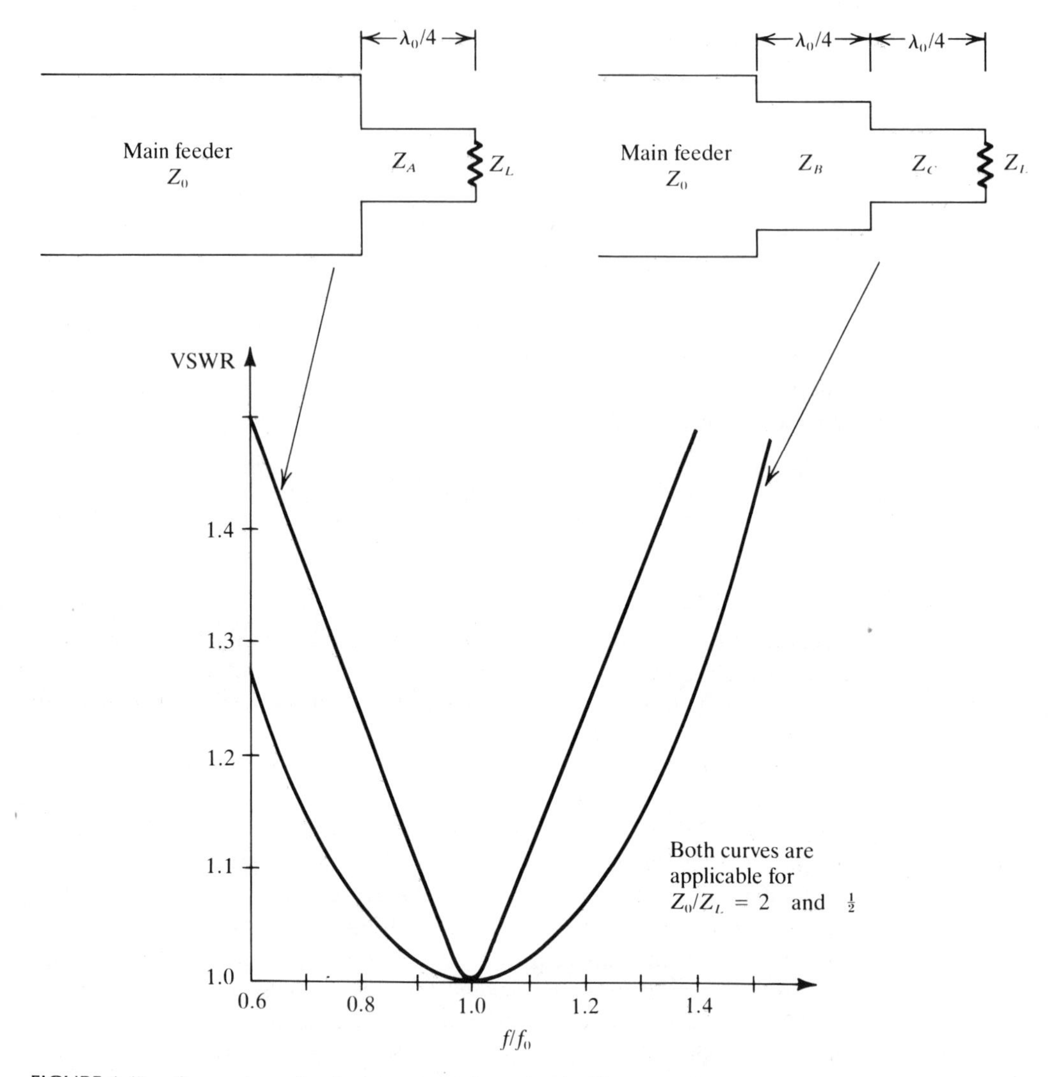

FIGURE 1.17. Comparison of a single section $\lambda/4$, and a double section binomial $\lambda/4$ transformers.

manner that is mathematically similar to the coefficients of the binomial expansion. In Fig. 1.17 a comparison is made between a single $\lambda/4$ section and a double binomial transformer. The main feeder impedance is denoted by Z_0 in each case, and the load impedance by Z_L. The characteristic impedance of the single $\lambda/4$ section is denoted by Z_A, and as already shown by Eq. (1.93),

$$Z_A = Z_0^{1/2} Z_L^{1/2} \tag{1.94}$$

The design equations for the binomial transformer are

$$Z_B = Z_0^{3/4} Z_L^{1/4} \tag{1.95}$$

$$Z_C = Z_0^{1/4} Z_L^{3/4} \tag{1.96}$$

The curves show the variation of VSWR as a function of frequency ratio, where f_0 is the frequency at which the sections are exactly one-quarter wavelength. The VSWR depends also on the Z_0/Z_L ratio, and for purposes of illustration this ratio is taken as 2:1. The curves are also applicable for a ratio of 1:2. Here Z_L is assumed to be purely resistive.

The variation of the input impedance of a quarter-wave section terminated in a resistive load, with variations in frequency, may be determined from Eq. (1.92) by noting that the angle βX is given by

$$\begin{aligned} \beta X &= \frac{2\pi}{\lambda} \frac{\lambda_0}{4} \\ &= \frac{\pi}{2} \frac{f}{f_0} \end{aligned} \tag{1.97}$$

Here f_0 is the frequency at which the section is exactly one-quarter wavelength and f is the frequency in general. Let $f = f_0 \pm \Delta f$; then Eq. (1.97) can be written as

$$\beta X = \frac{\pi}{2}(1 + \delta) \tag{1.98}$$

where $\delta = \pm \Delta f / f_0$ is the fractional frequency change.

Substitution of Eq. (1.98) in Eq. (1.92) and carrying out some straightforward algebraic manipulation, which is left as an exercise for the reader, results in, for the single section,

$$Z_{in} = Z_A \frac{Z_L \sin(\delta\pi/2) - jZ_A \cos(\delta\pi/2)}{Z_A \sin(\delta\pi/2) - jZ_L \cos(\delta\pi/2)} \tag{1.99}$$

The input impedance Z_{in} now takes the place of Z_L in Eq. (1.61), enabling the reflection coefficient for the main feeder to be calculated. The VSWR can then be calculated using Eq. (1.82).

EXAMPLE 1.10

A quarter-wave section is used to match a 90-Ω purely resistive load to a 50-Ω line. Calculate (a) the characteristic impedance of the matching section, and (b) the VSWR on the main feeder when the frequency is increased by 10% from the resonant value.

SOLUTION (a) From Eq. (1.94),

$$Z_A = \sqrt{50 \times 90} = \mathbf{67.1\ \Omega}$$

(b)

$$\frac{\delta\pi}{2} = \frac{0.1 \times \pi}{2} \text{ rad} = 9°$$

From Eq. (1.99),

$$Z_{\text{in}} = 67.1\left(\frac{90 \sin 9 - j67.1 \cos 9}{67.1 \sin 9 - j90 \cos 9}\right)$$

$$= 50.79 \angle 5.26°\ \Omega$$

$$\Gamma_{(\text{main feed})} = \frac{50.79 \angle 5.26 - 50}{50.79 \angle 5.26 + 50}$$

$$= \frac{0.58 + j4.66}{100.58 + j4.66}$$

Therefore,

$$\rho = 0.0466$$

From Eq. (1.82),

$$\text{VSWR} = \frac{1 + \rho}{1 - \rho} \simeq \mathbf{1.1}$$

For the binomial transformer, the input impedance to the first section is given by Eq. (1.99) modified to

$$Z_{\text{in1}} = Z_C \frac{Z_L \sin(\delta\pi/2) - jZ_C \cos(\delta\pi/2)}{Z_C \sin(\delta\pi/2) - jZ_L \cos(\delta\pi/2)} \tag{1.100}$$

The input impedance to the second section is given by a similar equation with Z_{in1} replacing Z_L:

$$Z_{\text{in2}} = Z_B \frac{Z_{\text{in1}} \sin(\delta\pi/2) - jZ_B \cos(\delta\pi/2)}{Z_B \sin(\delta\pi/2) - jZ_{\text{in1}} \cos(\delta\pi/2)} \tag{1.101}$$

Calculation of VSWR on the main feeder then follows the same procedure as for the single section, using Z_{in2} in place of Z_L in Eq. (1.61) to calculate the voltage reflection coefficient for the main feeder, and Eq. (1.82) to calculate VSWR. It will be seen that considerable computation is involved, which is best carried out on a computer or programmable calculator. The curves shown in Fig. 1.17 were determined using the procedure given here.

1.15 TRANSMISSION LINES AS CIRCUIT ELEMENTS

Inductive and capacitive elements can be realized using transmission-line sections, these being special cases of the impedance relationship given by Eq. (1.92). Referring to Fig. 1.15, at position l, the impedance $Z_{in} = V/I$ is given by eq. (1.92) with $\theta = 360\ l/\lambda$ degrees or $2\pi l/\lambda$ radians.

1.15.1 Inductance

In this case, the load impedance is made a short circuit; therefore, $Z_L = 0$ in Eq. (1.92) and this becomes

$$Z = jZ_0 \tan \frac{2\pi l}{\lambda} \tag{1.102}$$

The short-circuited line is sketched in Fig. 1.18(a), and the variation of impedance Z with length l in Fig. 1.18(b). It will be seen that the line is inductive, as signified by the $+j$ reactance values, for certain ranges of l only, and in practice the first range, $0 \leqslant l \leqslant \lambda/4$, is usually chosen. It is also of interest to know how Z varies with frequency. Let f_0 represent the frequency at which the line is exactly one-quarter wavelength, then since $l = \dfrac{\lambda_0}{4}$:

$$\frac{2\pi l}{\lambda} = \frac{\pi}{2}\frac{\lambda_0}{\lambda} = \frac{\pi}{2}\frac{f}{f_0} \tag{1.103}$$

Therefore,

$$Z = jZ_0 \tan \frac{\pi f}{2f_0} \tag{1.104}$$

The curve of Z as a function of frequency is sketched in Fig. 1.18(c). It will be seen that this is similar to Fig. 1.18(b). For comparison the reactance line for a constant inductance is also shown and it will be seen that it is only over a small range of frequencies starting at zero that the two curves coincide.

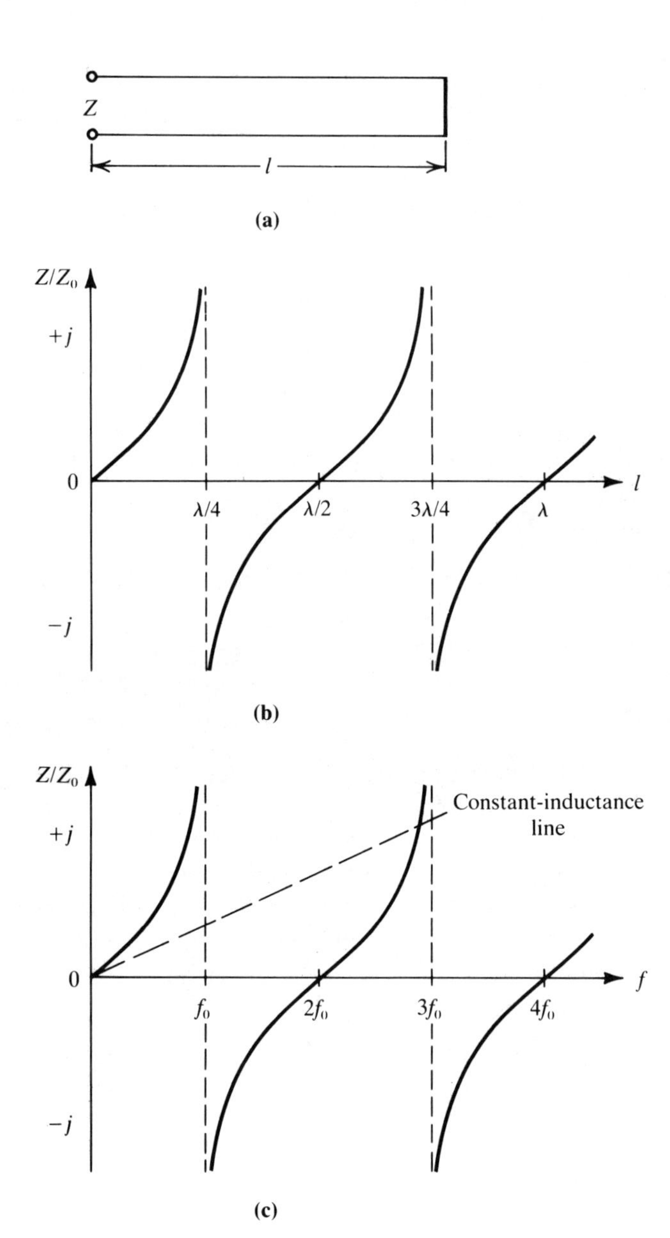

FIGURE 1.18. (a) Short-circuited line section; (b) variation of Z with l; (c) variation of Z with f.

EXAMPLE 1.11

A transmission section is 0.2λ long at a frequency of 500 MHz, and its characteristic impedance is 50 Ω. Calculate (a) the $\frac{1}{4}\lambda$ resonant frequency, (b) the input impedance at 500 MHz when the line is short circuited, and (c) the input impedance at 700 MHz when the line is short circuited.

SOLUTION

(a) $f_0 = 500 \times \dfrac{0.25\lambda}{0.2\lambda} = \mathbf{625\ MHz}$

(b) From Eq. (1.102),

$$Z = j50 \tan 2\pi \times 0.2$$
$$\simeq \boldsymbol{j154\ \Omega}$$

(c) From Eq. (1.104),

$$Z = j50 \tan \frac{\pi \times 700}{2 \times 625}$$
$$\simeq -\mathbf{j262\ \Omega}$$

1.15.2 Capacitance

As seen from Fig. 1.18 a short-circuited line behaves as a capacitive element for lengths in the range λ/4 to λ/2 and for higher multiples of this range. The capacitive reactance occurs when the j coefficient is negative. Capacitive reactance can also be achieved using an open-circuited line, as will now be shown. Equation (1.92) can be written as

$$Z = Z_0 \frac{\cos \beta l + j\,(Z_0/Z_L) \sin \beta l}{(Z_0/Z_L) \sin + j \cos \beta l} \tag{1.105}$$

Now, with an open-circuit termination, $Z_L = \infty$ and Eq. (1.105) reduces to

$$Z = -jZ_0 \cot \frac{2\pi l}{\lambda} \tag{1.106}$$

The open-circuited line is shown in Fig. 1.19(a), and the variation of impedance with length in Fig. 1.19(b). It will be seen that the line is capacitive, as signified by the $-j$ reactance values, for certain ranges of l only, and in practice the first range, $0 \leq l \leq \lambda/4$, is usually chosen. The variation of

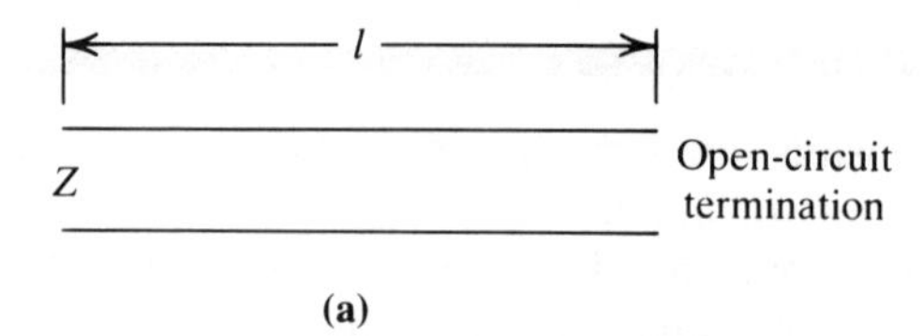

(a)

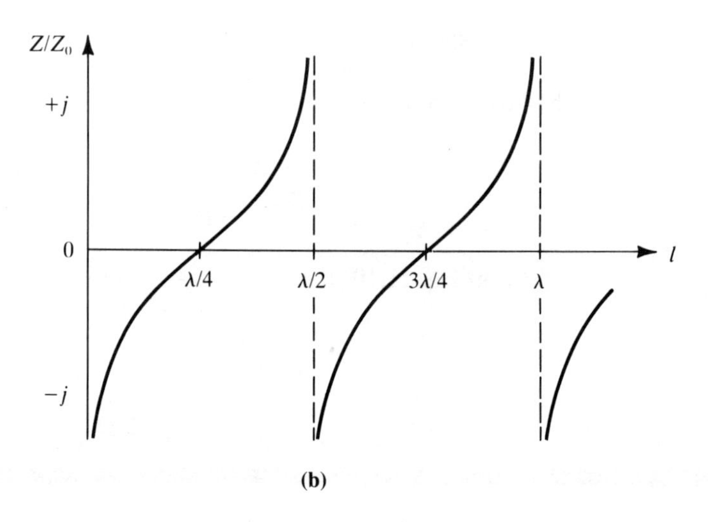

(b)

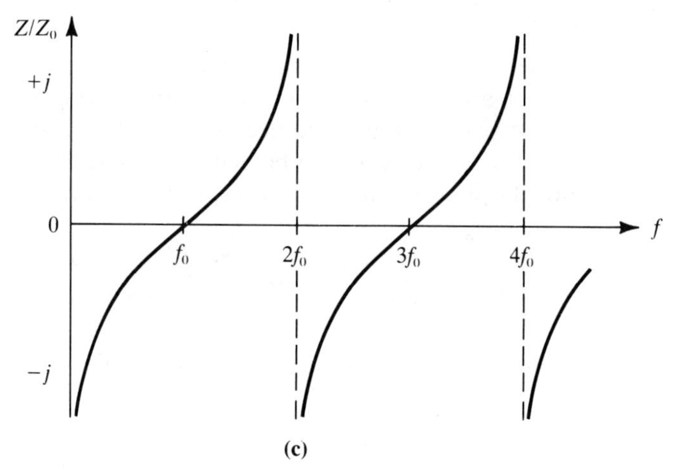

(c)

FIGURE 1.19. (a) Open-circuit line section; (b) variation of impedance with length; (c) variation of impedance with frequency.

impedance with frequency can be found in a similar way to that used to derive Eq. (1.104) and results in

$$Z = -jZ_0 \cot \frac{\pi f}{2f_0} \tag{1.107}$$

The curve of Z as a function of frequency is sketched in Fig. 1.19(c). A more direct comparison of the frequency response, with that for a constant capacitance C, may be made in terms of admittance Y. For the open-circuited line section the admittance is

$$Y = \frac{1}{Z} = jY_0 \tan \frac{\pi f}{2f_0} \tag{1.108}$$

while that for the constant capacitance is $j\omega C$. In Eq. (1.108) $Y_0 = 1/Z_0$. Equation (1.108) will be seen to be similar in form to Eq. (1.104), and this is shown in Fig. 1.20, which shows generalized curves for impedance or admittance.

1.15.3 Resonators

Figure 1.18 and Eq. (1.102) show that for a short-circuited line, the impedance is infinite at $l = \lambda/4$. In fact, the impedance exhibits an infinite discontinuity at this point. This is for an ideal line, which has zero losses. In practice, line losses will limit the impedance to some finite, but high value. The frequency at which this occurs is known as the *resonant frequency* of the line section, denoted by f_0 in Fig. 1.18(c). For frequencies lower than f_0 the impedance is inductive, and for frequencies higher than f_0, up to $2f_0$, it is capacitive. The $\lambda/4$ short-circuited line therefore behaves in a manner similar to that of a parallel tuned circuit in the vicinity of

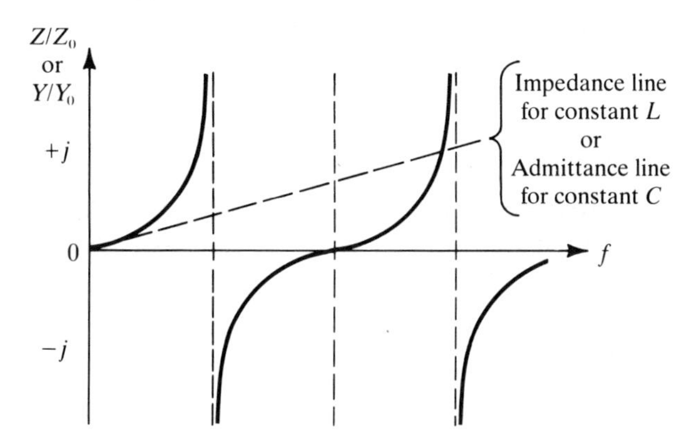

FIGURE 1.20. Generalized curves for impedance equations (1.104) and (1.108).

resonance, having a high impedance at resonance, an inductive impedance below resonance, and a capacitive impedance above resonance. Note, however, that unlike the tuned circuit, the transmission line exhibits multiple resonances, and furthermore, these alternate between very high and very low (ideally zero) impedance values, as shown in Fig. 1.18. The first zero impedance value occurs at $f = 2f_0$, the next high impedance value at $f = 3f_0$, and so on. The behavior at the first zero impedance, where $l = \lambda/2$, is of particular importance also. The line behaves as a series resonant circuit around this point, being capacitive below $2f_0$ down to f_0, and inductive for frequencies above $2f_0$ up to $3f_0$.

To summarize, ths short-circuited line behaves as a parallel tuned circuit around odd multiples of $\lambda/4$ and as a series tuned circuit around even multiples of $\lambda/4$. In applications it is desirable to keep the physical length of the line small, so where possible, operation would be confined to the first parallel resonant region, at $l = \lambda/4$. However, operation is not confined to short-circuited lines, and as shown in the next paragraph, open-circuited lines can also be used to provide tuned-circuit characteristics.

For an open-circuited line, Fig. 1.19 and Eq. (1.106) show that the characteristics are the duals of the short-circuited line characteristics; that is, the open-circuited line behaves as a series resonant circuit around odd multiples of $\lambda/4$, and as a parallel tuned circuit around even multiples of $\lambda/4$. Again, as with the shorted line, it is desirable to keep the physical length small, and this usually means confining operation to around the first $\lambda/4$ point, as a series resonator. However, with microstrip and stripline, it is difficult to achieve a good short circuit at the higher microwave frequencies (e.g., above about 3 GHz), so the open circuit line is often preferred for microstrip and stripline applications. These applications are described more fully in Section 5.8.

The Q-factor of the transmission-line resonant circuit is considerably higher than the L-C tuned-circuit counterpart. The Q-factor is given by [see, e.g., Marcuvitz (1971)]

$$Q = \frac{\beta_0}{2\alpha} \tag{1.109}$$

where β_0 is the phase-shift coefficient evaluated at resonance and α is the attenuation coefficient. Q-factors as high as 10^4 have been realized.

1.16 COUPLED LINES

The electromagnetic field from one transmission line can be used to couple energy into a second line, one such arrangement being illustrated in Fig. 1.21.

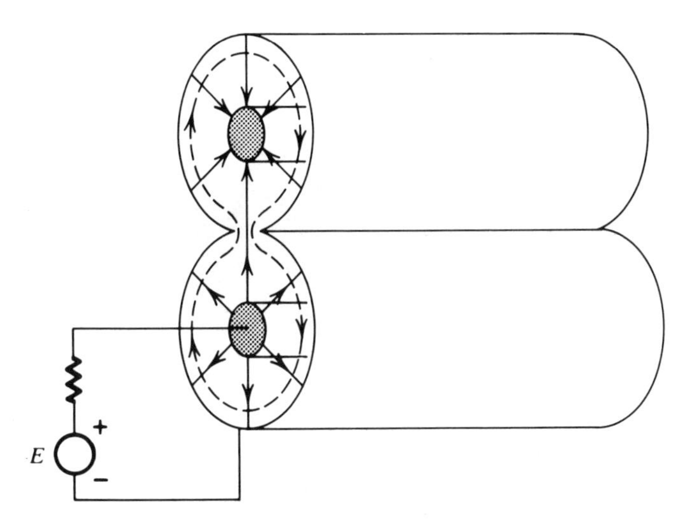

FIGURE 1.21. Coupled coaxial lines.

As shown, the lower coaxial line is energized by an external source, and a narrow slot allows some of the field to penetrate into the upper coaxial line. The electric field will induce an equal and opposite charge on the center conductor, giving rise to an electric field in the upper line which is reversed in direction from that in the lower line. The same magnetic field encloses both center conductors. As discussed in Section 1.9, reversal of the E field must be accompanied by a reversal of direction of propagation. The wave in the lower line propagates from left to right, while that in the upper line propagates from right to left. The coupler can be analyzed through the use of even- and odd-symmetry modes (Reed and Wheeler 1956). The plane of symmetry in this case is the plane containing the coupling slot, shown schematically in Fig. 1.22. The characteristic impedance of the individual lines is not specified at this point, and as will be shown, even-mode and odd-mode impedances must be specified. The even-symmetry circuits are shown in Fig. 1.23(a) and the odd-symmetry circuits are shown in Fig. 1.23(a) and the odd-symmetry circuits in Fig. 1.23(b). Subscript e denotes even and subscript o denotes odd and the number subscripts refer to the port numbers shown in Fig. 1.22. These four circuits are similar and can be represented by the single circuit of Fig. 1.23(c). Z_S stands for either Z_{0e} or Z_{0o}, depending on the symmetry mode being considered. For the top circuit of Fig. 1.23(b), E_S stands for $-\frac{1}{2}E$, and for the other circuits it stands for $+\frac{1}{2}E$ In practice, the coupled section is made equal to $\lambda/4$ in length, as this gives maximum coupling [see, e.g., Edwards (1981)]. The main properties of the coupler can be found from an analysis of the quarter-wave circuit shown in Fig. 1.23.

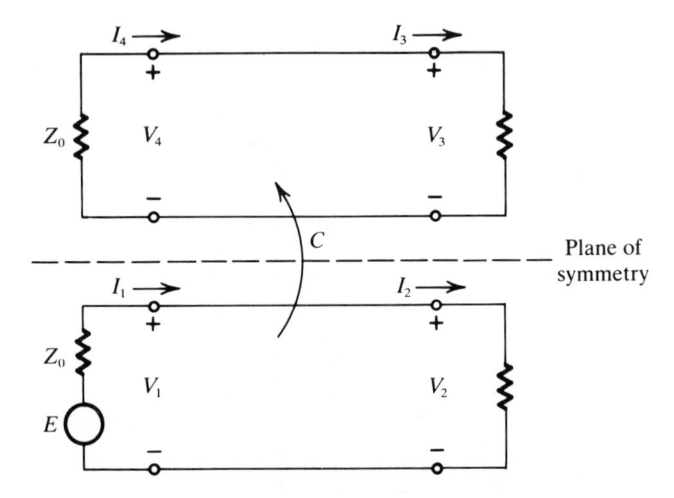

FIGURE 1.22. Coupled system with ports terminated in Z_o.

Referring to Fig. 1.23(c) and applying Eq. (1.93), the input impedance is

$$Z_{\text{in}} = \frac{Z_S^2}{Z_0} \tag{1.110}$$

Hence the input current I is

$$I_A = \frac{E_S}{Z_0 + Z_S^2/Z_0} \tag{1.111}$$

The voltage V_A is

$$\begin{aligned} V_A &= I_A Z_{\text{in}} \\ &= \frac{E_S(Z_S^2/Z_0)}{Z_0 + (Z_S^2/Z_0)} \\ &= \frac{E_S Z_S^2}{Z_0^2 + Z_S^2} \end{aligned} \tag{1.112}$$

The voltage V_B can be found by application of Eq. (1.90) on noting that V_B replaces V_L in this, Z_S replaces Z_0, I_A replaces I_S, and that $\theta = 90°$. Hence, from Eq. (1.90),

$$\begin{aligned} V_B &= -jI_A Z_S \\ &= -jE_S \frac{Z_S Z_0}{Z_0^2 + Z_S^2} \end{aligned} \tag{1.113}$$

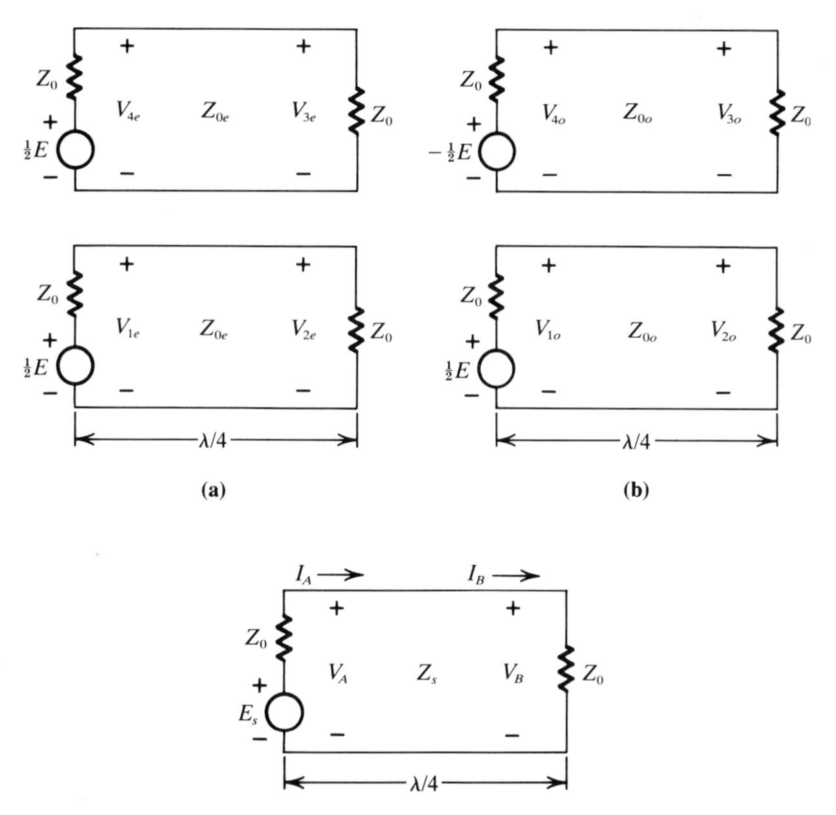

FIGURE 1.23. (a) Even-mode symmetry and (b) odd-mode symmetry for the circuit of Fig. 1.22. (c) Single equivalent circuit.

The voltages at the four ports can now be found. V_1 is obtained from V_A with E_S being replaced by $\frac{1}{2}E$ for both even and odd mode circuits.

$$
\begin{aligned}
V_1 &= V_{1e} + V_{1o} \\
&= \frac{\frac{1}{2}EZ_{0e}^2}{Z_0^2 + Z_{0e}^2} + \frac{\frac{1}{2}EZ_{0o}^2}{Z_0^2 + Z_{0o}^2} \\
&= \frac{E}{2}\left(\frac{Z_{0e}^2}{Z_0^2 + Z_{0e}^2} + \frac{Z_{0o}^2}{Z_0^2 + Z_{0o}^2}\right)
\end{aligned}
\tag{1.114}
$$

For the system to be matched the quantity in parentheses must equal unity since this makes $V_1 = \frac{1}{2}E$. It is left as an exercise (Problem 24) for the student to show that this requires

$$Z_0 = \sqrt{Z_{0e}Z_{0o}} \tag{1.115}$$

The voltage at port 4 is also obtained from V_A but in this case E_S is replaced by $\frac{1}{2}E$ for the even mode and by $-\frac{1}{2}E$ for the odd mode circuit:

$$\begin{aligned}
V_4 &= V_{4e} + V_{4o} \\
&= \left(\frac{\frac{1}{2}EZ_{0e}^2}{Z_0^2 + Z_{0e}^2} - \frac{\frac{1}{2}EZ_{0o}^2}{Z_0^2 + Z_{0o}^2}\right) \\
&= \frac{1}{2}E\left(\frac{Z_{0e}^2}{Z_{0e}Z_{0o} + Z_{0e}^2} - \frac{Z_{0o}^2}{Z_{0e}Z_{0o} + Z_{0o}^2}\right) \\
&= \frac{E}{2}\frac{Z_{0e} - Z_{0o}}{Z_{0e} + Z_{0o}}
\end{aligned} \quad (1.116)$$

The coupling factor C is defined as

$$C \triangleq \frac{V_4}{V_1} \quad (1.117)$$

Since $V_1 = \dfrac{E}{2}$ it is seen that

$$C = \frac{Z_{0e} - Z_{0o}}{Z_{0e} + Z_{0o}} \quad (1.118)$$

The voltage at port 2 is obtained from V_B with $\frac{1}{2}E$ replacing E_S for both even and odd mode circuits

$$\begin{aligned}
V_2 &= V_{2e} + V_{2o} \\
&= -j\frac{E}{2}\left(\frac{Z_{0e}Z_0}{Z_0^2 + Z_{0e}^2} + \frac{Z_{0o}Z_0}{Z_0^2 + Z_{0o}^2}\right) \\
&= -j\frac{E}{2}\left(\frac{Z_{0e}Z_o}{Z_{0e}Z_{0o} + Z_{0e}^2} + \frac{Z_{0o}Z_0}{Z_{0e}Z_{0o} + Z_{0o}^2}\right) \\
&= -j\frac{E}{2}\frac{2Z_0}{Z_{0e} + Z_{0o}}
\end{aligned} \quad (1.119)$$

It is left as an exercise (Problem 25) for the student to show that this can be written as

$$V_2 = -j\frac{E}{2_s}\sqrt{1 - C^2} \quad (1.120)$$

The transmission factor T is defined as

$$T \triangleq \frac{V_2}{V_1} \tag{1.121}$$

Again, with $V_1 = \frac{E}{2}$

$$T = -j\sqrt{1 - C^2} \tag{1.122}$$

Finally, the voltage at port 3 is obtained also from V_B, but in this case with $\frac{1}{2}E$ replacing E_S for the even mode, and $-\frac{1}{2}E$ replacing E_S for the odd mode, circuit

$$\begin{aligned} V_3 &= V_{3e} + V_{3o} \\ &= -j\tfrac{1}{2}E\left(\frac{Z_{0e}Z_0}{Z_o^2 + Z_{oe}^2} - \frac{Z_{0o}Z_0}{Z_0^2 + Z_{0o}}\right) \\ &= -j\tfrac{1}{2}E\left(\frac{Z_0}{Z_{0o} + Z_{0e}} - \frac{Z_0}{Z_{0e} + Z_{0o}}\right) \\ &= 0 \end{aligned} \tag{1.123}$$

These equations show that some of the incident energy entering at port 1 is transmitted to port 2, and some is coupled to port 4, while none is coupled to port 3. The incident wave, traveling from port 1 to port 2, couples energy into port 4 and none into port 3. A reflected wave in the lower line, traveling from port 2 to port 1, would couple energy into port 3 and none into port 4. Thus the coupler is capable of separating out incident and reflected waves, and this principle is used as the basis for directional couplers. Practical values for a microstrip coupler are given, and other types of coupler are described in Section 5.6. Some applications of directional couplers are discussed in Chapter 14.

In practical couplers a small amount of energy reaches port 3, and V_3 has some small but finite value. The directivity factor D is defined as

$$D \triangleq \frac{V_3}{V_1} \tag{1.124}$$

It will be seen that coupling factor, directivity factor, and transmission factor are all defined for the maximum coupling condition, that is, for $\theta = 90°$. Also, because V_2 lags on V_1 by 90°, the coupler is sometimes referred to as a 90° or quadrature coupler. In practice, the decibel values are usually specified and these are

$$\text{coupling dB} = 20 \log \frac{|V_4|}{|V_1|} = 10 \log \frac{P_4}{P_1} \tag{1.125}$$

$$\text{directivity dB} = 20 \log \frac{|V_3|}{|V_1|} = 10 \log \frac{P_3}{P_1} \tag{1.126}$$

$$\text{transmission dB} = 20 \log \frac{|V_2|}{|V_1|} = 10 \log \frac{P_2}{P_1} \tag{1.127}$$

Power P_i is the power delivered to the Z_0 load at port i.

EXAMPLE 1.12

The specifications for a directional coupler are coupling 10 dB and directivity 60 dB. Assuming that the system is matched as shown in Fig. 1.22, determine, for an input power of 1 mW, the power at the other ports.

SOLUTION The power at port 3 is -60 dBm or 1 nW; the power at port 4 is -10 dBm or 0.1 mW. The power delivered to port 3 is small enough to be neglected, and therefore the power transmitted to port 2 is $1 - 0.1 = 0.99$ mW. This follows from the power balance, and may also be deduced from Eq. (1.122).

In this example, the coupling, the directivity, and the transmission are seen to be negative quantities when expressed in decibels. This must always be the case since each of these quantities expressed as a power ratio must be less than unity. Care must be taken when using specification sheets for directional couplers as common practice is to specify the magnitudes only of these decibel values.

1.17 PROBLEMS

1. Determine the product LC in terms of permittivity and permeability for both the two-wire line and the coaxial cable.
2. Derive Eq. (1.24) from Eq. (1.20).
3. A 500-MHz sine wave is propagated over a matched transmission line that has an air dielectric. The phase velocity in air is 3×10^8 m/s. Calculate the wavelength of the signal on the line.
4. A solid dielectric has an attenuation coefficient of 0.001 m^{-1}. Calculate the factor by which the amplitude is reduced for a 1-m length of line.

5. For coaxial cable type RG-141/U, determine, for a frequency of 1.0 GHz, (a) the attenuation coefficient, and (b) the dielectric constant.
6. A 1.0-GHz sinusoidal signal is transmitted along an RG-141/U cable. Determine the wavelength on the line.
7. Use the values specified in Fig. 1.10 for cable type RG-58/U to determine (a) the dielectric constant, (b) the phase-shift coefficient at 1.0 GHz, and (c) the attenuation coefficient at 1.0 GHz.
8. For the RG-58/U find the power loss per unit length at a frequency of 1.0 GHz when the input power is at the maximum rated value.
9. Use the values specified in Fig. 1.10 for the semirigid cable, code 8, to determine (a) the dielectric constant, (b) the phase-shift coefficient at 10.0 GHz, (c) the attenuation coefficient at 10.0 GHz, and (d) the maximum power loss allowed per unit length at 10.0 GHz.
10. The semirigid cable, code 8, of Fig. 1.10 is to be used in an aircraft, the maximum altitude and temperature specifications being 35,000 ft and 120°C. Determine the maximum allowable power input and the corresponding power loss per unit length.
11. A voltage reflection coefficient is measured as 0.75, and the corresponding load voltage is 1 V. Determine (a) the incident voltage component at the load, (b) the reflected voltage component at the load, and (c) the current reflection coefficient.
12. The voltage reflection coefficient of 0.75 is measured on a 50-Ω line system. Calculate the load impedance.
13. A load impedance of $30\underline{/70^\circ}$ Ω is connected to a 50-Ω line. Calculate the voltage reflection coefficient.
14. For the system described in Problem 13, V_S is determined as 6 V, and this is used as the reference voltage. Given that the line is lossless and $\lambda/4$ long, calculate (a) V_{IS}, (b) V_{RS}, (c) V_{IL}, and (d) V_{RL}.
15. The voltage reflection coefficient at the load, as measured on a 50-Ω system, is $0.7\underline{/25^\circ}$. Determine the load impedance.
16. The voltage reflection coefficient at a load is known to be $0.6\underline{/30^\circ}$. Determine the corresponding voltage reflection coefficient at a distance of 0.1λ along the line from the load. The line attenuation may be assumed negligible.
17. A transmission-line section is 0.1λ long and has a characteristic impedance of 50 Ω. It is terminated in a 100-Ω purely resistive load. Calculate the input impedance.
18. Determine the impedances of the quarter-wave sections for a binomial transformer used to match a 90-Ω purely resistive load to a 50-Ω line.
19. A short-circuited transmission line section is exactly one-quarter wavelength at a frequency of 1250 MHz. Calculate (a) its inductive react-

ance, and (b) the equivalent inductance, at a frequency of 500 MHz. The characteristic impedance of the line is 50 Ω.

20. Repeat Problem 19 for a frequency of 1000 MHz.

21. A 50-Ω open-circuit line section is 0.1λ in length. Determine the input impedance.

22. An open-circuit line section is exactly $\lambda/4$ in length. Determine the input impedance, assuming that the line is lossless.

23. Determine the input admittance for the line section of Problem 21.

24. Show by substitution that Eq. (1.115) fulfills the matching condition for Eq. (1.114).

25. Using Eqs. (1.115) and (1.118), show that Eq. (1.120) is equivalent to Eq. (1.119).

REFERENCES

EDWARDS, T. C., 1981. *Foundations for Microstrip Circuit Design*. New York: John Wiley & Sons, Inc.

GLAZIER, E. V. D., and H. R. L. LAMONT, 1958. *The Services' Textbook of Radio*, Vol 5, *Transmission and Propagation*. London: Her Majesty's Stationery Office.

MARCUVITZ, NATHAN, 1971. *Microwave Fields and Circuits*. Englewood Cliffs, N.J.: Prentice-Hall, Inc.

REED, J., and G. J. WHEELER, 1956, "A method of analysis of symmetrical four-port networks," *IRE Trans. Microwave Theory Tech.*

RODDY, DENNIS, and JOHN COOLEN, 1984. *Electronic Communications*. Reston, Va.: Reston Publishing Company, Inc.

two
Scattering Parameters

2.1 INTRODUCTION

Scattering variables and scattering parameters (S-parameters) are especially useful in dealing with microwave circuits because they relate to signal flow rather than to voltages and currents directly. In many situations in microwave circuits, electric and magnetic field strengths, rather than voltages and currents, are measured, and scattering variables allow a generalized approach to be used. A major reason why S-parameters are preferred for characterizing microwave circuits is that they are measured in a matched impedance system, in contrast to the open- and short-circuit-type measurements required for other available network parameters, measurements that can be very difficult to make at microwave frequencies.

2.2 N-PORT VOLTAGE AND CURRENT CONVENTIONS

Figure 2.1(a) shows a very simple two-port network, consisting of a length of transmission line. The indicated voltage polarities and current directions are all important. As defined, positive current flows in at the assumed positive voltage terminal at both ports. Should the current direction be the opposite to that assumed the current will have a negative value. Reversal of voltage polarity will also show up as a negative numerical value. Figure 2.1(b) shows these conventions generalized to an n-port network.

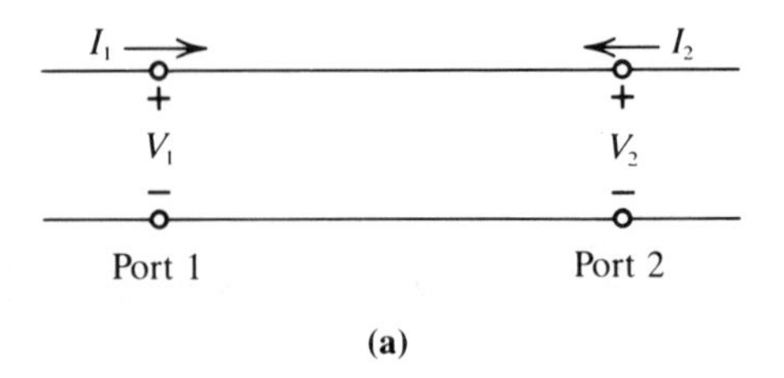

(a)

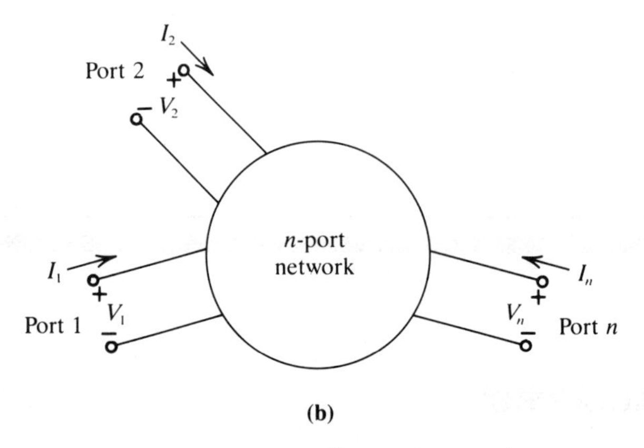

(b)

FIGURE 2.1. (a) Transmission-line section as a two-port network; (b) representation of an n-port network.

The n-port network is assumed to be linear; that is, the voltage and current at any one port are linearly related to the sum of the voltages and currents at the other ports. In effect, the network does not introduce nonlinear distortion such as squaring the voltage or current. Furthermore, unless otherwise stated, the voltages and currents are rms values of sinusoidal functions. At the nth port, therefore, the voltage and current may have their own phase angle relative to some previously established reference phase, and can be described by

$$V_n = |V_n| \angle\theta \tag{2.1}$$

$$I_n = |I_n| \angle\phi \tag{2.2}$$

The average power flowing into the nth port is then given by

$$P_n = |V_n|\,|I_n| \cos(\theta - \phi) \tag{2.3}$$

For the work in hand, a mathematically more convenient form of equation results when the voltage and current are kept as complex quantities, and the real part of the product of one with the complex conjugate of the other used. Denoting the complex conjugate by means of * and the real part by Re, then

$$P_n = \text{Re}\,(V_n I_n^*) \tag{2.4}$$

An alternative form is

$$P_n = \text{Re}(V_n^* I_n) \tag{2.5}$$

2.3 SCATTERING VARIABLES

Scattering variables at a port are defined in terms of the port voltage V_n, port current I_n, and a normalizing number Z_0. Because this normalizing number turns out to have dimensions of impedance, and also because it can be identified with the characteristic or wave impedance of transmission lines and waveguides, the symbol Z_0 is used. It must be kept in mind, however, that Z_0 can be applied to networks in general and is not restricted to use with transmission lines and waveguides. At any port n, the voltage and current can be synthesized by an incident scattering variable a_n and a reflected scattering variable b_n, defined such that

$$a_n + b_n \triangleq \frac{V_n}{\sqrt{Z_0}} \tag{2.6}$$

$$a_n - b_n \triangleq I_n \sqrt{Z_0} \tag{2.7}$$

Now although these definitions may appear to be unduly abstract, their significance will become apparent when applied to power flow and to the measurable quantities readily obtained at microwave frequencies. The scattering parameters are indicated on diagrams as shown in Fig. 2.2(a). It will be seen that all the *a*'s are directed into the ports, and all the *b*'s out of the ports. An interesting and essential point to grasp is that the role of a variable can change depending on the port with which it is associated. This is illustrated in Fig. 2.2(b), which shows port n of a network connected to a one-port network, the load impedance Z_L, designated port L. As can be seen, the reflected scattering variable from the load forms the incident scattering variable for the port n, and the reflected scattering variable from port n forms the incident scattering variable to the load. Thus, for Fig. 2.2(b),

$$a_n = b_L \tag{2.8}$$

$$b_n = a_L \tag{2.9}$$

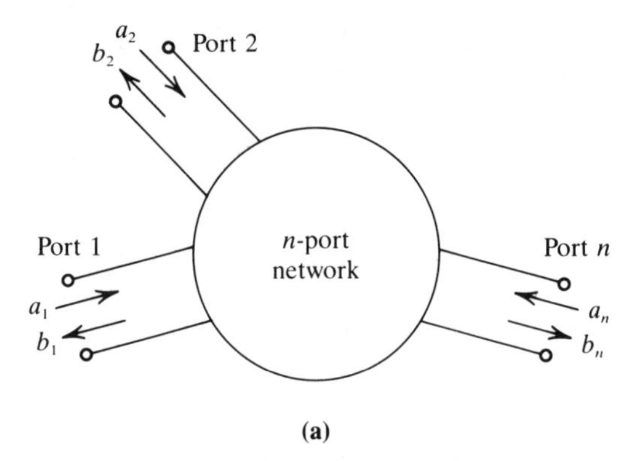

(a)

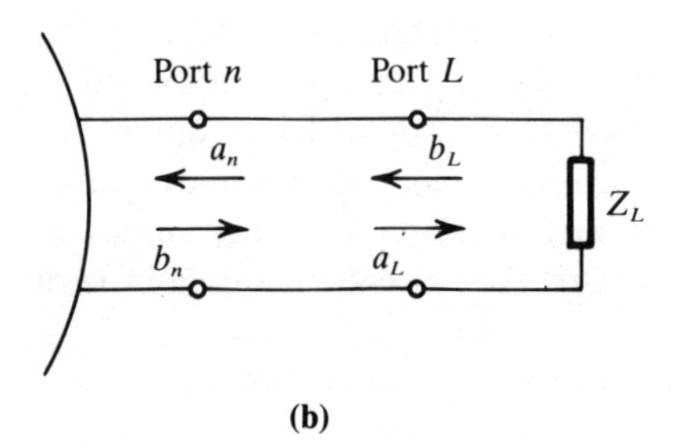

(b)

FIGURE 2.2. (a) Scattering variables for an n-port network; (b) interchange of scattering variables between two ports.

2.4 AVERAGE POWER IN TERMS OF SCATTERING VARIABLES

Equation (2.4) gives the average power in terms of voltage V_n and the complex conjugate of current I_n^*. Substituting Eqs. (2.6) and (2.7) therefore gives

$$P_n = \mathrm{Re}\left[\sqrt{Z_0}\,(a_n + b_n)\,\frac{a_n^* - b_n^*}{\sqrt{Z_0}}\right]$$

$$= \mathrm{Re}\,[a_n a_n^* - b_n b_n^* + b_n a_n^* - a_n b_n^*] \tag{2.10}$$

Now, the real part of $b_n a_n^* - a_n b_n^*$ is equal to zero, and the imaginary part may be ignored since the power equation requires the real part (Re) only. Also, $a_n a_n^* = |a_n|^2$ and $b_n b_n^* = |b_n|^2$, so that

$$P_n = |a_n|^2 - |b_n|^2 \tag{2.11}$$

Equation (2.11) shows that the average power flow into a port is that carried by the incident scattering variable minus that carried by the reflected scattering variable.

Although most of the analytical work of networks may be carried out in terms of the scattering variables, there will come a point where it will be necessary to determine these in terms of equivalent voltages and currents at the ports. This is easily done, since, adding Eqs. (2.6) and (2.7) gives

$$a_n = \frac{V_n}{2\sqrt{Z_0}} + \frac{I_n\sqrt{Z_0}}{2} \tag{2.12}$$

and substracting them gives

$$b_n = \frac{V_n}{2\sqrt{Z_0}} - \frac{I_n\sqrt{Z_0}}{2} \tag{2.13}$$

EXAMPLE 2.1

The rms voltage at a port is 3 $\angle 0$ V, and the rms current is 2 $\angle 30°$ A. Calculate (a) the scattering variables for a normalizing number of 100 Ω, and (b) the power flow into the port.

SOLUTION

(a) From Eq. (2.12),

$$\begin{aligned} a_n &= \frac{3}{2\sqrt{100}} + \frac{2\ \angle 30°\ \sqrt{100}}{2} \\ &= 0.15 + 8.66 + j5 \\ &= \mathbf{8.81 + \mathit{j}5} \end{aligned}$$

From Eq. (2.13),

$$\begin{aligned} b_n &= \frac{3}{2\sqrt{100}} - \frac{2\ \angle 30°\ \sqrt{100}}{2} \\ &= 0.15 - 8.66 - j5 \\ &= \mathbf{-8.51 - \mathit{j}5} \end{aligned}$$

(b) From Eq. (2.11),

$$\begin{aligned} P_n &= [(8.81)^2 + (5)^2] - [(8.51)^2 + (5)^2] \\ &= (8.81)^2 - (8.51)^2 \\ &= \mathbf{5.196\ W} \end{aligned}$$

2.5 ONE-PORT NETWORKS: REFLECTION COEFFICIENT

The simplest one-port network is the single impedance shown in Fig. 2.3(a). For this, the reflection coefficient Γ_n is defined as

$$\Gamma_n \triangleq \frac{b_n}{a_n} \tag{2.14}$$

Now, from Eqs. (2.6) and (2.7),

$$\frac{a_n + b_n}{a_n - b_n} = \frac{V_n}{I_L Z_0}$$

or

$$\frac{1 + \Gamma_n}{1 - \Gamma_n} = \frac{V_n}{I_n Z_0} \tag{2.15}$$

But since $Z_n = V_n/I_n$, Eq. (2.15) can be written as

$$\frac{1 + \Gamma_n}{1 - \Gamma_n} = \frac{Z_n}{Z_0} \tag{2.16}$$

The ratio Z_n/Z_0 occurs very frequently in this work and is referred to as the normalized impedance. By definition,

$$z_n \triangleq \frac{Z_n}{Z_0} \tag{2.17}$$

Equation (2.16) can therefore be written as

$$z_n = \frac{1 + \Gamma_n}{1 - \Gamma_n} \tag{2.18}$$

This in turn can be rearranged to give the further useful relationship:

$$\Gamma_n = \frac{z_n - 1}{z_n + 1}$$

$$= \frac{Z_n - Z_0}{Z_n + Z_0} \tag{2.19}$$

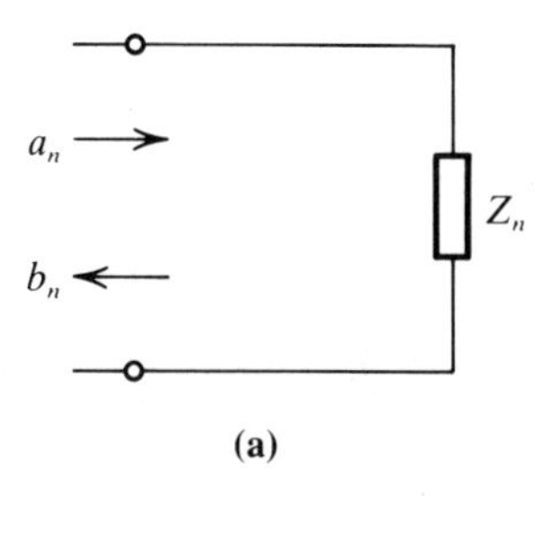

(a)

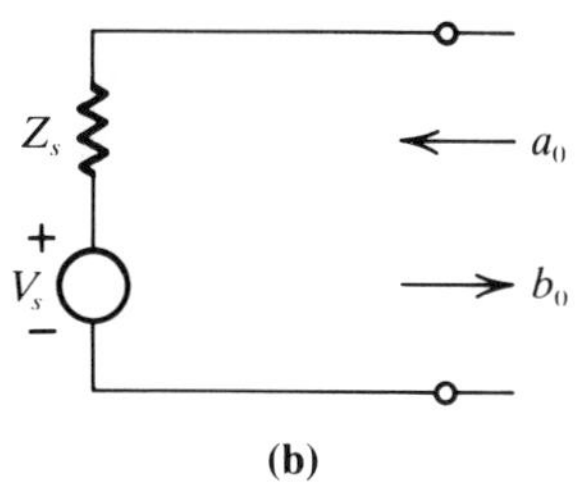

(b)

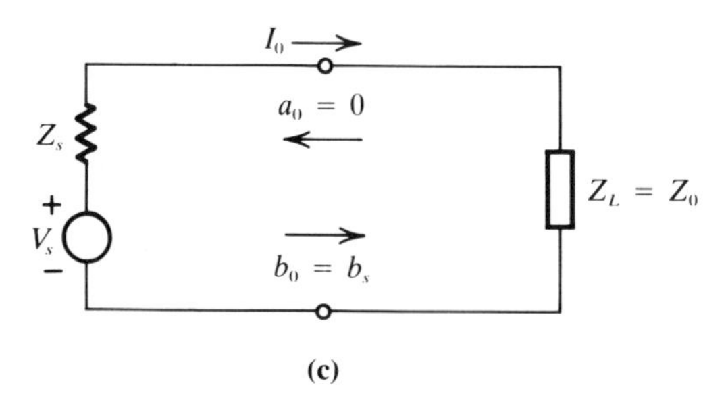

(c)

FIGURE 2.3. (a) Impedance as a one-port; (b) EMF source as a one-port; (c) circuit used for evaluating the source EMF contribution to the reflected variable.

Comparison of Eq. (2.19) with Eq. (1.61) shows that the reflection coefficient as defined in terms of scattering variables is identical to the voltage reflection coefficient defined previously for transmission lines, provided that the normalizing number Z_0 is identified with the characteristic impedance of the line.

An EMF source is also a one-port and here the scattering variables must take the source EMF V_s into account. Figure 2.3(c) shows an EMF source connected to a load Z_L. The incident variable for the source is a_0 and this will give rise to a reflected component $\Gamma_s a_0$, where Γ_s is the reflection coefficient of the source and is given by Eq. (2.19) on replacing the subscript n with the subscript s. In addition to this component, the source EMF will contribute a component b_s, so that the total reflected variable from the source is

$$b_0 = \Gamma_s a_0 + b_s \tag{2.20}$$

The component b_s may be found by arranging the circuit so that a_0 is zero, and this can be done by making $Z_L = Z_0$, or equivalently, $z_L = 1$. From Eq. (2.19) this would give $\Gamma_L = 0$, and $a_0 = b_L = \Gamma_L b_s = 0$. Making $Z_L = Z_0$ results in the circuit of Fig. 2.3(c), from which

$$I_0 = \frac{V_s}{Z_s + Z_0} \tag{2.21}$$

Now bearing in mind that $a_0 = 0$ and that the current flowing into the source is $-I_0$, Eq. (2.7) gives

$$a_0 - b_0 = -I_0\sqrt{Z_0}$$

Since $a_0 = 0$, Eq. (2.20) gives $b_0 = b_s$ and therefore

$$-b_s = -\frac{V_s\sqrt{Z_0}}{Z_s + Z_0}$$

or

$$b_s = \frac{V_s\sqrt{Z_0}}{Z_s + Z_0} \tag{2.22}$$

EXAMPLE 2.2

An EMF source has $V_s = 0.3$ V and $Z_s = 50\ \Omega$. Calculate the scattering parameters normalized to $Z_0 = 50\ \Omega$.

SOLUTION From Eq. (2.17),

$$z_s = \frac{50}{50}$$
$$= 1$$

From Eq. (2.19),

$$\Gamma_s = \frac{1 - 1}{1 + 1}$$
$$= \mathbf{0}$$

From Eq. (2.22),

$$b_s = \frac{0.3\sqrt{50}}{50 + 50}$$

$$= \mathbf{0.0212}$$

EXAMPLE 2.3

A load impedance of 100 $\angle -30°$ Ω is connected to the source of Example 2.2. Determine (a) the reflection coefficient for the load, and (b) the scattering variables existing at the common port, normalized to $Z_0 = 50$ Ω.

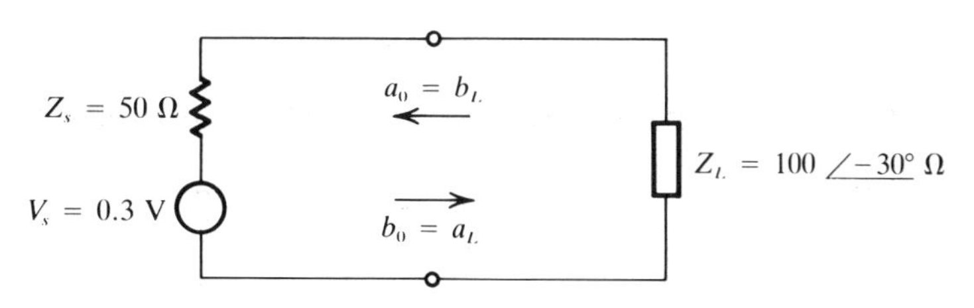

FIGURE 2.4. Circuit for Example 2.3.

SOLUTION The circuit is shown in Fig. 2.4.

(a) From Eq. (2.17),

$$z_L = \frac{100\angle -30°}{50}$$

$$= 2\angle -30°$$

From Eq. (2.19),

$$\Gamma_L = \frac{2\ \angle -30° - 1}{2\ \angle -30° + 1}$$

$$= \frac{1.732 - j1 - 1}{1.732 - j1 + 1}$$

$$= \mathbf{0.426\ \angle -33.69°}$$

(b) From Example 2.2, $\Gamma_s = 0$ and $b_s = 0.0212$. Hence, from Eq. (2.20),

$$b_0 = 0 + 0.0212$$

$$= \mathbf{0.0212}$$

From Fig. 2.4 [and as discussed in connection with Eq. (2.8) and (2.9)],

$$a_L = b_0$$
$$= \mathbf{0.0212}$$

Also, $a_0 = b_L$, and from Eq. (2.14),

$$b_L = (0.426\angle -33.69^\circ) \times 0.0212$$

Hence

$$\boldsymbol{a_0 = b_L = 0.009\angle -33.69^\circ}$$

2.6 SCATTERING EQUATIONS FOR A TWO-PORT NETWORK

Figure 2.5(a) shows the general representations of a two-port network. Because the network is linear, the reflected variable at a given port will be the algebraic sum of two components, the component originating from the incident variable at that port, and the component transmitted through the network from the other incident variable. This allows the network to be modeled by two scattering equations:

$$b_1 = S_{11}a_1 + S_{12}a_2 \tag{2.23}$$

$$b_2 = S_{21}a_1 + S_{22}a_2 \tag{2.24}$$

where S_{11}, S_{12}, and so on, are known as the *scattering parameters*. These parameters can be measured when the network is impedance matched for zero reflections, in contrast to other system parameters (such as h-parameters and Y-parameters), which require open- and short-circuit terminations. Such extreme terminations at microwaves usually lead to instability and unwanted oscillation. Figure 2.5(b) shows how the network would be terminated in order to make $a_2 = 0$. That this is so follows from the fact that $\Gamma_L = 0$ and therefore $b_L = 0$. But $b_L = a_2$, as shown in Fig. 2.5(b), and therefore $a_2 = 0$. Under these conditions, Eqs. (2.23) and (2.24) reduce to

$$S_{11} = \left.\frac{b_1}{a_1}\right|_{a2=0} \tag{2.25}$$

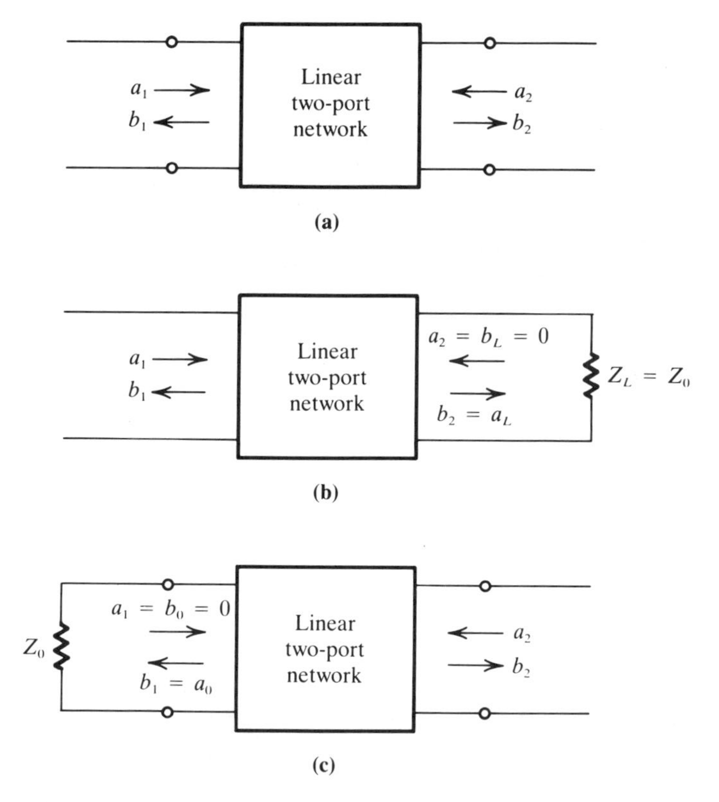

FIGURE 2.5. (a) Scattering variables for a linear two-port network; (b) Arrangement for measuring scattering parameters S_{11} and S_{21}; (c) arrangement for measuring scattering parameters S_{22} and S_{12}.

$$S_{21} = \left.\frac{b_2}{a_1}\right|_{a2=0} \tag{2.26}$$

Assuming, therefore, that the scattering variables a_1, b_1, and b_2 can all be measured, the scattering parameters as given by Eqs. (2.25) and (2.26) can be determined. A method of measuring the scattering variables is given in Chapter 14. S_{11} is termed the input reflection coefficient with $a_2 = 0$. Note carefully the similarity (and difference) between this and the reflection coefficient defined by Eq. (2.14). In the latter case the reflection occurs from a passive impedance, and therefore originates from the incident variables, whereas in the general two-port situation [Fig. 2.5(a)], the reflected component b_1 has two components, and the component originating from a_2 must be forced to zero, in order to enable S_{11} to be determined.

A similar procedure can be adopted in order to measure S_{22} and S_{12}, by terminating port 1 in Z_0 and measuring from port 2, as shown in Fig. 2.5(c).

TABLE 2.1
Scattering Parameters for a Two-Port Network

Parameter	*Meaning*	*Reflectionless Matched Condition*
S_{11}	Input reflection coefficient	$a_2 = 0$
S_{22}	Output reflection coefficient	$a_1 = 0$
S_{12}	Reverse transmission coefficient	$a_1 = 0$
S_{21}	Forward transmission coefficient	$a_2 = 0$

From this

$$S_{22} = \left.\frac{b_2}{a_2}\right|_{a_1=0} \tag{2.27}$$

$$S_{12} = \left.\frac{b_1}{a_2}\right|_{a_1=0} \tag{2.28}$$

Table 2.1 summarizes the scattering parameters. The scattering parameters for a low-cost gallium arsenide field-effect transistor (FET) are shown in Table 2.2. The transistor is manufactured by the Avanek Company. The use of the scattering parameters in determining transistor amplifier performance is covered in Chapter 8.

2.7 NETWORK ANALYSIS USING SCATTERING PARAMETERS

The analysis will be confined to a linear two-port network as shown in Fig. 2.6, but this covers a wide range of practical situations. The first objective of the analysis is to obtain expressions for the scattering variables a_1, b_1, a_2, and b_2 in terms of the network parameters S_{11}, S_{12}, S_{21}, and S_{22}, the source quantities Γ_s and V_s, and the load reflection coefficient Γ_L. These parameters and quantities can be assumed known, as they will either be specified or will be obtainable by measurement. Once the scattering variables have been found, they can be used to determine voltage and power gains for the network, and overall reflection coefficients for network with load, and with source connected. The scattering variables may be obtained as follows. From Fig. 2.6,

$$\Gamma_L = \frac{b_L}{a_L} = \frac{a_2}{b_2} \tag{2.29}$$

TABLE 2.2
Typical Scattering Parameters, Common Source* for Avantek AT-12570-5, 2-10 GHz Small Signal Gallium Arsenide FET

AT·12570-5 V_{DS} = 3 V, I_{DS} = 20 mA

FREQUENCY (GHz)	S11 (MAG)	S11 (ANG)	S21 (MAG)	S21 (ANG)	S12 (MAG)	S12 (ANG)	S22 (MAG)	S22 (ANG)
1.00	.969	−32	4.02	150	.036	68	.514	−23
2.00	.928	−64	3.91	121	.068	45	.483	−49
3.00	.869	−98	3.61	93	.093	23	.434	−75
4.00	.782	−126	3.19	66	.106	3	.389	−97
5.00	.741	−155	2.84	41	.115	−14	.352	−119
6.00	.719	179	2.54	20	.117	−29	.339	−140
7.00	.720	156	2.27	−1	.118	−43	.338	−159
8.00	.692	139	2.02	−21	.116	−57	.349	−174
10.00	.687	107	1.72	−59	.114	−81	.381	153

AT·12570-5 V_{DS} = 5 V, I_{DS} = 50 mA

FREQUENCY (GHz)	S11 (MAG)	S11 (ANG)	S21 (MAG)	S21 (ANG)	S12 (MAG)	S12 (ANG)	S22 (MAG)	S22 (ANG)
1.00	.963	−34	5.07	148	.027	69	.559	−22
2.00	.907	−69	4.82	118	.051	47	.526	−45
3.00	.841	−104	4.34	89	.067	26	.474	−63
4.00	.753	−133	3.77	63	.076	7	.433	−87
5.00	.713	−161	3.30	38	.081	−6	.400	−107
6.00	.697	173	2.91	17	.082	−19	.392	−125
7.00	.700	151	2.58	−4	.084	−30	.399	−145
8.00	.671	135	2.28	−24	.085	−40	.416	−160
10.00	.668	104	1.93	−62	.089	−60	.464	169

*As measured in the Avantek TF-070 test fixture. Fixtures available, factory inquiry. (Courtesy Avantek Inc., Santa Clara, Ca.)

and from this, $b_2 = \Gamma_L a_2$. This relationship for b_2 may be substituted in the scattering equation (2.24) to give

$$0 = S_{21}a_1 + \left(S_{22} - \frac{1}{\Gamma_L}\right)a_2 \tag{2.30}$$

Also, from Fig. 2.6 it is seen that $a_1 = b_0$ and $b_1 = a_0$ and these relationships may be substituted into Eq. (2.20) to give

$$a_1 = \Gamma_s b_1 + b_s \tag{2.31}$$

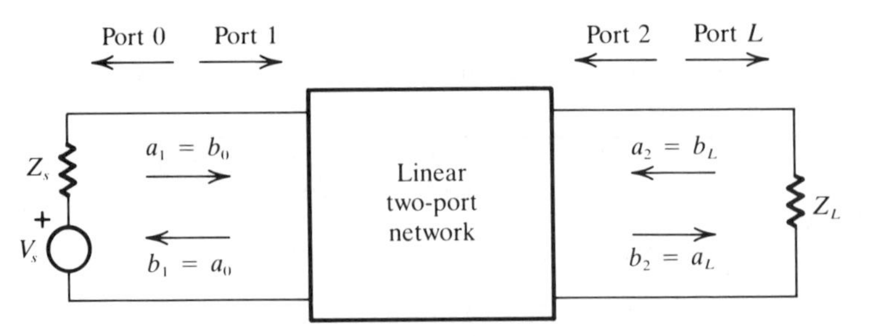

FIGURE 2.6. Scattering variables associated with a linear two-port network.

Equation (2.31) may now be substituted into the scattering equation (2.23) to get:

$$\frac{-b_s}{\Gamma_s} = \left(S_{11} - \frac{1}{\Gamma_s}\right)a_1 + S_{12}a_2 \tag{2.32}$$

Equations (2.30) and (2.32) can be solved for a_1 and a_2, but they are not in the most convenient form as written because they contain terms $1/\Gamma$; since often Γ is zero, the resulting term in infinity makes the mathematics more difficult to handle. Multiplying through by the respective reflection coefficients clears the reciprocal terms from the equations, and these can be rewritten as

$$\text{[Eq. (2.30)]:} \quad 0 = A_{21}a_1 + A_{22}a_2 \tag{2.33}$$

$$\text{[Eq. (2.32)]:} \quad -b_s = A_{11}a_1 + A_{12}a_2 \tag{2.34}$$

where, for convenience, $A_{21} = \Gamma_L S_{21}$, $A_{12} = \Gamma_s S_{12}$, $A_{11} = \Gamma_s S_{11} - 1$, and $A_{22} = \Gamma_L S_{22} - 1$. The determinant of the system is

$$D = \begin{vmatrix} A_{21} & A_{22} \\ A_{11} & A_{12} \end{vmatrix} = A_{21}A_{12} - A_{11}A_{22} \tag{2.35}$$

Solving for a_1 gives

$$a_1 = \frac{\begin{vmatrix} 0 & A_{22} \\ -b_s & A_{12} \end{vmatrix}}{D} = \frac{b_s A_{22}}{D} \tag{2.36}$$

Solving for a_2 yields

$$a_2 = \frac{\begin{vmatrix} A_{21} & 0 \\ A_{11} & -b_s \end{vmatrix}}{D} = \frac{-b_s A_{21}}{D} \tag{2.37}$$

Once a_1 and a_2 have been found, b_1 and b_2 can be determined from Eqs. (2.23) and (2.24), this being left as an exercise for the reader. The complete set of equations for the scattering variables are collected together here for easy reference:

$$a_1 = \frac{b_s A_{22}}{D} \tag{2.36}$$

$$a_2 = \frac{-b_s A_{21}}{D} \tag{2.37}$$

$$b_1 = \frac{b_s(A_{22}S_{11} - S_{12}A_{21})}{D} \tag{2.38}$$

$$b_2 = \frac{b_s(A_{22}S_{21} - S_{22}A_{21})}{D} \tag{2.39}$$

The auxiliary equations are

$$A_{11} = \Gamma_s S_{11} - 1 \tag{2.40}$$

$$A_{22} = \Gamma_L S_{22} - 1 \tag{2.41}$$

$$A_{12} = \Gamma_s S_{12} \tag{2.42}$$

$$A_{21} = \Gamma_L S_{21} \tag{2.43}$$

$$D = A_{21}A_{12} - A_{11}A_{22} \tag{2.35}$$

Note that although it may be possible to use the simpler relationships, $b_1 = (a_1 - b_s)/\Gamma_s$, from Eq. (2.31), and $b_2 = a_2/\Gamma_L$, from Eq. (2.29), instead of Eqs. (2.38) and (2.39), these "simpler" equations lead to the indeterminate expression 0/0 when the reflection coefficients are zero, which is often the case in practice. This difficulty is not encountered with the forms given in Eqs. (2.38) and (2.39).

EXAMPLE 2.4

The scattering parameters for transistor AT-12570-5 are shown in Table 2.2. The transistor is used in a common-source amplifier at $V_{DS} = 3$ V, $I_{DS} = 20$ mA, and $f = 4.00$ GHz. The signal source can be represented by a 10 μV EMF in series with a 50-Ω resistance, and the load is 100 Ω purely resistive. Calculate the scattering variables normalized to $Z_0 = 50\ \Omega$.

SOLUTION

From Table 2.2,

$$S_{11} = 0.782\ \underline{/-126^\circ} \qquad S_{21} = 3.19\underline{/66^\circ}$$

$$S_{12} = 0.106\ \underline{/3^\circ} \qquad S_{22} = 0.389\underline{/-97^\circ}$$

From Eq. (2.19),

$$\Gamma_s = \frac{50 - 50}{50 + 50} = 0$$

$$\Gamma_L = \frac{100 - 50}{100 + 50} = \frac{1}{3}$$

From Eq. (2.40),

$$A_{11} = -1$$

From Eq. (2.41),

$$A_{22} = \frac{0.389\ \underline{/-97^\circ}}{3} - 1 = 1.024\ \underline{/-172.78^\circ}$$

From Eq. (2.42),

$$A_{12} = 0$$

From Eq. (2.43),

$$A_{21} = \frac{3.19}{3}\ \underline{/66^\circ} = 1.063\ \underline{/66^\circ}$$

From Eq. (2.35),

$$D = 0 - (-1)(1.024\ \underline{/-172.78^\circ}) = 1.024\ \underline{/-172.78^\circ}$$

From Eq. (2.22),

$$b_s = \frac{10\sqrt{50}}{50 + 50} = 0.707\ \mu\text{V}/\sqrt{\Omega}$$

At this point, all the constants of the system have been determined, and it now remains to calculate the scattering variables.

From Eq. (2.36),

$$a_1 = \frac{0.707 \times (1.024\ \underline{/-172.78°})}{1.24\ \underline{/-172.78°}}$$

$$= \mathbf{0.707\ \underline{/0°}}$$

From Eq. (2.37),

$$a_2 = \frac{-0.707 \times 1.063\ \underline{/66°}}{1.024\ \underline{/-172.78°}}$$

$$= \mathbf{0.734\ \underline{/58.78°}}$$

From Eq. (2.38),

$$b_1 = \frac{0.707[(1.024\ \underline{/-172.78})(0.782\ \underline{/-126°}) - (0.106\ \underline{/3°})(1.063\ \underline{/66°})]}{1.024\ \underline{/-172.78°}}$$

$$= \mathbf{0.476\ \underline{/-127.3°}}$$

From Eq. (2.39),

$$b_2 = \frac{0.707[(1.024\ \underline{/-172.78°})(3.19\ \underline{/66°}) - (0.389\ \underline{/-97°})(1.063\ \underline{/66°})]}{1.024\ \underline{/-172.78°}}$$

$$= \mathbf{2.203\ \underline{/58.78°}}$$

In this example, the units for the scattering variables are $\mu V/\sqrt{\Omega}$.

Note that in this example, b_2 can be evaluated using the simpler formula $b_2 = a_2/\Gamma_L$, but trying to use the similar relationship, Eq. (2.31) to evaluate b_1 leads to difficulties.

2.8 VOLTAGE GAIN

The terminal voltage gain of the two-port network of Fig. 2.6 can be defined as

$$A_v \triangleq \frac{V_2}{V_1} \tag{2.44}$$

where V_2 is the terminal voltage at port 2 and V_1 the terminal voltage at port 1. Applying Eq. (2.6) to each port results in

$$A_v = \frac{a_2 + b_2}{a_1 + b_1} \tag{2.45}$$

Now using the Eqs. (2.36) through (2.39) for the scattering variables results in

$$A_v = \frac{-A_{21} + (A_{22}S_{21} - S_{22}A_{21})}{A_{22} + (A_{22}S_{11} - S_{12}A_{21})} \tag{2.46}$$

The factor A_{11} does not enter into Eq. (2.46), so a knowledge of the source parameters is not required.

EXAMPLE 2.5

Calculate the terminal voltage gain for the amplifier in Example 2.4.

SOLUTION

Since in Example 2.4 the scattering variables are determined directly, Eq. (2.45) is the most convenient to use in this case. The values are $a_1 = 0.707$, $a_2 = 0.734 \angle 58.78°$, $b_1 = 0.476 \angle -127.78°$, and $b_2 = 2.203 \angle 58.78°$.

Hence

$$A_v = \frac{0.734 \angle 58.78° + 2.203 \angle 58.78°}{0.707 + 0.476 \angle -127.3°}$$

$$= \frac{2.937 \angle 58.78°}{0.565 \angle -42.1}$$

$$= \mathbf{5.2 \angle 100.9°}$$

EXAMPLE 2.6

A small-signal amplifier for use at 10 GHz utilizes a GaAs FET, type HFET-2202, for which

$$S_{11} = 0.67 \angle 143^\circ \qquad S_{22} = 0.66 \angle -142^\circ$$

$$S_{21} = 1.65 \angle -20^\circ \qquad S_{12} = 0.06 \angle 20^\circ$$

Given that $\Gamma_L = 0$, calculate the voltage gain.

SOLUTION Note that in this example the scattering variables cannot be calculated since the source parameters are not given, but Eq. (2.46) can be used to determine voltage gain. From Eq. (2.41) with $\Gamma_L = 0$, $A_{22} = -1$, and from Eq. (2.43), $A_{21} = 0$. Hence, applying Eq. (2.46), we have

$$\begin{aligned} A_v &= \frac{0 + (-1.65 \angle -20^\circ)}{-1 + (-0.67 \angle 143^\circ)} \\ &= \frac{1.65 \angle -20^\circ}{0.615 \angle 40.9^\circ} \\ &= \mathbf{2.68} \angle -60.9^\circ \end{aligned}$$

2.9 TRANSDUCER POWER GAIN

Although the word *transducer* is more usually associated with devices that convert a signal from one form to another (e.g., a pressure-to-voltage transducer), the word is also used to describe the power gain of a linear two-port network, which converts a source power to a load power. The transducer power G_T gain is defined as

$$\begin{aligned} G_T &= \frac{\text{power delivered to load}}{\text{power available from source}} \\ &= \frac{P_L}{P_{av}} \end{aligned} \tag{2.47}$$

The power available from a source is a reference power and is that power which the source can deliver to a conjugately matched load connected directly to the source. This means that if the source impedance is $Z_s = R$

$+ jX$, the load impedance must be $R - jX$ for the available power to be realized. This situation is shown in Fig. 2.7(a), in which the complex conjugate of Z_s is denoted by Z_s^*. It should be noted that the power available from the source is a *reference power* which does not depend in any way on the two-port network nor the load connected to this. It is determined entirely by the source parameters b_s as given by Eq. (2.22), and Γ_s, which can be shown as follows.

From Fig. 2.7(a) it is seen that $a_0 = b_1$. From the definition of reflection coefficient, $b_1 = \Gamma_s^* a_1$. Substituting these relationships in Eq. (2.20) gives

$$\begin{aligned} b_0 &= \Gamma_s \Gamma_s^* a_1 + b_s \\ &= \rho_s^2 a_1 + b_s \end{aligned} \tag{2.48}$$

where ρ_s is the modulus of Γ_s. Also, from Fig. 2.7(a), $b_0 = a_1$ and therefore Eq. (2.48) becomes

$$a_1 = \rho_s^2 a_1 + b_s$$

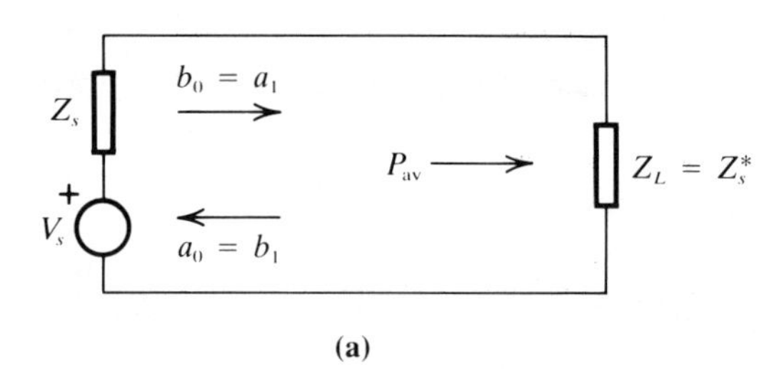

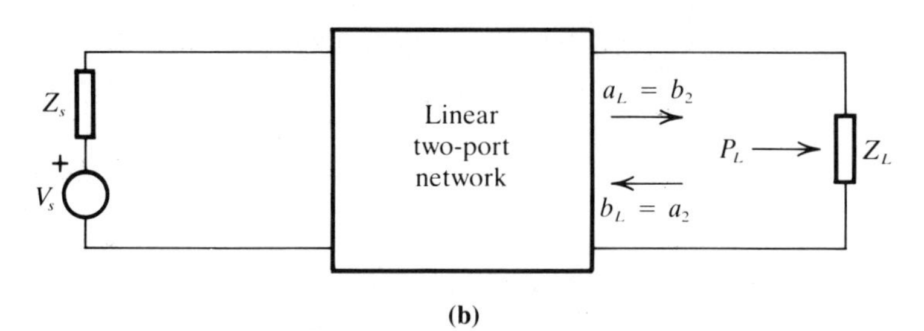

FIGURE 2.7. (a) Source conjugately matched so that the load power is by definition the available power from the source; (b) source feeding a linear two-port network, delivering power to an arbitrary load Z_L.

or

$$a_1 = \frac{b_s}{1 - \rho_s^2} \tag{2.49}$$

Using the relationship $b_1 = \Gamma_s^* a_1$ in Eq. (2.11) gives for the available power,

$$P_{av} = |a_1|^2(1 - \rho_s^2) \tag{2.50}$$

Equation (2.49) can now be used to substitute for a_1 in this expression for available power to give

$$P_{av} = \frac{|b_s^2|}{1 - \rho_s^2} \tag{2.51}$$

Next, an expression for load power in terms of the known parameters must be obtained. Applying Eq. (2.11) to the load power for Fig. 2.7(b) yields

$$P_L = |a_L|^2(1 - \rho_L^2) \tag{2.52}$$

$$= |b_2|^2(1 - \rho_L^2) \tag{2.53}$$

Using Eqs. (2.51) and (2.53), the transducer power gain, as defined by Eq. (2.47), becomes

$$G_T = \frac{|b_2^2|\,(1 - \rho_L^2)(1 - \rho_S^2)}{|b_s^2|} \tag{2.54}$$

With substitution of Eq. (2.39) for b_2, the b_s term cancels out, and the full expression for transducer power gain is

$$G_T = \frac{|A_{22}S_{21} - S_{22}A_{21}|^2(1 - \rho_L^2)(1 - \rho_s^2)}{|A_{21}A_{12} - A_{11}A_{22}|^2} \tag{2.55}$$

Bearing in mind that the A factors are complex, as given by Eqs. (2.40) through (2.43), evaluation of the power gain in general, from Eq. (2.55), will be quite difficult unless carried out by machine.

EXAMPLE 2.7

For the amplifier in Example 2.6, given that $\Gamma_s = 0$ also, determine the transducer power gain.

SOLUTION With $\Gamma_s = 0$, $A_{12} = 0$ and $A_{11} = -1$. Using these values, together with the values obtained in Example 2.6 in Eq. (2.55), gives

$$G_T = \frac{|-S_{21}|^2(1)}{|-1|^2}$$
$$= 1.65^2$$
$$= \mathbf{2.723\ (or\ 4.35\ dB)}$$

2.10 EFFECTIVE REFLECTION COEFFICIENTS

The reflection coefficient at the input port of a two-port network may be defined as

$$\Gamma_{11} \triangleq \frac{b_1}{a_1} \tag{2.56}$$

using Eqs. (2.36) and (2.38), this can be written as

$$\Gamma_{11} = \frac{A_{22}S_{11} - S_{12}A_{21}}{A_{22}}$$
$$= S_{11} - \frac{S_{12}A_{21}}{A_{22}} \tag{2.57}$$

Substituting for A_{21} from Eq. (2.43) and for A_{22} from Eq. (2.41) gives

$$\Gamma_{11} = S_{11} - \frac{S_{12}S_{21}\,\Gamma_L}{\Gamma_L S_{22} - 1} \tag{2.58}$$

Equation (2.58) shows that the effective input reflection coefficient contains both the input reflection coefficient S_{11} defined for matched output conditions, and a term that is dependent on the transmission coefficients and the load reflection coefficient. As can be seen, with $\Gamma_L = 0$, Γ_{11} reduces to S_{11}. Also, if S_{12} is very small (much less than unity), as is sometimes the case, the contribution of the output circuit to the input reflection coefficient is negligible.

An effective output reflection coefficient can be defined in a similar manner. In this case, the source voltage V_s is zero, and a signal applied at the output port will have two components a_2 and b_2, which are related by

$$\Gamma_{22} \triangleq \frac{b_2}{a_2} \tag{2.59}$$

By symmetry, the output reflection coefficient is obtained from Eq. (2.58) as

$$\Gamma_{22} = S_{22} - \frac{S_{12}S_{21}\Gamma_s}{\Gamma_s S_{11} - 1} \tag{2.60}$$

2.11 PROBLEMS

1. By expressing V_n and I_n in exponential form, show that Eqs. (2.4) and (2.5) are equivalent to Eq. (2.3).
2. Verify that Eq. (2.3) gives the same power flow as calculated in Example 2.1.
3. Express the following voltages and currents, in rectangular form. Assuming rms values, find the power in each case: (a) $V = 30\angle 10°$ V, $I = 2\angle -20°$ A; (b) $V = 10\angle 60°$ V, $I = 3\angle -30°$ A; (c) $V = 100\angle 0$ V, $I = 2\angle 25°$ A.
4. Express the following voltages and currents in polar form. Assuming rms values, find the power in each case: (a) $V = 5 + j3$ V, $I = 0.1 + j0.2$ A; (b) $V = 7 + j5$ V, $I = 0.1 + j0.3$ A.
5. Scattering variables measured at a port are: $a = 5 - j3$, $b = 2 + j5$. The normalizing impedance is $Z_0 = 50\ \Omega$. Calculate the voltage and current.
6. Scattering variables measured at a port are: $0.9\angle 35°$ and $0.3\angle 70°$. The normalizing impedance is $Z_0 = 50\ \Omega$. Calculate the voltage and current.
7. Calculate the power flow into a port in each case, for the following sets of scattering variables at the port: (a) $a = 2 + j3$, $b = 0.1 - j0.1$; (b) $a = 30 - j40$, $b = 5 - j5$.
8. Calculate the power flow into a port in each case, for the following sets of scattering variables at the port: (a) $a = 3\angle -30°$, $b = 2\angle 45°$; (b) $a = 100\angle 0°$, $b = 10\angle 0°$.
9. Calculate the scattering variables in each of the following cases for $Z_0 = 100\Omega$: (a) $V = 3 - j2$ V, $I = 0.3 + j0.4$ A; (b) $V = 10\angle 30°$ V, $I = 0.5\angle 70°$ A.

10. Show, using Eqs. (2.12) and (2.13) and for Z_0 real, that Eq. (2.11) is equivalent to Eq. (2.4).

11. Find the reflection coefficient for each of the following sets of scattering variables: (a) $a = 0.3 + j0.4$, $b = 0.1 - j0.2$; (b) $a = -0.5 + j0.2$, $b = 0 - j$; (c) $a = 0.7\angle 30°$, $b = 0.5\angle -20°$; (d) $a = 3\angle 0°$, $b = 0.2\angle 90°$.

12. For the reflection coefficients and characteristic impedances given, find the reflecting impedance in each case: (a) $\Gamma = 0.7\angle 30°$, $Z_0 = 50\ \Omega$; (b) $\Gamma = 0.9\angle -35°$, $Z_0 = 100\ \Omega$; (c) $\Gamma = 0.1 - j.2$, $Z_0 = 50\ \Omega$; (d) $\Gamma = 0.5 + j0$, $Z_0 = 600\ \Omega$.

13. For the impedances given, find the reflection coefficient in each case. All impedance values are in ohms: (a) $Z_0 = 50$, $Z_L = 75 - j50$; (b) $Z_0 = 100$, $Z_L = 100 - j100$; (c) $Z_0 = 50$, $Z_L = 100\angle 40°$; (d) $Z_0 = 50$, $Z_L = 25\angle 0°$; (e) $Z_0 = 100$, $Z_L = 10\angle 90°$.

14. An EMF source has an internal impedance of $100 + j0$ ohms, and EMF of 3 V. Using $Z_0 = 50\ \Omega$, find the reflection coefficient and scattering variable for the source.

15. An EMF source has an internal impedance of $100 + j50\ \Omega$ and EMF of 10 mV. Using $Z_0 = 50\ \Omega$, find the reflection coefficient and scattering variable for the source.

16. Show that b_s of Eq. (2.22) can be expressed as

$$b_s = \frac{V_s(1 - \Gamma_s)}{2\sqrt{Z_0}}$$

17. Show, using Eqs. (2.14) and (2.20), that for a source connected to a load, the scattering variables are given by

$$b_0 = \frac{b_s}{1 - \Gamma_s\Gamma_L} \quad \text{and} \quad a_0 = \frac{\Gamma_L}{1 - \Gamma_s\Gamma_L}\, b_s$$

18. Using the results of Problem 16, solve for the scattering variables when $V_s = 1$ V, $Z_s = 80\ \Omega$, $Z_L = 30 + j40\ \Omega$, and $Z_0 = 50\ \Omega$.

19. Using the scattering parameters listed in Table 2.2 for $V_{ds} = 3$ V and $f = 10$ GHz, determine the scattering variables a_1, a_2, b_1, b_2 for the following load conditions: (a) 50 Ω; (b) 100 Ω; (c) 200 Ω. The source is 3 mV EMF in series with a 50-Ω internal impedance. $Z_0 = 50\ \Omega$. All impedances are real.

20. Using the scattering parameters listed in Table 2.2 for $V_{ds} = 5$ V and $f = 1$ GHz, determine the scattering variables a_1, a_2, b_1, b_2 for the following source and load impedances: (a) source $(30 + j10)$, load (50

$+ j0$); (b) source ($50 + j0$), load ($30 + j60$); (c) source ($20 - j20$), load ($40 - j5$). All impedance values are in ohms, $Z_0 = 50\ \Omega$ and the source EMF is 3 mV.

21. Calculate the voltage gain A_v for each set of conditions specified in Problems 19 and 20.

22. Using the scattering parameters listed in Table 2.2 for $V_{ds} = 3$ V and $f = 6$ GHz, determine the voltage gain A_v for a load impedance of 70 Ω, and a characteristic impedance of 50 Ω, both purely resistive.

23. Calculate the power gain G_T for each set of conditions specified in Problems 19 and 20.

24. Calculate the effective input reflection coefficient for each set of conditions specified in Problems 19 and 20.

25. Calculate the effective output reflection coefficient for each set of conditions specified in Problems 19 and 20.

FURTHER READING

Hewlett-Packard Application Note 95-1, S-Parameter Techniques for Faster, More Accurate Network Design.

Hewlett-Packard Application Note 154, S-Parameter Design.

three
The Smith Chart

3.1 INTRODUCTION

In Chapters 1 and 2 it is seen that important interrelationships exist between the voltage reflection coefficient, impedance, and voltage standing-wave ratio, and calculations leading from one to the other are complex and difficult in general. The fundamental relationship between reflection coefficient and impedance at a pair of terminals is given by Eqs. (1.61) and (2.19), which in terms of normalized impedance may be written as

$$\Gamma = \frac{z - 1}{z + 1} \tag{3.1}$$

Subscripts will only be shown where it is necessary to distinguish specific impedance levels, so the normalized impedance in general is

$$z = \frac{Z}{Z_0} \tag{3.2}$$

In microwave applications Z_0 may be identified with the characteristic impedance of the transmission medium.

Since the impedance will be complex in general, the reflection coefficient will also be complex, and may be represented by a magnitude ρ and a phase angle ϕ by either of the forms

$$\Gamma = \rho \angle \phi = \rho e^{j\phi} \tag{3.3}$$

The Smith chart provides a graphical means of mapping between Eqs. (3.1) and (3.3).

3.2 CONSTRUCTION OF THE SMITH CHART

In the Smith chart, Eqs. (3.1) and (3.3) are plotted on an Argand diagram, known as the $t = p + jq$-plane. Considering first Eq. (3.3) when this is represented on the t-plane, then

$$\rho e^{j\phi} = p + jq \tag{3.4}$$

from which

$$\rho = \sqrt{p^2 + q^2} \tag{3.5}$$

and

$$\phi = \tan^{-1}\frac{q}{p} \tag{3.6}$$

This relationship is shown in Fig. 3.1 for two specific values of reflection coefficient.

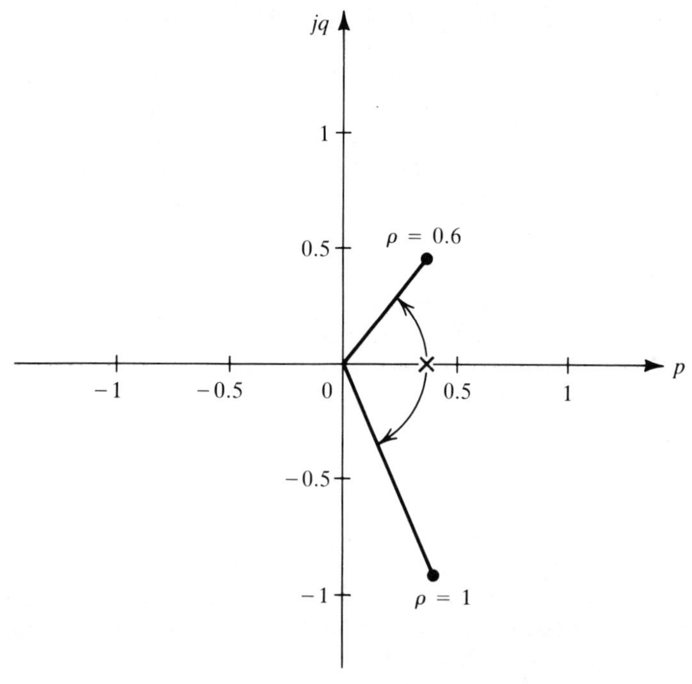

FIGURE 3.1. Two reflection coefficients $\Gamma = 0.6 \angle 53.13°$ and $= 1 \angle 67.38°$ drawn on the $t = p + jq$ plane.

To show Eq. (3.1) it is first necessary to split this into its real and imaginary parts. The normalized impedance z will have a normalized resistive component r, and a normalized reactance component x, where

$$z = r + jx \tag{3.7}$$

Therefore, Eq. (3.1) can be written as

$$\frac{(r - 1) + jx}{(r + 1) - jx} = p + jq \tag{3.8}$$

When this is separated into its real and imaginary parts, the two equations that result are

$$\left(p - \frac{r}{r + 1}\right)^2 + q^2 = \left(\frac{1}{r + 1}\right)^2 \tag{3.9}$$

$$\left(q - \frac{1}{x}\right)^2 + (1 - p)^2 = \left(\frac{1}{x}\right)^2 \tag{3.10}$$

Equation (3.9) represents a family of circles of radius $1/(r + 1)$, and centered along the p-axis at positions $r/(r + 1)$. Three such circles are shown in Fig. 3.2(a). Later it will be seen that negative values of r need not be shown, and the regular Smith chart extends only to the $r = 0$ circle.

Equation (3.10) also represents a family of circles. These have radius $1/x$ and are centered along a line given by $p = 1$. The centers are at positions $1/x$ on this line. Figure 3.2(b) shows four such circles. Now, because the Smith chart does not extend beyond the $r = 0$ circle, only those segments of the x-circles that lie within the $r = 0$ circle are shown on the regular Smith chart. Figure 3.3 shows the t-plane with Figures 3.1 and 3.2 combined.

On an actual Smith chart, the p-axis, instead of being calibrated linearly, is calibrated in r-values. The calibration is obtained from Eq. (3.8) on setting x equal to zero. (At $x = 0$, the radius of the x-circle becomes ∞ and the circle degenerates into the p-axis.) This gives

$$\frac{r - 1}{r + 1} = p \tag{3.11}$$

Some sample values are

$$p = -1 \qquad r = 0$$

$$p = 0 \qquad r = 1$$

$$p = 1 \qquad r = \infty$$

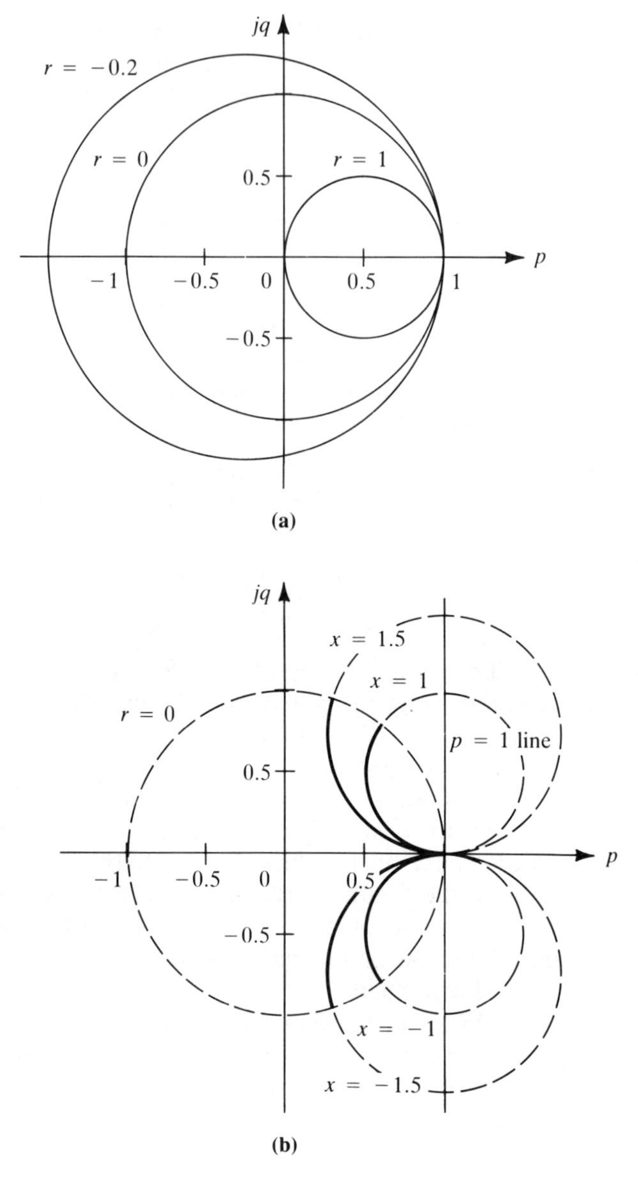

FIGURE 3.2. (a) Constant r-circles [Eq. (3.9)] and (b) constant x-circles [Eq. (3.10)] on the $t = p + jq$ plane.

The r-scale is seen to extend from zero to infinity for p-values ranging from -1 to $+1$. Also on the Smith chart, the q-axis need not be shown and the segments of x-circles are marked directly in the x-values. An additional scale is added around the perimeter of the chart, showing the reflection coefficient angle. The basic Smith chart with these scales is shown in Fig. 3.4.

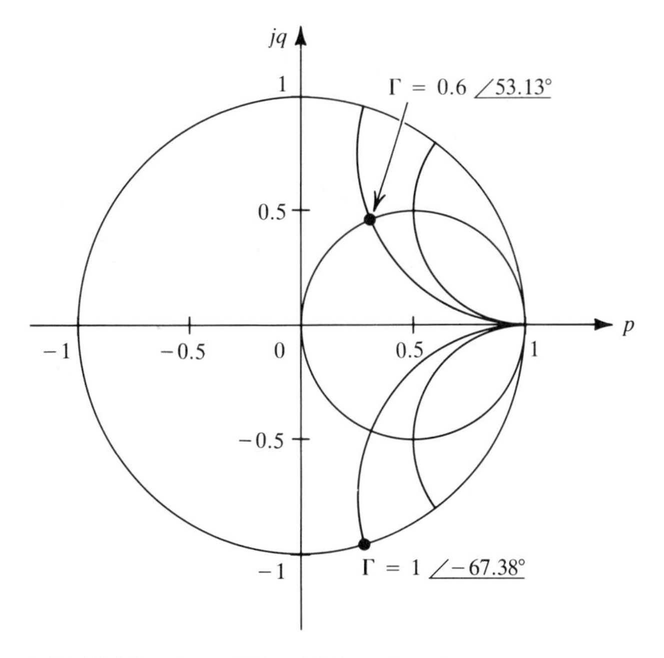

FIGURE 3.3. Figures 3.1 and 3.2 combined.

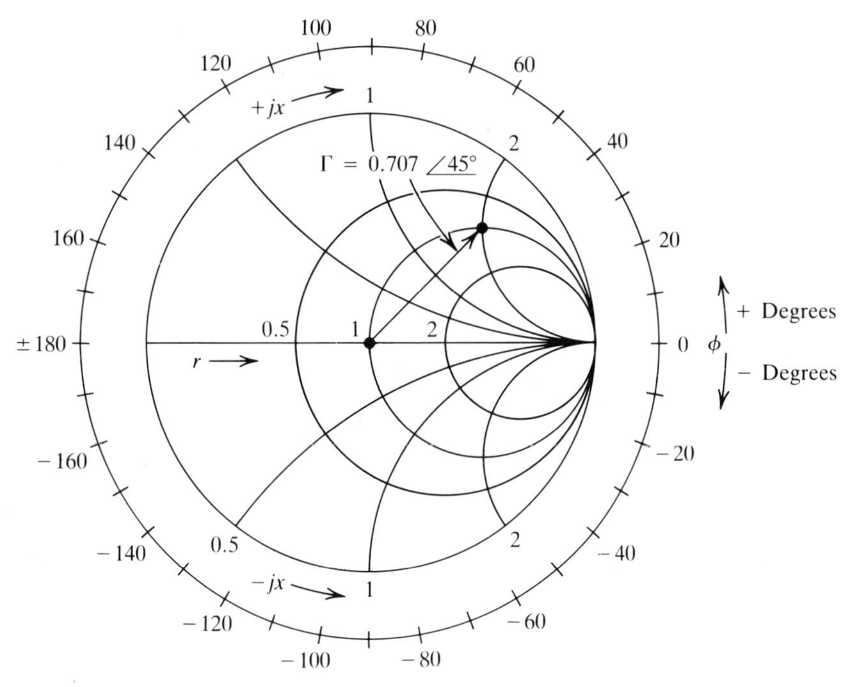

FIGURE 3.4. Basic Smith chart. A reflection coefficient of $0.707\angle 45^\circ$, corresponding to a normalized impedance of $z = 1 + j2$, is also shown.

Also shown on Fig. 3.4 is a reflection coefficient of $0.707\ \underline{/45^\circ}$. That this corresponds to a normalized impedance value of $z = 1 + j2$ is easily verified by applying Eq. (3.1):

$$\begin{aligned}\Gamma &= \frac{z - 1}{z + 1} \\ &= \frac{1 + j2 - 1}{1 + j2 + 1} \\ &= 0.5 + j0.5 \\ &= 0.707\ \underline{/45^\circ}\end{aligned}$$

3.3 REFLECTION COEFFICIENT AND NORMALIZED ADMITTANCE

By defining normalized admittance as

$$y \triangleq \frac{1}{z} \tag{3.12}$$

Equation (3.1) can be rewritten as

$$\Gamma = \frac{1 - y}{1 + y} \tag{3.13}$$

Therefore,

$$-\Gamma = \frac{y - 1}{y + 1}$$

or

$$\rho\ \underline{/\phi - 180^\circ} = \frac{y - 1}{y + 1} \tag{3.14}$$

Equation (3.14) shows that y can be represented by the point on the chart which is on the same ρ circle and diametrically opposite z.

In terms of the real and imaginary parts of y,

$$y = g + jb \tag{3.15}$$

Thus, what was originally the r-axis on the Smith chart can also be calibrated in g-values; the $+jx$-axis can also be calibrated in $+jb$-values, and the $-jx$-axis in $-jb$-values.

EXAMPLE 3.1

Use the Smith chart to find the normalized admittance corresponding to a normalized impedance of $z = 0.5 + j0.5$. Verify the result by direct calculation.

SOLUTION From the Smith chart (Fig. 3.5),

$$\mathbf{y = 1 - j1}$$

By direct calculation,

$$y = \frac{1}{0.5 + j0.5} = \frac{0.5 - j0.5}{(0.5)^2 + (0.5)^2} = \mathbf{1 - j1}$$

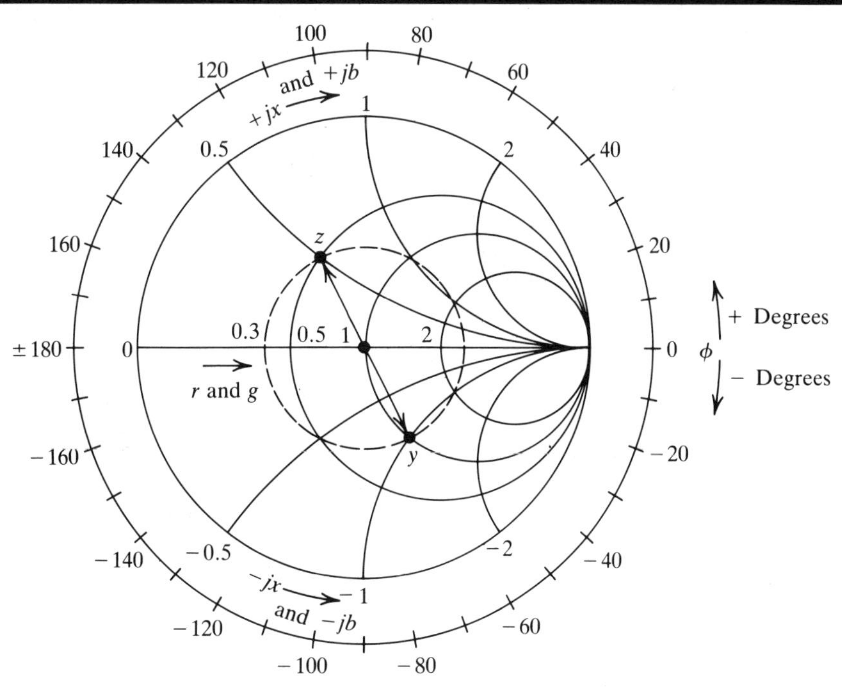

FIGURE 3.5. Example 3.1: $z = 0.5 + j0.5$ and $y = 1 - j1$. Also shown is that the chart calibration applies for normalized admittance as well as normalized impedance.

3.4 USE OF THE SMITH CHART WITH SLOTTED LINE MEASUREMENTS

The rather complex calculations described in Section 1.14 may be carried out quickly and efficiently by means of the Smith chart. The reflection coefficient is, from Eq. (1.69), for a lossless line in which $\gamma = j\beta$,

$$\Gamma = \Gamma_L e^{-j2\beta l} \tag{3.16}$$

and since $\Gamma_L = \rho_L \angle\phi$, this can be rewritten as

$$\Gamma = \rho_L e^{j(\phi - 2\beta l)} \tag{3.17}$$

Now this can be shown on the Smith chart in the same manner as described for Eq. (3.3), it only being necessary to use the angle $(\phi - 2\beta l)$ rather than ϕ alone. This is shown in Fig. 3.6(a) for arbitrary l. The particular case for $l = l_{\min}$ corresponding to the position of a voltage minimum is shown in Fig. 3.6(b). As shown in Fig. 1.14(b), the angle in this particular case is given by

$$2\beta l_{\min} - \phi = \pi \tag{3.18}$$

The reflection coefficient at $l = l_{\min}$ is, from Eqs. (3.17) and (3.18),

$$\begin{aligned} \Gamma_{\min} &= \rho_L e^{-j\pi} \\ &= -\rho_L \end{aligned} \tag{3.19}$$

The impedance at a voltage minimum can now be evaluated using Eq. (1.60) normalized:

$$\begin{aligned} z_{\min} &= \frac{1 + \Gamma_{\min}}{1 - \Gamma_{\min}} \\ &= \frac{1 - \rho_L}{1 + \rho_L} \end{aligned} \tag{3.20}$$

But Eq. (1.82) shows that VSWR $= (1 + \rho_L)/(1 - \rho_L)$ and hence

$$z_{\min} = \frac{1}{\text{VSWR}} \tag{3.21}$$

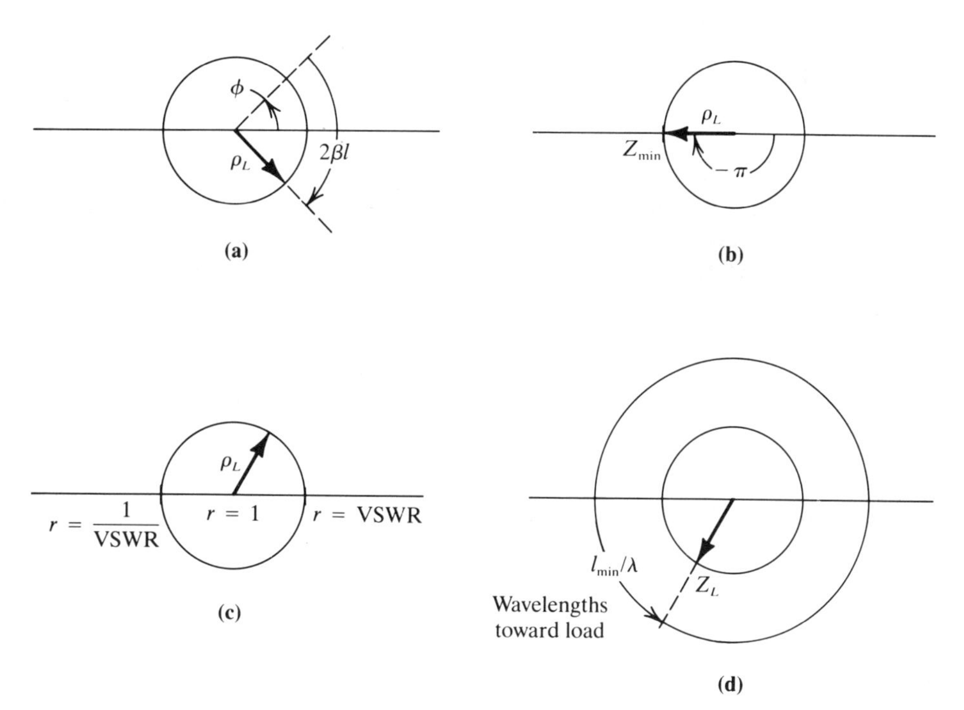

FIGURE 3.6. (a) Representation of reflection coefficient [Eq. (3.17)]; (b) reflection coefficient and impedance at a voltage minimum; (c) VSWR circle; (d) location of load impedance point referenced to the voltage minimum.

Therefore, the point labeled z_{min} on Fig. 3.6(b) is equal numerically to 1/VSWR. A circle centered on $r = 1$ (i.e., at the zero origin of the t-plane) and of radius ρ_L must pass through the z_{min} point. Also, as shown in Fig. 3.5, the point diametrically opposite must be $y_{min} = 1/z_{min}$, which of course is numerically equal to VSWR. Hence, for a given VSWR, a circle may be drawn on the Smith chart as shown in Fig. 3.6(c).

It follows that the z_L point must be located at a point which is advanced in phase by amount $2\beta l_{min}$ from the z_{min} point, as shown in Fig. 3.6(d). To make it easy to locate the z_L point, the Smith chart has a distance scale in addition to the angle scale. The distance scale is calibrated in fractions of wavelength and is referenced from the z_{min} axis. Both directions are shown, *wavelengths toward load* and *wavelengths toward generator*. Figure 3.7 shows the Smith chart with these scales, and the following example illustrates its use.

EXAMPLE 3.2

Measurements on a 50-Ω line yielded the following results: VSWR = 2:1, l_{min1} = 6 cm, and l_{min2} = 21 cm. Determine (a) the components of the

Impedence or Admittance Coordinates

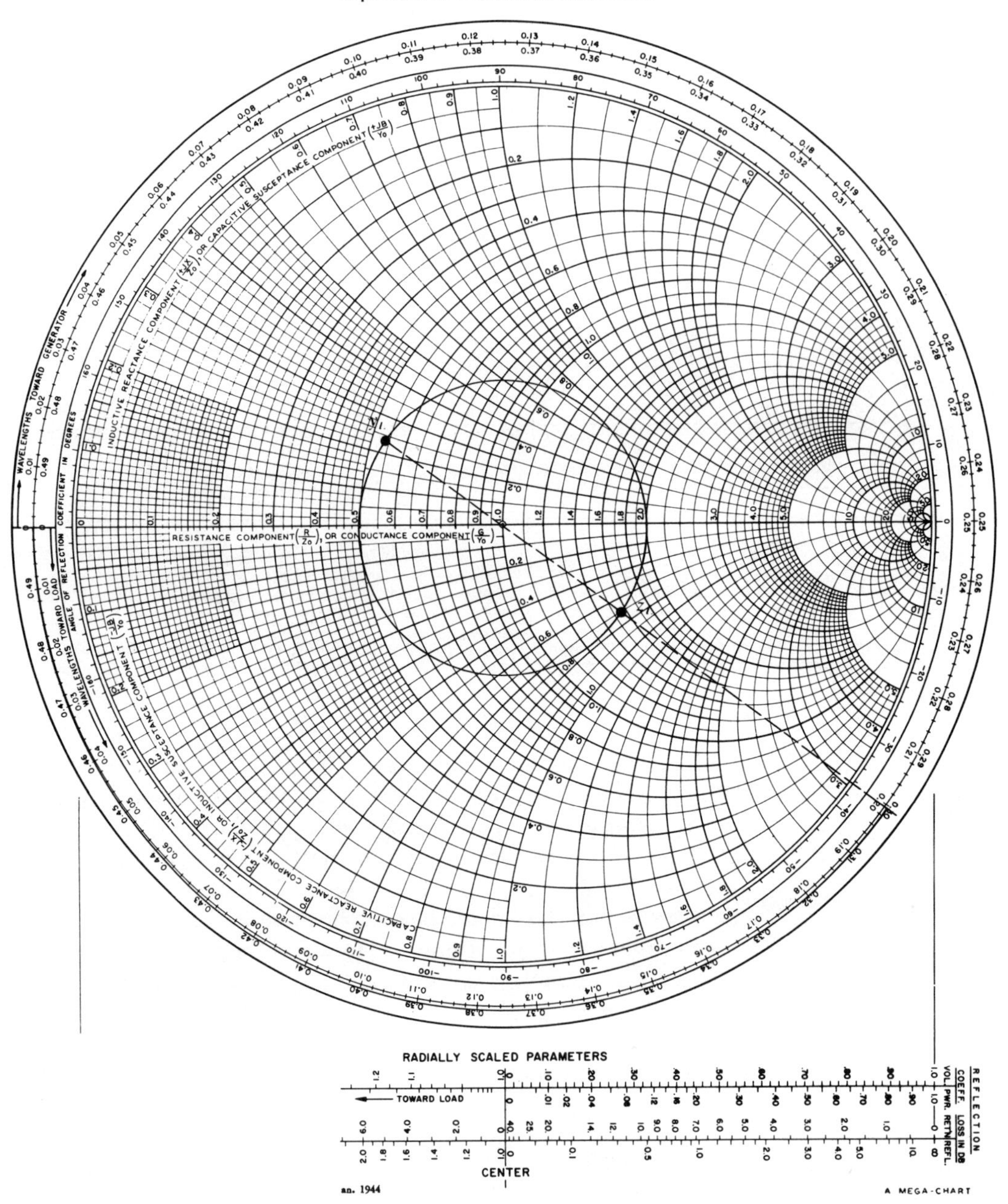

FIGURE 3.7. Solution to Example 3.2.

equivalent series load impedance, and (b) the components of the equivalent parallel load impedance.

SOLUTION From the data given, VSWR = 2. Therefore, draw a circle centered on $r = 1$, cutting the r-axis at 2 and $\frac{1}{2}$ (see Fig. 3.7). From the data given,

$$\frac{l_{min}}{\lambda} = \frac{6}{2 \times (21 - 6)}$$

$$= 0.2$$

(a) On the chart draw a radius that is displaced by distance 0.2λ along the *wavelengths toward load* scale. Point z_L is located at the point where this radius cuts the VSWR circle, which from Fig. 3.7, is

$$z_L = 1.55 - j0.68$$

and

$$Z_L = 50 z_L$$

$$= \mathbf{77.5 - j34\ \Omega}$$

(b) Using the theory developed in connection with Fig. 3.5, the equivalent admittance is located diametrically opposite z_L and as shown on Fig. 3.7 is

$$y_L = 0.54 + j0.24$$

and

$$Y_L = \frac{y_L}{50}$$

$$= 0.0108 + j0.0048\ \mathrm{S}$$

Therefore,

$$R_p = \frac{1}{0.0108}$$

$$= \mathbf{92.59\ \Omega}$$

and

$$X_p = \frac{-1}{0.0048}$$

$$= \mathbf{-208\ \Omega}$$

3.5 IMPEDANCE TRANSFORMATION

Impedance transformation as described by Eq. (1.92) in Section 1.14 may be solved using a Smith chart. The load impedance point is first located on the chart. Next, a circumferential distance equal to the line length in wavelengths is marked off, in the direction of *wavelengths toward generator*, starting at the z_L point. This arc terminates at the required z_{in} point. The method is illustrated in the following example.

EXAMPLE 3.3

A $Z_0 = 100\text{-}\Omega$ line is terminated in an impedance $Z_L = 60 - j130\ \Omega$. For a line length of $\frac{3}{8}\lambda$, determine the input impedance.

SOLUTION

$$z_L = \frac{60 - j130}{100}$$

$$= 0.6 - j1.3$$

This point is located on the Smith chart as shown in Fig. 3.8. Following the circular arc of length $\frac{3}{8}\lambda$ along the *wavelengths toward generator* scale leads to the input impedance value of

$$z_{in} = 2.7 + j2.3$$

Hence

$$Z_{in} = 100 z_{in}$$

$$= \mathbf{270 + j230\ \Omega}$$

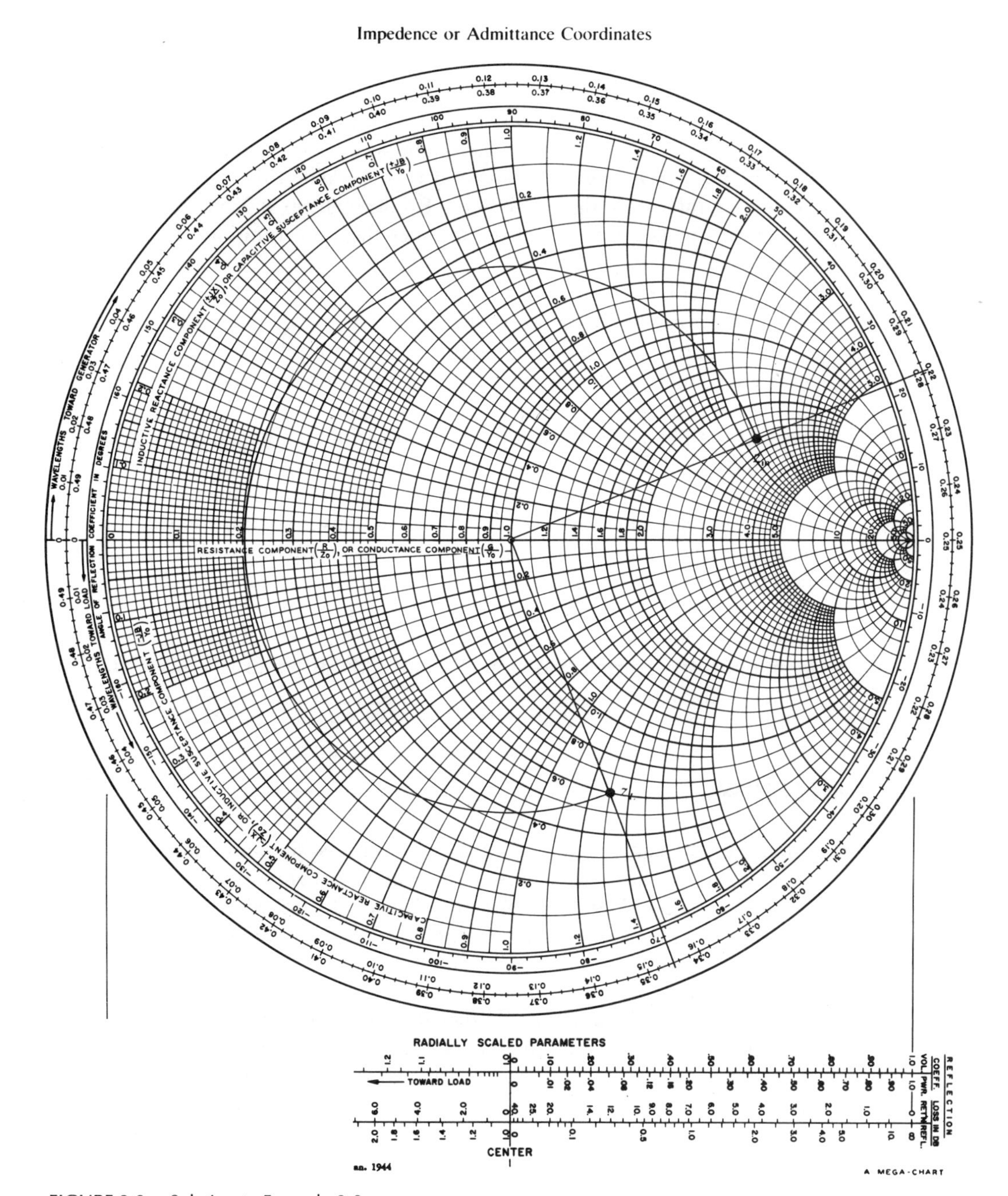

FIGURE 3.8. Solution to Example 3.3.

A number of points may be noted from Example 3.3. First, the load impedance is transformed from a capacitive impedance to an inductive impedance at the line input. Second, normalization to 100 Ω is used, illustrating the fact that the Smith chart can be used with any real-valued normalizing number, not just the commonly encountered value of 50 Ω. Third, the *wavelengths* scale in total is 0.5λ. When the z_L radius is projected, it intersects this at 0.345λ. Rotation from 0.335λ to 0.5λ gives 0.155λ; a further rotation of 0.22λ gives the total rotation of 0.375λ (or $\frac{3}{8}\lambda$).

In Section 1.15 it is shown that short-circuited and open-circuited line sections can be used as capacitive and inductive elements. Such sections are frequently used in parallel with line transforming sections to provide impedance matching. In these situations a knowledge of admittance values rather than impedance values is of more use. As already shown in connection with Fig. 3.7, the admittance corresponding to a given impedance is found diametrically opposite the impedance point on the same VSWR circle on the Smith chart. Transformation of the admittance by means of a line section then follows exactly the method given for impedance, starting from the admittance point. On the Smith chart it will be seen that a common scale is provided for normalized admittance and impedance. If the load admittance value is known, instead of impedance, this point may be entered directly on the chart, which obviates the need for the initial impedance to admittance transformation step. Some of these points are illustrated in detail in the next section.

3.6 STUB MATCHING

Line sections connected in parallel with a main line and load system are referred to as *stubs*. In the design of small-signal amplifiers constructed in microstrip, a considerable range of characteristic impedance values can be realized, so the design is not restricted to 50-Ω systems. Microstrip and stripline design methods are described in more detail in Chapter 5.

For certain applications coaxial line sections may have to be used, for example where power-handling capacity requires this. To facilitate use of common connectors and readily available coaxial cables and components, it is usual to maintain a constant characteristic impedance throughout the system, for example 50 Ω.

3.6.1 Single-Stub Matching

Single-stub matching makes use of an adjustable stub (usually shorted), and adjustable position in relation to the load, as illustrated in Fig. 3.9(a). The object is to transform the Z_L which will be complex in general, to a

$Z_0 \angle 0°$ value that matches the main feeder. Knowing Y_L and therefore y_L (normalized to Z_0), length l_1 is found which converts y_L to some intermediate value $1 + jb$. Stub l_2 provides a susceptance $-jb$ which tunes out the $+jb$ component, leaving a conductance component of unity which therefore matches the main feeder. The problem is best solved using the Smith chart, as illustrated in the following example.

EXAMPLE 3.4

A load impedance of $73 - j80\ \Omega$ has to be matched to a $50 \angle 0°$-Ω line using single-stub matching, as illustrated in Fig. 3.9(a). Determine the required lengths l_1 and l_2 in fractions of a wavelength.

SOLUTION

Normalize Z_L to 50 Ω:

$$z_L = \frac{73 - j80}{50}$$

$$= 1.46 - j1.6$$

The following steps are shown on the Smith chart of Fig. 3.9(b).

Step 1: Locate z_L on the chart.

Step 2: Draw a circle through z_L centered on $r = 1$.

Step 3: Locate y_L on the circle, diametrically opposite z_L.

Step 4: Move l_1/λ *towards generator* until $y = 1 + jb$ is reached. From the chart,

$$\frac{l_1}{\lambda} = \mathbf{0.117}$$

and

$$jb = j1.38$$

Step 5: The stub must provide a susceptance of $-j1.38$. From the chart, the required l_2/λ *toward load* (the *stub load*, y_s, is a short circuit, or $y_s = \infty$) is $l_2/\lambda = 0.1$.

Thus the required lengths are $l_1 = \mathbf{0.11\lambda}$ and $l_2 = \mathbf{0.1\lambda}$.

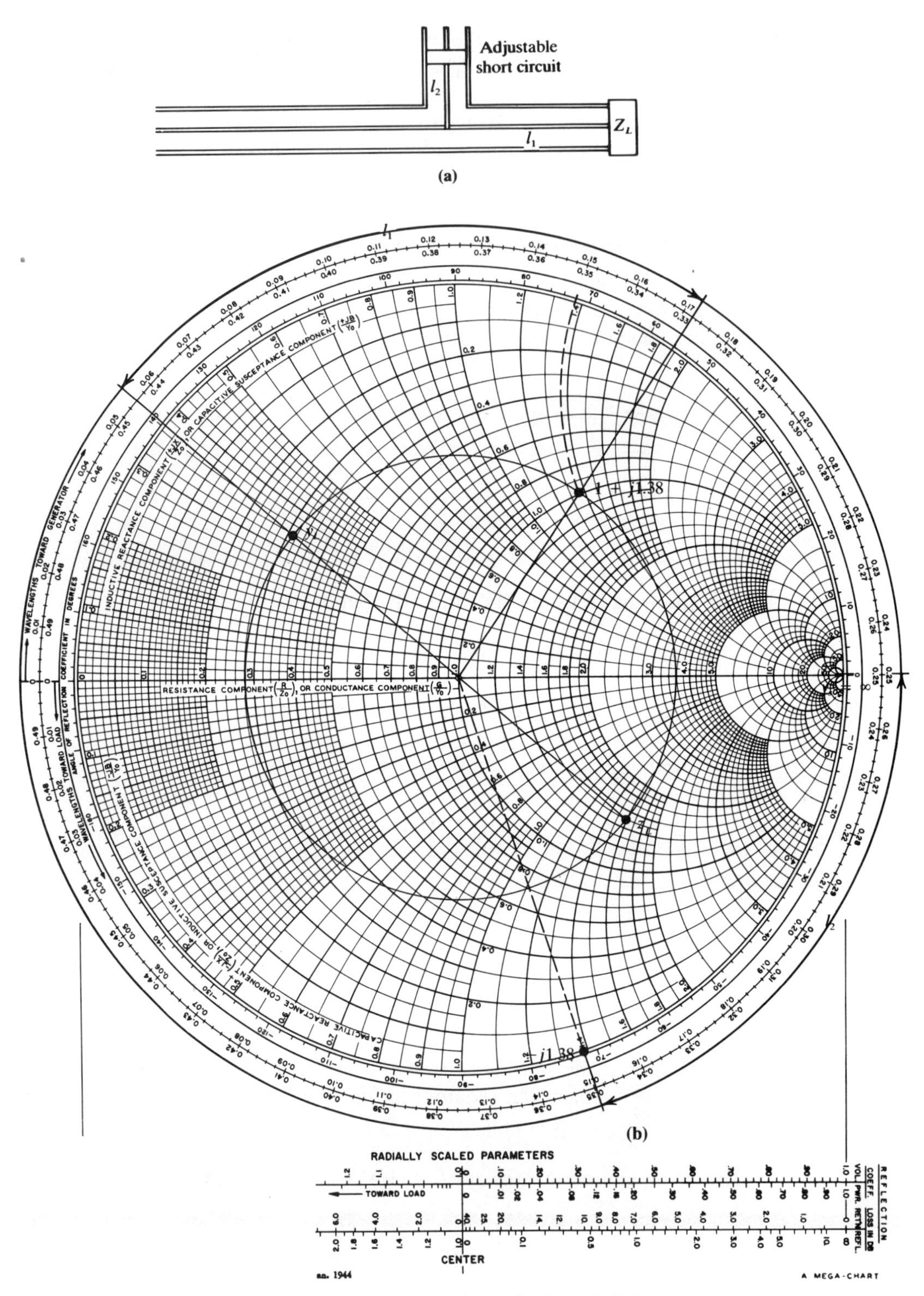

FIGURE 3.9. (a) Single-stub matcher; (b) Smith chart solution for Example 3.4.

3.6.2 Double-Stub Matching

One difficulty with the single-stub matcher is that the position l_1 is not easily adjustable. A double-stub matcher overcomes this difficulty in that its position is fixed and the two variables are the lengths of the stubs l_1 and l_2, as shown in Fig. 3.10(a). Figure 3.10(b) shows the Smith chart solution for the particular problem in Example 3.5, but can also be used to illustrate the general principles. Length l, which is arbitrarily fixed, transforms the load admittance y_L to some new value y'_L. Stub 1 is adjusted to alter the susceptance component of y'_L, the conductance component remaining constant, until point y_A is reached on the circle A. This circle is chosen because the $\frac{3}{8}\lambda$ displacement of the fixed length between stubs, transforms the circle A into circle B, which is the unity conductance circle. Thus admittance y_A is transformed into admittance y_B. Stub 2 is then adjusted to tune out the susceptance of y_B, leaving a normalized conductance component of unity, so the system is matched. Stub 1 could also be adjusted to bring admittance y'_L to the y'_A position, which in turn would be transformed to y'_B by the $\frac{3}{8}\lambda$ section, and this would also be a solution.

Not all values of y'_L can be matched successfully. If y'_L falls within the $g = 2$ circle [shown crosshatched in Fig. 3.10(b)], none of the constant-conductance circles cuts the circle A, so matching cannot be achieved. However, this small excluded range is generally accepted as a reasonable price to pay for the practical convenience afforded by the double-stub arrangement.

As shown, the spacing between stubs is $\frac{3}{8}\lambda$, which is the usual practical arrangement. Other spacings can be used, and of course when the matcher is used at frequencies other than the design frequency, the spacing, as a fraction of the wavelength, will alter. This will be true also of the spacing between the load and the first stub. (And, of course, the excluded area will alter as the frequency of operation is changed.) In practical terms, then, the Smith chart solution would be used only to establish the approximate values required for l_1 and l_2. Stub 1 would then be set at a number of points close to the calculated value, and at each setting, stub 2 would be adjusted to minimize the VSWR on the main line (which therefore has to be measured during the setting-up procedure). The combined settings of stub 1 and stub 2 that give the lowest measured VSWR are the practical values used.

EXAMPLE 3.5

Rework the matching problem of Example 3.4 using a double-stub matching unit for which the spacing between stubs is $\frac{3}{8}\lambda$ and the spacing between first stub and load is 0.042λ.

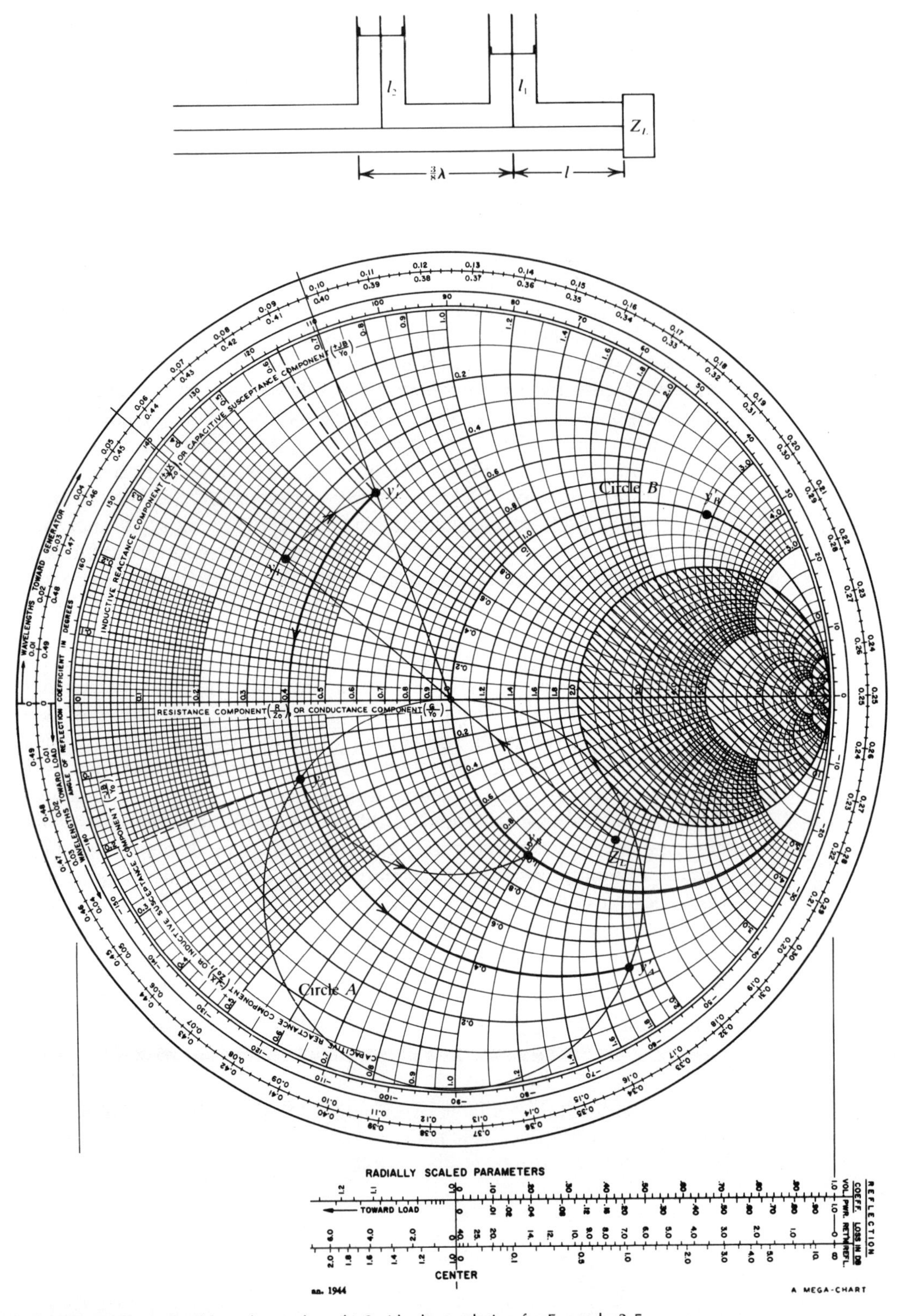

FIGURE 3.10. (a) Double-stub matcher; (b) Smith chart solution for Example 3.5.

SOLUTION Refer to the Smith chart of Fig. 3.10(b). Steps 1, 2, and 3 are all the same as in Example 3.4, and enable the load admittance point y_L to be located.

Step 4: Move distance 0.042λ *toward generator*. This locates point y'_L. From the chart this is $y'_L = 0.4 + j0.63$.

Step 5: Move along the constant conductance 0.4 circle until circle A is reached. This gives $y_A = 0.4 - j0.21$. The total susceptance change is seen to be $(-j0.21) - j0.63 = -j0.84$. From the chart, a shorted stub of length 0.139λ [measured from the infinite susceptance (short circuit) point to the $-j0.84$ point on the *wavelengths toward generator* scale] will do this. Alternatively, the length l_1 can be calculated from Eq. (1.102):

$$\frac{1}{-j0.84} = j \tan\left(2\pi \frac{l_1}{\lambda}\right)$$

Therefore,

$$\frac{l_1}{\lambda} = \frac{\arctan (1/0.84)}{2\pi}$$
$$= \mathbf{0.139}$$

Step 6: The $\frac{3}{8}\lambda$ section transforms circle A into circle B and hence point y_A into point y_B. From the chart

$$y_B = 1 - j1$$

Hence stub 2 must provide a susceptance of $+j1$ to tune out the $-j1$ component. From the chart, or from Eq. (1.102), this gives $l_2 = \frac{3}{8}\lambda$ for a shorted stub.

Hence the required stub lengths are $l_1 = \mathbf{0.139\lambda}$ and $l_2 = \mathbf{0.375\lambda}$.

The range of matching can be increased by using a three-stub matcher, the spacing between stubs in practice being $\frac{3}{8}\lambda$. The tuning procedure is similar to that for the double stub unit but obviously will be more involved because there are three variables instead of two.

3.7 PROBLEMS

1. Using a Smith chart and with $Z_0 = 50\ \Omega$, determine the reflection coefficients for the following impedances, all in ohms: (a) $100\angle 0°$; (b) $50 - j50$; (c) $0 + j75$. Verify your answers by calculation.
2. Using a Smith chart, and with $Z_0 = 50\ \Omega$, determine the impedance for the following reflection coefficients: (a) $0.7\angle 45°$; (b) $1\angle 180°$; (c) $0.5\angle 0°$. Verify your answers by calculation.
3. Find using a Smith chart, and by calculation, the admittances corresponding to the following impedances, all in ohms: (a) $10 - j30$; (b) $100\angle 45°$; (c) $0 - j50$.
4. Find using a Smith chart, and by calculation, the impedances corresponding to the following admittances all in millisiemens: (a) $1 - j2$; (b) $3\angle 30°$; (c) $0 - j5$.
5. Measurements on a 50-Ω slotted line yielded the following results: distance of first voltage minimum from load, 10 cm; distance of second voltage minimum from load, 40 cm; VSWR = 2.3:1. Determine the termination (a) as an impedance and (b) as an admittance. (c) What are the values of the equivalent parallel resistance and reactance for part (b)?
6. A 50-Ω slotted line has its distance scale calibrated with the zero at the source end and an unknown fixed length of line between the slotted section and load. With the unknown load as the termination, the first voltage minimum occurs at 40 cm on the distance scale and the VSWR is 3:1. With a short circuit in place of the load, the first voltage minimum occurs at 47 cm and the second at 32 cm. (See Fig. 14.10.) Determine the value of the load termination (a) as an impedance and (b) as an admittance. (c) What are the equivalent parallel resistance and reactance values for part (b)?
7. A 100-Ω transmission line 0.2 wavelength in length is terminated by a 50-Ω resistor. Find, using a Smith chart, the impedance at the line input.
8. A 50-Ω transmission line 1.3 wavelengths in length is terminated in an impedance of $30 - j50\ \Omega$. Using a Smith chart, determine the impedance at the input to the line.
9. A 50-Ω transmission line 0.4 wavelengths in length is terminated in an admittance of $0.01 - j0.04$ S. Using a Smith chart determine (a) the input impedance, (b) the input admittance, and (c) the equivalent parallel values of resistance and reactance for part (b).
10. A transmission-line section of characteristic impedance 100 Ω is terminated in a load consisting of a 30-Ω resistor in parallel with a 2-pF capacitor. The line length is 7 cm. Determine the impedance at the line input when the frequency is 2.65 GHz.

11. In a 50-Ω transmission line system, a load impedance of 75 − j40 Ω has to be matched to a 50-Ω line through a single-stub matcher. Using a Smith chart, determine the distance of the stub from the load, the length of the stub, and the stub termination.
12. Repeat Problem 11 for a load impedance of 75 Ω purely resistive.
13. Repeat Problem 11 for a load consisting of a 100-Ω resistor in parallel with a +j50-Ω reactance.
14. In a 50-Ω system, a two-stub matcher is used to match the load to the 50-Ω line. The separation between stubs in the matcher is 3λ/8 and the distance to the nearest stub from the load is 0.01λ. Determine which of the following loads can be matched (all values in ohms): (a) $10\angle 0°$; (b) $250\angle 0°$; (c) 30 + j40; (d) a 100-Ω resistor in parallel with a −j70-Ω reactance.
15. For the two-stub matching system of Problem 14, the load impedance is 75 − j60 Ω. Determine the lengths of the matching stubs required.

FURTHER READING

GLAZIER, E. V., and H. R. L. LAMONT, 1958. *The Services' Textbook of Radio*, Vol. 5, *Transmission and Propagation*. London: Her Majesty's Stationery Office.

KENNEDY, GEORGE, 1970. *Electronic Communication Systems*. New York: McGraw-Hill Book Company.

LIAO, SAMUEL Y., 1980. *Microwave Devices and Circuits*. Englewood Cliffs, N.J.: Prentice-Hall, Inc.

four
Waveguides and Waveguide Components

4.1 INTRODUCTION

At frequencies higher than about 3000 MHz, transmission of electromagnetic waves along lines and cables becomes difficult, mainly because of the losses that occur both in the solid dielectric needed to support the conductors, and in the conductors themselves. It is possible to transmit an electromagnetic wave down a metallic tube, this being known as a *waveguide*. The commonest form of waveguide is rectangular in cross section, as shown in Fig. 4.1(a). Induced currents in the walls of the waveguide give rise to power losses and to minimize these losses, the waveguide wall resistance is made as low as possible. Because of the skin effect, the currents tend to concentrate near the inner surface of the guide walls, and these are sometimes specially plated to reduce resistance.

Apart from determination of losses, the walls of a waveguide may be considered to be perfect conductors. Two important boundary conditions result, which determine the mode of propagation of an electromagnetic wave along a guide. These are: (1) electric fields must terminate normally on the conductor [i.e., the tangential component of electric field must be zero, Fig. 4.1(b)]; and (2) magnetic fields must lie entirely tangentially along the wall surface [i.e., the normal component of magnetic field must be zero, Fig. 4.1(c)]. Knowing these boundary conditions provides a simple way of visualizing how the various modes of waveguide transmission occur, and these will be discussed in the following sections.

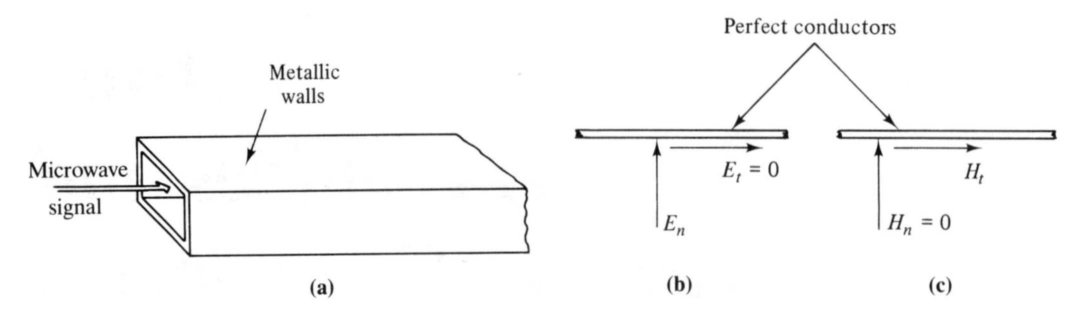

FIGURE 4.1. (a) Rectangular waveguide; (b) Electric field boundary conditions; (c) Magnetic field boundary conditions.

4.2 DOMINANT MODE IN A RECTANGULAR WAVEGUIDE

The boundary conditions already referred to preclude the possibility of a waveguide supporting a *transverse electromagnetic* (TEM) wave [see Fig. 1.11(c)] propagation, since the magnetic field is at right angles to the direction of propagation (along the axis of the waveguide) and therefore would have to terminate normally to the sidewalls, which cannot happen. One possible solution is for the magnetic field to form loops along the

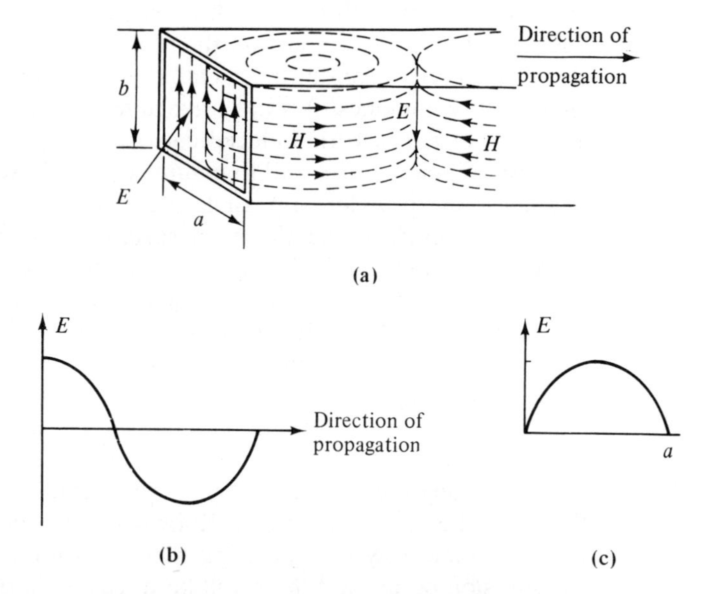

FIGURE 4.2. (a) One possible field configuration for a waveguide; (b) Electric field amplitude along guide axis; (c) Electric field amplitude along guide width.

direction of propagation, as shown in Fig. 4.2(a), these lying parallel to the top and bottom walls and tangentially to the sidewalls.

The variation of electric field as a function of distance along the direction of propagation is shown in Fig. 4.2(b) and along the cross section in Fig. 4.2(c). The propagation mode sketched in Fig. 4.2 is known as a *transverse electric* (TE) *mode* because the electric field is entirely transverse to the direction of propagation. (It is also known as an *H mode*, signifying that part of the magnetic field lies along the direction of propagation.)

Subscripts are used to denote the number of half-cycles of variation which occur along the a and b sides. As shown in Fig. 4.2, one half-cycle (i.e., one maximum) occurs along the a side and none along the b side; this mode is therefore referred to as the *TE_{10} mode*. The TE_{10} mode is the dominant mode in waveguide transmission, for, as will be shown, it supports the lowest-frequency waveguide mode.

In the following sections, operation in the TE_{10} mode will be assumed unless otherwise stated.

4.3 WAVELENGTH RELATIONSHIPS

Figure 4.3(a) shows how a TE_{10} wave may be formed as the resultant of two TEM waves crossing each other. At points of intersection shown, the individual waves add vectorially. At those points where the electric fields reinforce each other, shown by double crosses and double dots, the magnetic field is directed up and down; at the points where the electric fields cancel, shown by dots and crosses together, the magnetic field is directed left and right. Metallic walls may be placed along the L–R direction as shown in Fig. 4.3(b) without violating the boundary conditions, as it will be seen that the electric field is zero, and the magnetic field tangential at these walls. Metallic top and bottom walls can be put in position as the electric field will terminate normally on these, and the magnetic field lies parallel to them.

The direction of the magnetic field loops alternates as shown in Fig. 4.3(b), one loop occupying a distance of a half wavelength in the guide. This will be different from a half-wavelength of the individual TEM waves and is shown as $\frac{1}{2}\lambda_g$, where λ_g is termed the *guide wavelength*.

The frequency of the TE wave will be the same as that of the TEM waves (i.e., it will have the same number of cycles in a second). Let v_p be the phase velocity of the TE wave; then, applying the general relationship given by Eq. (1.29) to the TE wave and the individual TEM waves

TE wave:

$$\lambda_g f = v_p \tag{4.1}$$

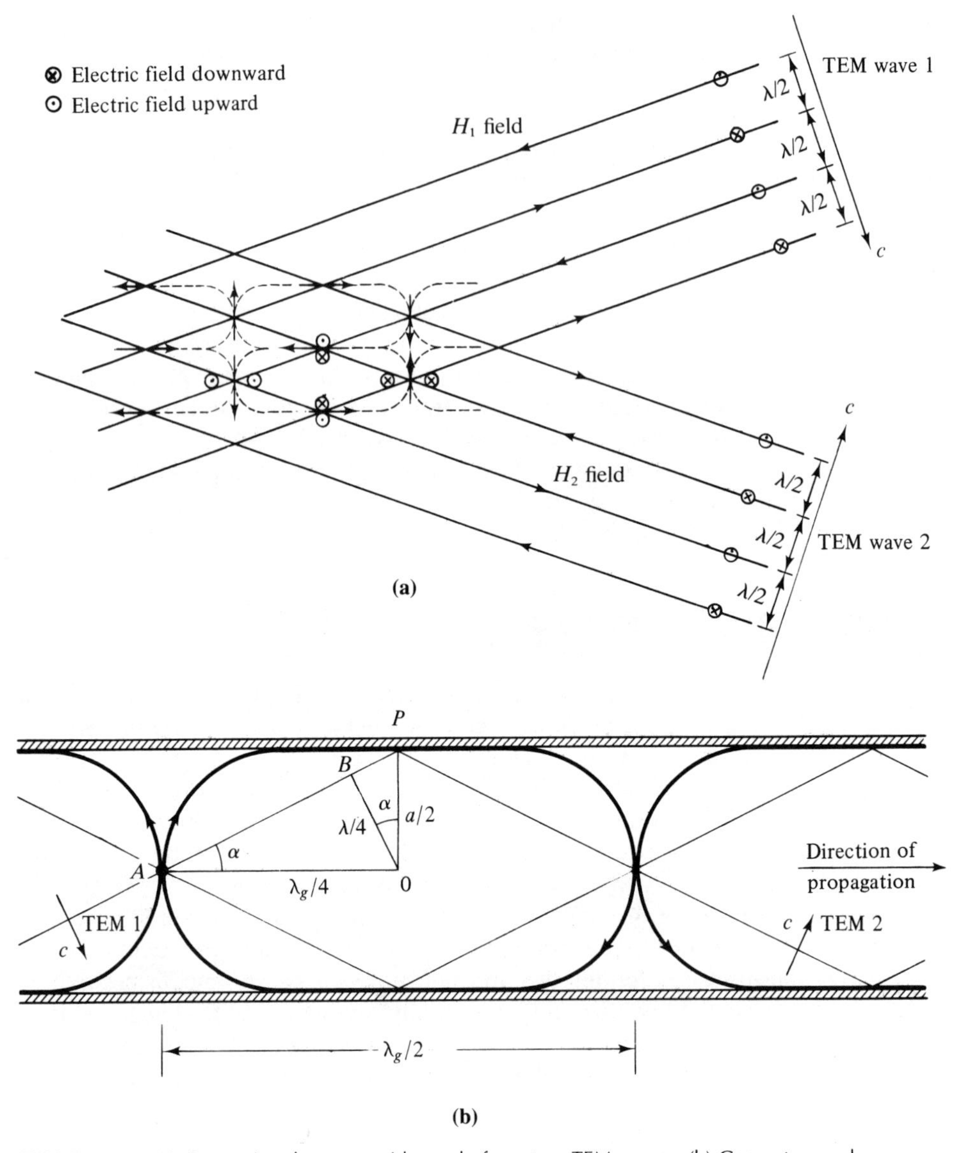

FIGURE 4.3. (a) Generating the waveguide mode from two TEM waves; (b) Geometry used in determining TE_{10} properties.

TEM wave (in air):

$$\lambda f = c \tag{4.2}$$

from which

$$v_p = c\frac{\lambda_g}{\lambda} \tag{4.3}$$

From the geometry of Fig. 4.3(b), it will be seen that:
From right-angled triangle OBP:

$$\cos\alpha = \frac{\lambda}{2a} \tag{4.4}$$

From right-angled triangle ABO:

$$\sin\alpha = \frac{\lambda}{\lambda_g} \tag{4.5}$$

and since $\sin^2\alpha + \cos^2\alpha = 1$, therefore

$$\left(\frac{\lambda}{\lambda_g}\right)^2 + \left(\frac{\lambda}{2a}\right)^2 = 1$$

or

$$\frac{1}{\lambda_g^2} = \frac{1}{\lambda^2} - \frac{1}{(2a)^2} \tag{4.6}$$

where λ_g is the guide wavelength of the TE_{10} mode, λ the free-space wavelength of either TEM wave, and a the broad dimension of the guide.

From Eq. (4.6) it is seen that when $\lambda = 2a$, the guide wavelength becomes zero; this corresponds to the individual TEM waves bouncing from side to side with no velocity component directed along the guide. Clearly, then, $\lambda = 2a$ represents the longest-wavelength TEM wave that can be induced into a guide, as a TE mode will not be generated for longer wavelengths. For shorter TEM wavelengths, Eq. (4.6) shows that the guide wavelength is real and positive, so that TE-mode propagation takes place.

The term $2a$ is referred to as the cutoff wavelength of the TE_{10} mode, λ_c, and, rewriting Eq. (4.6), we have

$$\frac{1}{\lambda_g^2} = \frac{1}{\lambda^2} - \frac{1}{\lambda_c^2} \tag{4.7}$$

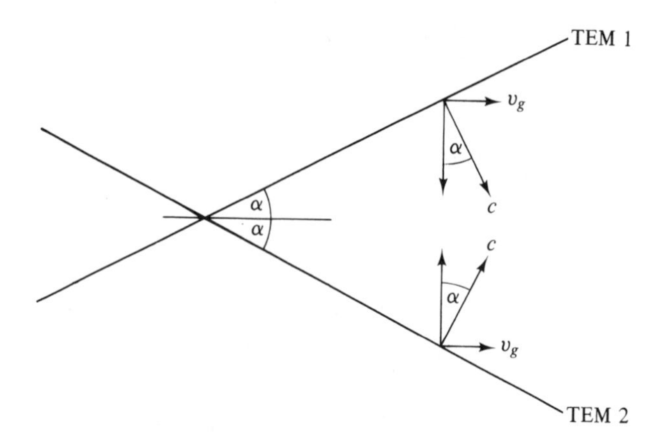

FIGURE 4.4. Group velocity as the horizontal component of c.

4.4 PHASE AND GROUP VELOCITIES

From Eqs. (4.3) and (4.5) it is seen that

$$v_p = \frac{c}{\sin \alpha} \tag{4.8}$$

Since the sine of an angle is never greater than unity, v_p can never be less than c; it can, of course, be greater.

The individual TEM waves have to travel a zigzag path and therefore the velocity component along the guide, which is the velocity at which the energy in the waves is conveyed, is less than c. From Fig. 4.4 this component, known as the group velocity v_g, is seen to be

$$v_g = c \sin \alpha \tag{4.9}$$

Combining Eqs. (4.8) and (4.9) gives

$$v_g v_p = c^2 \tag{4.10}$$

Also, Eq. (4.9) can be rewritten in terms of wavelength, since these are the quantities that are normally known:

$$v_g = c \frac{\lambda}{\lambda_g} \tag{4.11}$$

Unless otherwise stated, it is assumed that the waveguide dielectric is dry air, for which free-space permittivity and permeability apply.

EXAMPLE 4.1

A rectangular waveguide has a broad wall dimension of 0.900 in., and is fed by a 10-GHz carrier from a coaxial cable as shown in Fig. 4.5. Determine if a TE_{10} wave will be propagated and, if so, its guide wavelength, phase, and group velocities.

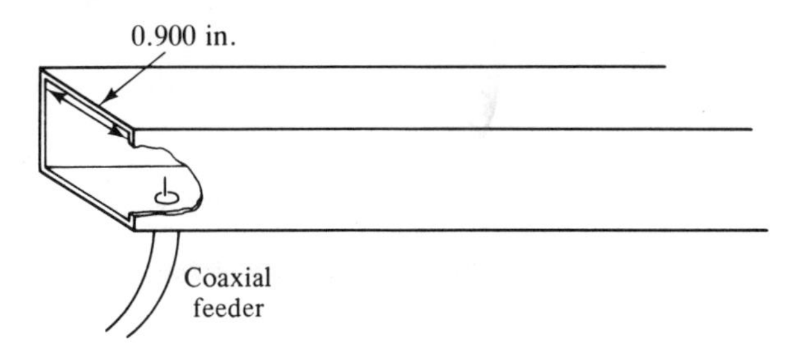

FIGURE 4.5. Example 4.1.

SOLUTION

$$a = 0.900 \text{ in.}$$
$$= 2.29 \text{ cm}$$

Therefore,

$$\lambda_c = 2 \times 2.29$$
$$= \mathbf{4.58\ cm}$$
$$\lambda = \frac{3 \times 10^8}{10^{10}}$$
$$= \mathbf{3\ cm}$$

(Note that it is the free-space wavelength and not the wavelength along the coaxial cable which is used.) Therefore, $\lambda_c > \lambda$ and a TE_{10} wave will propagate.

$$\frac{1}{\lambda_g^2} = \frac{1}{\lambda^2} - \frac{1}{\lambda_c^2}$$

Therefore,

$$\lambda_g = \frac{\lambda}{\sqrt{1 - (\lambda/\lambda_c)^2}}$$
$$= \frac{3}{\sqrt{1 - 0.042}}$$
$$= \mathbf{3.97\ cm}$$

$$v_p = c \times \frac{\lambda_g}{\lambda}$$

$$= 3 \times 10^8 \times 1.32$$

$$= \mathbf{3.97 \times 10^8 \ m/s}$$

$$v_g = c \times \frac{\lambda}{\lambda_g}$$

$$= \frac{3 \times 10^8}{1.32}$$

$$= \mathbf{2.27 \times 10^8 \ m/s}$$

4.5 WAVE IMPEDANCE

The *wave impedance* is defined as the ratio of the transverse components of electric to magnetic fields. In synthesizing the TE_{10} mode the two TEM waves are of equal amplitude and in time phase; thus $E_1 = E_2 = E$ and $H_1 = H_2 = H$. The entire electric field E is transverse, along dimension b, while from Fig. 4.3 it is easily ascertained that the transverse component of H lies along dimension a and is

$$H_a = H \sin \alpha \tag{4.12}$$

Hence the wave impedance for the TE_{10} mode is

$$Z_w = \frac{E_b}{H_a}$$

$$= \frac{E}{H \sin \alpha} \tag{4.13}$$

The wave impedance of a TEM wave in free space is known to be

$$Z_0 = \frac{E}{H}$$

$$= \sqrt{\frac{\mu_0}{\varepsilon_0}} \tag{4.14}$$

Thus substituting Eqs. (4.14) and (4.5) in Eq. (4.13) gives

$$Z_w = Z_0 \frac{\lambda_g}{\lambda} \tag{4.15}$$

$$= 120\pi \frac{\lambda_g}{\lambda} \quad \Omega \tag{4.16}$$

when the numerical values for μ_0 and ε_0 are substituted.

The importance of the wave impedance concept is that it can be used in an analogous manner to the characteristic impedance of a transmission line, so that the theory of reflections, standing waves, and the Smith chart developed in Chapter 3 can also be applied to waveguides.

4.6 STANDING WAVES

Consider a section of waveguide closed by a perfectly conducting sheet at the end. The boundary conditions require that the fields adjust so that the electric field is zero, and the magnetic field entirely tangential at the closure (i.e., the field patterns are similar to those at the walls of the guide). The resultant wave pattern can be accounted for in terms of the incident TE wave, and a reflected TE wave, the combination of which sets up a standing-wave pattern along the guide, similar to that described in Section 1.12 for transmission lines. It is important to realize that the resultant wave pattern is stationary in space and varies in time, whereas the traveling wave shown in Fig. 4.3(b) is time invariant but of course moves along the guide. Figure 4.6 emphasizes this point. On the left are shown the conditions in a short-circuited guide at time intervals of one-fourth the periodic time of the wave, the reference time being chosen at a maximum field condition. The sequence on the right shows how a single traveling wave would vary over the same time intervals. (For clarity, only the magnetic field loops are shown.)

It will be observed that at $\frac{1}{4}\lambda_g$ from the closed end the transverse magnetic field is zero while the electric field is a maximum. Therefore, the wave impedance is infinite at this section across the guide. At the $\frac{1}{2}\lambda_g$ section, the electric field is zero while the transverse magnetic field component is maximum, so that the wave impedance is zero. Thus the quarter-wavelength and half-wavelength sections of guide have the same transforming properties as those of the transmission-line sections described in Section 1.14.

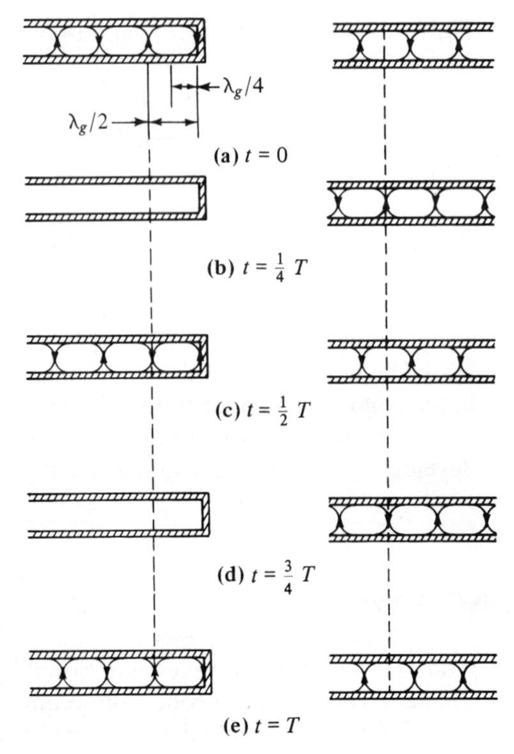

FIGURE 4.6. Comparing standing waves (left) to traveling waves (right).

4.7 WAVEGUIDE TERMINATIONS

To terminate a waveguide without causing reflections, the terminating impedance must equal the wave impedance of the incident wave. Consider first a thin sheet of resistive material closing the end of the waveguide, Fig. 4.7(a). Between the top and bottom of the guide, the sheet presents a resistance of path length b and cross-sectional area ta, where t is the thickness of the sheet; a and b are the guide dimensions defined previously. Only the sheet resistance between top and bottom needs to be considered, since the current direction is at right angles to the H-field at the sheet. Let ρ be the resistivity of the sheet material; then the terminating resistance is

$$R_T = \rho \frac{b}{ta} \tag{4.17}$$

The *sheet resistivity* R_s is defined as

$$R_s = \frac{\rho}{t} \ \Omega \text{ square} \tag{4.18}$$

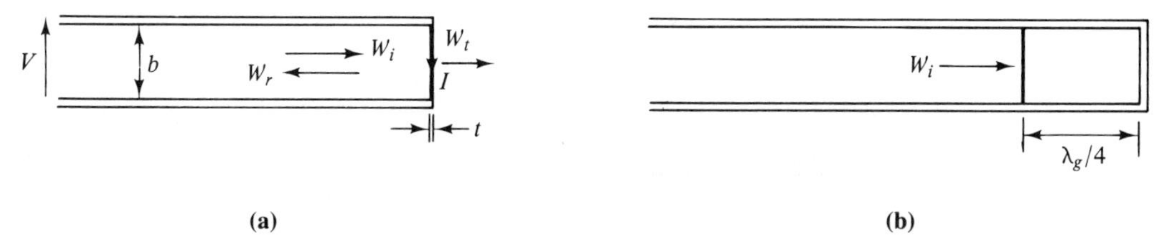

FIGURE 4.7. (a) Waveguide terminated with sheet resistance; (b) $\frac{1}{4}\lambda_g$ section added to termination.

The units of the square do not matter. R_s will have the same value for a square inch, a square centimeter, and so on. What does matter is that the units of ρ and t should be consistent. If ρ is in Ω-m, then t must be in meters and R_s will be in ohms per square. If ρ is in $\mu\Omega$-in., t must be in inches and R_s will be in $\mu\Omega$ per square.

Now, the current I in the sheet supports the tangential component of magnetic field H_T, where

$$H_T = \frac{I}{a} \tag{4.19}$$

Also, the voltage across the sheet gives rise to an electric field component:

$$E_T = \frac{V}{b} \tag{4.20}$$

Therefore,

$$\frac{E_T}{H_T} = \frac{Va}{Ib} \tag{4.21}$$

But $V/I = R_T = R_s(b/a)$, and substituting this in Eq. (4.21) gives

$$\frac{E_T}{H_T} = R_s \frac{b}{a} \frac{a}{b}$$

$$= R_s \tag{4.22}$$

E_T and H_T will consist of three components: incident, reflected, and transmitted waves, shown in Fig. 4.7(a) as W_i, W_r, and W_t, respectively. The transmitted wave can be eliminated by extending the guide $\frac{1}{4}\lambda_g$ beyond the terminating sheet, as shown in Fig. 4.7(b), and short-circuiting the end. The $\frac{1}{4}\lambda_g$ shorted section presents an infinite impedance; therefore, no wave

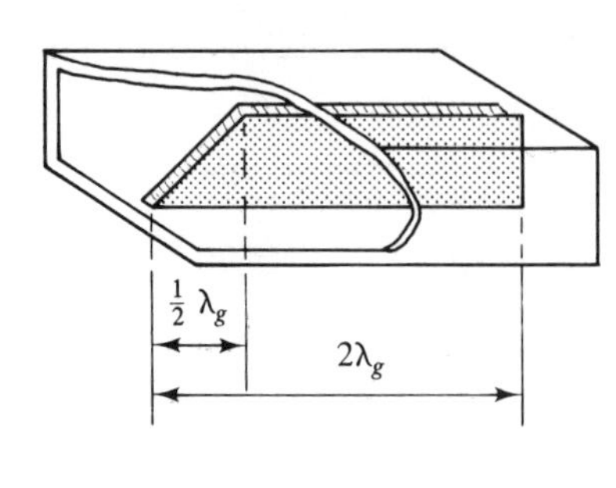

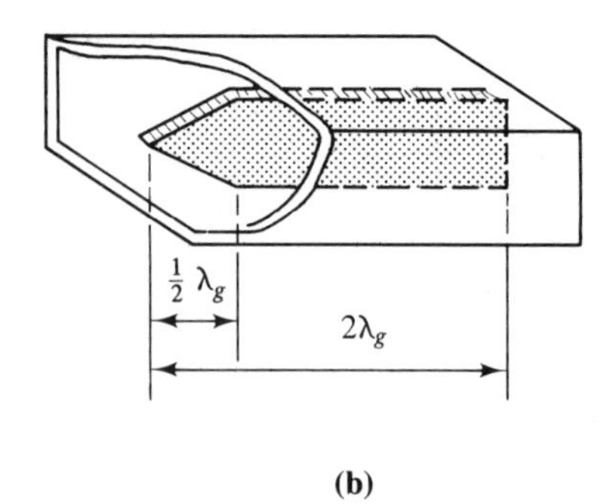

FIGURE 4.8. Practical waveguide terminations.

can be transmitted into it. Next, the sheet resistivity R_s is made equal to the incident-wave impedance Z_w:

$$Z_w = R_s \tag{4.23}$$

This ensures that $E_T/H_T = E_i/H_i = E/H$; that is, the incident wave is absorbed by the sheet just as though it were continuing down an infinitely long waveguide in which the wave impedance was Z_w (or R_s). Thus the reflected wave is eliminated.

The arrangement discussed illustrates two important principles, that of sheet resistivity and the idea of using a $\frac{1}{4}\lambda_g$ shorted section to isolate the load from external impedances. The arrangement is not satisfactory in practice, however, largely because it requires an adjustable short circuit on the $\frac{1}{4}\lambda_g$ section which is rather critical to adjust. More practical arrangements for terminating a guide are shown in Fig. 4.8. The resistive strip is set in the plane of the maximum electric field, and the taper ensures a gradual change in electric field, which reduces reflections to a negligible level. Either form of taper shown in Fig. 4.8 is satisfactory, and the dimensions shown are typical of those used in practice. A carbon-coated strip may be used, or for more stable operation, a glass strip covered with a thin film of metal which, in turn, has a thin dielectric coating for protection. Surface resistivities of the order of 500 Ω per square are typical.

4.8 ATTENUATORS

An arrangement similar to that shown in Fig. 4.8 can be used to provide attenuation in a waveguide. Two common methods are shown in Fig. 4.9. In Fig. 4.9(a), the thin resistive sheet can be moved from the sidewall, where it produces minimum attenuation, to the center of the guide, where it produces maximum attenuation. The mechanical drive for the sheet is

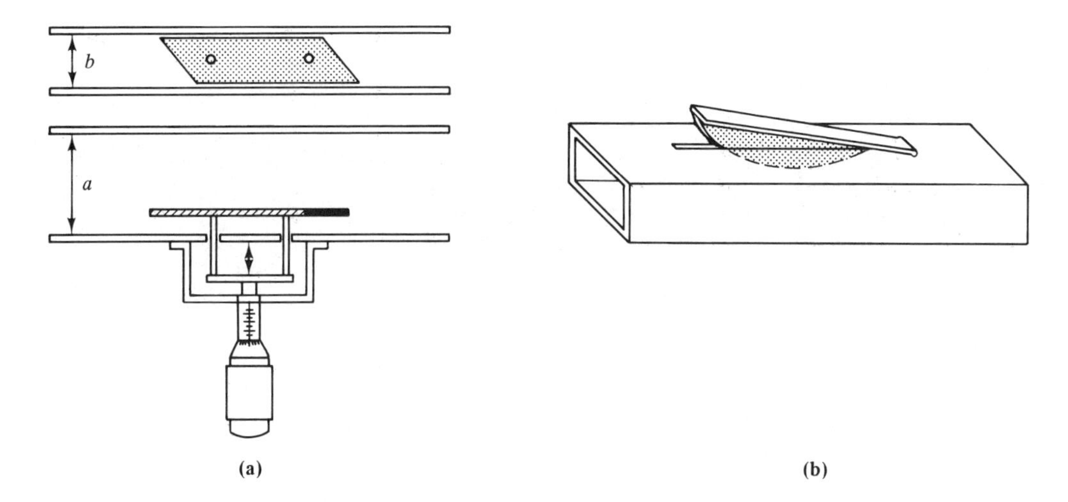

FIGURE 4.9. Waveguide attenuators.

often fitted with a micrometer control so that fine adjustment of attenuation can be made and accurately calibrated. The flap attenuator of Fig. 4.9(b) is simple to construct, and, as shown in the next section, the slot position is such that radiation is minimized. However, some radiation does occur, and this type is not used for accurate work.

4.9 CURRENTS IN WALLS

As already stated, the traveling-wave pattern is time-invariant; that is, the pattern appears to keep its shape as it moves down the guide. At any given cross section of the guide, however, the magnetic field (and the electric field) appears to vary in time as the loops of alternate polarity sweep by. This gives rise to induced currents in the walls of the guide, these currents being at right angles to the magnetic field. The currents for the TE_{10} mode are as shown in Fig. 4.10(a). This pattern moves down the guide along with the field pattern at the phase velocity. It is emphasized that the pattern moves at the phase velocity, not the current (since that would imply electrons moving faster than the speed of light!). In effect, the current pattern builds up and decays as the TE wave sweeps by. An analogy can be drawn with a sea wave moving along a sea wall, creating a splash, Fig. 4.10(b). The splash travels at the phase velocity (which is faster than the velocity at which the wave approaches the wall), and people standing at the edge of the sea wall will move back as the splash passes them, so the ripple in the line also moves at the phase velocity.

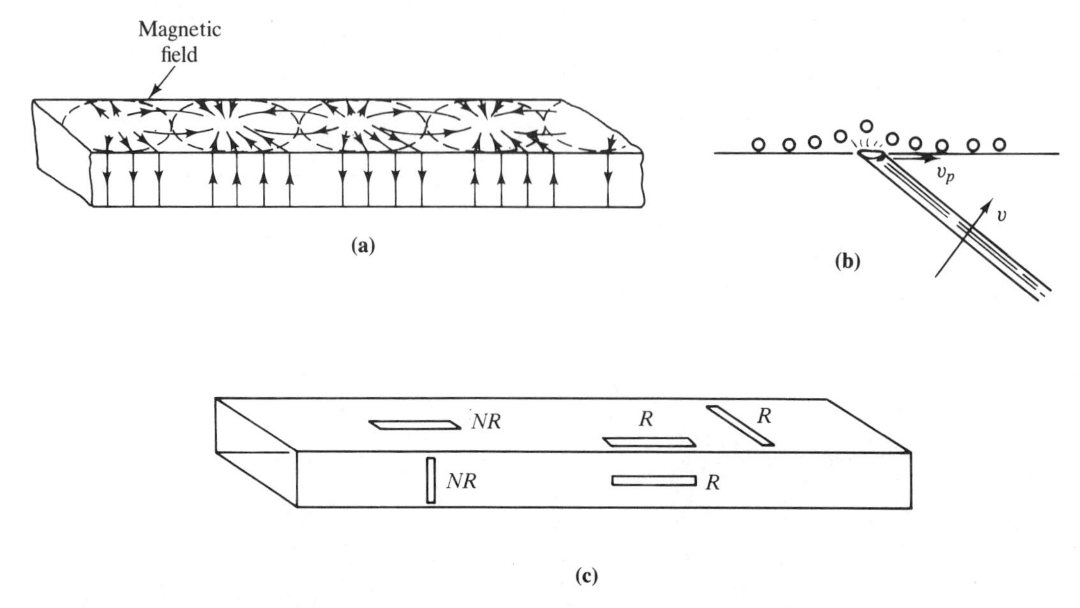

FIGURE 4.10. (a) Currents in walls for the TE_{10} mode; (b) Sea wave analogy illustrating phase velocity; (c) Nonradiating (NR) and radiating (R) slots.

Knowing the current pattern is important, as it enables slots to be correctly positioned in the walls, to serve various purposes. Slots that do not noticeably interrupt the current flow are termed *nonradiating*, since they result in minimum disturbance of the internal fields, and therefore little electromagnetic energy leaks through them. The positions of two nonradiating slots are shown in Fig. 4.10(c), these being labeled NR. It will be seen that a nonradiating slot can be placed along the center of the broad wall of the guide and use is made of this in the attenuator of Fig. 4.9(b). Another application is the slotted waveguide, which enables standing waves to be measured, the technique being similar to that described in Section 14.5 for transmission lines.

Slots that do interrupt the current flow, such as those labeled *R* in Fig. 4.10(c), are termed *radiating* slots. These produce maximum disturbance of the internal fields, which results in energy being radiated. Radiating slots form the basis of slot antennas, described in Section 10.12.

4.10 CONTACTS AND JOINTS

The properties of a short-circuited section can be used to provide an electrical short circuit without the necessity of providing a solid mechanical contact at the point of short circuit. This principle is incorporated into the

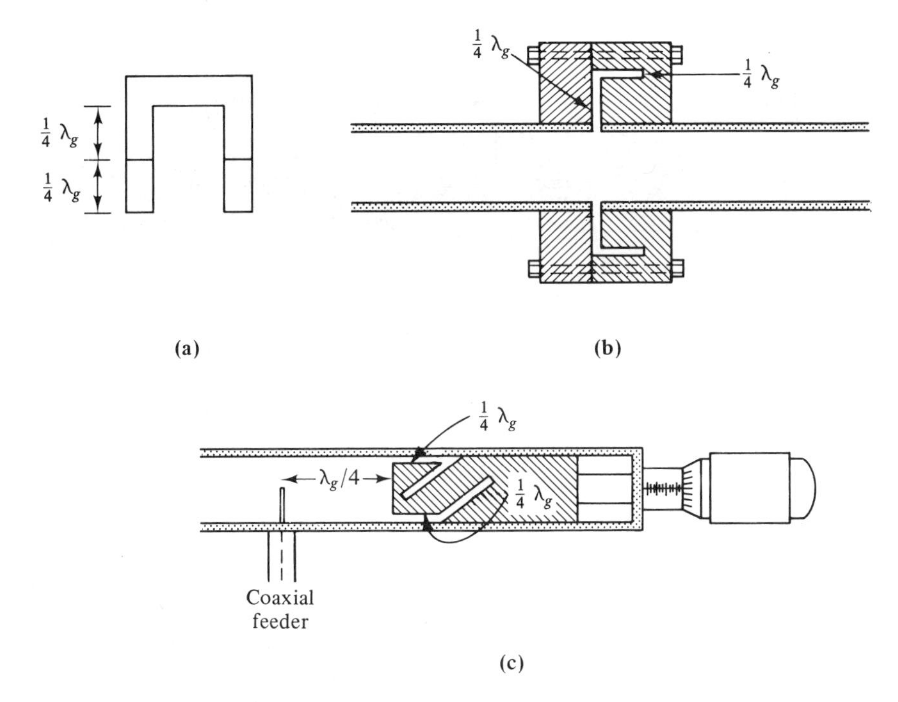

FIGURE 4.11. (a) Two $\frac{1}{4}\lambda_g$ transforming sections in series; (b) A coupling flange utilizing (a); (c) A movable short-circuiting contact utilizing (a); the contact illustrated provides a $\frac{1}{4}\lambda_g$ shorted section for the coaxial feeder.

design of some types of flanges used for coupling guide sections together and in the design of movable short-circuiting contacts. Two $\frac{1}{4}\lambda_g$ transformations take place, as shown in Fig. 4.11(a). The top $\frac{1}{4}\lambda_g$ section transforms the solid short circuit at the top to an open circuit at the junction of the two $\frac{1}{4}\lambda_g$ sections. Here a mechanical joint occurs, but since this is a high impedance point the currents are small and the joint resistance is not important. The second $\frac{1}{4}\lambda_g$ section transforms the open circuit back to a short circuit at the entry point. Application of the principle to a coupling flange is illustrated in Fig. 4.11(b) and to a movable short circuit in Fig. 4.11(c). In the latter figure, a coaxial feeder is shown by way of example, the $\frac{1}{4}\lambda_g$ short-circuited section performing the same function as that shown in Fig. 4.7(b).

4.11 REACTIVE STUBS

Equivalent reactive elements can be introduced into a waveguide in a variety of ways, a common method being to use an adjustable screw stub as shown in Fig. 4.12. When the stub is only a short way in, as shown in

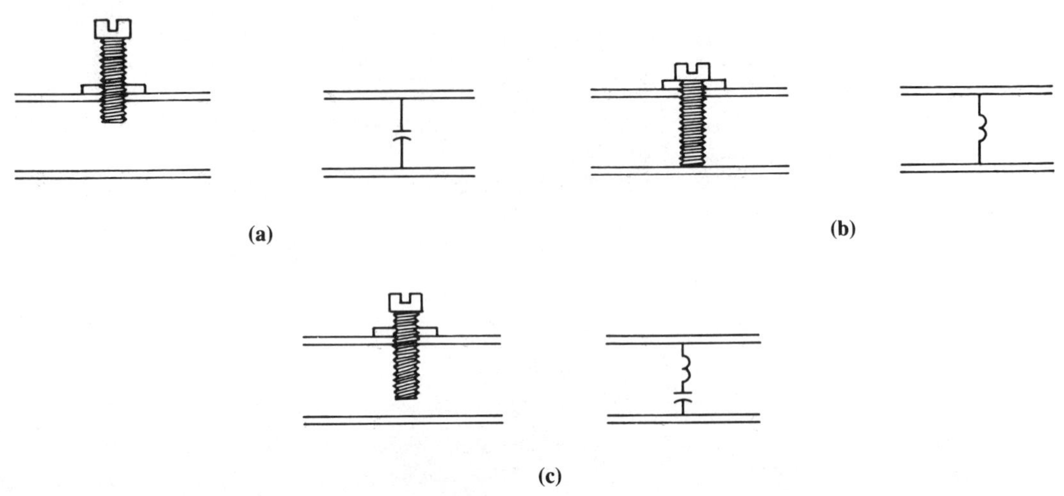

FIGURE 4.12. Reactive stubs: (a) Capacitive stub; (b) Inductive stub; (c) Series LC stub.

Fig. 4.12(a), it acts as a capacitor as it produces an increase in the electric flux density of the wave in the vicinity of the stub.

With the screw all the way in, such that it forms a post between the top and bottom of the guide, Fig. 4.12(b), a path is provided for induced currents which set up a magnetic field; such a post therefore acts as an inductor. An equivalent series *LC* circuit is formed when the screw is sufficiently far in for both components to be significant but neither dominant, such as is shown in Fig. 4.12(c). Screw stubs are used singly, and in groups of two and three, to provide matching devices between a waveguide and load.

4.12 WAVEGUIDE T-JUNCTIONS

A waveguide T-junction provides a means of splitting, and also of combining, power in a waveguide system. Two basic types of T-junctions are available, the *E*-plane junction and the *H*-plane junction.

The *E*-plane junction is illustrated in Fig. 4.13(a). Because the electric field plane, which is parallel to the narrow walls of the junction, is common to the three arms of the junction as shown in Fig. 4.13(b), the junction is referred to as an *E*-plane junction. It will be seen that the field splits in such a manner that, with the input to port 2, the electric fields in arms 1 and 3 are in antiphase. Also, referring to Fig. 4.13(b), $E_2 = E_1 + E_3$. This is analogous to the voltage relationship in a series circuit and for this reason the *E*-plane junction is also referred to as a *series junction*.

The *H*-plane junction is illustrated in Fig. 4.14(a). The name arises

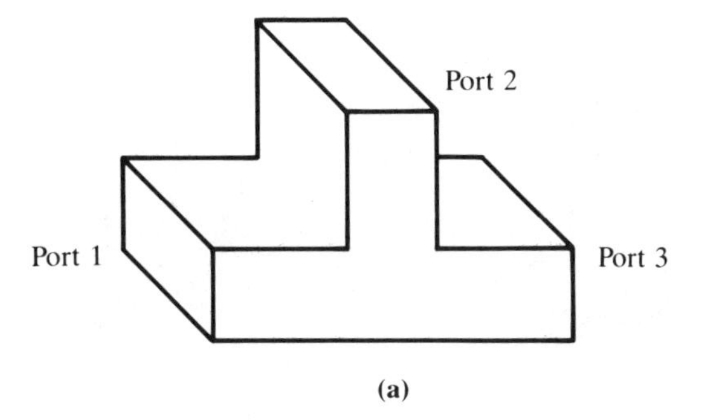

(a)

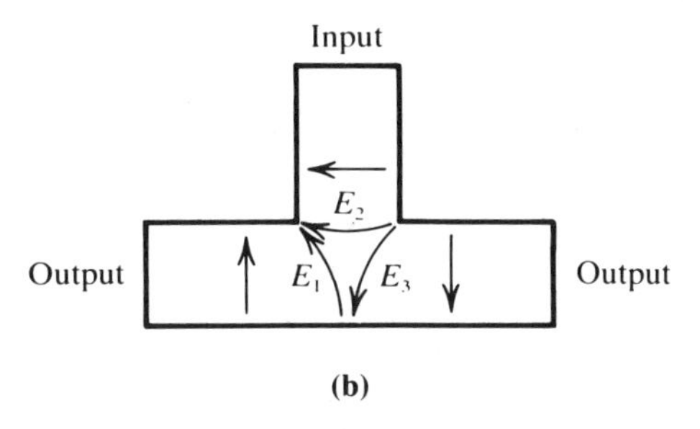

(b)

FIGURE 4.13. (a) *E*-plane junction, also known as a series junction; (b) electric field distribution with input applied to port 2, and outputs taken from ports 1 and 3.

in this case because the plane of the magnetic field is common to all three arms, as illustrated in Fig. 4.14(b). Because the magnetic field loops divide in parallel between arms 1 and 3 in a manner similar to current division between branches in a parallel circuit, the junction is also referred to as a *shunt junction*. Also, the outputs at ports 1 and 3 are in phase for the *H*-plane junction, in contrast to the antiphase relationship for the *E*-plane junction.

The transmission properties of the T-junction can be described using scattering variables. In Chapter 2 the scattering equations (2.23) and (2.24) are developed for a two-port network, and the extension of these to a three-port network is straightforward. Referring to Fig. 4.15(a), the scattering equations are

$$b_1 = S_{11}a_1 + S_{12}a_2 + S_{13}a_3 \tag{4.24}$$

$$b_2 = S_{21}a_1 + S_{22}a_2 + S_{23}a_3 \tag{4.25}$$

$$b_3 = S_{31}a_1 + S_{32}a_2 + S_{33}a_3 \tag{4.26}$$

In these equations, the b term on the left-hand side is the reflected variable

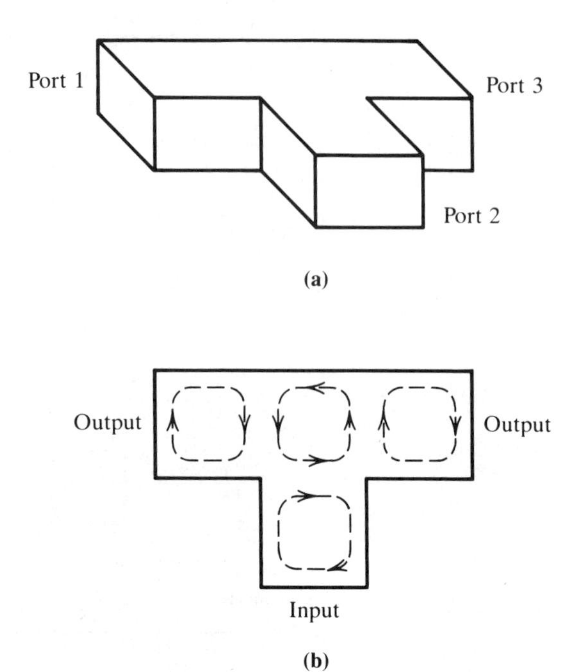

FIGURE 4.14. (a) *H*-plane junction, also known as a shunt junction; (b) magnetic field distribution with input applied to port 2 and outputs taken from ports 1 and 3.

at each port, consisting of the reflected component originating at the port itself, plus the components transmitted through the junction from the other ports. For example, b_2, the reflected variable at port 2, consists of the reflected component $S_{22}a_2$ plus the transmitted components $S_{21}a_1$ and $S_{23}a_3$.

For practical use, numerical values must be found for the scattering parameters, and as a first step toward this, symmetrical properties of the junction may be used to reduce the number of scattering parameters in-

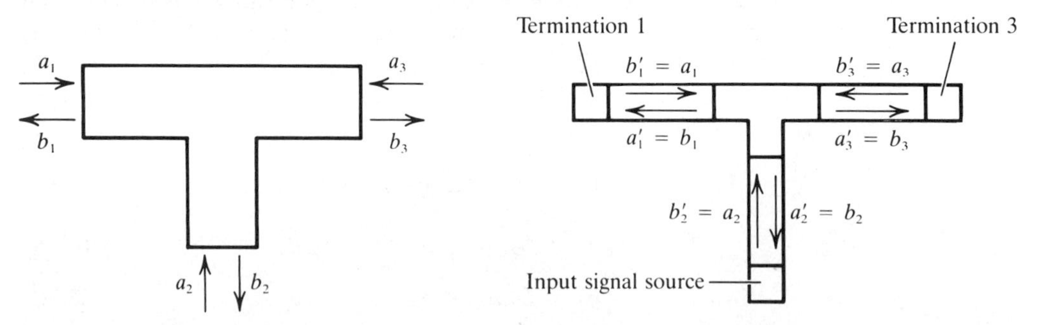

FIGURE 4.15. (a) Scattering variables for a three-port network; (b) a T-junction fed at port 2, used to split power between ports 1 and 3. The primed variables are the scattering variables referred to the source and terminations.

volved. Because there are no nonlinear elements in the junction, then in general $S_{ij} = S_{ji}$ and $S_{11} = S_{33}$. Furthermore, for the H-plane junction, $S_{21} = S_{23}$, and for the E-plane junction, $S_{21} = -S_{23}$. The negative sign is necessary here to account for the phase inversion described previously. Making use of these properties reduces the problem to that of finding values for the scattering parameters S_{11}, S_{12}, S_{13}, and S_{22}.

As shown in Chapter 2 [Eqs. (2.8) and (2.9), and Fig. 2.2(b)], the role of a scattering variable depends on its direction relative to the "port." For example, the incident variable a_1 at port 1 is also the reflected variable b_1' originating from the termination connected to port 1. Figure 4.15(b) shows the situation with input power to port 2, and terminations connected to ports 1 and 3. In each arm of the circuit, the scattering variables may be interpreted as incident or reflected depending on whether one is looking towards the T-junction port or the connection at the other end. Terminations 1 and 3 can be chosen independent of the T-junction parameters to provide matched terminations, so that b_1' and b_3' both become zero. This in turn means that a_1 and a_3 are zero. Making use of the power equation (2.11), the net input power to port 2 is $|a_2|^2 - |b_2|^2$ or $|a_2|^2(1 - |S_{22}|^2)$, and the output power is $|b_1|^2 + |b_3|^2$ or $2|S_{12}|^2|a_2|^2$, since $|S_{21}| = |S_{23}|$ by symmetry. For a lossless T-junction the input power must equal the output power and therefore

$$(1 - |S_{22}|^2) = 2|S_{12}|^2 \tag{4.27}$$

Now S_{22} is the reflection coefficient at port 2, and it can be reduced to zero by use of a suitable matching iris. Assuming that this is done, Eq. (4.27) gives $|S_{12}| = 1/\sqrt{2}$, and from the symmetry arguments presented earlier the following values can be deduced:

$$S_{12} = S_{21} = \frac{1}{\sqrt{2}} \tag{4.28}$$

$$S_{23} = S_{32} = \pm\frac{1}{\sqrt{2}} \tag{4.29}$$

where the + sign is used with the H-plane junction and the − sign with the E-plane junction. It must be kept in mind that these values are for the condition that $S_{22} = 0$.

Once S_{22} has been made equal to zero it is not possible to make either S_{11} or S_{33} equal to zero by similar iris matching methods. This is because the irises will interact with each other and either S_{22} will no longer be zero, or, if this is readjusted, mismatch again occurs at S_{11} and S_{33}. A more detailed argument based on power considerations which is too lengthy to include here shows that $S_{11} = S_{33} = \frac{1}{2}$ and $S_{13} = S_{31} = \frac{1}{2}$, for $S_{22} = 0$. Thus the scattering Eqs. (4.24) through (4.26) can be rewritten as

$$b_1 = \frac{a_1}{2} + \frac{a_2}{\sqrt{2}} + \frac{a_3}{2} \tag{4.30}$$

$$b_2 = \frac{a_1}{\sqrt{2}} + 0 \pm \frac{a_3}{\sqrt{2}} \tag{4.31}$$

$$b_3 = \frac{a_1}{2} \pm \frac{a_2}{\sqrt{2}} + \frac{a_3}{2} \tag{4.32}$$

The + signs are used for the *H*-plane junction and the − signs for the *E*-plane junction.

One important consequence which this analysis shows up is that when S_{22} is designed to be zero, then S_{11} and S_{33} are $\frac{1}{2}$, and a mismatch will exist at these ports. For example, if power is applied to port 1, and ports 2 and 3 are correctly terminated (matched) so that a_2 and a_3 are both zero, the reflection coefficient at port 1 will still be $\frac{1}{2}$. The VSWR on the connecting guide to port will then be, from Eq. (1.82),

$$\text{VSWR} = \frac{1 + 0.5}{1 - 0.5} = 3$$

The coefficients on the right-hand side of Eqs. (4.24) through (4.26) form what is known as the *scattering matrix* for the T-junction, and this is formally written as

$$S = \begin{bmatrix} S_{11} & S_{12} & S_{13} \\ S_{21} & S_{22} & S_{23} \\ S_{31} & S_{32} & S_{33} \end{bmatrix} \tag{4.33}$$

When the numerical values are substituted the scattering matrix becomes

$$S = \begin{bmatrix} \frac{1}{2} & \frac{1}{\sqrt{2}} & \frac{1}{2} \\ \frac{1}{\sqrt{2}} & 0 & \pm\frac{1}{\sqrt{2}} \\ \frac{1}{2} & \pm\frac{1}{\sqrt{2}} & \frac{1}{2} \end{bmatrix} \tag{4.34}$$

It will be seen that the scattering matrix gives all the important information about the junction.

EXAMPLE 4.2

A 10-mW signal is fed into port 2 of a lossless *H*-plane junction for which $S_{22} = 0$. If the other ports are terminated in matched loads, calculate the power delivered through each of the other arms.

SOLUTION The power flow through arm 2 is 10 mW. By symmetry, this divides equally between the other two arms; therefore, arms 1 and 3 each carry 5 mW.

EXAMPLE 4.3

For the junction in Example 4.2 the incident power fed to port 1 is 10 mW. If the other ports are terminated in matched loads, determine the power delivered through each arm.

SOLUTION The matched terminations result in $a_2 = a_3 = 0$. The net power input to port 1 is

$$\begin{aligned} P_1 &= a_1^2(1 - |S_{11}|^2) \\ &= 10 \times [1 - (0.5)^2] \\ &= \mathbf{7.5\ mW} \end{aligned}$$

The power transferred to port 2 is

$$\begin{aligned} P_2 &= |a_1|^2 |S_{21}|^2 \\ &= 10 \times \left(\frac{1}{\sqrt{2}}\right)^2 \\ &= \mathbf{5\ mW} \end{aligned}$$

The power transferred to port 3 is

$$\begin{aligned} P_3 &= |a_1|^2 |S_{31}|^2 \\ &= 10 \times \left(\frac{1}{2}\right)^2 \\ &= \mathbf{2.5\ mW} \end{aligned}$$

Note that $P_1 = P_2 + P_3$.

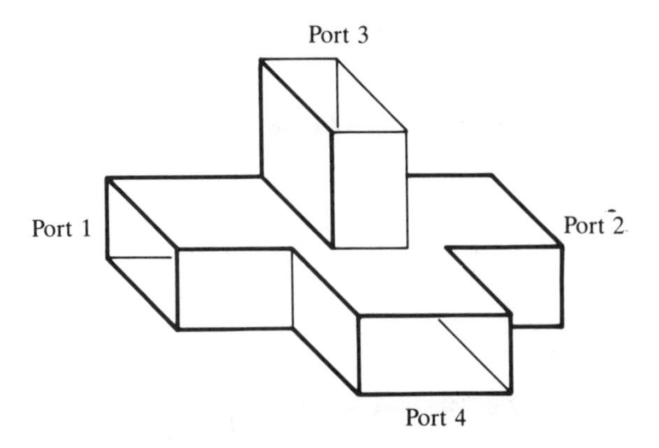

FIGURE 4.16. Hybrid-T, also known as the magic-T.

4.13 THE HYBRID-T JUNCTION

The hybrid-T junction is a combined E-plane and H-plane arrangement of junctions as illustrated in Fig. 4.16. One of the main advantages of this type of junction is that it is possible to design all reflection coefficients, S_{11}, S_{22}, S_{33}, and S_{44} to be zero, and therefore the hybrid-T provides much better matching than either the H-plane or E-plane junctions alone. An equally important property of this junction is that power fed in any one port will not appear at the opposite port but will divide between the two other ports. This is shown in Table 4.1.

TABLE 4.1
Power Distribution in a Hybrid-T

Power-In Port	*Power-Out Ports*
1	3 and 4
2	3 and 4
3	1 and 2
4	1 and 2

A qualitative picture of the operation of the hybrid-T can be obtained by considering the electric field distribution in each of the arms when the input is applied to each of the ports in turn. Figure 4.17(a) shows the situation when the input is applied to port 3. The signal divides between ports 1 and 2 as in the case of the E-plane junction of Fig. 4.13, and the antiphase signals cancel in arm 4. No output appears therefore at port 4. Figure 4.17(b) shows the input applied at port 4. The signal divides between arms 1 and 2, as for the H-plane junction of Fig. 4.14, and the antiphase signals cancel in arm 3. Thus no output appears at port 3. Figure 4.17(c)

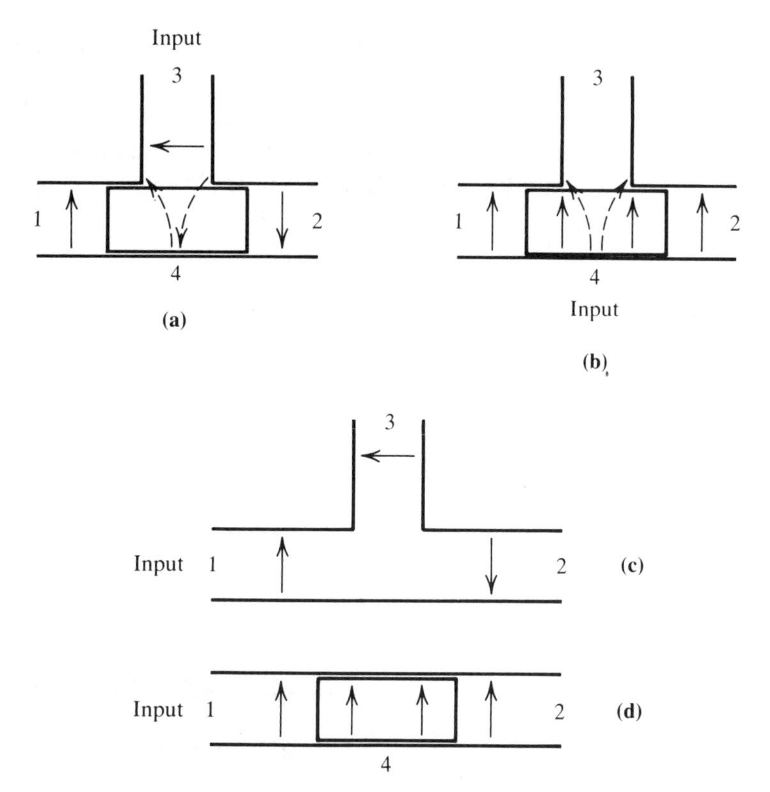

FIGURE 4.17. Electric field distribution with the input to (a) port 3 and (b) port 4. The input to port 1 showing (c) *E*-plane distribution and (d) *H*-plane distribution.

and (d) show the electric fields for the individual *E*-plane and *H*-plane transfers when the signal is applied to port 1. Comparing these it will be seen that the signals in arm 2 are antiphase, and therefore they cancel out, and no output appears at port 2. Because of symmetry, an input at port 2 will not produce an output at port 1.

When the individual reflection coefficients S_{ii} have been adjusted to zero the scattering matrix for the hybrid-T is found to be

$$S = \frac{1}{\sqrt{2}} \begin{bmatrix} 0 & 0 & 1 & 1 \\ 0 & 0 & -1 & 1 \\ 1 & -1 & 0 & 0 \\ 1 & 1 & 0 & 0 \end{bmatrix} \tag{4.35}$$

It will be noted that because there are four ports, four scattering equations are required to describe the transfer properties of the hybrid-T. Because of the property that no output appears at the port opposite the input port, the hybrid-T is sometimes referred to as a *magic-T*.

EXAMPLE 4.4

A hybrid-T has a scattering matrix given by Eq. (4.35), the ports being numbered as shown in Fig. 4.16. A 10-mW signal is applied to each port in turn, the remaining ports being terminated in matched loads each time. Calculate the power delivered to the matched loads in each case.

SOLUTION When the input is applied to port j, the output scattering variable at port i is

$$b_i = a_j S_{ij}$$

All other a variables are zero because of the matched load conditions. Inspection of Eq. (4.35) shows that S_{ij} is equal to either zero or $\pm 1/\sqrt{2}$ and therefore P_i will either be zero or $10 \times |\pm 1/\sqrt{2}|^2 = 5$ mW. Hence

With $P_1 = 10$ mW input, $P_2 = 0$, and $P_3 = P_4 = 5$ mW.
With $P_2 = 10$ mW input, $P_1 = 0$, and $P_3 = P_4 = 5$ mW.
With $P_3 = 10$ mW input, $P_4 = 0$, and $P_1 = P_2 = 5$ mW.
With $P_4 = 10$ mW input, $P_3 = 0$, and $P_1 = P_2 = 5$ mW.

4.14 APPLICATIONS OF THE HYBRID-T JUNCTION

The hybrid-T may be used to couple two equal power transmitters together such that the power to the load is doubled while the transmitters are isolated from each other. Figure 4.18(a) shows a schematic of the arrangement. For equal powers from the transmitters, $P_3 = P_4 = P$ and $a_3 = a_4 = a$. As previously shown, when ports 3 and 4 are used as input ports, no coupling exists between them. The antenna matched to port 1 makes a_1 equal to zero. The second row of the scattering matrix, eq. (4.35) gives $b_2 = (0 + 0 - a + a)\sqrt{2} = 0$ and hence $a_2 = 0$ also. With a_1 and a_2 both equal to zero the third and fourth rows of the scattering matrix give $b_3 = 0$ and $b_4 = 0$, and the scattering equations reduce to

$$\begin{aligned} b_1 &= S_{13}a_3 + S_{14}a_4 \\ &= \frac{a_3}{\sqrt{2}} + \frac{a_4}{\sqrt{2}} \\ &= \sqrt{2}a \end{aligned} \tag{4.36}$$

Hence the output power is at port 1 and is equal to

$$P_1 = |b_1|^2 = 2|a|^2 = 2P \tag{4.37}$$

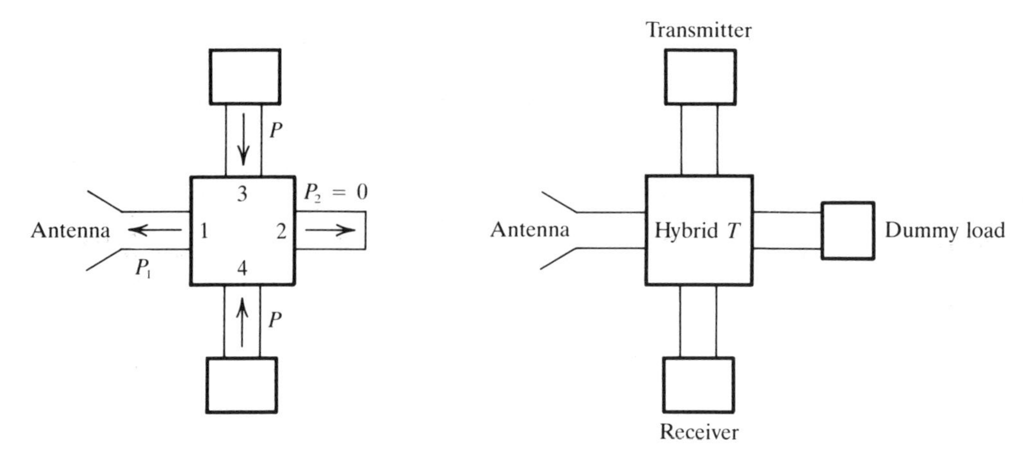

FIGURE 4.18. (a) Hybrid-T used to couple two transmitters to a single antenna; (b) hybrid-T used to couple a transmitter and a receiver to a common antenna.

Another application is where it is required to work a common antenna with a transmitter and a receiver without these interfering with each other. The transmitter and receiver are connected to opposite ports (3 and 4 or 1 and 2) and therefore they are isolated from each other. The antenna is connected to one of the remaining ports, and a matched dummy load to the other, as shown in Fig. 4.18(b). Power from the transmitter divides equally between the dummy load and antenna, with none going to the receiver. Signal power from the antenna divides equally between the receiver and the transmitter, the latter power being dissipated as heat in the transmitter output resistance. Note therefore that a 3-dB power loss occurs each way.

Figure 4.19 shows how the hybrid-T is used to form a low-noise mixer circuit. The incoming signal is fed to arm 4 and utilizes the H-plane properties of the junction so that the received signal components arriving at the mixer diodes are in phase. The local oscillator signal is fed to arm 3, and utilizes the E-plane properties of the junction, so that the local oscillator components arriving at the mixer diodes are in antiphase. The 1F signals will therefore be in antiphase but these add in the push-pull output transformer. Now one of the problems encountered with a single-diode mixer is that noise from the local oscillator can mask weak received signals. With the hybrid-T arrangement the noise components at the diodes are in antiphase. These mix with the main oscillator signals which are also antiphase to produce in-phase 1F components of noise which tend to cancel at the output of the balanced transformer.

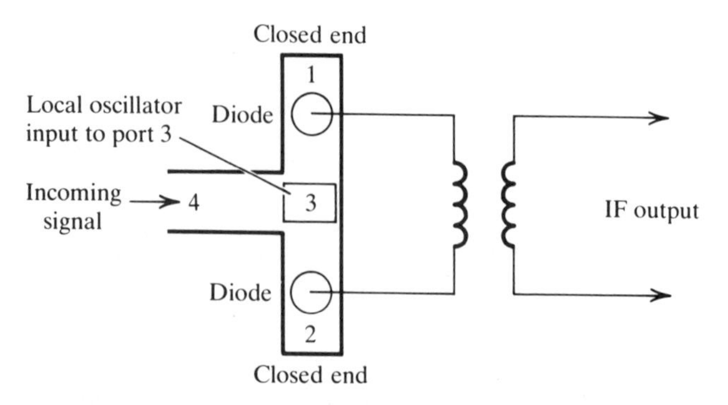

FIGURE 4.19. Low-noise mixer circuit utilizing a hybrid-T.

4.15 THE HYBRID RING

The hybrid ring is another form of four-port junction with properties similar to those of the hybrid-T. One arrangement for the hybrid ring is shown in Fig. 4.20. A signal applied at any one port is seen to have two paths available, and provided that the other ports are reflectionless matched, the applied signal will divide evenly between these two paths. The difference in length between the two paths will determine the phase difference between the two signals when they recombine at any given port. For example, the two paths between port 1 and port 2 are $\lambda_g/4$ and $5\lambda_g/4$. The path difference is therefore $5\lambda_g/4 - \lambda_g/4 = \lambda_g$, and this corresponds to a phase difference of 360°, or equivalently 0°. The differences in path lengths

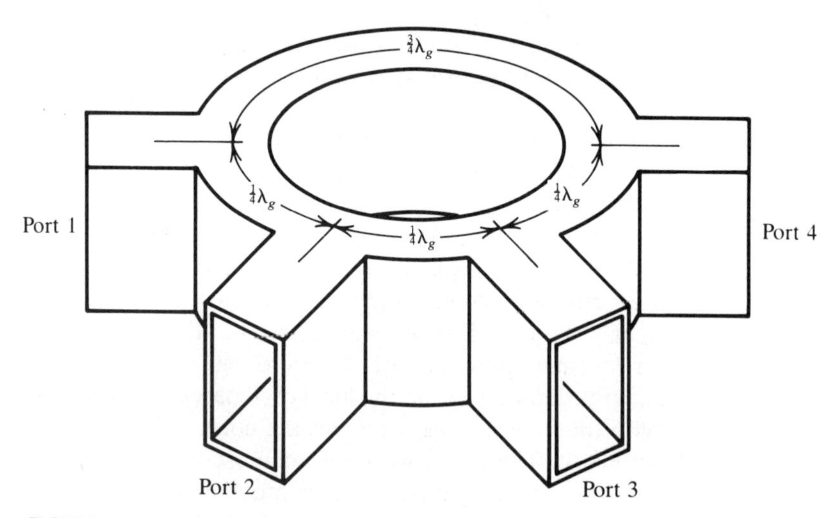

FIGURE 4.20. Hybrid ring utilizing *E*-plane junctions.

TABLE 4.2
Differential Path Length and Phase for the Hybrid Ring

Input Port	*Output Port*			
	1	*2*	*3*	*4*
1	—	λ_g (360°)	$\frac{1}{2}\lambda_g$ (180°)	0
2	λ_g (360°)	—	λ_g (360°)	$\frac{1}{2}\lambda_g$ (180°)
3	$\frac{1}{2}\lambda_g$ (180°)	λ_g (360°)	—	λ_g (360°)
4	0	$\frac{1}{2}\lambda_g$ (180°)	λ_g (360°)	—

between any two ports are shown in Table 4.2, together with the corresponding phase differences.

It will be seen that a signal applied at port 1 will result in outputs at ports 2 and 4 but not at port 3, the 180° differential phase shift at this port resulting in cancellation of the two branch signals. Similar conclusions can be drawn for the other combinations shown in Table 4.2.

Hybrid rings may be constructed as *E*-plane or *H*-plane rings, the *E*-plane type being the most common. The *E*-plane ring is illustrated in Fig. 4.20. Hybrid rings are also constructed in microstrip and stripline, and a more detailed analysis of these types is given in Section 5.6.

4.16 DIRECTIONAL COUPLERS

Just as it is possible to have coupled lines, as described in Section 1.16, so waveguide sections may be coupled to form directional couplers. The quantities defined by Eqs. (1.125) through (1.127) are also used for waveguide couplers, but because powers and not voltages are normally measured in waveguides, the power equations are the most convenient, and these are repeated here:

$$\text{Coupling dB} = 10 \log \frac{P_4}{P_1} \tag{4.38}$$

$$\text{Directivity dB} = 10 \log \frac{P_3}{P_1} \tag{4.39}$$

$$\text{Transmission dB} = 10 \log \frac{P_2}{P_1} \tag{4.40}$$

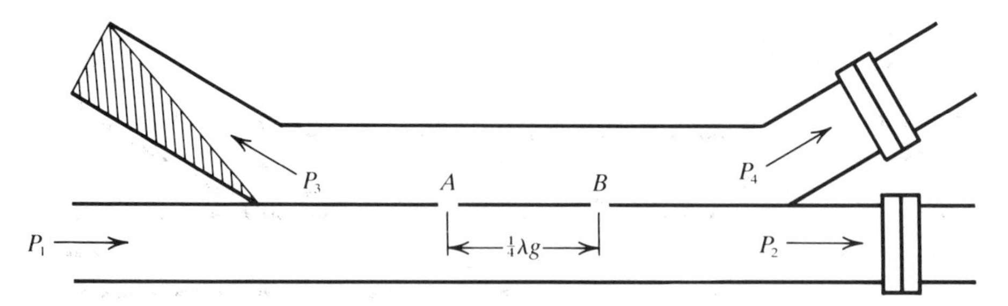

FIGURE 4.21. Directional couplers: (a) two hole; (b) three hole; (c) single hole.

The subscripts 1 through 4 refer to the ports shown in Fig. 4.21, which illustrates one simple form of waveguide coupler. All ports are assumed to be matched, and most of the power fed in at port 1 will pass through to port 2. A small amount of power is coupled into the top guide through the two coupling holes, spaced $\lambda_g/4$. The coupled waves propagate in both directions. Those traveling toward port 4 travel the same distance and therefore add in phase to provide an output at port 4. For port 3, the wave coupled in through hole A leads that from hole B by 180°. This is because, taking hole A as reference, the wave going to hole B has to travel a further $\lambda_g/4$ in the main guide, and then $\lambda_g/4$ back again. Hence the waves going to port 3 cancel. In practice some small amount of power will reach port 3 because of imperfections in the coupling, and the directivity given by Eq. (4.39) is a measure of this.

The two-hole coupler described is narrowband, because it depends on the $\lambda_g/4$ spacing between holes. The frequency response can be broadened by using more than two coupling holes.

Single-hole couplers are also available. These work on the principle that the electric and magnetic fields coupled through a single hole give rise to independent electromagnetic waves in the branch guide. By orienting the branch guide at a specific angle to the main guide (rather than having it lying parallel as shown in Fig. 4.21), the waves can be made to add at port 4 and cancel at port 3. The position and shape of the coupling hole are also important factors (Levy 1966; Cohn and Levy 1984).

4.17 CAVITY RESONATORS

A cavity resonator may be thought of as a section of waveguide which permits electromagnetic fields to exist in it only at certain sharply defined frequencies, or resonant frequencies. A cavity resonator may be constructed by closing off both ends of a rectangular guide section with metallic end plates, as shown in Fig. 4.22(a). Apart from the small coupling loop, the cavity is totally enclosed and the boundary conditions described in

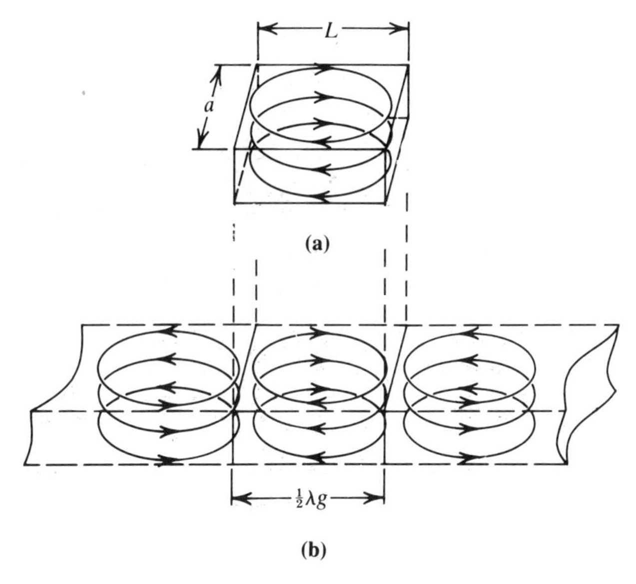

FIGURE 4.22. (a) Rectangular cavity resonator; (b) how part (a) may be considered as a section of a TE_{10} guide.

Section 4.1 must be satisfied at all six walls. By comparing parts (a) and (b) of Fig. 4.22, it will be seen that the cavity mode is equivalent to one half-wavelength of variation of a TE_{10} mode, and for this reason the cavity mode is referred to as a TE_{101} mode. The boundary conditions are automatically satisfied at the end plates if they are positioned $\frac{1}{2}\lambda_g$ apart for the TE_{10}. In effect, the TE_{10} wave is reflected from the end plates with a reflection coefficient of -1, and the mode within the cavity is a result of the standing waves generated as described in Section 4.6.

The current in the small coupling loop at the end of the coaxial feed sets up the magnetic field in the correct direction, but for the TE_{101} mode to exist, the frequency of the input signal must satisfy Eq. (4.7) with the additional constraint that $\lambda_g = 2L$. If the length L is made equal to λ_g the cavity mode becomes TE_{102}, and in general, the cavity can support TE_{mnp} and TM_{mnp} modes where m and n have the meanings attached in Section 4.18, and p represents the number of half-wavelengths in the direction of L.

For the TE_{101} mode, $\lambda_c = 2a$ and $\lambda_g = 2L$ and therefore applying Eq. (4.7) gives, for the free-space wavelength,

$$\lambda = \frac{2}{\sqrt{(1/a)^2 + (1/L)^2}} \tag{4.41}$$

Figure 4.23 shows how a cavity resonator may be incorporated as a filter in a waveguide run. The irises form the closed ends of the cavity and

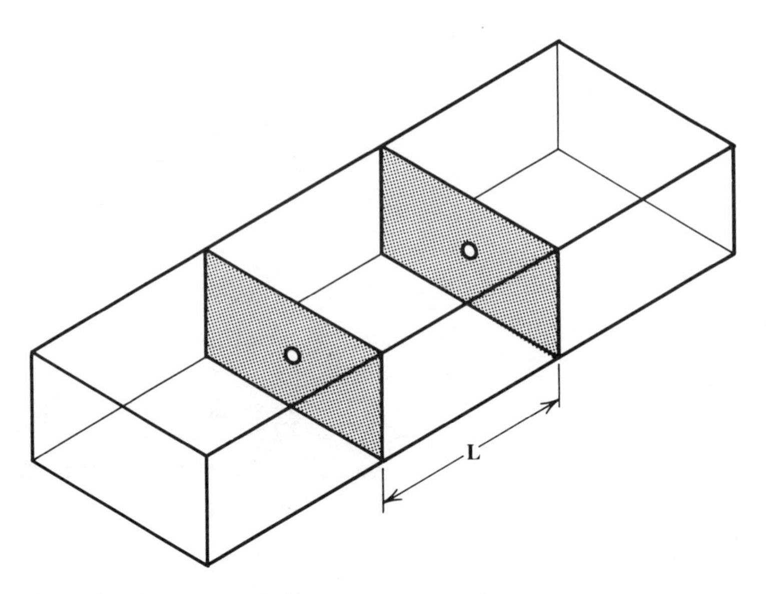

FIGURE 4.23. Waveguide filter using a cavity resonator.

the small holes in the centers allow coupling into and out of the cavity. The magnetic field along one side of the iris will couple through the hole to the other side, and provided that the length L is exactly λ_g, a TE_{101} mode will be set up which couples through to the output section of the guide. The coupling drops off rapidly at frequencies on either side of the resonant frequency and therefore the cavity acts as a narrow bandpass filter.

The cavity resonator is also widely used as a *cavity wavemeter*, which is an instrument for measuring wavelength, and therefore of course frequency. Figure 4.24 illustrates the principle of operation of the absorption wavemeter. Here, hole coupling is used to couple the electric field from the main guide into the cavity. Since the electric field of the TE_{10} mode in the main guide is normal to the broad wall, it will be normal to the plane of the hole and pass through this and will be in the correct direction to set up the TE_{101} mode providing the length L of the cavity is equal to λ_g. The length L of the cavity is adjustable through the micrometer drive at one end, and this can be calibrated directly in guide wavelength. When the cavity is resonant with the wave in the main guide it will absorb energy from this, and as a result, a sharp dip will be observed in the output power from the main guide. Alternatively, a crystal detector may be coupled into the cavity, and this will indicate a peak when the cavity is resonant with the wave in the main guide.

Cylindrical resonators are also available, and Fig. 4.25 shows the TE_{111} and the TM_{010} cylindrical modes. In general, TM_{mnp} and TE_{mnp}

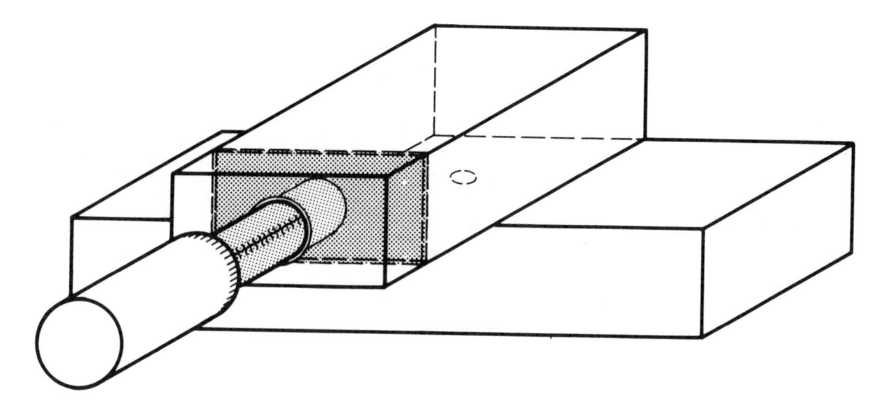

FIGURE 4.24. Cavity absorption wavemeter.

modes may be excited where m and n have the meanings attached for circular waveguides in Section 4.18, and p represents the number of half-length variations along the axis of the cavity. Cylindrical resonators, and resonators of other shapes, are fundamental to the operation of microwave tubes, described in Chapter 9. The reentrant cavity shown in Fig. 4.26 is used as the frequency-determining element in the klystron tube (see Section 9.3). It is termed reentrant because the cavity wall structure folds back inside itself as shown. An expression for the resonant frequency can be found by considering the cavity as a coaxial line section, short circuited at the reentrant end, and terminated in a capacitor C_g at the other end, the capacitance C_g of the gap being given approximately by

$$C_g = \frac{\varepsilon_0 A}{d} = \frac{\varepsilon_0 \pi a^2}{d} \tag{4.42}$$

The impedance of the short-circuited, coaxial line section is given by

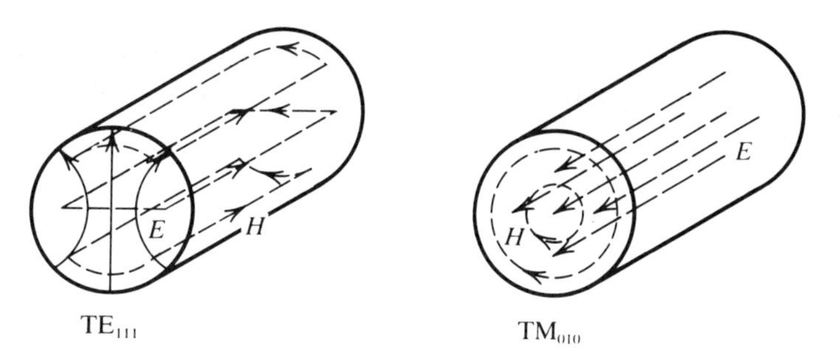

FIGURE 4.25. Cylindrical resonator modes.

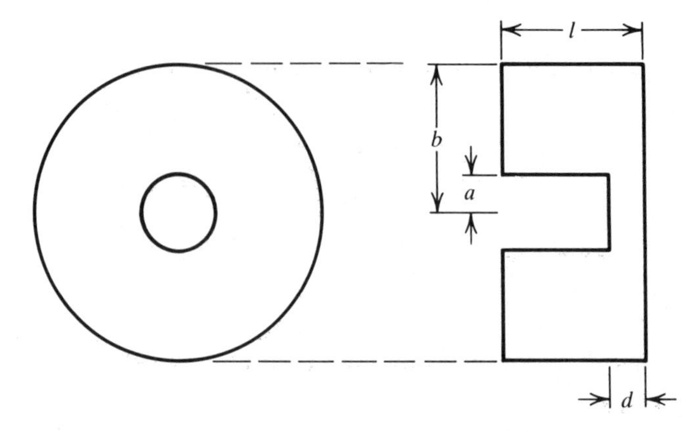

FIGURE 4.26. Reentrant cavity.

Eq. (1.102) as

$$Z_L = jZ_0 \tan\frac{2\pi l}{\lambda}$$

$$= jZ_0 \tan\frac{\omega l}{c} \tag{4.43}$$

and the impedance of C_g is

$$Z_C = \frac{-j}{\omega C_g} \tag{4.44}$$

These two impedance–frequency relationships are plotted in Fig. 4.27(a). Resonance occurs at the points labeled f_{c1}, f_{c2}, and so on, where the total impedance goes to zero. Thus it will be seen that the resonant frequency is many valued, but usually in practice the cavity would be operated at the first resonance, f_{c1}.

When the gap d is very small compared to the other dimensions of the cavity, the capacitance C_g is large and the situation is as illustrated in Fig. 4.27(b). It will be seen that the first resonance occurs at low values of Z_L such that $\tan(\omega l/c) \sim \omega l/c$ and the resonant condition becomes

$$\frac{1}{\omega_{c1} C_g} = Z_0 \frac{\omega_{c1} l}{c}$$

from which

$$\omega_{c1}^2 = \frac{c}{C_g Z_0 l} \tag{4.45}$$

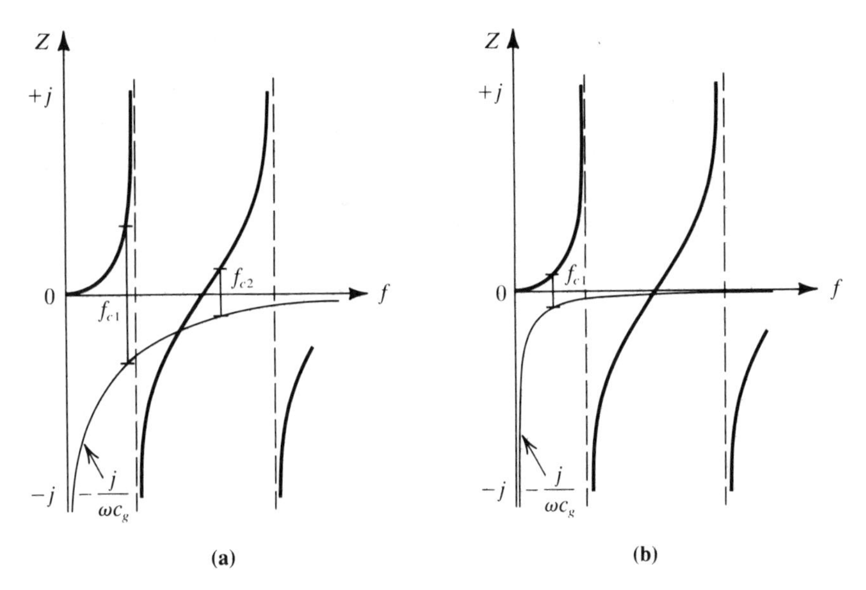

FIGURE 4.27. (a) Multiple resonances f_{c1}, f_{c2}, and so on, for a reentrant cavity; (b) resonant conditions for large C_g.

Since $(c/\omega_{c1})^2 = (\lambda_{c1}/2\pi)^2$, then using Eq. (4.45) gives

$$\lambda_{c1}^2 = 4\pi^2 C_g Z_0 lc \tag{4.46}$$

Finally, substituting for Z_0 from Eq. (1.33), and for C_g from Eq. (4.42) and simplifying yields

$$\lambda_{c1} = 2\pi a \sqrt{\frac{l}{2d} \ln \frac{b}{a}} \tag{4.47}$$

4.18 OTHER MODES

Higher modes can occur in waveguides, an example of the TE_{20} mode being sketched in Fig. 4.28(a). *Transverse magnetic* (TM) *modes* can also occur, the TM_{11} mode being sketched in Fig. 4.28(b). It can be shown that the cutoff wavelength for TE_{mn} and TM_{mn} modes in general is given by

$$\left(\frac{1}{\lambda_c}\right)^2 = \left(\frac{m}{2a}\right)^2 + \left(\frac{n}{2b}\right)^2 \tag{4.48}$$

where m and n are integers. Equation (4.48) is seen to reduce to $\lambda_c = 2a$

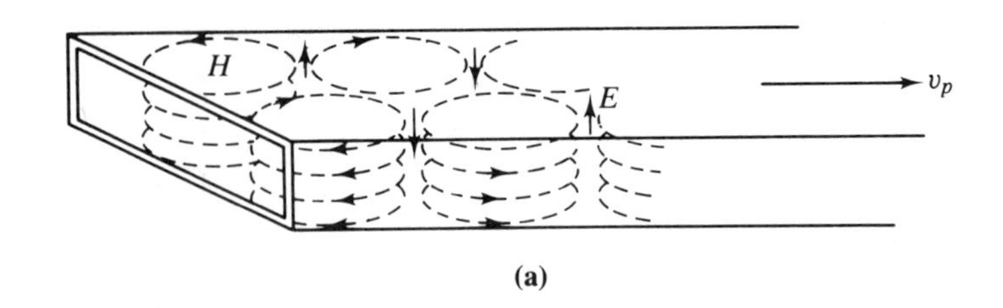

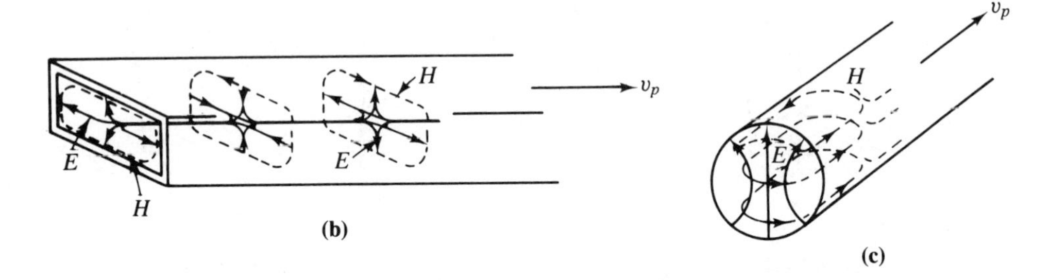

FIGURE 4.28. Other waveguide modes: (a) TE_{20} mode; (b) TM_{11} mode; (c) TE_{11} circular waveguide mode.

for $m = 1$ and $n = 0$, which is the condition already introduced in Eq. (4.7) for the TE_{10} mode.

The TM_{11} mode is the lowest TM mode that can occur as the boundary conditions exclude the TM_{10} mode. For transmission purposes only the TE_{10} mode is used, and the guide dimensions are chosen, in conjunction with the input frequency, to cut off all but the dominant mode (TE_{10}). For example, for a standard guide WR-90 the dimensions are:

Outside walls:

$$1.000 \times 0.500 \text{ in.}$$

Wall thickness:

$$0.050 \text{ in.}$$

The inside wall dimensions are, therefore,

$$a = 2.286 \text{ cm}$$

$$b = 1.016 \text{ cm}$$

The cutoff wavelengths for some of the various modes are:

TE_{10}:

$$\lambda_c = 2a$$

$$= 4.572 \text{ cm}$$

TE_{20}:

$$\lambda_c = a$$

$$= 2.286 \text{ com}$$

TE_{01}:

$$\lambda_c = 2b$$

$$= 2.032 \text{ cm}$$

TE_{11}, TM_{11}:

$$\left(\frac{1}{\lambda_c}\right)^2 = \left(\frac{1}{2a}\right)^2 + \left(\frac{1}{2b}\right)^2$$

Therefore,

$$\lambda_c = 1.857 \text{ cm}$$

Tabulating these results, we have the following:

Mode	λ_c *(cm)*	f_c *(GHz)*
TE_{10}	4.572	6.56
TE_{20}	2.286	13.1
TE_{01}	2.032	14.8
TE_{11}, TM_{11}	1.857	16.2

The recommended operating frequency range for the WR-90 guide is 8.20 to 12.40 GHz, and it will be seen that only the TE_{10} mode will be excited. Of course, if a higher frequency is fed into the guide, higher modes will be excited; for example, if a frequency of 15 GHz is used, the first three modes may exist simultaneously in the guide. For all these modes, including the circular mode described in the next paragraph, Eq. (4.48) applies.

Circular guides also support waveguide modes, the TE_{11} circular mode being sketched in Fig. 4.28(c). (The subscripts here denote circumferential and radial variations of the fields which are much more complex than the modes in the rectangular guide; only the TE_{11} circular mode will be considered here.) The cutoff wavelength is related to the diameter of the guide, the value derived from theory being

$$\lambda_c = 1.71d \tag{4.49}$$

where d is the diameter of the circular guide.

Circular guides have special properties which enable them to be used for rotating joints, and the TE_{01} mode has the unusual characteristic of attenuation becoming less as the frequency is increased, which is attractive for transmissions at the higher microwave frequencies. However, the mechanical problems of making and maintaining precise dimensions in a circular guide are much more formidable than in a rectangular guide, and the latter finds greater use for transmission purposes.

4.19 PROBLEMS

1. Determine λ_c, λ_g, λ, v_p, and v_g for standard waveguide WR-62 for which the outside dimensions are 0.702 × 0.391 in., the wall thickness is 0.040 in., when the exciting frequency is 10 GHz.
2. Determine the TE_{10} wave impedance for the WR-62 guide of Problem 1.
3. Explain how standing waves can be produced in a waveguide. What is the spacing, in terms of wavelength, between successive minima along the standing-wave pattern?
4. By analogy with the transmission-line theory presented in Section 1.10 determine the reflection coefficient for the electric field of a TE_{10} wave of wave impedance 680 Ω terminated as shown in Fig. 4.7(b), the sheet resistivity of the load being 500 Ω/square.
5. Also by analogy to the transmission theory presented in Section 1.13, determine the electric field standing-wave ratio for the guide conditions of Problem 4.
6. What is meant by sheet resistivity, and how does this differ from the resistivity of a material? What are the units for sheet resistivity?
7. Illustrate on a Smith chart the transformations that take place in the $\frac{1}{2}\lambda_g$ short-circuited section shown in Fig. 4.11(a).
8. Explain the difference between radiating and nonradiating slots in a TE_{10} mode waveguide.

9. For the T-junction shown in Fig. 4.13, the S_{22} parameter is made equal to zero. A power of 5 mW is incident at port 2 and the other ports are reflectionless matched. Determine the power flow through the junction.

10. Repeat Problem 9, but with the 5 mW incident at port 1, and the other two ports matched. Would you expect the same results with ports 1 and 3 interchanged?

11. Write out in full the scattering equations associated with the scattering matrix equation (4.34). Hence show that when ports 1 and 3 are reflectionless matched, and power input is applied to port 2, the scattering equations reduce to $b_1 = b_3 = a_2/\sqrt{2}$ for the *H*-plane junction, and $b_1 = -b_3 = a_2/\sqrt{2}$ for the *E*-plane junction. From this result deduce that the input power splits equally between ports 1 and 3.

12. From the scattering equations obtained in Problem 11, show that when the input is applied to port 1 and the other two are reflectionless matched, the scattering equations reduce to $b_1 = b_3 = a_1/2$; $b_2 = a_1/\sqrt{2}$. Hence show that if the net power input to port 1 is P_{in}, the power flow to port 2 is $\frac{2}{3}P_{in}$ and to port 3 $\frac{1}{3}P_{in}$.

13. Write out in full the scattering equations associated with the scattering matrix, Eq. (4.35). Hence show that power input to either port 1 or port 2 results in power out at ports 3 and 4, and power in at either port 3 or 4 results in power out at ports 1 and 2. All ports may be assumed to be reflectionless matched.

14. Given that the power input to port 1 of Problem 13 is 5 mW, and that all other ports are reflectionless matched, determine the power delivered at each of the other ports.

15. The specifications for a directional coupler give the directivity as 40 dB, the coupling as 10 dB, and the transmission loss as 1 dB. With reference to Fig. 4.21, for a power input of 5 mW at port 1, determine the power at each of the other ports.

16. A range of directional couplers have available couplings of 3, 6, 10, and 20 dB. Determine the percentage of primary power that is coupled to the auxiliary arm in each case.

17. The specifications for a directional coupler are: coupling 20 dB; directivity 40 dB; transmission loss, including coupling, 0.4 dB. For a 1-mW power input at port 1, determine the power delivered at each of the other ports, and determine the power dissipated in the primary arm. Express the latter in decibels relative to the input power.

18. A reentrant cavity as shown in Fig. 4.26 has dimensions $a = 3$ cm, $b = 9$ cm, $l = 9$ cm, and $d = 5$ mm. Determine its lowest resonant frequency, assuming that the conditions of Fig. 4.27(b) apply.

19. Using Eqs. (4.42) and (4.43), solve for the first and second resonant frequencies for the resonator of Problem 18.

20. Determine, for a WR62 rectangular guide, the cutoff wavelengths for all modes up to $n = m = 2$.

21. A standard waveguide WR187, has the following dimensions: outside walls 2.000×1.000 in., wall thickness 0.064 in. Determine the highest frequency that can be transmitted if only the TE_{10} mode is permitted. What modes can exist in it if it is excited by a wave of frequency 6.00 GHz?

REFERENCES

COHN, S. B. and R. LEVY, 1984. "History of microwave passive components with particular attention to directional couplers," *IEEE Trans. Microwave Theory Tech.*, Vol. MTT-32 No. 9.

LEVY, R., 1966. "Directional couplers," in *Advances in Microwaves,* Vol. 1. Ed. by Leo Young. New York: Academic Press, Inc.

FURTHER READING

GLAZIER, E. V., and H. R. L. LAMONT, 1958. *The Services' Textbook of Radio*, Vol. 5, *Transmission and Propagation*. London: Her Majesty's Stationery Office.

KENNEDY, GEORGE, 1970. *Electronic Communication Systems*. New York: McGraw-Hill Book Company.

LIAO, SAMUEL Y., 1980. *Microwave Devices and Circuits.* Englewood Cliffs, N.J.: Prentice-Hall, Inc.

five
Substrate Lines and Components

5.1 INTRODUCTION

Miniaturization of microwave circuits has taken place through the development of conducting systems constructed on solid dielectric substrates. Such systems are used mainly in low-power applications, for example, front-end circuits for microwave receivers. Although much of the circuitry could, in principle, be realized in coaxial line construction, the planar approach allows for greater flexibility and compactness of design. A number of different types of structures are currently in use, and the principal ones are described in this chapter.

5.2 STRIPLINE

The basic *stripline* structure, also known as *triplate*, is shown in Fig. 5.1(a). It consists of a flat central conductor embedded in a dielectric material, with ground planes on upper and lower surfaces. The stripline can be thought of as a highly distorted form of coaxial line, and the electric and magnetic field distribution shown in Fig. 5.1(b) should be compared with those for the coaxial line shown in Fig. 1.11(b). The TEM mode may be assumed for wave propagation along the stripline.

Practical stripline is constructed from two substrates as shown in Fig. 5.2. Once the desired circuit is etched on one of the copper planes, the two boards are bolted together to form the completed stripline circuit as shown in Fig. 5.2(b). It is important that the two sections be clamped

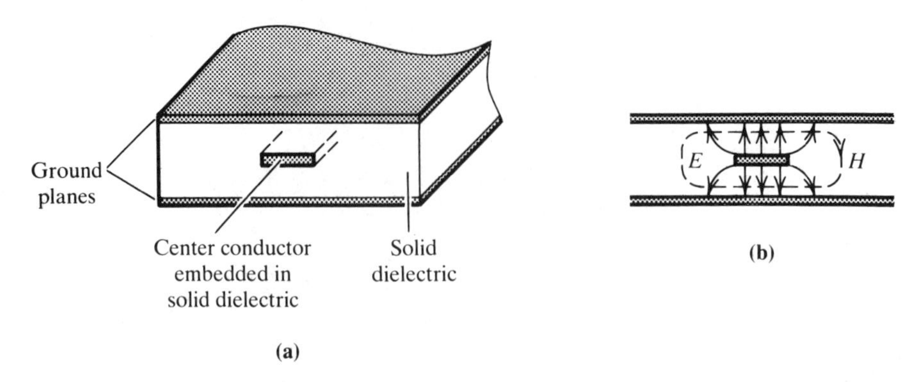

FIGURE 5.1. (a) Basic stripline; (b) electric (*E*) and magnetic (*H*) field distributions.

tightly together so that the dielectric material closes in around the etched circuit lines. A residual air gap will exist at the edges of the etched conductor pattern. This air gap may be left "as is" or it may be filled in with silicone grease, or better, by means of a polyethylene laminate. Connections to the stripline are preferably made by means of edge connectors as shown in Fig. 5.2(c), although broadwall connectors are also available. Connectors are usually of the 3-mm subminiature type.

Thermal properties and mechanical stability are major factors that must be taken into account, in addition to the electrical properties. The most widely used materials as listed by Howe (1974) are woven Teflon fiberglass, microfiber Teflon glass, polyolefin, polyphenelene oxide, cross-linked polystyrene, and glass-filled polystyrere. Polyguide (see *Electronized Chemicals Co. Polyguide Handbook*), a commercially available material, is an irradiated polyolefin. Exposure to radiation results in cross-linking of molecules in the polyolefin, which gives the material high mechanical stability and uniform electrical properties. Polyguide is available with copper cladding on one or both sides, or unclad; it also comes with copper on one side and aluminum on the ground-plane side, which has certain advantages in circuit design and fabrication. The copper cladding is specified in ounces, this referring to the ounces of copper used per square yard of surface. Common weights are 1 oz and 2 oz, and the corresponding copper thicknesses as specified for polyguide being 0.0014 in. $\pm 10\%$ and 0.0028 in. $\pm 10\%$. Boards come in standard thicknesses, these being designated by the nominal thicknesses, which typically are $\frac{1}{32}$, $\frac{1}{16}$, and $\frac{1}{8}$ in. The dielectric constant (relative permittivity) for polyolefin is 2.32, and this may be assumed independent of frequency at least up to X-band frequencies (10 GHz). The material has been used successfully over the frequency range 100 MHz to 30 GHz.

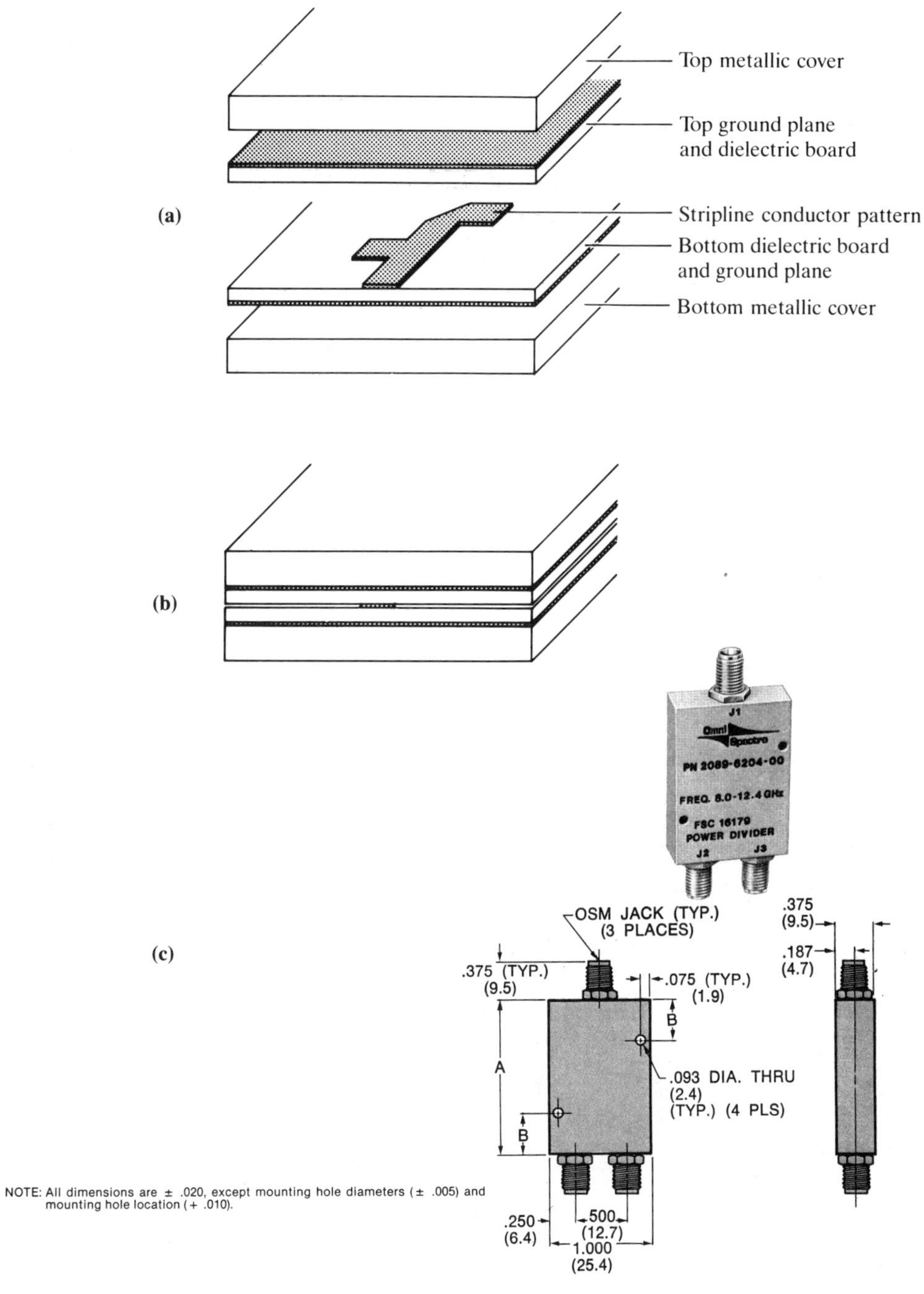

FIGURE 5.2. (a) Exploded view of stripline assembly; (b) stripline assembled; (c) four-way divider made in stripline by Americon, Inc. [*Part (c) courtesy of Stripline Components Catalog, Americon, Inc., Waltham, Mass.*]

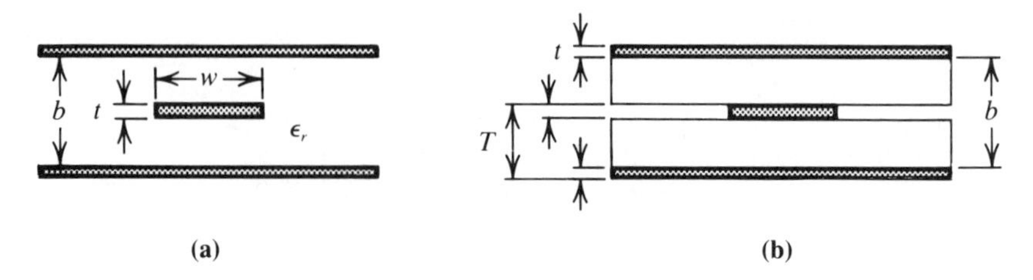

(a) (b)

FIGURE 5.3. (a) Dimensions used in determining characteristic impedance; (b) dimensions of a practical line (see Example 5.1).

5.2.1 Characteristic Impedance of Stripline

The characteristic impedance of any line is a function of the geometry of the line and the dielectric constant, for example as shown by Eqs. (1.34) and (1.35) for the two-wire line and the coaxial cable. With stripline the impedance as a function of line geometry is much more complex than the lines discussed in Chapter 1. Many of the design formulas are given in Howe (1974), and the equations for characteristic impedance are summarized here. Dimensions used are shown in Fig. 5.3(a). It is necessary to consider the characteristic impedance in two separate regions defined as follows:

High-impedance region:

$$\frac{w}{b-t} \leqslant 0.35 \qquad \text{and} \qquad \frac{t}{b} \leqslant 0.25 \tag{5.1}$$

$$Z_0 = \frac{60}{\sqrt{\varepsilon_r}} \ln \frac{4b}{\pi d_0} \quad \Omega \tag{5.2}$$

This should be compared with Eq. (1.35) for the coaxial line. In Eq. (5.2), d_0 represents an equivalent center conductor diameter given by

$$d_0 = \frac{w}{2}\left[1 + \frac{t}{\pi w}\left(1 + \ln \frac{4\pi w}{t} + 0.51\pi \left(\frac{t}{w}\right)^2\right)\right] \tag{5.3}$$

Low-impedance region:

$$\frac{w}{b-t} > 0.35 \tag{5.4}$$

$$Z_0 = \frac{94.15}{\sqrt{\varepsilon_r}\,[(w/ba) + C]} \tag{5.5}$$

where

$$a = \left(1 - \frac{t}{b}\right) \tag{5.6}$$

and

$$C = \frac{1}{\pi}\left[\frac{2}{a}\ln\left(\frac{1}{a} + 1\right) - \left(\frac{1}{a} - 1\right)\ln\left(\frac{1}{a^2} - 1\right)\right] \tag{5.7}$$

The use of these equations are illustrated in the following examples.

EXAMPLE 5.1

Determine the characteristic impedance of a polyguide stripline of width 0.014 in., made on 1-oz $\frac{1}{16}$-in. board. Assume that $\varepsilon_r = 2.32$.

SOLUTION

For 1-oz board t is specified as 0.0014 in. The nominal board thickness T includes dielectric and both ground planes and therefore the separation between ground planes is, from Fig. 5.3(b),

$$b = 2T - 3t$$

$$= 2 \times \frac{1}{16} - 3 \times 0.0014 = 0.1208 \text{ in.}$$

Hence

$$\frac{w}{b - t} = \frac{0.014}{0.1208 - 0.0014} = 0.1173$$

and

$$\frac{t}{b} = \frac{0.0014}{0.1208} = 0.0116$$

These values show that the high-impedance approximation applies. From Eq. (5.3),

$$d_0 = \frac{0.014}{2}\left[1 + \frac{0.0014}{\pi \times 0.014}\left(1 + \ln\frac{4\pi \times 0.014}{0.0014} + 0.51\pi\left(\frac{0.0014}{0.014}\right)^2\right)\right]$$

$$= 0.00830$$

From Eq. (5.2),

$$Z_0 = \frac{60}{\sqrt{2.32}}\ln\frac{4 \times 0.1208}{\pi \times 0.00830}$$

$$= \mathbf{115\ \Omega}$$

EXAMPLE 5.2

Determine the characteristic impedance of a polyguide stripline of width 0.056 in., made on 1-oz, $\frac{1}{16}$-in. board.

SOLUTION

From Example 5.1, $\varepsilon_r = 2.32$, $b = 0.1208$ in., and $t = 0.0014$ in. A calculation will show that $w/(b - t) > 0.35$ and therefore the low-impedance approximation applies.

$$a = 1 - \frac{0.0014}{0.1208} = 0.9884$$

From Eq. (5.7),

$$C = \frac{1}{\pi}\left[\frac{2}{0.9884}\ln\left(\frac{1}{0.9884} + 1\right) - \left(\frac{1}{0.9884} - 1\right)\ln\left(\frac{1}{0.9884^2} - 1\right)\right]$$

$$= 0.4642$$

From Eq. (5.4),

$$Z_0 = \frac{94.15}{\sqrt{2.32}\left(\dfrac{0.056}{0.1208 \times 0.9884} + 0.4642\right)}$$

$$= \mathbf{66.24\ \Omega}$$

The accuracy of Eqs. (5.2) and (5.5) is quoted by Howe (1974) as being better than 1.3%, the worst value occurring at $w/(b - t) = 0.35$.

The problem of finding line dimensions for a given Z_0 is more difficult. As a guide to which approximation to use, the product $\sqrt{\varepsilon_r}Z_0$ is equal to 119 Ω for a line of zero thickness ($t = 0$) and $w/b = 0.35$. For example, to make a 50-Ω line in polyguide, the product $\sqrt{2.32} \times 50 = 76.16$ Ω. Since this is less than 119 Ω, the low-impedance approximation must be used. Now, knowing $\sqrt{\varepsilon_r}Z_0$, a board size must be chosen that will fix the values of b and t, and working back from Eqs. (5.5) and (5.7), the line width can be determined.

EXAMPLE 5.3

Determine the line width for a 50-Ω stripline on $\frac{1}{32}$-in. 1-oz polyguide.

SOLUTION

Since $\sqrt{\varepsilon_r}Z_0 = \sqrt{2.32} \times 50 = 76.16$ Ω, the low-impedance approximation should be used.

$$b = 2 \times \frac{1}{32} - 3 \times 0.0014 = 0.583 \text{ in.}$$

$$a = 1 - \frac{0.0014}{0.0583} = 0.9760$$

Substituting this value of a in Eq. (5.7) gives for C,

$$C = 0.4836$$

From Eq. (5.5),

$$w = ba\left(\frac{\sqrt{\varepsilon_r} Z_0}{94.15} - C\right)$$

$$= 0.0583 \times 0.9760 \times \left(\frac{76.16}{94.15} - 0.4836\right)$$

$$= \mathbf{0.0185\ in.}$$

Howe (1974) quotes the lower practical limit to line width as about 0.012 in. The manufacturer's design guide for polyguide shows Z_0 values ranging from about 5 to 110 Ω.

5.3 MICROSTRIP

Microstrip line employs a single ground plane, the conductor pattern on the top surface being open as shown in Fig. 5.4(a). Although microstrip and stripline cover a similar range of applications, in practice each one is best suited to certain areas. An obvious advantage of microstrip is that discrete components are easily connected into circuit. The open structure means that the electromagnetic field is not confined to the solid dielectric, but is partly in the air space as shown in Fig. 5.4(b). This makes shielding of the circuit more difficult, and also gives rise to more complicated modes of propagation.

A range of substrate materials is available, but for production runs alumina is the most widely used, for frequencies up to about 20 GHz. At higher frequencies, quartz, and sometimes sapphire, is used. Circuit patterns and ground planes are usually deposited on the substrates using thin-film or thick-film techniques (Edwards 1981, 1982). As with stripline, connections to external coaxial lines are usually made through 3-mm subminiature connectors.

The relative permittivity of alumina depends on its purity and is specified (Edwards 1982) as $8.2 \leqslant \varepsilon_r \leqslant 10.2$. An important parameter in microstrip design is the *effective relative permittivity* ε_{eff}, which takes into

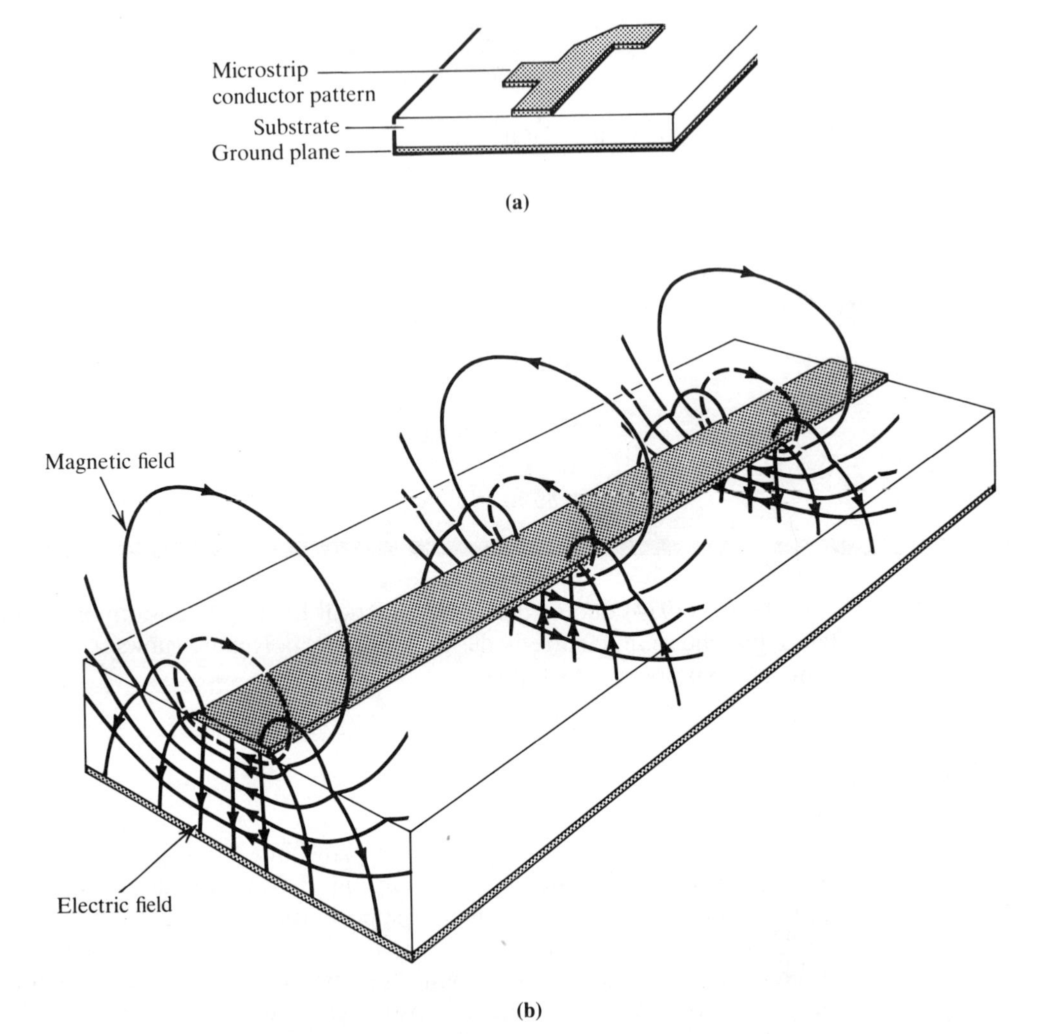

FIGURE 5.4. (a) Microstrip line; (b) pattern of the electromagnetic field of a microstrip line. In this figure both the electric field and the magnetic field are traverse: this is an approximation which is valid only at the lower microwave frequencies. The lines of force go partly through air and partly through dielectric. The dashed line shows the magnetic field close to the strip. [*Part (b) courtesy of Philips Tech. Rev., Vol. 32, 1971.*]

account the air–solid dielectric combination. From a knowledge of ε_{eff}, the phase velocity can be determined using Eq. (1.42) modified to

$$v_p = \frac{c}{\sqrt{\varepsilon_{eff}}} \tag{5.8}$$

Line wavelength is then determined for a given frequency using the relationship $\lambda = v_p/f$.

5.3.1 Characteristic Impedance and ε_{eff} for Microstrip

As with stripline, the impedance must be considered in two separate areas. Design details are given in Edwards (1981), and only the main results can be summarized here. In the following, w is the line width, and h the substrate thickness. For convenience in applying the design formulas, the following parameters are defined:

$$A = \frac{119.9}{\sqrt{2(\varepsilon_r + 1)}} \tag{5.9}$$

$$B = \frac{1}{2}\left(\frac{\varepsilon_r - 1}{\varepsilon_r + 1}\right)\left(\ln\frac{\pi}{2} + \frac{\ln 4/\pi}{\varepsilon_r}\right) \tag{5.10}$$

$$C = \ln\left(\frac{4h}{w} + \sqrt{\left(\frac{4h}{w}\right)^2 + 2}\right) \tag{5.11}$$

$$D = \frac{59.95\pi}{\sqrt{\varepsilon_r}} \tag{5.12}$$

It will be seen that A, B, and D, depend on ε_r, while C depends on w/h. Narrow lines, for which $w/h < 3.3$, result in high characteristic impedance values given by

$$Z_0 = A(C - B) \tag{5.13}$$

The effective relative permittivity for narrow lines requires a different crossover value for w/h this being $w/h < 1.3$

$$\varepsilon_{eff} = \frac{\varepsilon_r + 1}{2\left(1 - \frac{B}{C}\right)^2} \tag{5.14}$$

These equations enable the characteristic impedance and effective relative permittivity to be determined from a knowledge of the w/h ratio and dielectric relative permittivity. The inverse problem, that of determining a w/h ratio for a known line impedance can be solved for narrow lines, for which $Z_0 > (44 - 2\varepsilon_r)$, by use of the equation

$$\frac{h}{w} = \frac{e^C}{8} - \frac{e^{-C}}{4} \tag{5.15}$$

C in this case is determined from Eq. (5.13) as

$$C = \frac{Z_0}{A} + B \tag{5.16}$$

The changeover value for effective relative permittivity is $Z_0 > (63 - 2\varepsilon_r)$, (i.e., $w/h < 1.3$).

Wide lines have low characteristic impedance values. For $w/h > 3.3$

$$Z_0 = \frac{D}{\left[\dfrac{w}{2h} + 0.44127 + \dfrac{0.08226(\varepsilon_r - 1)}{\varepsilon_r^2} + \dfrac{(\varepsilon_r + 1)}{2\pi\varepsilon_r}\left(1.4516 + \ln\left(\dfrac{w}{2h} + 0.94\right)\right)\right]} \tag{5.17}$$

Also, for $w/h > 1.3$, the effective relative permittivity is given by

$$\varepsilon_{\text{eff}} = \frac{\varepsilon_r + 1}{2} + \frac{\varepsilon_r - 1}{2\left(1 + \dfrac{10h}{w}\right)^{0.555}} \tag{5.18}$$

Again, if the characteristic impedance and relative permittivity are known, the line dimensions can be found for $Z_0 < (44 - 2\varepsilon_r)$, from

$$\frac{w}{h} = \frac{2}{\pi}\left[\frac{\pi D}{Z_0} - 1 - \ln\left(\frac{2\pi D}{Z_0} - 1\right)\right] + \frac{\varepsilon_r - 1}{\pi\varepsilon_r}\left[\ln\left(\frac{\pi D}{Z_0} - 1\right) + 0.293 - \frac{0.517}{\varepsilon_r}\right] \tag{5.19}$$

The changeover value for relative permittivity is $Z_0 < (63 - 2\varepsilon_r)$ (i.e., $w/h > 1.3$) as before.

EXAMPLE 5.4

Determine the characteristic impedance and the effective relative permittivity for a microstrip line for which $w/h = 4$ and $\varepsilon_r = 9$.

SOLUTION Since $w/h > 3.3$ the low impedance approximation, Eq. (5.17) applies

$$Z_0 = \frac{62.779}{(2 + 0.4412) + \dfrac{0.08226 \times 8}{81} + \dfrac{10}{18\pi}(1.4516 + \ln 2.94)}$$

$$= \mathbf{21.67\ \Omega}$$

Application of Eq. (5.18) gives

$$\varepsilon_{\text{eff}} = \frac{10}{2} + \frac{8}{2(1 + 2.5)^{0.555}}$$

$$= \mathbf{6.7}$$

5.4 ATTENUATION AND POWER LOSSES

For both microstrip, and stripline, losses occur in the dielectric and in the conductors. In addition, radiation losses occur with microstrip, and because of its mixed dielectric, a surface propagation mode can exist along with the main TEM mode, which results in additional losses.

For both types of line the conductor and dielectric losses are complicated functions of line geometry, conductivity, dielectric constant, and frequency. Figure 5.5 shows the variation of conductor losses and dielectric losses for a 50-Ω microstrip line on alumina substrate. It will be seen that in this case the dielectric loss is negligible compared to conductor loss. However, this is not true in general. The losses are comparable where plastic substrates are used, and with semiconductor substrates the dielectric loss can exceed the conductor loss. For example, for a 50-Ω line on a 0.254-mm substrate, silicon exhibits 0.4 dB/cm dielectric loss and about 0.13 dB/cm conductor loss at 10 GHz. Gallium arsenide exhibits a similar conductor loss and a dielectric loss of about 0.044 dB/cm. For comparison, at 10 GHz, the values for alumina from Fig. 5.5 are 0.044 dB/cm conductor loss and 0.005 dB/cm dielectric loss. All data are obtained from Fig. 5.5.

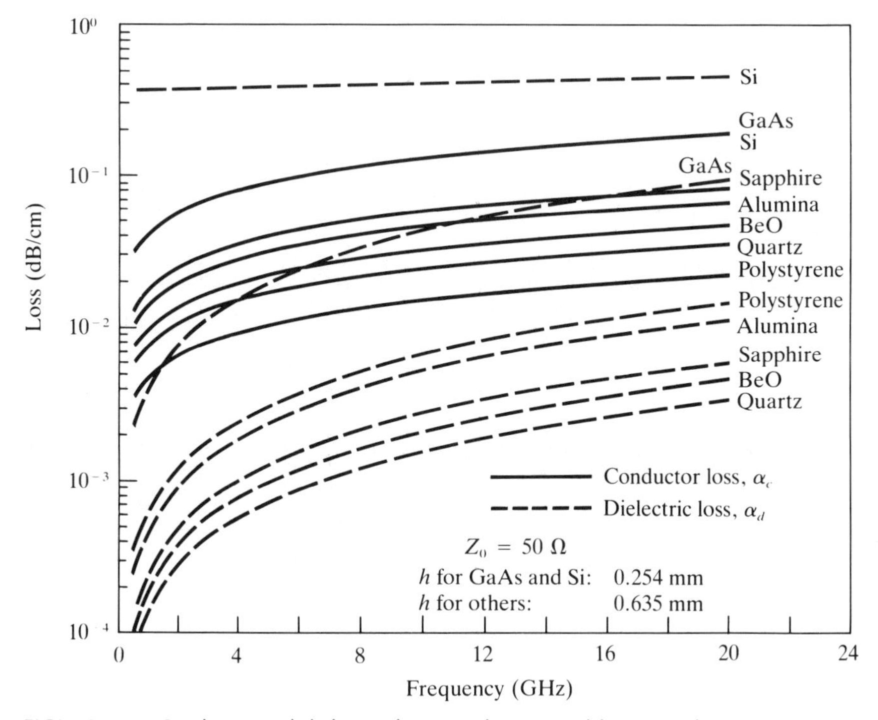

FIGURE 5.5. Conductor and dielectric losses as functions of frequency for microstrip lines on various substrates. [*From K. C. Gupta, Ramash Garg, and I. J. Bahl, Microstrip Lines and Slotlines (Dedham, Mass.: Artech House, 1979).*]

Although stripline and microstrip find widest use in small-signal applications, they are capable of handling powers up to a few kilowatts in some circumstances. Continuous-wave (CW) rating and peak power rating must be considered separately.

The CW power limit is set mainly by the thermal conductivity of the substrate, as this determines how fast the heat can be removed. Figure 5.6 shows the average power/frequency curve as specified for polyguide. It will be seen that at 1 GHz the average power is limited to about 420 W, and it drops to about 125 W at 10 GHz. By contrast, alumina and the semiconductor substrates silicon and gallium arsenide have much better thermal conductivities and the average power-handling capability is of the order of a few kilowatts, relatively independent of frequency.

Peak power is limited by breakdown in the line assembly. The dielectric strength of the dielectric is one of the important limiting factors, as is the air breakdown along the edges of the etched line conductors. Sharp edges and corners cause a concentration of electric field which can precipitate breakdown. Laminations are sometimes used in stripline to fill in the air gap at the conductor edges, and use of dielectric paint for the same purpose has been suggested (Howe 1974). Internal mismatches that result in high electric fields can also precipitate breakdown, and in many instances the limiting factor is the breakdown in the external connector, which is usually of the 3-mm subminiature variety.

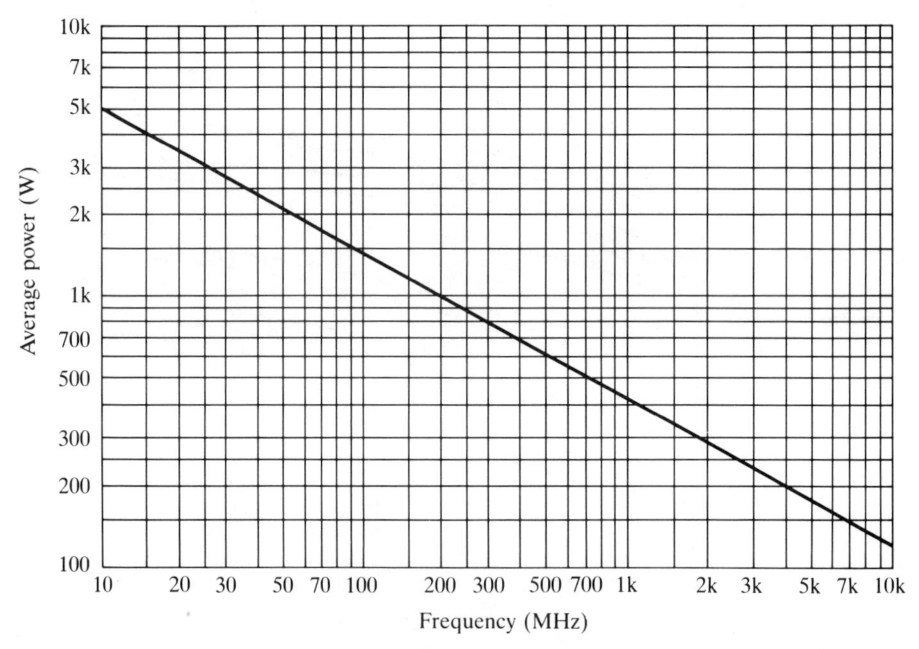

FIGURE 5.6. Power-handling capability of polyguide. (*Courtesy of Electronized Chemicals Corp., Burlington, Mass.*)

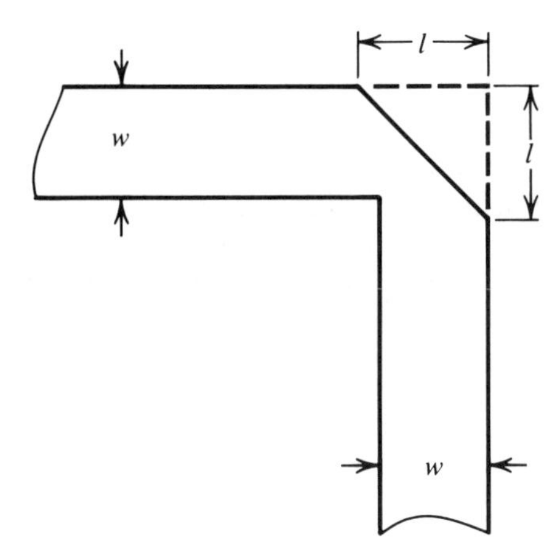

FIGURE 5.7. Mitered bend.

For stripline, peak powers up to 25 kW have been reported, but Howe (1974) suggests a safe limit of 5 kW. Edwards (1981) has calculated a peak power limit of 200 kW for the transition between microstrip and connector, and suggests a safe limit of 100 kW. These figures are for the connection alone, and do not take into account possible internal breakdown. When derating for this is taken into account, it would seem that a safe limit for both types of line is around 5 kW.

5.5 BENDS

As mentioned in the preceding section, sharp corners must be avoided, as they result in impedance mismatch and in charge concentration, both of which can precipitate breakdown. Where bends are required, the corners are mitered as shown in Fig. 5.7. The design of such bends is discussed in Howe (1974) and Edwards (1981). Typically, the mitered length is about equal to the line width w. Specific values obtained from the references quoted are: for 50-Ω stripline, $l = 1.131w$, and for 50-Ω microstrip, $l = 1.194w$.

5.6 MICROSTRIP AND STRIPLINE COUPLERS

5.6.1 Electromagnetic Field Couplers

The basic principles of coupled transmission lines are outlined in Section 1.16, and although illustrated with reference to a coaxial system, the principles are applicable generally to any system of coupled lines.

The electric fields for the even- and odd-symmetry modes on coupled microstrip lines are shown in Fig. 5.8(a) and (b). With stripline, the conductors are embedded in uniform dielectric, whereas with microstrip the dielectric is discontinuous at the air–substrate interface. This creates a difference in relative permittivity for even and odd modes which becomes significant at the higher microwave frequencies, resulting in different wavelengths for the modes. For the present analysis, however, the mode wavelengths will be assumed equal. As mentioned in Section 1.16, each mode has its own characteristic impedance and these impedance values are basic design parameters. In a design situation, the coupling factor C, and the system characteristic impedance Z_0, are the known parameters, and from Eqs. (1.115) and (1.118), it may be shown that (see Problem 11)

$$Z_{0e} = Z_0 \sqrt{\frac{1 + C}{1 - C}} \tag{5.20}$$

$$Z_{0o} = Z_0 \sqrt{\frac{1 - C}{1 + C}} \tag{5.21}$$

Special design procedures are then required to find line dimensions which simultaneously fit the even- and odd-mode impedance values. The detailed design for a coupler is given in Edwards (1981), the values for which are: coupling, 10 dB; single microstrip feed line impedance, 50 Ω; center fre-

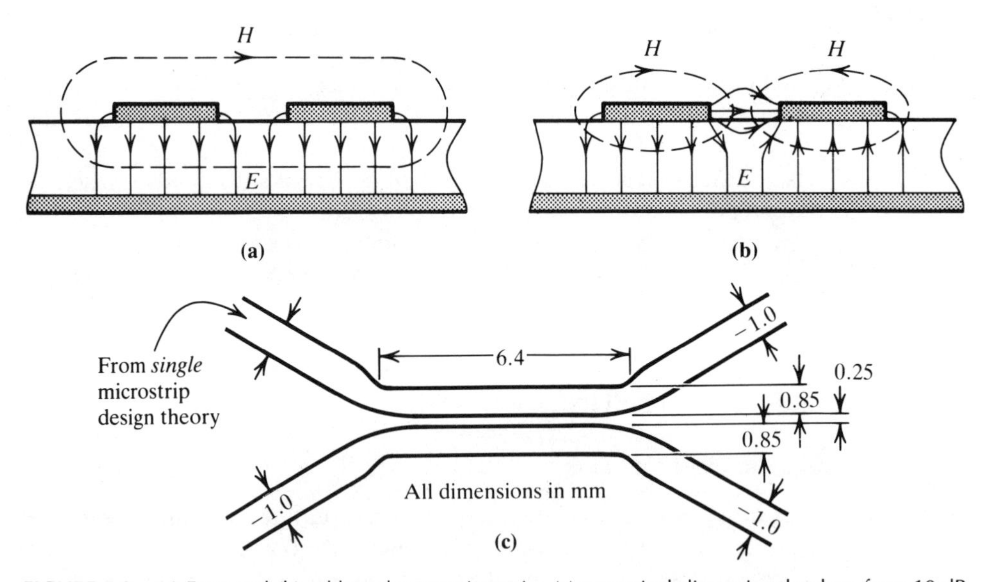

FIGURE 5.8. (a) Even and (b) odd modes on microstrip; (c) numerical dimensional values for a 10-dB coupler design. [*Part (c) from T. C. Edwards, Foundations for Microstrip Design; reprinted by permission of John Wiley & Sons Ltd., © 1981.*]

quency, 5 GHz; Z_{0e}, 69.5 Ω; Z_{0o} 36 Ω. The dimensions in millimeters for the coupler are reproduced in Fig. 5.8(c).

5.6.2 Branch Couplers

Transmission lines may also be coupled through connecting branches, such couplers being known as branch-line couplers. Compared to the field coupler, the branch-line coupler can provide tighter coupling and can handle higher powers. The branches may consist of chokes or filters which provide added flexibility in design.

A basic single-section branch-line coupler is illustrated in Fig. 5.9(a). Here it is more convenient to work in terms of characteristic admittances. The system characteristic admittance is Y_0 ($= 1/Z_0$). The direct lines, one from port 1 to port 2, and one from port 3 to port 4, each have a characteristic admittance of $\sqrt{2}\ Y_0$, and the branch lines coupling these two direct lines each have characteristic admittance Y_0. The even-mode symmetry circuits are shown in Fig. 5.9(b), and the odd-mode symmetry circuits in Fig. 5.9(c). For the even-symmetry mode, the λ/8 stubs are open circuited, and therefore from Eq. (1.108) these can be represented by a $+jY_0$ susceptance. For the odd-mode symmetry, the λ/8 stubs are short circuited, and therefore from Eq. (1.102) these can be represented by a $-jY_0$ susceptance. As with Fig. 1.23, the circuits shown in Fig. 5.9(b) and (c) can be represented by a single circuit, shown in Fig. 5.10. In this analysis, where double signing occurs, e.g. $\pm$ or $\mp$, the upper sign is used for even-mode analysis, and the lower sign for odd-mode analysis. The EMF source E_S represents the $-\frac{1}{2}E$ source in the top circuit in Fig. 5.9(c), and otherwise it represents the $+\frac{1}{2}E$ sources for the other half-circuits.

For Fig. 5.10, the load admittance is

$$Y_B = Y_0(1 \pm j) \tag{5.22}$$

When this is transferred to the input through the λ/4 transformer, it becomes

$$\begin{aligned} Y_A &= \frac{(\sqrt{2}\ Y_0)^2}{Y_B} \\ &= Y_0(1 \mp j) \end{aligned} \tag{5.23}$$

The input admittance is therefore

$$\begin{aligned} Y_{\text{in}} &= \pm jY_0 + Y_A \\ &= \pm jY_0 + Y_0(1 \mp j) \\ &= Y_0 \end{aligned} \tag{5.24}$$

Thus the input voltage V_A is

$$V_A = \tfrac{1}{2} E_S \tag{5.25}$$

Using Eq. (1.90) with $\theta = 90°$, I_A replacing I_S, V_B replacing V_L,

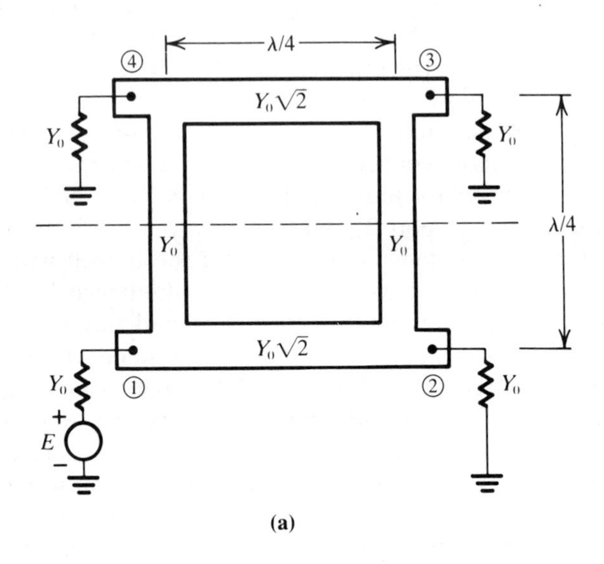

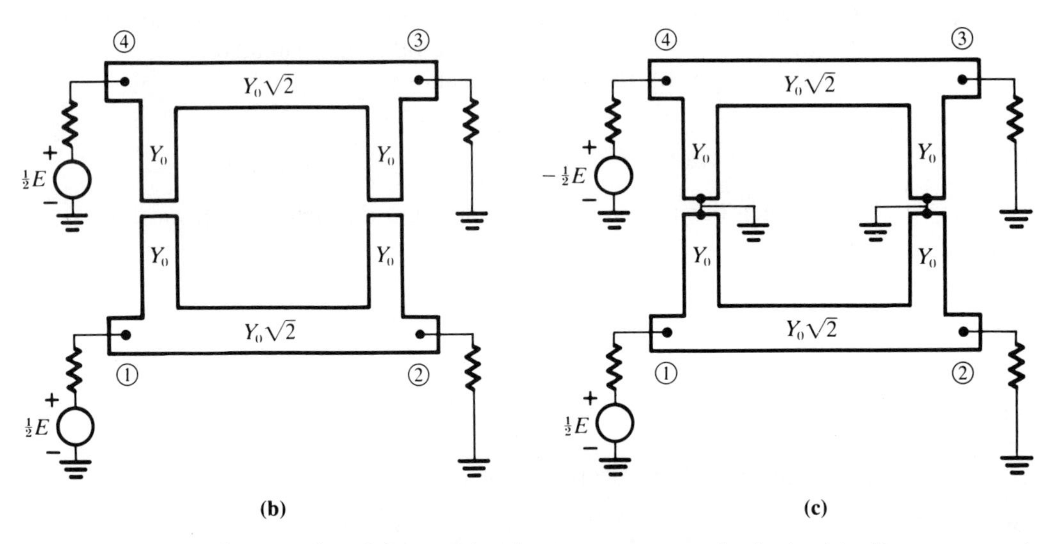

FIGURE 5.9. (a) Single-section branch-line coupler; (b) even-symmetry-mode circuits; (c) odd-symmetry-mode circuits.

and $\sqrt{2}\ Y_0$ replacing $1/Z_0$ gives V_B as

$$V_B = -j\frac{I_A}{\sqrt{2}\ Y_0}$$

$$= -j\frac{V_A Y_A}{\sqrt{2}\ Y_0}$$

$$= \frac{-jE_S(1 \mp j)}{2\sqrt{2}} \tag{5.26}$$

Following the procedure described in Section 1.16, the port voltages can now be found as follows:

V_1 is obtained from V_A with $E/2$ replacing E_s for both even and odd mode circuits

$$V_1 = V_{Ae} + V_{Ao}$$

$$= \tfrac{1}{2}(\tfrac{1}{2}E) + \tfrac{1}{2}(\tfrac{1}{2}E) \tag{5.27}$$

$$= \frac{E}{2}$$

V_4 is also obtained from V_A but in this case E_s is replaced by $E/2$ for the even mode, and by $-E/2$ for the odd mode, circuit

$$V_4 = V_{Ae} + V_{Ao}$$

$$= \tfrac{1}{2}(\tfrac{1}{2}E) + \tfrac{1}{2}(-\tfrac{1}{2}E) \tag{5.28}$$

$$= 0$$

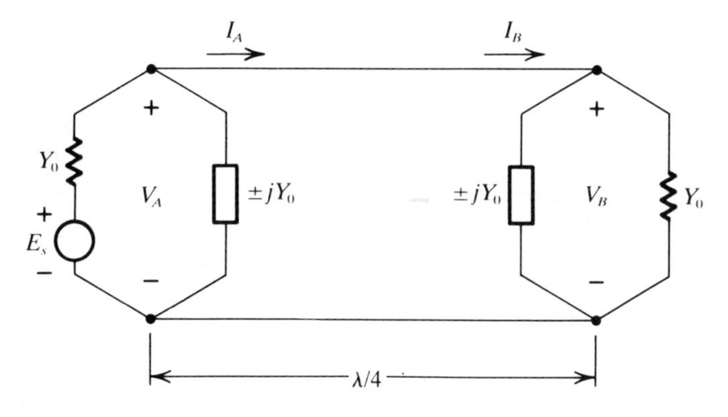

FIGURE 5.10. Single equivalent circuit for Fig. 5.9(b) and (c).

V_2 is found from V_B with $E/2$ replacing E_s for both even and odd mode circuits

$$V_2 = V_{Be} + V_{Bo}$$
$$= \frac{-j(\frac{1}{2}E)(1 - j)}{2\sqrt{2}} + \frac{-j(\frac{1}{2}E)(1 + j)}{2\sqrt{2}} \tag{5.29}$$
$$= \frac{-jE}{2\sqrt{2}}$$

V_3 is also obtained from V_B but in this case E_s is replaced by $E/2$ for the even mode and $-E/2$ for the odd mode, circuit

$$V_3 = V_{Be} + V_{Bo}$$
$$= \frac{-j(\frac{1}{2}E)(1 - j)}{2\sqrt{2}} + \frac{-j(-\frac{1}{2}E)(1 + j)}{2\sqrt{2}} \tag{5.30}$$
$$= \frac{-E}{2\sqrt{2}}$$

The coupling factor C, the transmission factor T, and the directivity factor D, as defined in Section 1.16, are, for the single-section branch coupler,

$$C = \frac{V_3}{V_1} = -\frac{1}{\sqrt{2}} \tag{5.31}$$

$$T = \frac{V_2}{V_1} = \frac{-j}{\sqrt{2}} \tag{5.32}$$

$$D = \frac{V_4}{V_1} = 0 \tag{5.33}$$

In practice, the directivity will have some small but finite value.

In decibels the coupling is seen to be 3 dB, and the transmission is also 3 dB. The coupler therefore divides the input power to port 1 evenly between ports 2 and 3, with none going to port 4. A reflected wave at port 2 would divide evenly between ports 1 and 4, with none of the power going to port 3. The device is termed a 3-dB directional coupler.

5.6.3 The Hybrid Ring

The hybrid ring (sometimes referred to as the rat-race ring) is shown in Fig. 5.11(a). The characteristic admittance of the system is Y_0, and the characteristic admittance of the connecting ring is $Y_0/\sqrt{2}$. The ring circum-

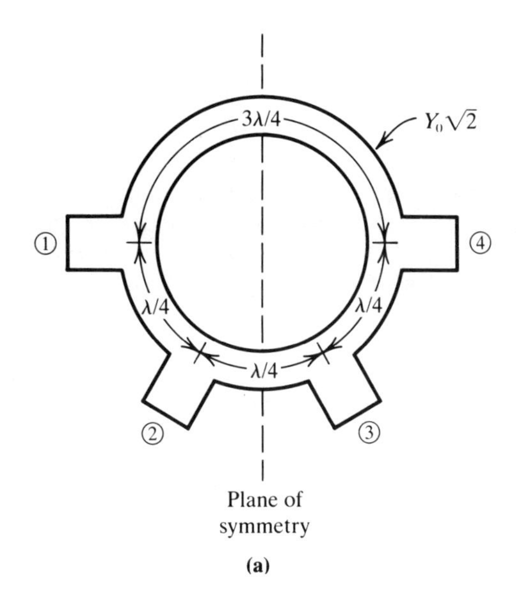

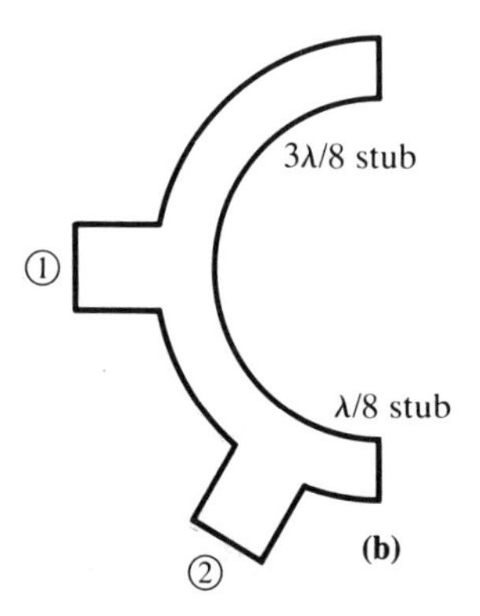

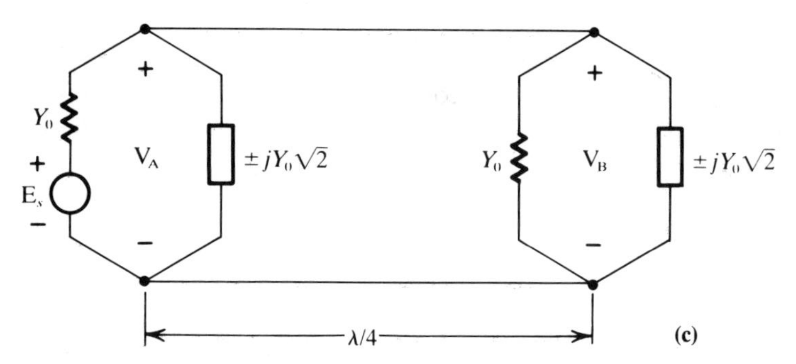

FIGURE 5.11. (a) Ring coupler; (b) half-circuit about the plane of symmetry; (c) equivalent circuit for part (b).

ference is divided by the ports into the length shown in Fig. 5.11(a). The symmetrical half-circuit is shown in Fig. 5.11(b), and the general equivalent circuit in Fig. 5.11(c). With the input applied to port 1, the half-circuit shows a $3\lambda/8$ stub directly across port 1 (and also one across port 4), a $\lambda/4$ transformer between ports 1 and 2 (also one between ports 4 and 3), and a $\lambda/8$ stub across port 2 (and one across port 3). The susceptance for the $\lambda/8$ stub is $+jY_0/\sqrt{2}$ even, $-jY_0/\sqrt{2}$ odd, and for the $3\lambda/8$ stub, $-jY_0/\sqrt{2}$ even, $+jY_0/\sqrt{2}$ odd. As in the previous analysis, the upper sign, whether $\pm$ or $\mp$, is associated with the even mode circuits and the lower sign with the odd mode circuits.

Referring to Fig. 5.11(c), the equivalent load admittance is

$$Y_B = Y_0\left(1 \pm \frac{j}{\sqrt{2}}\right) \tag{5.34}$$

This referred to the input is

$$\begin{aligned} Y_A &= \frac{(Y_0/\sqrt{2})^2}{Y_B} \\ &= \frac{Y_0}{2(1 \pm j/\sqrt{2})} \\ &= \frac{Y_0}{3}\left(1 \mp \frac{j}{\sqrt{2}}\right) \end{aligned} \tag{5.35}$$

The input admittance is

$$\begin{aligned} Y_{\text{in}} &= \mp j\frac{Y_0}{\sqrt{2}} + Y_A \\ &= \mp j\frac{Y_0}{\sqrt{2}} + \frac{Y_0}{3}\left(1 \mp \frac{j}{\sqrt{2}}\right) \\ &= \frac{Y_0}{3}(1 \mp j2\sqrt{2}) \end{aligned} \tag{5.36}$$

The voltage V_A is therefore

$$\begin{aligned} V_A &= \frac{E_S Y_0}{Y_0 + Y_{\text{in}}} \\ &= \frac{E_S Y_0}{Y_0 + (Y_0/3)(1 \mp j2\sqrt{2})} \\ &= \frac{E_S}{2}\left(1 \pm \frac{j}{\sqrt{2}}\right) \end{aligned} \tag{5.37}$$

(The algebraic manipulation is left as an exercise for the student; see Problem 13.) As in the previous analyses, the voltage V_B is found from Eq. (1.90) on replacing I_S with I_A, V_L with V_B, and $1/Z_0$ with $Y_0/\sqrt{2}$:

$$\begin{aligned} V_B &= -jI_A \frac{\sqrt{2}}{Y_0} \\ &= -jV_A Y_A \frac{\sqrt{2}}{Y_0} \\ &= -j\frac{E_S}{2}\left(1 \pm \frac{j}{\sqrt{2}}\right)\frac{Y_0}{2(1 \pm j/\sqrt{2})}\frac{\sqrt{2}}{Y_0} \\ &= -j\frac{E_S}{2\sqrt{2}} \end{aligned} \tag{5.38}$$

The port voltages can now be found as follows:

$V_1 = V_{Ae} + V_{Ao}$ with $E_s = \frac{1}{2}E$ for both even and odd mode circuits.

$$\begin{aligned} &= \frac{E}{4}\left(1 + \frac{j}{\sqrt{2}}\right) - \frac{E}{4}\left(1 - \frac{j}{\sqrt{2}}\right) \\ &= \frac{E}{2} \end{aligned} \tag{5.39}$$

$V_4 = V_{Ae} + V_{Ao}$ with $E_s = \frac{1}{2}E$ for the even, and $-\frac{1}{2}E$

for the odd mode circuit.

$$\begin{aligned} &= \frac{E}{4}\left(1 + \frac{j}{\sqrt{2}}\right) - \frac{E}{4}\left(1 - \frac{j}{\sqrt{2}}\right) \\ &= j\frac{E}{2\sqrt{2}} \end{aligned} \tag{5.40}$$

$V_2 = V_{Be} + V_{Bo}$ with $E_s = \frac{1}{2}E$

for both even and odd mode circuits.

$$\begin{aligned} &= -j\frac{\frac{1}{2}E}{2\sqrt{2}} - j\frac{\frac{1}{2}E}{2\sqrt{2}} \\ &= -j\frac{E}{2\sqrt{2}} \end{aligned} \tag{5.41}$$

$V_3 = V_{Be} + V_{Be}$ with $E_s = \frac{1}{2}E$

for the even, and $-\frac{1}{2}E$ for the odd mode circuit.

$$= 0 \tag{5.42}$$

Thus the power input to port 1 divides evenly between ports 2 and 4, with no output at port 3. V_2 lags V_1 by 90°, and V_4 leads V_1 by 90°. It will also be seen that since $V_1 = E_S = \frac{1}{2}E$, the input admittance of port 1 must equal the source admittance, or the source is matched.

A similar analysis with the input applied to port 2 shows that the power divides evenly between ports 1 and 3, and V_1 is in phase with V_3. Some applications of the hybrid ring are described in Chapter 6.

5.7 POWER JUNCTIONS

The layout for a simple power junction is shown in Fig. 5.12(a). Inputs applied at ports 2 and 3 appear as a combined output at port 1, ports 2 and 3 being isolated from each other. In this mode, the device is termed a *power combiner*. The device can also be operated in the reverse direction, power applied at port 1 dividing equally between ports 2 and 3.

The equivalent circuit for the power combiner is shown in Fig. 5.12(b). To illustrate the transmission properties, input is applied to port 3 only, the input to port 2 being set equal to zero. This equivalent circuit may be split into even- and odd-symmetry circuits as was done in the preceding section for directional couplers, and these half-circuits are shown in Fig. 5.13(a) and (b).

For the even-mode circuits, the $\lambda/4$ transformer transforms the $2Z_0$ load to an input impedance of $(\sqrt{2}Z_0)^2/2Z_0$ or Z_0. For the odd-mode circuits, the short circuit at the output is transformed to an open circuit at the input, and therefore the input impedance is simply the Z_0 impedance connected at the input. Thus the input voltage V_A for both even and odd circuits is

$$V_A = \tfrac{1}{2}E_S \tag{5.43}$$

where E_S is the generalized EMF source [Fig. 5.13(c)]. E_S represents $-\frac{1}{2}E_3$ for the top half-circuit of Fig. 5.13(b), and $+\frac{1}{2}E_3$ for the other half-circuits. The load impedance Z_B represents $2Z_0$ for the even-mode circuits and a short-circuit for the odd-mode circuits. The voltage V_B for the odd-mode circuits is therefore zero. For the even-mode circuits V_B may be found using Eq. (1.90), on substituting I_A for I_S, V_B for V_L, and $\sqrt{2}Z_0$ for Z_0:

$$\begin{aligned} V_{Be} &= -jI_A(\sqrt{2}Z_0) \\ &= -j\frac{E_S}{2Z_0}(\sqrt{2}Z_0) \\ &= -j\frac{E_S}{\sqrt{2}} \end{aligned} \tag{5.44}$$

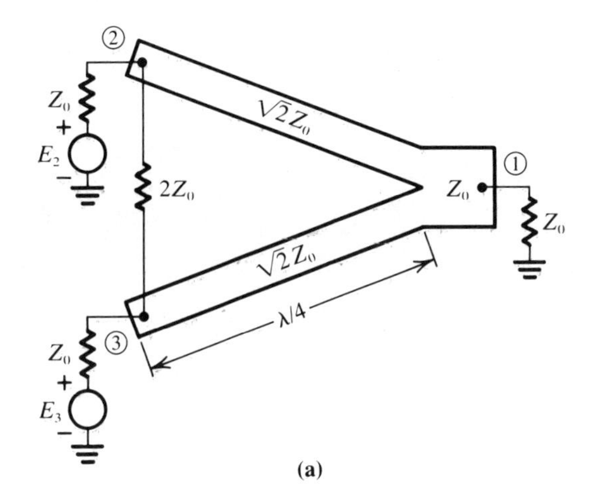

(a)

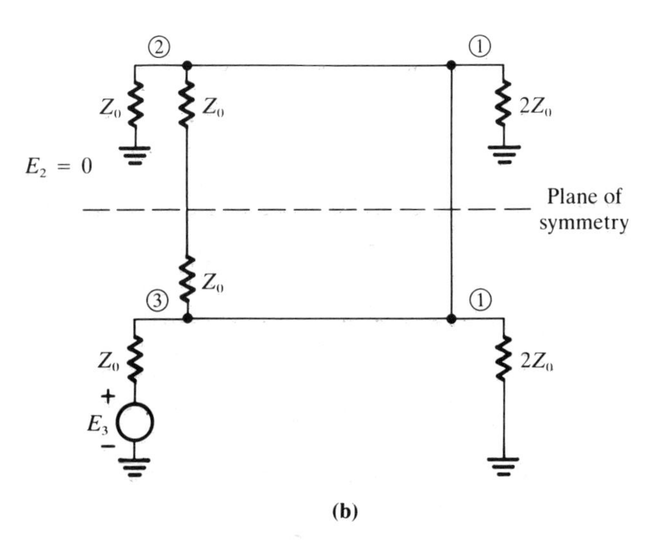

(b)

FIGURE 5.12. Power combiner: (a) microstrip layout; (b) equivalent circuit.

Following the procedure used in Section 5.6, the port voltages are found as

$V_3 = V_{Ae} + V_{Ao}$ with $E_s = \frac{1}{2}E_3$ for both even and odd mode circuits.

$$= \tfrac{1}{2}(\tfrac{1}{2}E_3) + \tfrac{1}{2}(\tfrac{1}{2}E_3)$$

$$= \tfrac{1}{2}E_3 \tag{5.45}$$

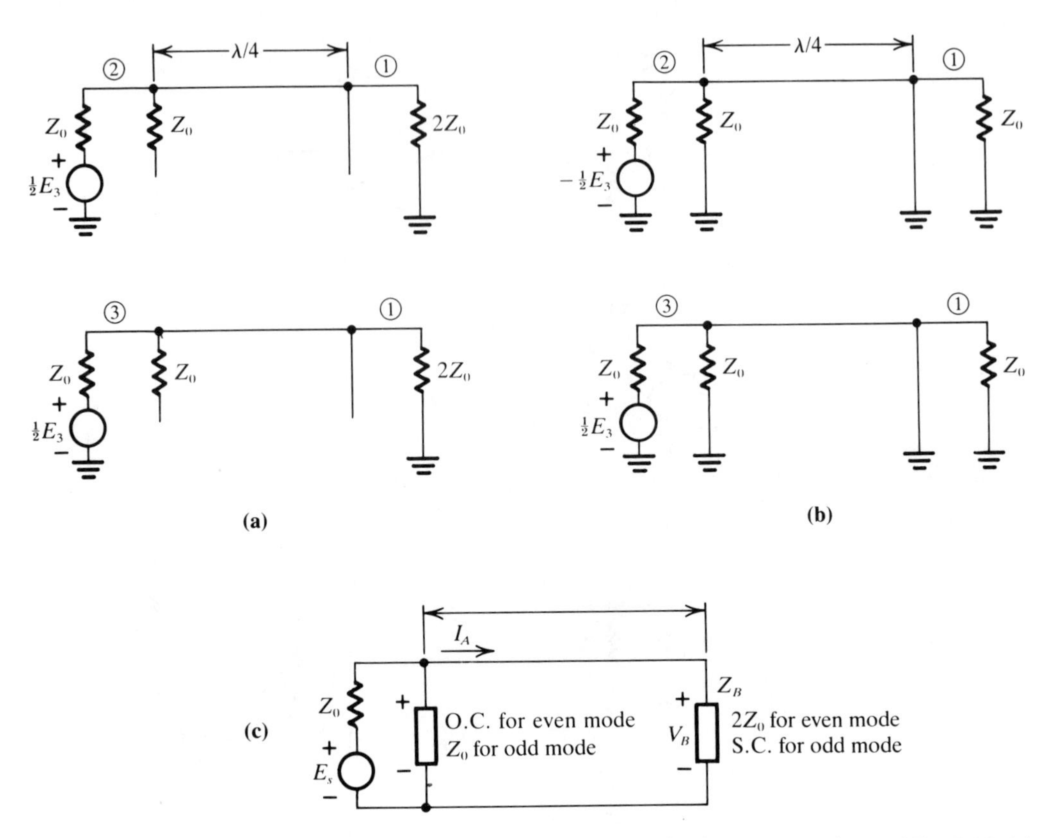

FIGURE 5.13. (a) Even-mode half-circuits; (b) odd-mode half-circuits for the power combiner of Fig. 5.12; (c) generalized equivalent half-circuit.

$$V_2 = V_{Ae} + V_{Ao} \text{ with } E_s = \tfrac{1}{2}E_3 \text{ for the even, and } -\tfrac{1}{2}E_3 \text{ for the odd mode circuit.} \tag{5.46}$$

$$= 0$$

$$V_1 = V_{Be}$$

$$= -j\frac{E_S}{\sqrt{2}}$$

$$= -j\frac{E_3}{2\sqrt{2}} \tag{5.47}$$

Since $V_3 = \frac{1}{2}E_3$, the E_3 source is matched. The output voltage V_1

lags V_3 by 90° and shows a 3-dB loss. The voltage V_2 is zero, which means that no power is transferred from port 3 to port 2.

The 3-dB power loss occurs in the balancing resistor connected between ports 2 and 3. Since V_2 is zero, the voltage across the $2Z_0$ balancing resistor is V_3. It is left as an exercise (Problem 16) for the student to show that the power dissipated in this balancing resistor is equal to the power delivered to the load at port 1.

The device can be used as a power divider. Power applied to port 1 divides equally between ports 2 and 3 (with zero loss in the balancing resistor), and the voltage at either output port lags that at the input port by 90° (see Problem 18).

The two-way power junction is a particular case of the N-way hybrid power divider described by Wilkinson (1960).

5.8 MICROSTRIP AND STRIPLINE FILTERS

As shown in Section 1.15.3, transmission-line sections can be used as resonators. With microstrip and stripline, the open-circuit line is preferred at frequencies above about 3 GHz, because of the difficulty in achieving a good short circuit at the higher frequencies. Figure 5.14(a) shows two simple $\lambda/4$ filter configurations. From the results of Section 1.15.3, the open circuit $\lambda/4$ is known to present a low impedance at the resonant frequency, and therefore it acts as a *band-stop filter*, for frequencies around the resonant value. The attenuation, in decibels, for the simple $\lambda/4$ filters is shown in Fig. 5.14(b). Here the attenuation refers to the number of decibels that the output signal is below the input signal. As shown, the attenuation is about 40 dB at the resonant frequency of 10 GHz.

Although the basic layouts for microstrip and stripline are similar, the detailed dimensions will differ because of the different effective dielectric permittivities in each case. Detailed design information will be found in Edwards (1981) and Howe (1974).

Bandpass filter characteristics can be achieved by using an open-circuit $\lambda/2$ line section. The passband peak around resonance may be broadened while maintaining relatively sharp sides to the response curve, by coupling a number of $\lambda/2$ sections together. A common method of coupling is to use the capacitive coupling between parallel sections as illustrated in Fig. 5.15(a). Here three open-circuit $\lambda/2$ sections can be seen capacitively coupled, and the two outer sections are also capacitively coupled to the input and output. The closer spacing seen at the input and output indicates that this coupling is tighter than the intersection coupling. The bandpass filter characteristic is shown in Fig. 5.15(b).

To reduce the overall length of filters employing $\lambda/2$ sections, the line sections may be folded or arranged in more compact geometries as

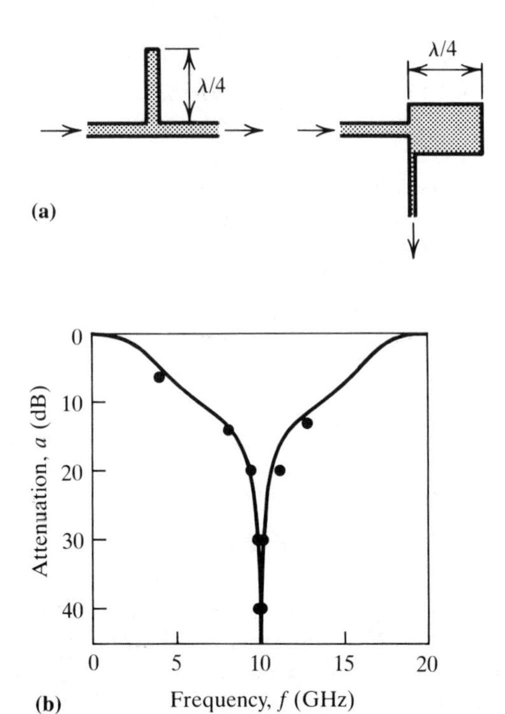

FIGURE 5.14. (a) Band-stop filters, consisting of quarter-wave resonators which have zero impedance at the resonant frequency at the junction point; (b) some measurements of the attenuation a of a filter as illustrated (top right) for various frequencies. (*Courtesy of Philips Tech. Rev., Vol. 32, 1971.*)

shown in Fig. 5.16. Figure 5.16(a) shows the pyramid type of fold, Fig. 5.16(b) the hairpin type of fold, and Fig. 5.16(c), what is termed a pseudo-interdigital type of filter.

5.9 OTHER TYPES OF SUBSTRATE LINES

Although microstrip, and stripline are the most commonly used types of substrate lines at the lower microwave frequencies, the dimensional tolerances of these lines become more critical, and losses increase as frequency increases. For millimeter wave applications, other types of line, notably the fin line, are preferred. Figure 5.17 shows in cross section a selection of some of the types available, and Table 5.1 lists the characteristic impedance ranges and typical unloaded Q-factors for these.

Fin line offers a number of advantages for use in millimeter systems. The impedance range covers all normal requirements, the unloaded Q-factor is relatively high (and hence losses low). Extreme miniaturization

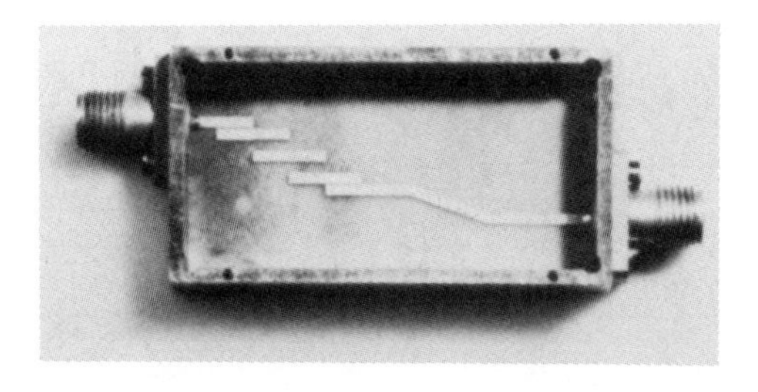

(a)

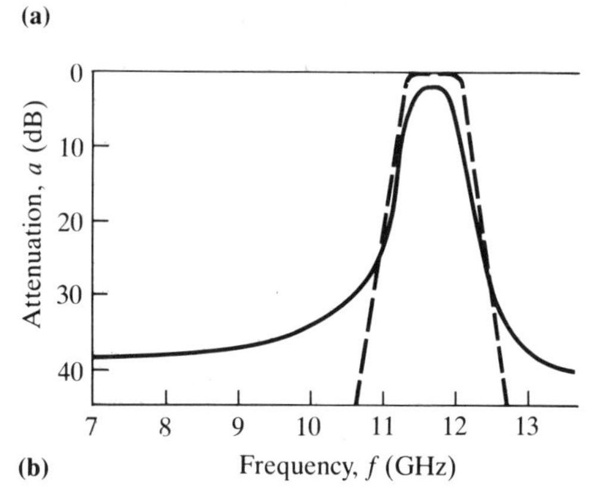

(b)

FIGURE 5.15. (a) Bandpass filter for 12 GHz, consisting of three coupled half-wave resonators; (b) bandpass response of this filter. Solid curve, measured response; dashed curve, calculated response, neglecting losses and strays. (*Courtesy of Philips Tech. Rev., Vol. 32, 1971.*)

can be avoided, and it is suitable for low-cost batch production of circuits (Meier 1974)

As the name suggests, and as illustrated in Fig. 5.17, the metallic lines forming the circuit are attached as fins to the broad walls of a waveguide, but the mode of propagation differs from the TE_{10} mode described

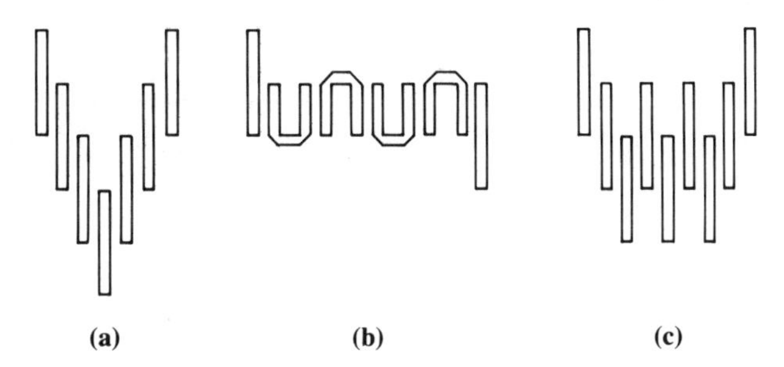

(a) (b) (c)

FIGURE 5.16. Common methods of folding side-coupled half-wave filters: (a) pyramid fold; (b) hairpin fold; (c) pseudo-interdigital. [*From H. Howe, Stripline Circuit Design (Dedham, Mass.:, Artech House, 1974); with permission.*]

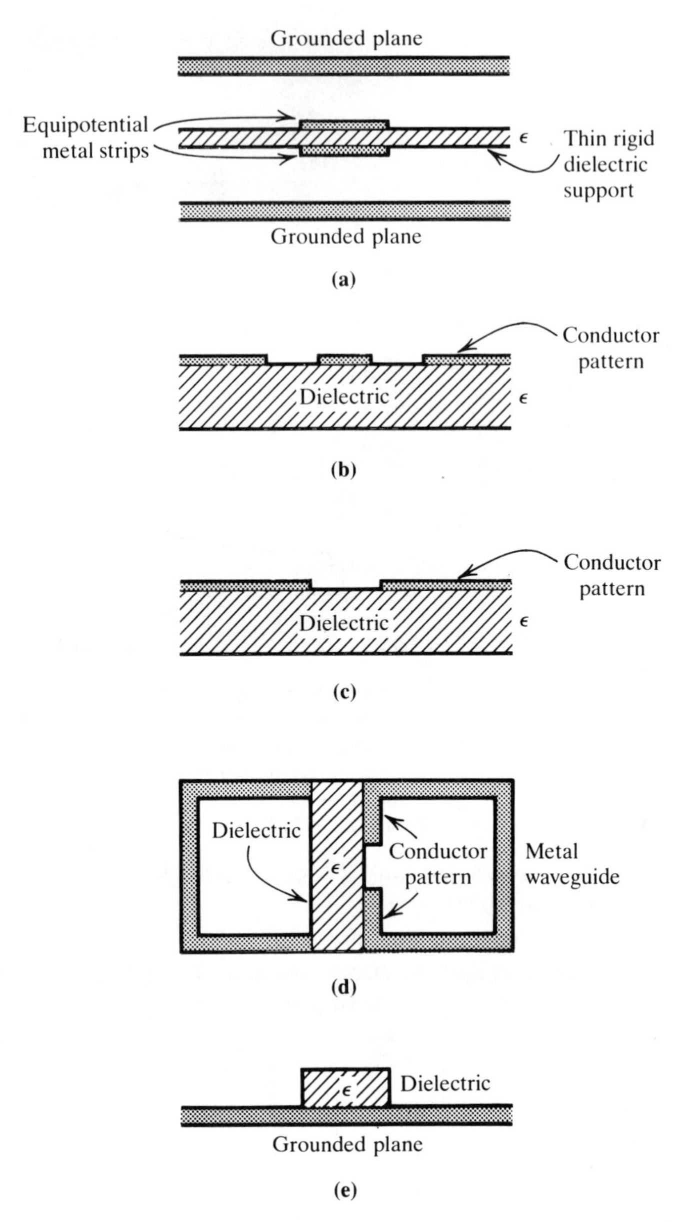

FIGURE 5.17. Cross sections of various substrate lines: (a) suspended substrate (high-Q) stripline; (b) coplanar waveguide; (c) slot line; (d) fin line; (e) image line.

TABLE 5.1
Characteristic Impedances and Typical Unloaded Q-Factors for Various Structures

Structure	Z (Ω)	*Q-factor*
Microstrip	20–125	150
Inverted microstrip	25–130	300
Trapped inverted microstrip	30–140	600
Suspended stripline	40–150	600
Coplanar waveguide	40–150	100
Slotline	60–200	120
Fin line	10–400	700
Image line	26	2500

Source: Edwards (1982).

in Section 4.2. The structure is shown in more detail in Fig. 5.18. The fins have to make good electrical contact with the metallic walls of the guide, and where passive components only are involved, the fins may be connected directly to the walls as shown. Where active devices are used, which require dc bias, the fins have to be dc isolated, which can be achieved through λ/4 choke coupling [see, e.g., Meier (1974)].

Figure 5.19(a) shows the fin-line layout for a magic-T (Gegemann 1978), which shows clearly the tapered transition used to couple from the waveguide to the fin line. Figure 5.19(b) shows the complete mixer circuit which utilizes the magic-T.

Because fin-line circuits lie in the plane of the electric field of the waveguide, they are also referred to as E-plane circuits.

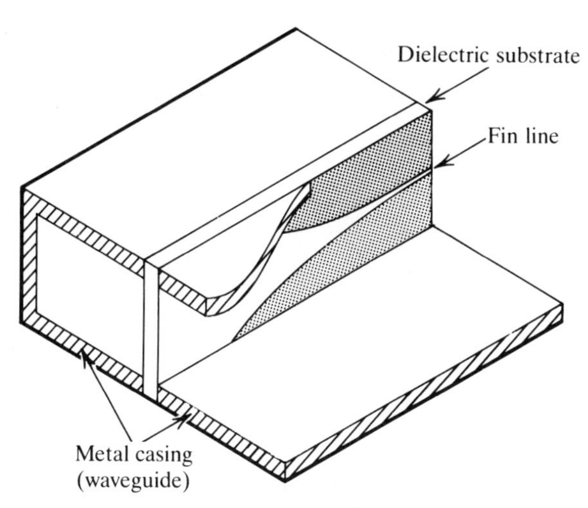

FIGURE 5.18. Cutaway view of fin line.

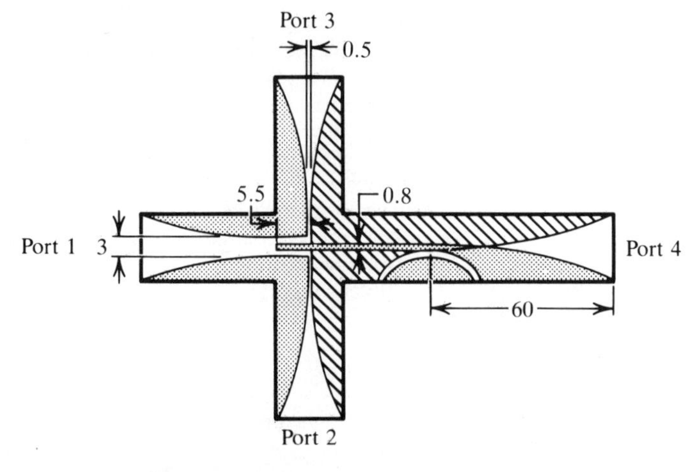

(a)

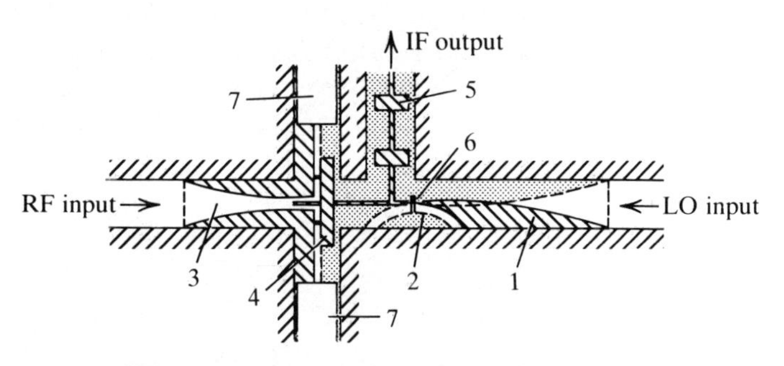

(b)

FIGURE 5.19. (a) Substrate of the complete fin-line magic-T with its waveguide input and output ports; typical dimensions are in millimeters; thickness of the substrate is 0.254 mm; permittivity is 2.22 (RT/Duroid 5880). (b) Complete mixer configuration: 1, waveguide to microstrip transition; 2, additional metalization; 3, fin-line taper; 4, microstrip stub; 5, low-pass filter; 6, block capacitor; 7, variable shorts. (*From Gunther Begemann, "An X-band balanced fin-line mixer," IEEE Trans. Microwave Theory Tech., © IEEE 1978; with permission.*)

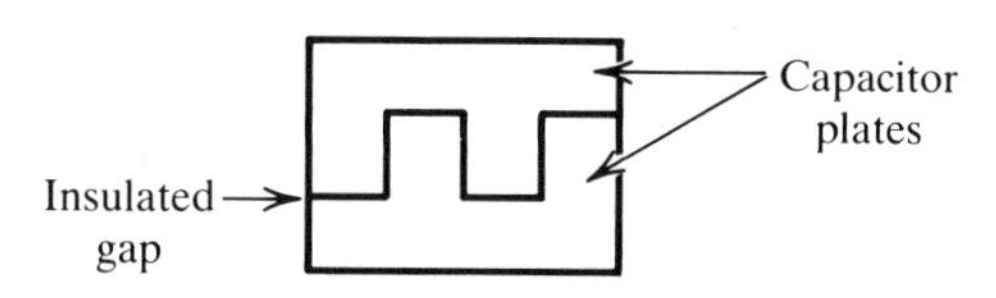

FIGURE 5.20. Coplanar capacitor.

5.10 LUMPED COMPONENTS

Resistors, inductors, and capacitors can all be made using either thin-film or thick-film techniques, although thin-film methods provide better control of circuit parameters. Resistors are made using standard thin-film techniques, and the range most commonly required is 20 to 200 Ω. These can be made using Nichrome, with simple rectangular geometries (i.e., non-meandered).

Capacitors, where comparatively high values are required (e.g., 10 pF), are made also as normal thin-film capacitors, with overlapping plates separated by the dielectric. Where smaller values of capacitance are required, an interdigitated-type construction is used (Fig. 5.20). Here both plates of the capacitor lie in the same plane, and capacitances in the range 0.1 to 0.4 pF can be obtained with very high Q-factors.

Inductors in the range 0.5 to 2.0 nH, and with Q-factors of the order of 50 can be achieved using the construction shown in Fig. 5.21(a). In constructing these, a 300-Å layer of Nichrome is first evaporated onto the substrate, followed by a 0.5-μm gold layer. This is plated up to a thickness in the range 3 to 5 μm, to improve the Q-factor. Figure 5.21(b) shows how inductance and Q-factor vary with width. Resonant circuits may be formed as shown in Fig. 5.22. Figure 5.22(a) shows a series resonant circuit, and Fig. 5.22(b) a parallel resonant circuit.

The advantages claimed for lumped component circuits are that:

1. They can be made small physically (in terms of wavelength).
2. They do not require high-dielectric-constant substrates because the circuits are deposited on one side of the substrate (compared with distributed circuits where a ground plane is used).
3. Resonant circuits have only one resonant frequency compared with multiple harmonic resonances associated with distributed circuits (e.g., transmission lines).

Regarding the last point, it should be realized, however, that at sufficiently high frequencies, lumped circuits will eventually appear as distributed circuits, and care would have to be taken to ensure that parasitic resonances occur outside the useful frequency range. Also, in regard to

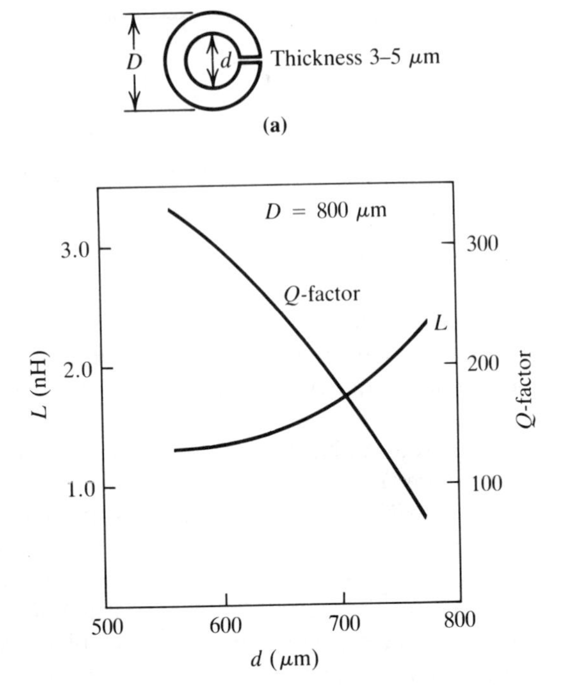

FIGURE 5.21. Film-type inductor: (a) physical configuration; (b) inductance and Q values versus width.

point 1, the limiting factor in physical size may well be proximity effects of boundaries (e.g., screening cans) rather than the physical size of the circuit. Also, a metallic enclosure can behave as a resonant cavity, which then requires the addition of absorbers to damp out unwanted resonances.

5.11 SAW DEVICES

In its basic form, the surface acoustic wave (SAW) device has two electrode structures, one for input and one for output, deposited on a piezoelectric substrate. The piezoelectric effect enables an electrical input signal to set up a surface acoustic wave on the substrate, which in turn couples with the output electrode structure to produce an output signal. In effect, the SAW device is a coupled circuit in which the coupling mechanism is the surface acoustic wave.

Two substrate materials are in widespread use, quartz and lithium niobate. The substrate has to be optically polished, and aluminum electrodes, of the order of 1000 A are deposited on this. Electrode dimensions

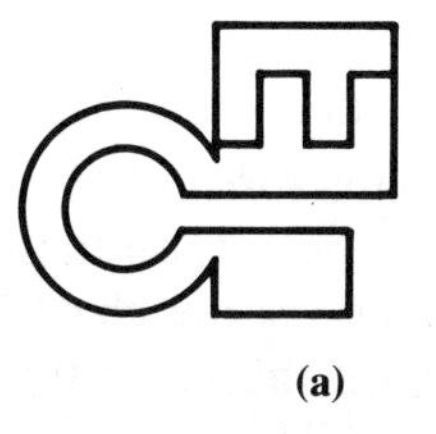

(a)

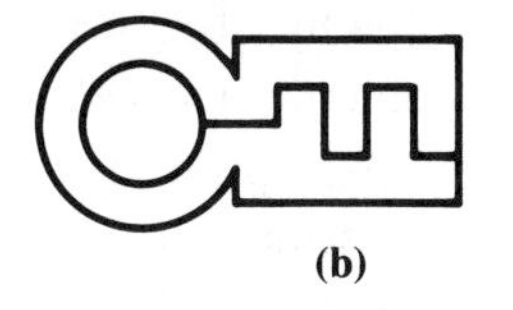

(b)

FIGURE 5.22. Film-type resonant circuits: (a) series resonant; (b) parallel resonant.

and spacing range from fractions of a wavelength to several wavelengths, but because of the low phase velocity of the surface wave these dimensions are small. The phase velocity of the surface acoustic wave is of the order of 3 km/s (compared to 3 Mm/s for an electromagnetic wave in free space), and for example, a signal frequency of 100 MHz would have a wavelength of (3 km/s)/(100MHz) = 30 um. Thus very compact devices are possible, which operate over a frequency range of about 10 MHz to 1.5 GHz. The electrodes are readily shaped to provide a wide range of useful transfer characteristics, and SAW devices are used as delay lines, bandpass, and bandstop filters. Programmable filters can also be constructed from SAW devices, and a powerful signal processing component known as a Fourier transformer is available in SAW device technology. The output of the Fourier transformer is a time analog of the spectrum of the input signal, and this allows the equivalent of frequency filtering to be carried out in the time domain.

5.12 PROBLEMS

1. A stripline is constructed in $\frac{1}{16}$-in. 2-oz Polyguide, the line width being 0.01 in. Determine the characteristic impedance.
2. A stripline is constructed in $\frac{1}{32}$-in. 2-oz Polyguide, the line width being 0.050 in. Determine the characteristic impedance.
3. It is required to construct a 40-Ω line on $\frac{1}{16}$-in. 2-oz Polyguide. Determine the line width needed.

4. It is required to construct a 100-Ω line on $\frac{1}{8}$-in. 2-oz Polyguide. Determine the line width needed.
5. For a given microstrip line, the ratio of line width to substrate thickness is $w/h = 3.5$, and the relative permittivity of the substrate material is 9. Determine (a) the characteristic impedance of the line, (b) the effective relative permeability, and (c) wavelength on the line for a frequency of 3 GHz.
6. Repeat Problem 5 for $w/h = 2$.
7. Repeat Problem 5 for $w/h = 1$.
8. A 10-Ω microstrip line is to be constructed on an alumina substrate, the relative permittivity of which is 9. Determine the required ratio of w/h and the effective relative permeability.
9. Repeat Problem 8 for a 75-Ω line.
10. Using Eqs. (1.115) and (1.118), derive Eqs. (5.20) and (5.21).
11. A directional coupler is designed for a coupling factor of 10 dB and a characteristic impedance of 50 Ω. Determine the even- and odd-mode characteristic impedances.
12. Referring to Fig. 5.9, describe the power flow through the coupler when the input is applied to port 3, the other ports being matched.
13. Carry out the algebraic simplification required in the final step to arrive at Eq. (5.37).
14. Referring to Fig. 5.11, carry out an analysis showing the signal flow when the input signal is connected to port 2.
15. Using the fact that with $E_2 = 0$ voltage V_3 appears across the $2Z_0$ resistor of Fig. 5.12(a), show that the power dissipated in this resistor is equal to the power delivered to the Z_0 load at port 1. Assume port 2 to be terminated in an impedance Z_0.
16. For the circuit of Fig. 5.12(a), develop an expression for the power delivered to the Z_0 resistor at port 1 in terms of the source voltages E_2 and E_3. Assume these have the same frequency but arbitrary phase difference.
17. A power divider is constructed according to the arrangement of Fig. 5.12(a), the input signal being applied to port 1, and output signals taken from ports 2 and 3. All ports may be assumed matched. Develop expressions for the output terminal voltages in terms of the source EMF, and hence show that the input power divides evenly betweeen the output ports.

REFERENCES

BEGEMANN, GUNTHER, 1978. "An X-band balanced fin-line mixer," *IEEE Trans. Microwave Theory Tech.*, Vol. MTT-26, No. 12, pp. 1007–1011.

EDWARDS, T. C. 1981. *Foundations for Microstrip Circuit Design*. New York: John Wiley & Sons, Inc.

EDWARDS, T. C., 1982. "Integrated wave-guiding media for microwaves and millimetre waves," *Electon. Power*, Vol. 28, No. 6, pp. 454–458.

GUPTA, K. C., and AMARJIT SINGH (eds.), 1974. *Microwave Integrated Circuits*. New York: John Wiley & Sons, Inc.

HOWE, JR., HARLAN, 1974. *Stripline Circuit Design*. Dedham, Mass.: Artech House, Inc.

MEIER, PAUL J., 1974. "Integrated fin-line millimeter components," *IEEE Trans. Microwave Theory Tech.,* Vol. MTT-22, No. 12, pp. 1209–1216.

WILKINSON, ERNEST J., 1960. "An *N*-way hybrid power divider," *IEEE Trans. Microwave Theory Tech.*, Vol. MTT-8, No. 1, pp. 116–118.

FURTHER READING

DE RONDE, F. C., 1970. "A new class of microstrip directional couplers," *IEEE G-MTT Int. Microwave Symp. Digest.*

FREY, JEFFREY (ed.), 1975. *Microwave Integrated Circuits*. Dedham, Mass.: Artech House, Inc.

JONES, E. M. T., and J. T. BOLLJAHN, 1956. "Coupled strip transmission line filters and directional couplers," *IEEE Trans. Microwave Theory Tech.*, Vol. MTT-4, No. 2.

SHELEG, BORIS, and BARRY E. SPIELMAN, 1974. "Broad band couplers using microstrip with dielectric overlays," *IEEE Trans. Microwave Theory Tech.*, Vol. MTT-22, No. 12.

SIMON, ERNST T., 1984. "Calculator program simplifies microstrip line computations," *Microwaves & RF*, July.

SURFACE ACOUSTIC WAVES. Proc IEEE May 1976, Vol. 64, No. 5.

VAN HEUVEN, J. H. C., and A. G. VAN NIE, 1971. "Microwave integrated circuits," *Philips Tech. Rev.*, Vol. 32, No. 9/10/11/12.

six
Microwave Ferrite Components

6.1 INTRODUCTION

Ferrites are *magnetic dielectrics*; that is, they exhibit both magnetic properties similar to ferromagnetic metals and high resistivity associated with dielectrics. The high resistivity prevents the flow of eddy currents in ferrites, and allows them to be used at microwave frequencies. By contrast, eddy currents in ferromagnetic metals increase directly with frequency, which restricts their use to low frequencies.

The magnetic properties of a ferrite can be controlled by means of an external magnetic field. This field can be supplied either by a permanent magnet, or where field variation is required, by an electromagnet. Control of the magnetic properties in this way gives rise to a number of widely used devices, including phase shifters, isolators, circulators, switches, tuned filters, and resonators. In these devices, the microwave signal level is kept low such that nonlinear effects do not occur in the ferrite; thus they are known as linear devices. At high-microwave levels, nonlinear effects occur which are made use of in ferrite power limiters, harmonic generators, and ferromagnetic amplifiers. Only linear devices are covered in this chapter.

6.2 FERRIMAGNETISM

Ferrites owe their magnetic properties to the magnetic dipole moments associated with electron spin. The quantum mechanical theory of matter shows that electrons have a property analogous to spin in the classical

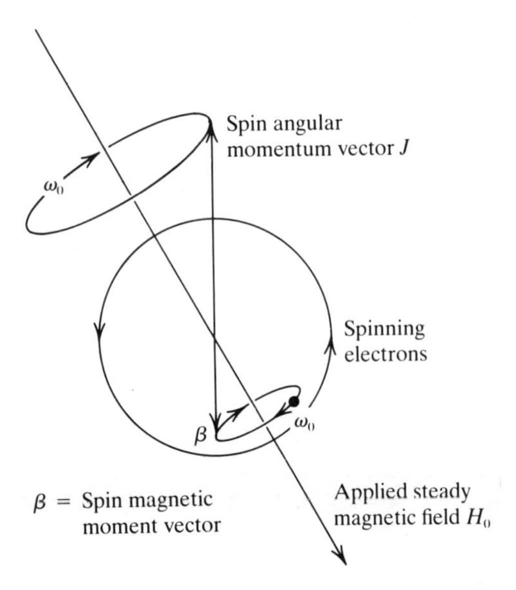

FIGURE 6.1. Electron spin angular momentum and magnetic moment vectors.

mechanical sense. An electron may be thought of as spinning about its own axis, and as a result, it possesses spin angular momentum. The spin angular momentum gives rise to a magnetic moment; that is, the spinning electron sets up its own magnetic field.

Spin angular momentum and spin magnetic moment are vector quantities that point in opposite directions, as shown in Fig. 6.1. The ratio of these vectors, known as the *gyromagnetic ratio*, is a constant given by $\mu_0 q_e/m_e$, where μ_0 is the permeability of free space, q_e is the charge on an electron, and m_e is the rest mass of an electron. The symbol for the gyromagnetic ratio is γ. In an atomic lattice, interaction of the spin momentum with orbital momentum can modify the value of γ to $\mu_0 g q_e/2m_e$, where g is known as the Landé splitting factor, or simply the g-factor. Thus

$$\begin{aligned}\gamma &= \frac{\mu_0 q_e g}{2m_e} \\ &= \frac{4\pi 10^{-7} \times (-1.61 \times 10^{-19})g}{2 \times 9.11 \times 10^{-31}} \\ &= -1.105 \times 10^5 g \, \frac{\text{rad/s}}{\text{A/m}} \qquad (6.1)\end{aligned}$$

The g-factor is approximately equal to 2, and for many ferrites it equals 2 exactly.

When a steady magnetic field is applied, shown as H_0 in Fig. 6.1, the magnetic moment interacts with this and the resultant torque causes the angular momentum vector to precess about the axis of the applied field. The rate of precession, or angular frequency ω rad/s, is given by

$$\omega_0 = |\gamma| H_0 \tag{6.2}$$

The frequency ω_0 is also known as the gyromagnetic resonant frequency. It is quite easy to adjust the magnetic field strength H_0 so that ω_0 falls within the microwave range, and as a result, microwave signals can interact strongly with a ferrite under certain circumstances, to be described later.

EXAMPLE 6.1

The magnetic flux density in the air gap of a permanent magnet is 0.1 tesla (T). Calculate the gyromagnetic resonant frequency of a ferrite placed in the air gap.

SOLUTION The magnetic field strength in the air gap is

$$H = \frac{B}{\mu_0}$$

$$= \frac{0.1}{4\pi \times 10^{-7}}$$

$$= 7.9 \times 10^4 \text{ A/m}$$

Hence using Eqs. (6.1) and (6.2) and assuming that $g = 2$ gives

$$\omega_0 = 2.21 \times 10^5 \times 7.96 \times 10^4 \text{ rad/s}$$

and

$$f = \frac{\omega_0}{2\pi} = \mathbf{2.8\ GHz}$$

In a ferrite, losses eventually reduce the precession to zero. The spin magnetic moment vectors then lie parallel to the applied field vector, some pointing in the same direction and some in opposition. The composition of ferrites is such that there is always a net magnetic moment in the direction of the applied field, and the ferrite is magnetized. This is termed *ferrimagnetism*.

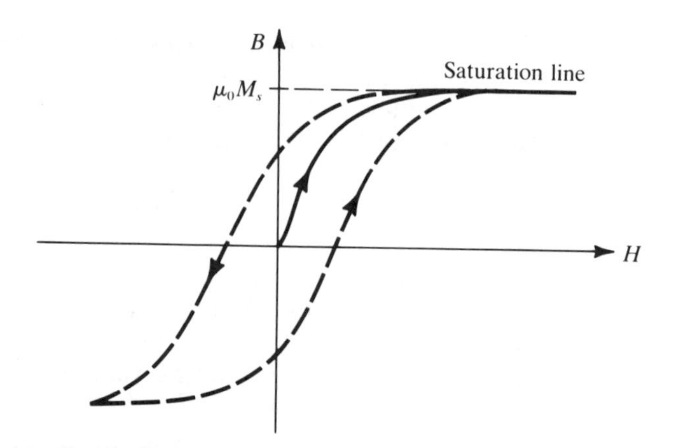

FIGURE 6.2. Magnetization curve, showing the saturation magnetization M_s.

In the bulk material, the spin magnetic moments are organized in domains, each domain being a small magnetized volume of material. These domains are randomly oriented in the unmagnetized material. In the presence of an external field they become aligned with the field, eventually producing magnetic saturation. In Fig. 6.2 is sketched the magnetization curve for a ferrite. This shows the relationship between magnetic flux density B (teslas) and magnetic field strength H (A/m), the relationship being expressed mathematically as

$$B = \mu_0(H + M) \tag{6.3}$$

M is the internal magnetization resulting from spin. As mentioned previously, increasing the external field H aligns more of the domains in the field direction, until eventually saturation sets in. The slope of the saturation line is μ_0. The intercept of this line on the B-axis gives the value of the saturation magnetization M_s. In SI units, M_s is measured in A/m, the same unit as magnetic field strength. For reference, the familiar B-H loop is shown dashed in Fig. 6.2.

Another frequency of importance, which will be used later in specifying ferrite properties, is given by

$$\omega_M = |\gamma| M_s \tag{6.4}$$

Typical values of M_s for some ferrites are given in Table 6.1.

It should be noted that in some texts [see, e.g., Roberts (1960) and Clarricoats (1961)], instead of Eq. (6.3), the relationship between B and H is given as $B = \mu_0 H + M$. This gives M the dimensions of flux density, and M/μ_0 appears in equations where here only M appears.

TABLE 6.1
Nominal Values for Some Ferrite Properties

Material	*g-Factor*	*Curie Temperature* T_c		*Magnetization*		*Line Width,* ΔH		*Anisotropy Const.* K_1	
		°C	*K*	$4\pi M_s(G)$	M_s *(A/m)*	*Oe*	*A/m*	*ergs/cm²*	*J/m³*
$MnFe_2O_4$	2 ± 0.01	300	573	5200	4.138×10^5	160	1.27×10^4	-2.2×10^4	-2.2×10^3
$NiFe_2O_4$	2.19	585	858	3400	2.71×10^5	75	5.97×10^3	-6.3×10^4	-6.3×10^3
$MgFe_2O_4$	2.01	440	713	1400	1.114×10^5	40	3.18×10^3	-5×10^3	-5×10^2
$5Fe_2O_3 \cdot 3Y_2O_3$ (yttrium iron garnet, YIG)	2.005	287	560	1725	1.373×10^5	30	2.39×10^3	-1.2×10^4	-1.2×10^3

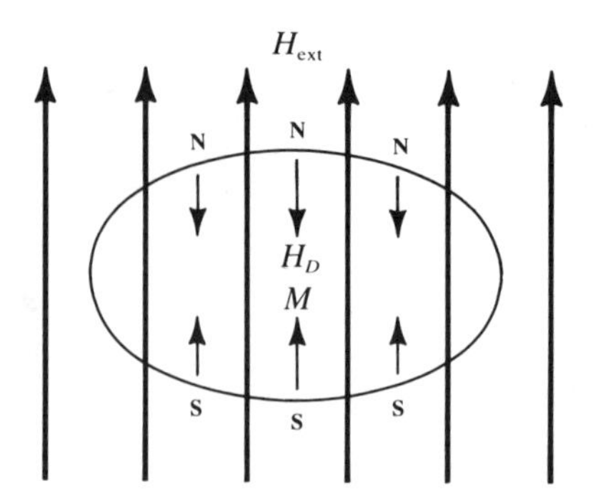

FIGURE 6.3. Ellipsoidal ferrite in an external magnetic field H_{sat}. H_D is the internal demagnetizing field, proportional to M.

6.3 THE BIASING FIELD

The magnetic field that is applied to a ferrite for control purposes sets up a biasing field within the ferrite. The biasing field, denoted by H_0, is related directly to the externally applied field, but is not in general equal to it. When the ferrite becomes magnetized, the magnetic poles on its surface set up an internal field which opposes the external field. This internal field is proportional to the magnetization M. Figure 6.3 shows the situation for an ellipsoidal ferrite (i.e., shaped like a symmetrically pointed egg). The external field is H_{ext}, the demagnetization field is H_D, and this is proportional to the magnetization M. The constant of proportionality is termed the demagnetization factor, N, and the effective internal field becomes

$$\begin{aligned} H_0 &= H_{ext} - H_D \\ &= H_{ext} - NM \end{aligned} \tag{6.5}$$

In general, separate demagnetization factors are required for the three Cartesian axes, and these are usually denoted as N_x, N_y, and N_z. These must satisfy the relationship

$$N_x + N_y + N_z = 1 \tag{6.6}$$

For the situation with the external biasing field applied along the z-axis sufficient to produce saturation, and with a microwave field having x-, y- and z-axis components of magnetic field as shown in Fig. 6.4, the

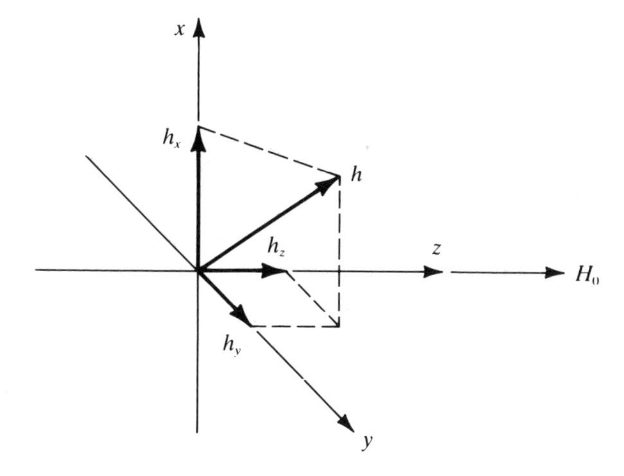

FIGURE 6.4. Magnetic field components. H_0 is the biasing field. Lowercase letters refer to the microwave field components.

equations are

$$H_z = H_0 = H_{ext} - N_z M_s \tag{6.7}$$

$$h_x = h_{x,ext} - N_x m_x \tag{6.8}$$

$$h_y = h_{y,ext} - N_y m_y \tag{6.9}$$

Lower case letters refer to the alternating (microwave) field components.

Because the ferrite is assumed to be saturated in the z-direction, the h_x component has negligible effect and can be ignored.

The equation for gyromagnetic resonant frequency under these conditions is given by

$$\omega_R = |\gamma| \sqrt{[H_{ext} + (N_x - N_z)M_s][H_{ext} + (N_y - N_z)M_s]} \tag{6.10}$$

Values for the demagnetizing factors have been worked out for a variety of shapes, some examples being:

Sphere:

$$N_x = N_y = N_z = 1/3$$

Under these conditions $H_{ext} = H_0$ and Eq. (6.10) yields

$$\omega_R = \omega_0 = |\gamma| H_0 \tag{6.11}$$

Thin disk with plane perpendicular to z-axis:

$$N_x = N_y = 0 \qquad N_z = 1$$

Substituting these values in Eq. (6.10) yields

$$\omega_R = |\gamma|(H_{\text{ext}} - M_s) \tag{6.12}$$

Thin circular rod, with axis of symmetry along the z-axis:

$$N_x = N_y = 0.5 \qquad N_z = 0$$

Substituting these values in Eq. (6.10) yields

$$\omega_R = |\gamma|\left(H_{\text{ext}} + \frac{M_s}{2}\right) \tag{6.13}$$

It will be seen that for the sphere, the effective internal biasing field is equal to the external field; for the thin disk, it is equal to $(H_{\text{ext}} - M_s)$; and for the thin rod, it is equal to $(H_{\text{ext}} + M_s/2)$. For simplicity in the diagrams that follow, the biasing field is shown as H_0, it being understood that this represents an effective internal biasing field of H_0.

6.4 MAGNETIC ANISOTROPY

Experimentally, it is known that single-crystal ferrites can be magnetized more easily along some axes than others. This arises because coupling exists between the electron spin momentum and electron orbital momentum, which is a function of direction in the crystal. For a given crystal type, certain axes can be identified as easy axes of magnetization, and others as hard axes of magnetization. The difference in magnetization energy required between the easy axis and the hard axis is known as the *anisotropy energy*.

The anisotropy energy is taken into account, to a first-order approximation, through an anisotropy constant, K_1, which has units of joules per cubic meter (J/m^3). K_1 is a measured parameter, and can be positive or negative depending on the material. Some nominal values are given in Table 6.1. An equivalent magnetic force can be evaluated for K_1, which depends on the direction of magnetization in the crystal. The cubic crystal is typical of ferrites, the main axes of which are shown in Fig. 6.5. With the biasing field along the [001] axis and the microwave magnetic field

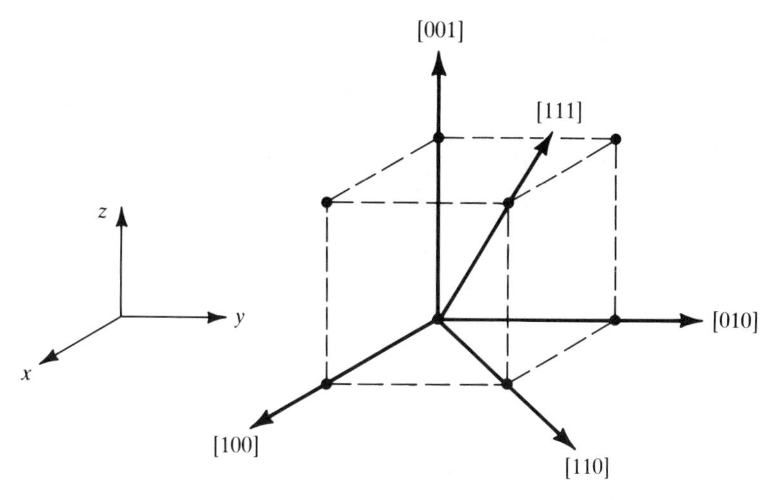

FIGURE 6.5. Axes of a cubic crystal. Miller indices are shown as $[m_1m_2m_3]$. The principal axes are [100], [110], and [111].

along the [100] direction, the effective anisotropy field strength is given by

$$H^a = \frac{2K_1}{\mu_0 M_s} \tag{6.14}$$

If K_1 is a negative quantity, H^a will oppose the demagnetizing field, H_D, in Fig. 6.3. In effect, it aids the external field. If K_1 is positive, the opposite happens.

Magnetic anisotropy affects the internal biasing field in a manner rather similar to that of the demagnetizing field discussed in the preceding section. The effect is particularly important in resonators utilizing gyromagnetic resonance. As shown by Eq. (6.10), the resonant frequency is determined not only by the external field, but by the demagnetizing factors N_x, N_y, and N_z. Magnetic anisotropy is often taken into account by calculating an equivalent demagnetizing factor, N^a, for the anisotropy field. Thus, for the case considered above, the demagnetizing factor would be determined as

$$N^a M_s = H^a = \frac{2K_1}{\mu_0 M_s}$$

Therefore,

$$N^a = \frac{2K_1}{\mu_0 M_s^2} \tag{6.15}$$

In general, there will be equivalent anisotropy demagnetizing factors N_x^a and N_y^a, corresponding to N_x and N_y of Eq. (6.10), and this equation can be rewritten as

$$\omega_R = |\gamma| \sqrt{(H_{\text{ext}} + (N_x - N_x^a - N_z)M_s(H_{\text{ext}} + (N_y - N_y^a - N_z)M_s)} \tag{6.16}$$

Recall that if k_1 is negative, N^a is negative and the anisotropy field aids H_{ext}. Applications are discussed in Section 6.15.

6.5 RESONANT LINE WIDTH

Resonance can be observed as a peak in absorbed power as either the frequency or the biasing field is varied through the resonant point. Normally, the frequency of the microwave source is held constant, and the external biasing field varied to produce a resonance curve as sketched in Fig. 6.6. The width of the peak gives a measure of the losses in the ferrite, a narrow peak signifying relatively lower loss.

The width of the peak is normally measured at the half-power point and is denoted by ΔH. This is what is meant by the line width. The corresponding angular frequency bandwidth is given by

$$\Delta\omega = |\gamma| \, \Delta H \tag{6.17}$$

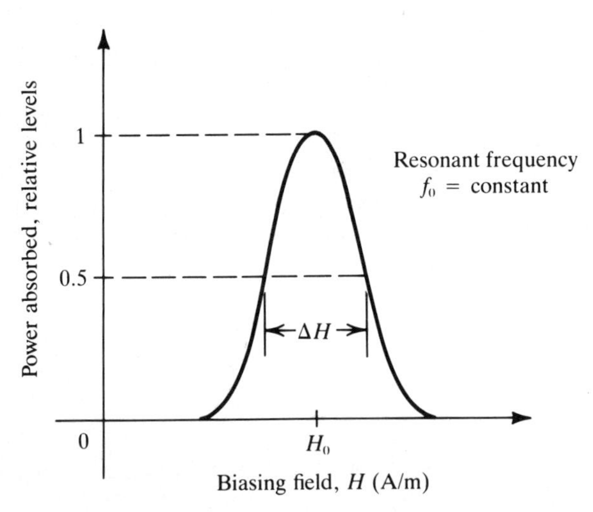

FIGURE 6.6. Resonant curve with magnetic field strength as the variable. The half-power line width is ΔH.

The Q-factor is given by the usual formula relating resonant frequency and bandwidth as

$$Q = \frac{\omega_0}{\Delta\omega}$$

$$= \frac{H_0}{\Delta H} \tag{6.18}$$

In measuring ΔH, care must be taken to minimize losses of the measuring system, or to be able to allow for these, if the loss component of the ferrite alone has to be determined. Some typical values of ΔH are given in Table 6.1.

6.6 TEMPERATURE EFFECTS

Magnetization is strongly temperature dependent. As temperature is increased, the thermal energy disrupts the magnetic alignment, and at some critical temperature the thermal energy is sufficient to completely destroy the internal magnetization. At this temperature the material becomes paramagnetic. For ferromagnetic materials the critical temperature is known as the *Curie temperature* T_c. For ferrites it is sometimes referred to as the *Neel temperature*. Common practice, which will be followed here, is to use the term "Curie temperature" for both.

In Fig. 6.7(a) is sketched the magnetization–temperature curve. This closely follows the relation (Say 1954)

$$\frac{M_s}{M_{s\max}} = \tanh\left(\frac{M_s}{M_{s\max}}\frac{T_c}{T}\right) \tag{6.19}$$

As can be seen from Fig. 6.7, the magnetization varies rapidly in the region of Curie temperature, and in microwave applications operating temperatures should be well below the Curie point. Some nominal values for Curie temperatures are given in Table 6.1.

Resonant line width also varies markedly with temperature, a sketch of the variation being shown in Fig. 6.7(b). Just below the Curie point, the line width decreases with decrease in temperature, but over most of the useful temperature range, the line width shows an increase with decrease in temperature.

Anisotropy decreases, in general, with increase in temperature below the Curie point. At the Curie point the anisotropy energy is zero. A sketch of anisotropy energy versus temperature for two ferrites is shown in Fig. 6.7(c).

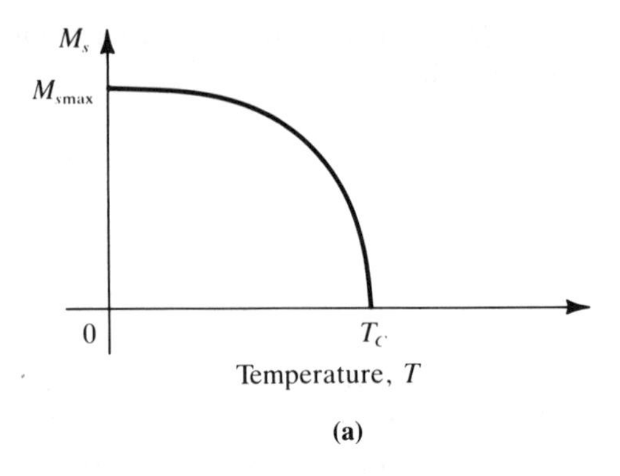

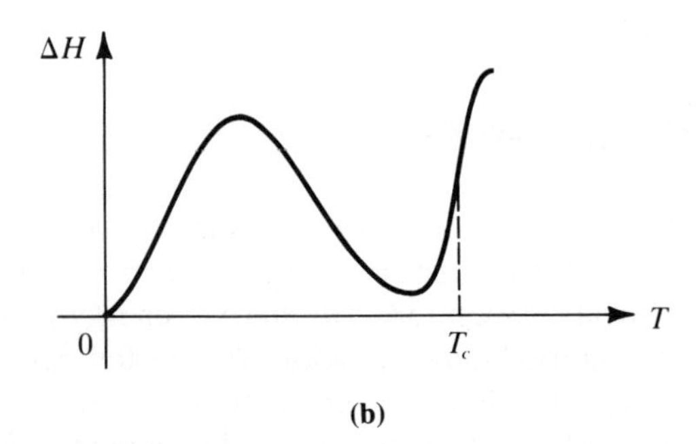

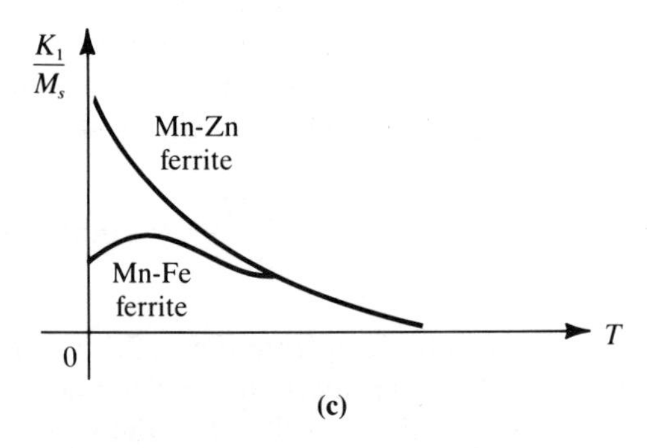

FIGURE 6.7. Temperature variations of (a) saturation magnetization, (b) line width, and (c) anisotropy. [*Parts (b) and (c) after Lax and Button, 1962.*]

6.7 UNITS

SI (Système International) units have been used so far in describing the properties of ferrites. However, it will be found that the centigrade-gram-second electromagnetic units (or CGSEMU) are widely used in practice. Many of the tables presented in handbooks dealing with ferrites use these units. In these, magnetic field strength is measured in oersteds and magnetic flux density in gauss. The conversion factors required between these and the SI units of amperes/meter and teslas are:

$$1 \text{ tesla} = 10^4 \text{ gauss} \tag{6.20}$$

$$1 \text{ oersted} = \frac{10^3}{4\pi} \text{ amperes/meter} \tag{6.21}$$

Use of these conversion factors is illustrated in the following example.

EXAMPLE 6.2

Convert the following: (a) 0.1 T to gauss, (b) 1400 G to teslas, (c) 10^5 A/m to oersteds, and (d) 400 oe to amperes/meter.

SOLUTION:

(a) $0.1 \times 10^4 = 10^3$ G

(b) $1400 \times 10^{-4} = 0.14$ T.

(c) $10^5 \times 4\pi \times 10^{-3} = 1256.6$ Oe.

(d) $400 \times \dfrac{10^3}{4\pi} = 3.183 \times 10^4$ A/m.

Sometimes the magnetic field strength in SI units is given in ampere-turns per meter. For example, a long solenoid of N turns carrying a current of I amperes sets up a magnetic field strength of NI/l ampere-turns per meter at the center, where l is the length of the solenoid, assumed very much larger than its diameter. Since number of turns is a dimensionless quantity, A-t/m is equivalent to A/m.

In the cgs units, $B = H + 4\pi M$ and the quantity $(4\pi M_s)$ gauss is tabulated rather than M_s. To convert to SI units, let $M_s' = 4\pi M_s$; then $M_s = (M_s' \times 10^{-4})/\mu_0$ (A/m) where M_s' is in gauss.

The energy constant K_1 is given in ergs/cm^3 in cgs emu, and in J/m^3 in SI units. The conversion factor is

$$1 \text{ J/m}^3 = 10 \text{ ergs/cm}^3 \tag{6.22}$$

In Table 6.1, some nominal values are given for the various properties discussed, the values being given in both SI units and cgs electromagnetic units. These are nominal values, for illustration purposes. The range of values encountered in practice will vary considerably around the nominal values shown, depending on the conditions of measurement and the actual ferrite specimen measured. The nominal values are for room-temperature conditions. The term "ferrite" is used somewhat generally here to cover garnets as well. The ferrimagnetic garnets are rare-earth garnets which have a different molecular makeup from that of ferrites. However, the microwave properties are similar to ferrites, and in particular, yttrium iron garnet is widely used in microwave devices.

6.8 MICROWAVE PERMEABILITY

In Eq. (6.3) the relationship between magnetic flux density B, magnetic field strength H, and magnetization M is given as $B = \mu_0(H + M)$, this being the form used when relating magnetic properties to the internal behavior of the medium. In applications, the relationship between B and H can be expressed more usefully as $B = \mu H$, where μ is a quantity known as the magnetic permeability. The permeability can be written as the product of two factors, $\mu = \mu_0\mu_r$, where $\mu_0 = 4\pi \times 10^{-7}$ H/m is the permeability of free space and μ_r is the relative permeability of the material. In an isotropic magnetic material (one that has identical magnetic properties in all directions) μ_r is a scalar quantity, the value being the same for all directions of the magnetic field. The situation is different in a biased ferrite. Application of a biasing field results in precession, as described in Section 6.2, which will interact with the microwave field. As can be visualized, precession is a directional effect, and the extent of the interaction will depend on the direction of propagation relative to the direction of the biasing field. The microwave permeability will be different for different directions of propagation, and values are defined for specific cases.

Where the magnetic field of the microwave signal is right-hand circularly (RHC) polarized as illustrated in Fig. 6.8(a), and the direction of propagation is the same as that of the biasing field, the relative permeability, denoted by μ_{r+}, is given by

$$\mu_{r+} = 1 + \frac{\omega_M}{\omega_0 - \omega} \tag{6.23}$$

Where the microwave field is left-hand circularly (LHC) polarized, as shown in Fig. 6.8(b), the relative permeability, denoted by μ_{r-}, is given by

$$\mu_{r-} = 1 + \frac{\omega_M}{\omega_0 + \omega} \tag{6.24}$$

Here ω is the frequency of the microwave signal, and ω_0 and ω_M are the frequency parameters introduced in Section 6.2 [see Eqs. (6.2) and (6.4)]. Derivation of these equations is relatively difficult and will be found in

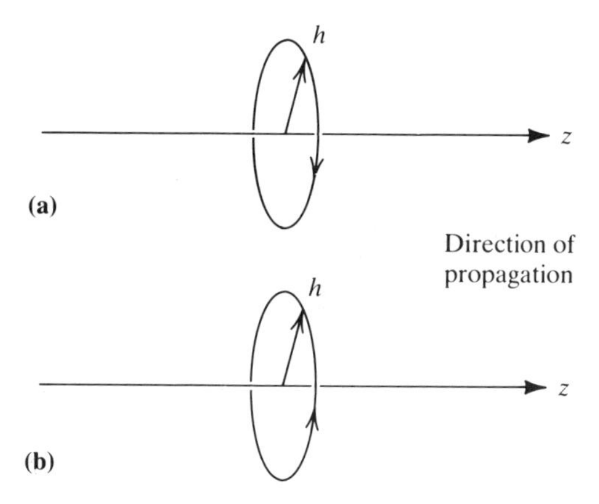

FIGURE 6.8. Microwave magnetic field vector h: (a) right-hand-circular (RHC) polarized; (b) left-hand-circular (LHC) polarized.

any of the more advanced textbooks listed at the end of the chapter. In the derivation, the ferrite medium is assumed to be infinite so that demagnetizing effects are absent (magnetic surface charges cannot accumulate), and the ferrite reaches saturation as soon as the biasing field is increased from zero.

Where the direction of propagation is the same as that of the biasing field, the ferrite is said to be *longitudinally magnetized*, and this is the situation described above. A practical situation of importance is where the ferrite fills completely the space between the conductors of a coaxial cable and is longitudinally magnetized. The relative permeability of the cable "dielectric," which of course is the ferrite, in this case is given by

$$\mu_{re} = \frac{2\mu_{r+}\mu_{r-}}{\mu_{r+} + \mu_{r-}} \tag{6.25}$$

When the direction of propagation is perpendicular to the direction of the biasing field, the ferrite is said to be *transversely magnetized*. Again, considering an infinite ferrite medium, two situations can arise. The first, of little interest in the present context, is where the magnetic field of the signal lies parallel to the biasing field. Since the ferrite is saturated in this direction, the relative permeability is unity, and the ferrite behaves as a normal dielectric. In the second situation, the electric vector of the signal lies parallel to the biasing magnetic field. If the biasing field is assumed to be along the z-axis, the microwave magnetic field will lie in the x-y plane. The relative permeability in this case is also given by Eq. (6.25).

In Fig. 6.9(a) is sketched the variation of μ_{r+}, μ_{r-}, and μ_{re} as functions of ω_0/ω. Recalling that ω_0 is directly proportional to H_0, the abscissa

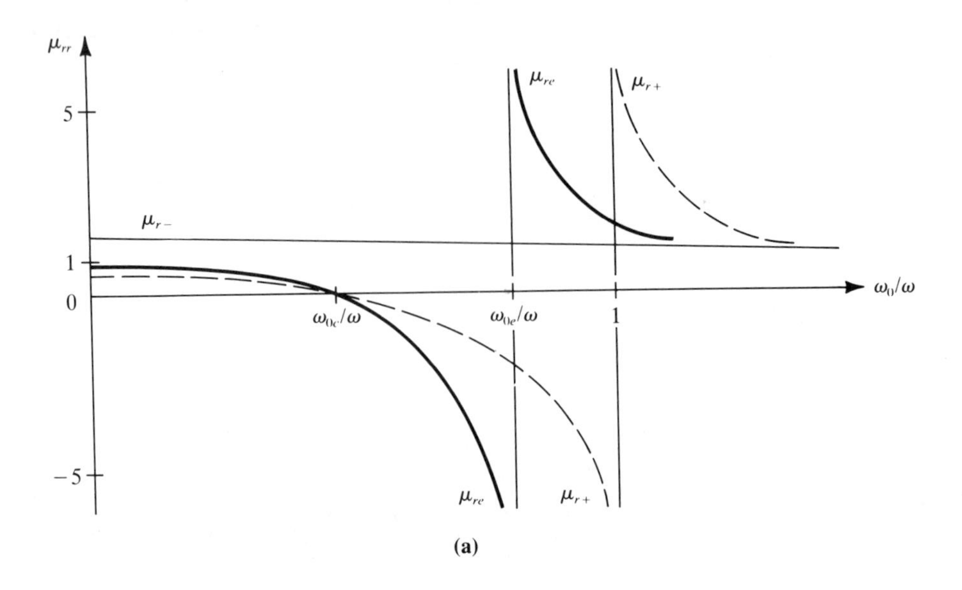

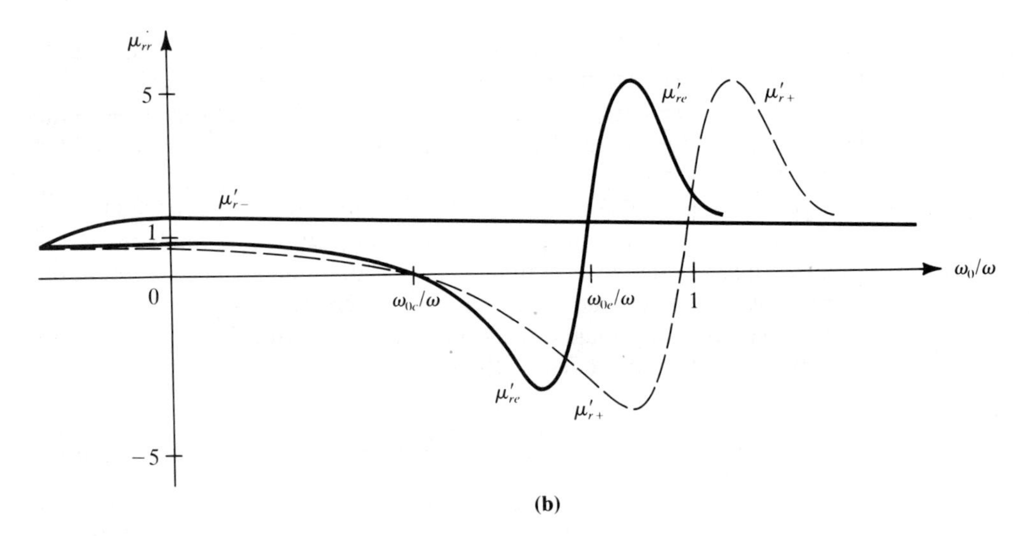

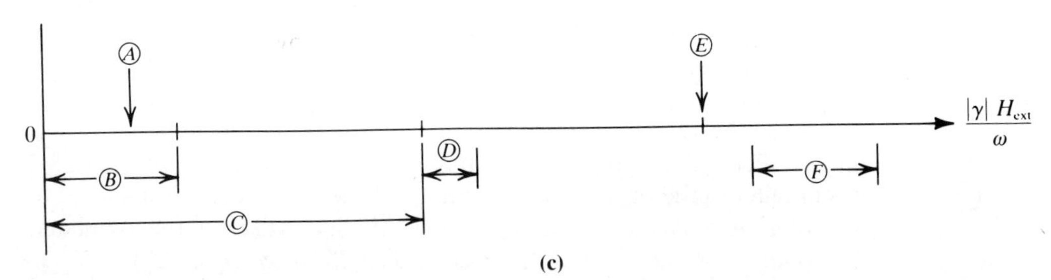

FIGURE 6.9. (a) Variation of u_{r+}, u_{r-}, and u_{re} as functions of ω_0/ω for a lossless ferrite; (b) as part (a) but taking losses into account; (c) extension of base to account for demagnetization $N_z\omega_m/\omega$. Letters indicate regions for specific device applications as listed in the text.

also shows how these quantities vary as the microwave frequency is held constant and H_0 is varied. Important points to note are that both μ_{r+} and μ_{re} are zero at some cutoff value of field H_{oc}, corresponding to ω_{oc}; μ_{re} shows a resonant point at ω_{oe}, and μ_{r+} shows a resonant point at $\omega_0/\omega = 1$.

It may be shown that (see problems 8, 9, and 10)

$$H_{0c} = \frac{\omega}{|\gamma|} - M_s \tag{6.26}$$

$$H_{0e} = \frac{M_s}{2}\left[\sqrt{1 + \left(\frac{2\omega}{\omega_m}\right)^2} - 1\right] \tag{6.27}$$

and the resonant condition for μ_{r+} is when

$$H_0 = \frac{\omega}{|\gamma|} \tag{6.28}$$

Keep in mind that ω is the angular frequency of the microwave signal, whereas ω_0 and ω_M are numbers which have dimensions of angular frequency and which are really measures of the biasing field through $H_0 = \omega_0/|\gamma|$ and of the saturation magnetization through $M_s = \omega_M/|\gamma|$.

So far, losses have been neglected. Conductive losses in ferrites are usually small enough to be ignored, and gyromagnetic damping losses account for the power dissipation in the ferrite. These losses can be taken into account by replacing the real number permeability discussed above with a complex number. The real part of this complex permeability (denoted in the literature by μ'_{r+}, etc.), determines the phase velocity in a similar manner to permeability for the lossless ferrite and the two are approximately equal except around the resonant points, as shown in Fig. 6.9(b). Around the resonant points, the infinite peaks are damped down to finite values, which is to be expected when losses are present. The power losses in the ferrite are related to the imaginary part of the complex permeability (denoted in the literature by μ''_{r+} etc.; for clarity, variations of the imaginary part have not been shown in Fig. 6.9).

As an alternative to varying H_0 while holding frequency constant, H_0 can be held constant and frequency varied. Figure 6.9 then shows the variations in permeability to be expected, but it must be kept in mind that frequency increases are to be read from right to left on the abscissa scale shown in Fig. 6.9.

The results above apply to an infinitely large ferrite medium. In an infinite medium demagnetization does not occur because there are no surfaces on which the demagnetizing field poles can accumulate (see Fig. 6.3). Saturation is reached as soon as the bias field is increased from zero. As discussed in Section 6.3, demagnetization will occur in a finite sample, and where the biasing field is in the z-direction, saturation is reached when

$$H_0 = H_{\text{ext}} - N_z M_s \tag{6.29}$$

Thus to the left of $H_0 = 0$ in a finite sample the ferrite is below saturation, and this is shown by the extension to the curves in Fig. 6.9(b). The abscissa scale for the external biasing field is shown in Fig. 6.9(c), this being normalized in terms of frequency, as in Fig. 6.9(a) and (b). In other words, the external field must exceed the demagnetizing field at saturation, or $H_{\text{ext}} > N_z M_s$, for the curves of Fig. 6.9(a) and (b) to apply.

The circled letters on Fig. 6.9(c) denote ranges for various practical devices [see von Aulock and Fay (1968)]. These are:

A: Latching devices
B: Reggia–Spencer phase shifter
C: Low-field phase shifters
D: Field displacement isolators
E: Resonant absorption devices
F: High-power devices

6.9 THE PROPAGATION COEFFICIENT

The propagation coefficient for an electromagnetic wave in ferrite is similar to the propagation coefficient for waves along a transmission line, introduced in Chapter 1. In the transmission-line situation the relative permeability could be taken as unity, whereas in the ferrite, the relative permeability has one of the forms given by Eq. (6.23), (6.24), or (6.25) depending on the propagation conditions as described in Section 6.8. Using the symbol μ_{rr} for any of these, the following relationships hold:

Phase velocity:

$$v_p = \frac{c}{\sqrt{\varepsilon_r \mu_{rr}}} \tag{6.30}$$

where c is the free-space velocity of light. Equation (6.30) should be compared with Eq. (1.42).

From Eq. (6.30), and using Eqs. (1.25) and (1.29), it is left as an exercise for the student to show:

Wavelength in ferrite:

$$\lambda = \frac{\lambda_0}{\sqrt{\varepsilon_r \mu_{rr}}} \tag{6.31}$$

Phase-shift coefficient:

$$\beta = \beta_0\sqrt{\varepsilon_r\mu_{rr}} \tag{6.32}$$

where β_0 is the free-space phase-shift coefficient, and λ_0 is the free-space wavelength.

It will be seen, therefore, that these quantities are determined by the particular form of μ_{rr}, which in turn depends on the mode of propagation and also on the value of the biasing field.

The attenuation coefficient, α in Eq. (1.9), will be determined by the damping losses in the ferrite.

6.10 PHASE SHIFT AND FARADAY ROTATION

A linearly polarized magnetic field can be resolved into RHC and LHC components as illustrated in Fig. 6.10. If, therefore, a linearly polarized wave is longitudinally propagated as described in Section 6.8, the RHC component will have a phase velocity determined by μ_{r+}, and the LHC component by μ_{r-}. The situation is illustrated in Fig. 6.11(a), which shows a ferrite rod of length l longitudinally magnetized. Figure 6.11(b) shows the RHC and LHC vectors at the input at the instant they are in-line. Figure 6.11(c) shows the vectors at the same instant at the output, each lagging, in phase, the RHC component by amount θ_+, the LHC component by amount θ_-, where:

$$\theta_+ = (\beta_0\sqrt{\varepsilon_r\mu_{r+}})l \tag{6.33}$$

$$\theta_- = (\beta_0\sqrt{\varepsilon_r\mu_{r-}})l \tag{6.34}$$

Since the two output vectors must rotate symmetrically about the line of polarization of the linearly polarized resultant, as shown in Fig. 6.10,

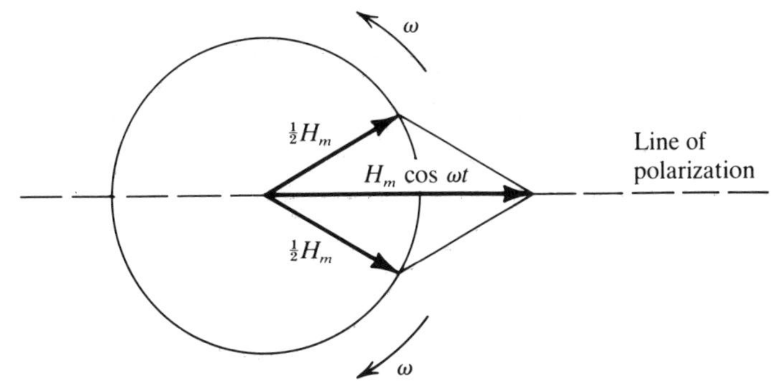

FIGURE 6.10. Resolution of a linear polarized vector H_m cos ωt into two constant-amplitude $\frac{1}{2}H_m$ vectors rotating in opposite directions at constant angular velocity.

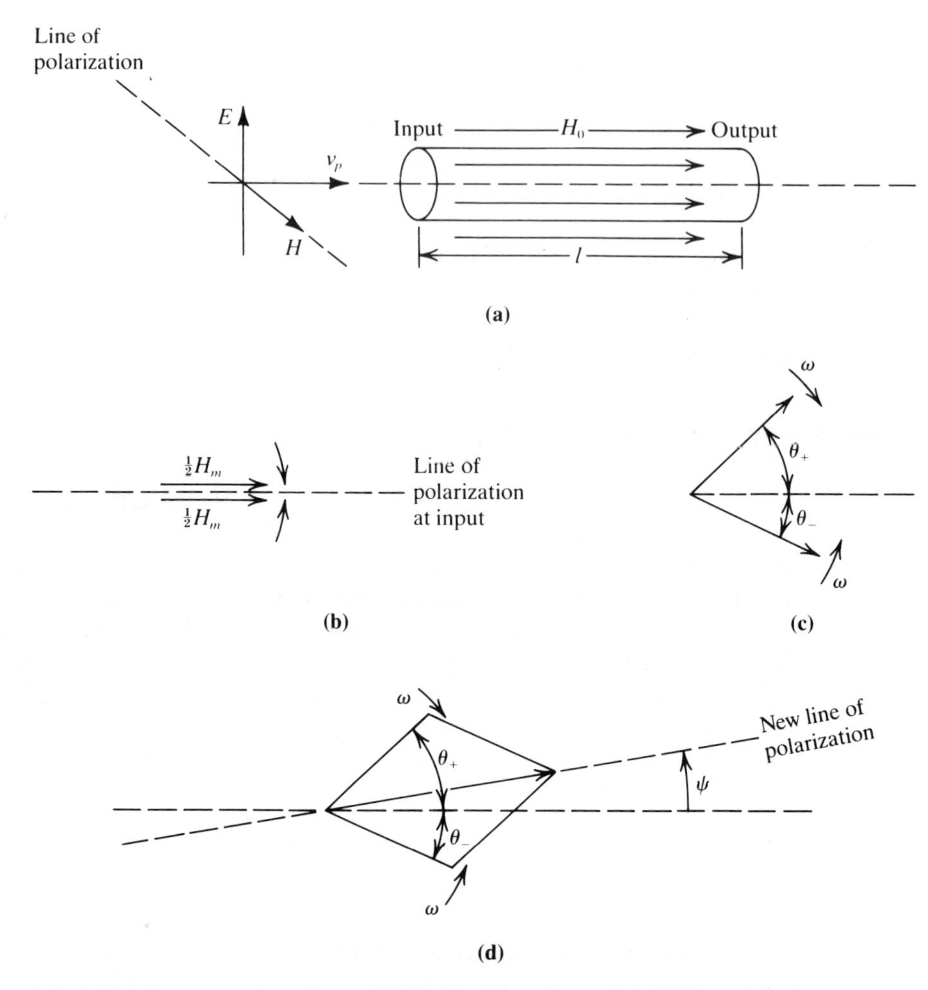

FIGURE 6.11. (a) Ferrite rod of length L with steady magnetic field B_o applied along length, and linear polarized wave propagating along length; (b) snapshot of rotating magnetic vectors at input; (c) snapshot of output vectors at same instant as part (b); (d) vector diagram showing new line of polarization for the resultant linear polarized magnetic vector at output.

the line of polarization must be shifted as shown in Fig. 6.11(d). From the geometry of Fig. 6.11(d),

$$\begin{aligned}\psi &= \frac{\theta_+ + \theta_-}{2} - \theta_- \\ &= \frac{\theta_+ - \theta_-}{2} \\ &= \tfrac{1}{2}\beta_0\sqrt{\varepsilon_r}\,(\sqrt{\mu_{r+}} - \sqrt{\mu_{r-}})l \end{aligned} \tag{6.35}$$

This is termed the Faraday polarization shift.

It is also seen from Fig. 6.11(d) that the phase lag of each component relative to the new line of polarization is

$$\theta = \frac{\theta_+ + \theta_-}{2}$$

$$= \frac{1}{2}\,\beta_0\sqrt{\varepsilon_r}\,(\sqrt{\mu_{r+}} + \sqrt{\mu_{r-}})l \qquad (6.36)$$

EXAMPLE 6.3

An 18-GHz linearly polarized wave is longitudinally propagated through a ferrite for which $\mu_{r+} = 0.955$, $\mu_{r-} = 1.015$, and $\varepsilon_r = 9$. Calculate (a) the length of rod required to produce a 45° polarization shift, and (b) the corresponding phase shift.

SOLUTION

$$\beta_0 = \frac{\omega}{c}$$

$$= \frac{2\pi \times 18 \times 10^9}{3 \times 10^8}$$

$$= 377 \text{ rad/m}$$

(a) Applying Eq. (6.35), and noting that since $\mu_{r-} > \mu_{r+}$, then ψ will be a negative angle, we obtain

$$-\frac{\pi}{4} = \frac{1}{2} \times 377 \times \sqrt{9}\,(\sqrt{0.955} - \sqrt{1.015})l$$

$$= -\,17.1l$$

Therefore,

$$l = \frac{\pi}{4 \times 17.1} = 0.0459 \text{ m} = \mathbf{4.59\ cm}$$

(b) Applying Eq. (6.36) gives

$$\theta = 377 \times 3 \times 0.0459 \times (\sqrt{0.955} + \sqrt{1.015})$$

$$= \mathbf{26.15°}$$

Faraday rotation is nonreciprocal. This means that the wave does not return to its original polarization line on being reflected back through the ferrite. This is illustrated in Fig. 6.12. The magnetic field of the forward wave is shown with a polarization shift of ψ in a clockwise direction. For

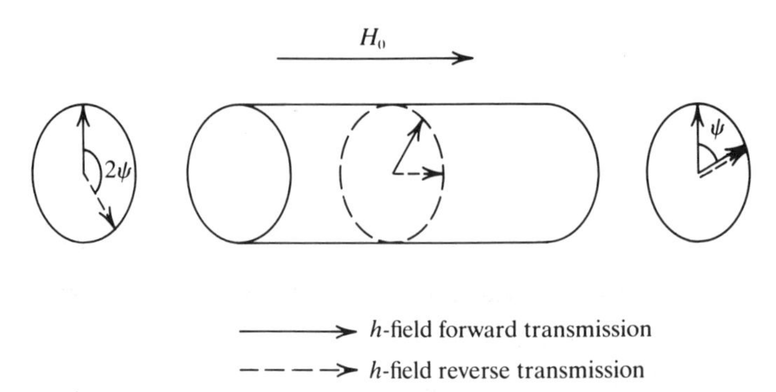

FIGURE 6.12. Nonreciprocal property of Faraday rotation.

purposes of illustration, the output wave is shown being reflected with no change of phase. On being transmitted back through the ferrite, the reflected wave will be rotated in a counterclockwise direction when viewed along the reverse direction of transmission, since the biasing field is $-H_0$ in this case. However, when viewed from input to output, this appears as a clockwise rotation, as shown in Fig. 6.12, and therefore the reflected wave is shifted by an amount 2ψ with respect to the direction of the original input h-vector.

The phase of the circularly polarized component will always lag the linearly polarized resultant. This is because it takes a finite time for transmission through the ferrite in either direction. Also, since the propagation time is the same for both directions, the phase shift is reciprocal. However, methods are available for achieving nonreciprocal phase shift, and some of these are described in later sections.

6.11 DEFINITIONS FOR NONRECIPROCAL DEVICES

As shown in Fig. 6.9, a variety of ferrite devices are available, many of which utilize the nonreciprocal property of ferrites. The definitions and symbols for these nonreciprocal devices, as given by Fox et al. (1955), are:

Differential phase shifter A transmission element for which the difference in phase shift (termed the differential phase shift) between the two directions of propagation is θ radians. The symbol is shown in Fig. 6.13(a). It is also known as a *directional phase shifter*.

Gyrator A special case of the differential phase shifter for which the differential phase shift is π radians. The symbol is shown in Fig. 6.13(b).

Isolator A device that permits power flow in one direction only, so named because it can be used to isolate an input port from reflections from other ports. Its operation depends on the absorption (rather than reflection) of power in the direction of no transmission. In a practical device a small amount of power will be transmitted in the reverse direction. The circuit symbol for the isolator is shown in Fig. 6.13(c).

Circulator A device that acts as a commutator for power. The circuit symbol for a four-port circulator is shown in Fig. 6.13(d). Referring to this, power entering at port 1 will be transmitted to port 2 only; power entering port 2 will be transmitted to port 3 only; power entering port 3 will be transmitted to port 4 only; and power entering port 4 will be transmitted to port 1 only. Circulator action is not restricted to four-port devices. Three-port circulators are widely used, and Fig. 6.13(e) shows how two four-port circulators may be combined to produce a six-port circulator.

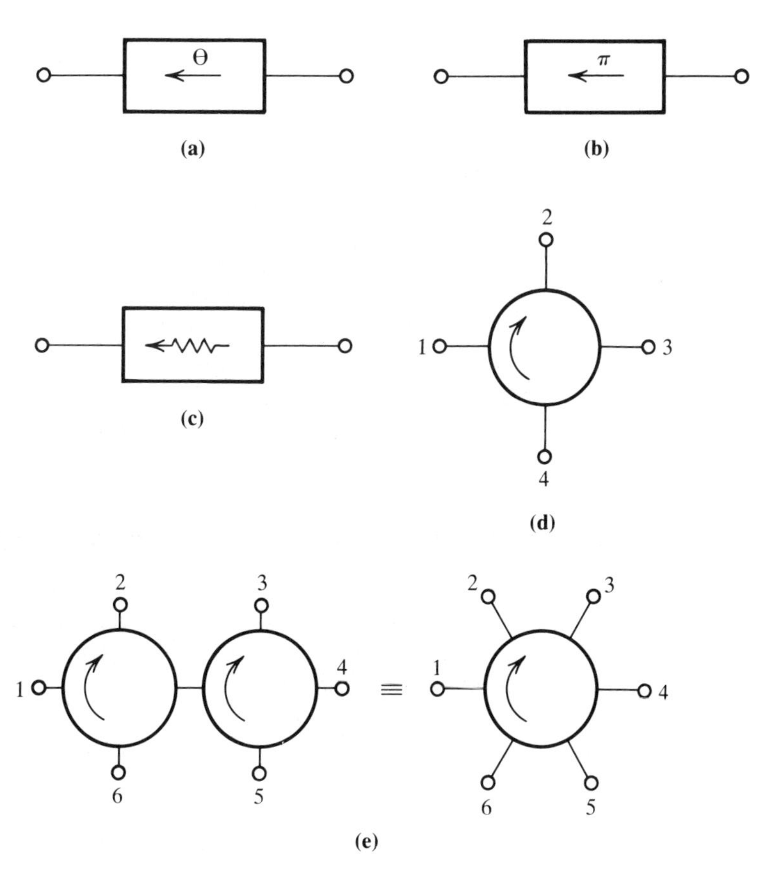

FIGURE 6.13. Circuit symbols for (a) differential phase shifter, (b) gyrator, (c) isolator, and (d) four-port circulator. (e) Showing how a six-port circulator may be realized from two four-port circulators.

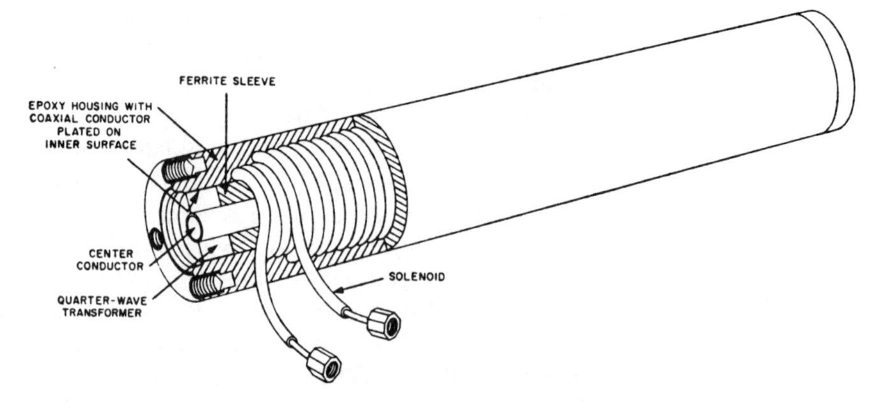

FIGURE 6.14. S-band coaxial phase shifter. [*From W. H. von Aulock and C. E. Fay, Linear-Ferrite Devices for Microwave Applications (New York: Academic Press, Inc., 1968).*]

6.12 PHASE SHIFTERS

The constructional features of a coaxial phase shifter are shown in Fig. 6.14. This particular device operates below resonance [in range C of Fig. 6.9(c)], and the saturation magnetization of the ferrite is 5.57×10^4 A/m. The dimensions of the coaxial system are 20.6 mm × 9.53 mm, and the length of the ferrite sleeve is 254 mm. The device operates over the frequency range 2.85 to 3.15 GHz. The solenoid is made of 3.175-mm-diameter copper tubing to permit the flow of cooling liquid, and the magnetic field strength can be varied from zero to about 8000 A/m. Quarter-wave matching transformers are fitted at each end.

Phase shift, which is reciprocal, is determined by μ_{re} as given by Eq. (6.25). The phase-shift coefficient is given by Eq. (6.32), where μ_{re} replaces μ_{rr}.

EXAMPLE 6.4

For the phase shifter described above, the relative permittivity of the ferrite is 12, and the initial (zero applied field) relative permeability is 0.65. Determine the change in phase shift at 3 GHz when the applied field is increased from zero to 6000 A/m.

SOLUTION Assuming that the Landé splitting factor is 2, then from Eq. (6.1),

$$|\gamma| = 1.105 \times 10^5 \times 2 = 2.21 \times 10^5$$

From Eq. (6.2), when $H = 6000$ A/m,

$$\omega_0 = 2.21 \times 10^5 \times 6000 = 1.326 \times 10^9 \text{ rad/s}$$

And from Eq. (6.4),

$$\omega_M = 2.21 \times 10^5 \times 5.57 \times 10^4 = 1.231 \times 10^{10} \text{ rad/s}$$

Hence, from Eqs. (6.23), (6.24), and (6.25), with $\omega = 2\pi \times 3 \times 10^9 = 1.885 \times 10^{10}$ rad/s:

$$\mu_{r+} = 1 + \frac{1.231}{0.1326 - 1.885} = 0.297$$

$$\mu_{r-} = 1 + \frac{1.231}{0.1326 + 1.885} = 1.61$$

$$\mu_{re} = \frac{2 \times 0.297 \times 1.61}{0.297 + 1.61} = 0.5$$

The effective permeability for this device is μ_{re}, and application of Eq. (6.32) gives

$$\Delta\theta = (\beta - \beta_i)l$$

$$= \beta_0 \sqrt{\varepsilon_r}(\sqrt{\mu_{re}} - \sqrt{\mu_{ri}})l$$

where β_i is the initial, or unbiased, phase-shift coefficient. Hence

$$\Delta\theta = \frac{2\pi \times 3 \times 10^9 \sqrt{12}}{3 \times 10^8}(\sqrt{0.5} - \sqrt{0.05}) \times \frac{254}{1000} \text{ rad}$$

$$= \mathbf{-5.48\ rad} \quad \text{or} \quad \mathbf{-314°}$$

Note that the minus sign occurs because the permeability is reduced from its initial value when the biasing field is applied.

Phase shift in a waveguide may be achieved by means of longitudinal magnetization of a ferrite rod as shown in Fig. 6.15. In this case, phase shift is somewhat similar to that described in Section 6.10, the magnetic field at the center of the guide being linearly polarized. However, away from the center of the guide the field is no longer linearly polarized, and the mode of operation is more complex. This device is known as a Reggia–Spencer phase shifter, after its originators.

Phase shift may also be achieved through use of a rectangular ferrite slab in a waveguide, transversely magnetized, as shown in Fig. 6.16(a). The presence of the ferrite slab alters the electric and magnetic field distributions (compared to the empty waveguide conditions) and furthermore,

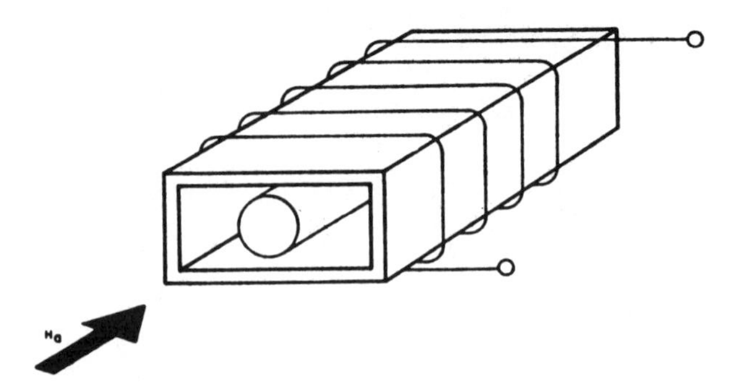

FIGURE 6.15. Reggia–Spencer phase-shifter configuration. [*From W. H. von Aulock and C. E. Fay, Linear-Ferrite Devices for Microwave Applications (New York: Academic Press, Inc., 1968).*]

the forward wave is affected differently from the reverse wave. The actual field distributions depend critically on the slab position and the slab thickness, and a typical result is shown in Fig. 6.16(b). Because the forward wave is affected differently from the reverse wave, a nonreciprocal phase shift can be achieved. Phase shift as a function of slab position is sketched in Fig. 6.17, for both forward and reverse waves. The differential phase shift, which is the difference between the forward and reverse phase shifts, is also shown in Fig. 6.17.

6.13 ISOLATORS

A sectional view of a Faraday rotation isolator is shown in Fig. 6.18. The end sections provide transitions from rectangular to round waveguide, and the ferrite rod, longitudinally magnetized, is contained in the round section of waveguide. The ferrite section is adjusted to produce a 45° Faraday rotation in a counterclockwise direction for a signal passing from port 1 to port 2. As discussed in Section 6.11, the effect is nonreciprocal, and therefore the polarization of an input signal at port 2 will also be rotated 45° in a counterclockwise direction on passing from port 2 to port 1.

An input signal at port 1 will have its electric field normal to the first resistance vane, and therefore no power will be absorbed from this signal by the vane. On passing through the center section, the polarization is rotated by 45°, and therefore the field is also normal to the second resistance vane, and the signal passes unattenuated to port 2. An input signal at port 2 will have its electric field normal to the resistance vane at this port, and again, the signal will pass unattenuated to the center section. However, the polarization of the electric field will now be rotated 45° counterclockwise and will lie parallel to the resistance vane in the first section. The

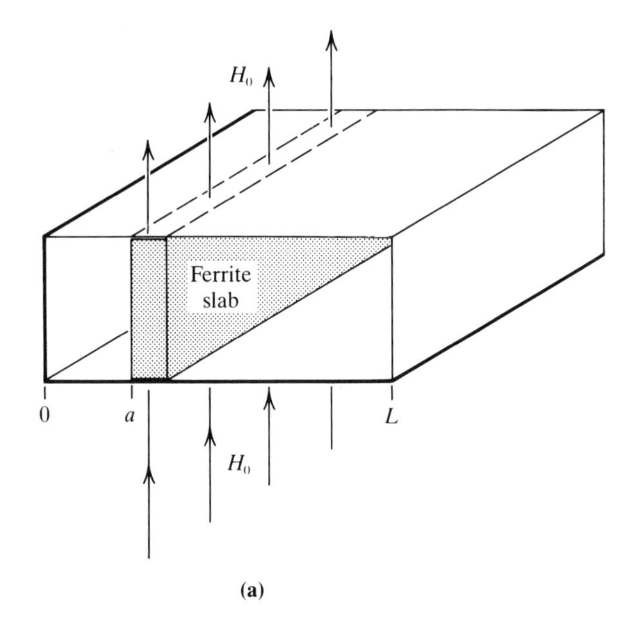

(a)

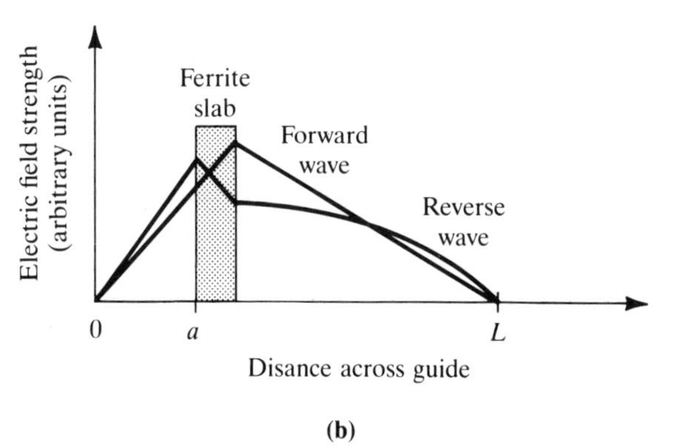

(b)

FIGURE 6.16. (a) Transversely magnetized ferrite slab in a rectangular waveguide; (b) electric field distribution for forward and reverse waves.

signal power will therefore be absorbed in this vane; in other words, a signal cannot be transmitted from port 2 to port 1. Should any reflection occur at the first resistance vane, the reflected signal will be rotated a further 45° on passing through the center section. The electric field will therefore lie parallel to the resistance vane at port 2, and will be absorbed in this.

Another type of isolator, known as a field displacement isolator, is illustrated in Fig. 6.19. As discussed in Section 6.12, the presence of a longitudinally magnetized ferrite slab alters the field distributions of the

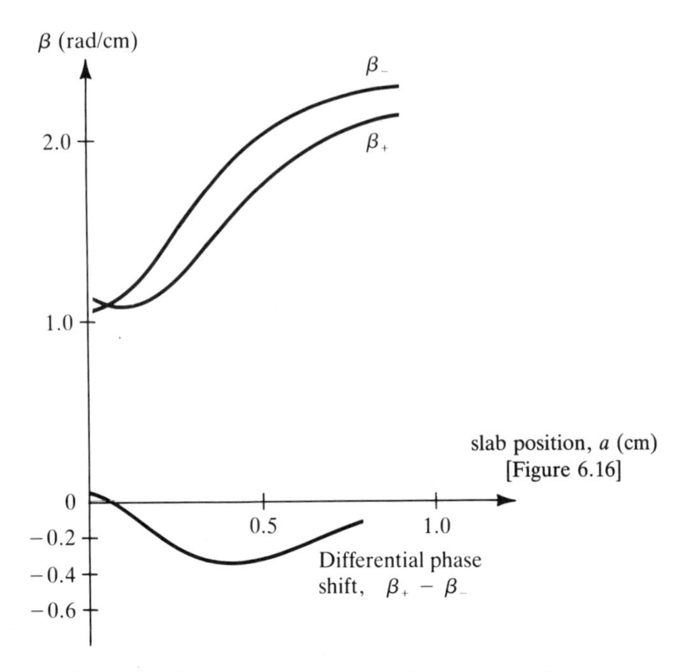

FIGURE 6.17. Phase shift as a function of slab position for the phase shifter shown in Fig. 6.16.

forward and reverse waves differently. A position can be found where the electric field of the forward wave is almost zero at the wall of the ferrite slab, while the electric field of the reverse wave has a relatively large value, as sketched in Fig. 6.19(b). By placing a resistance sheet on this wall as shown in Fig. 6.19(a), power is absorbed from the reverse wave but not from the forward wave.

6.14 CIRCULATORS

A sectionalized view of a Faraday rotation circulator is shown in Fig. 6.20. Comparing this with the isolator in Fig. 6.18, the structures are seen to be similar, but the circulator has two additional ports, shown as ports 3 and

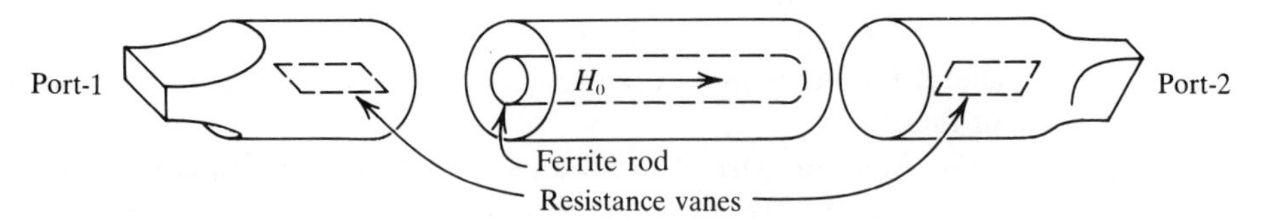

FIGURE 6.18. Sectionalized view of a Faraday rotation circulator.

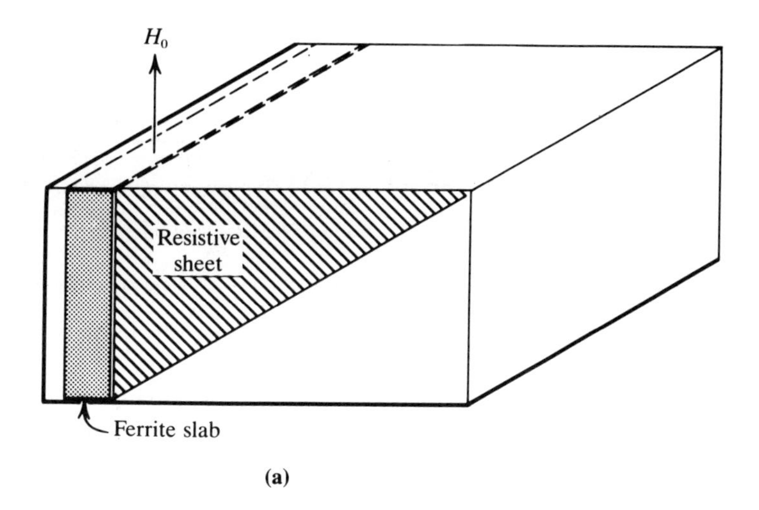

(a)

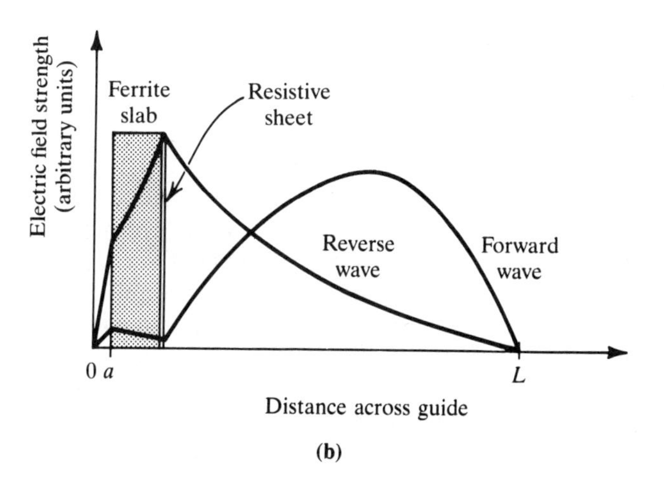

(b)

FIGURE 6.19. Field displacement isolator: (a) position of ferrite slab backed with resistive sheet; (b) electric field distribution for forward and reverse waves.

4. Also, the resistive vanes are replaced by metal septums. In studying the mode of operation of the circulator it should be kept in mind that the rectangular sections of guide carry TE waves, in which the electric field is normal to the broad walls of the guide. Thus a signal entering port 1 will not excite a wave in port 3, and the metal septum has little effect. This signal passes through the ferrite section to port 2 and in the process the polarization undergoes a 45° counterclockwise rotation. A signal entering port 2 will also bypass port 4 and will be rotated clockwise 45° (as viewed from port 2), so that it has the correct polarization to excite a TE wave in

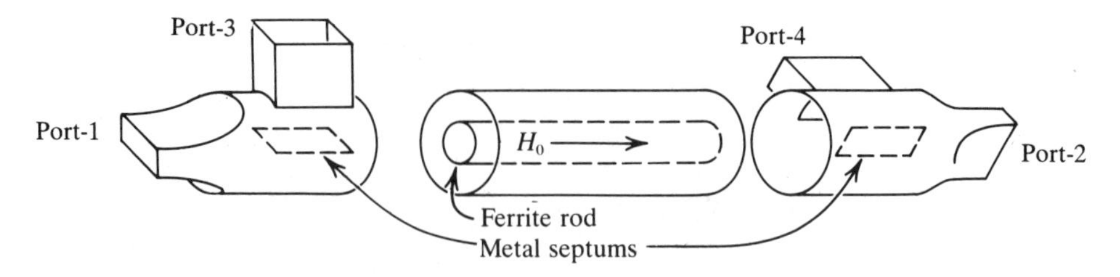

FIGURE 6.20. Sectionalized view of a Faraday rotation circulator.

port 3. The metal septum acts as a reflector and improves the signal transfer into port 3.

A signal entering port 3 will be rotated 45° counterclockwise on passing through the ferrite section and therefore will be correct for excitation of a TE wave in port 4. Finally, a signal entering port 4 will be rotatcd 45° clockwise (as viewed from the port 4 end of the ferrite) and will be transmitted out through port 1. The sequence of transmission is therefore seen to be port 1–2–3–4–1. . . , and this is the circulator action required.

Another type of circulator is the junction circulator, available in rectangular waveguide, in stripline, and for frequencies in the VHF-UHF ranges, as a lumped component device. In all these devices, a ferrite disk or post is used to rotate the electromagnetic field of the signal so that coupling occurs to the selected port only. The action can best be described with reference to the lumped component version.

Figure 6.21(a) shows a ferrite disk coupled to a resonant circuit. For clarity, only a single-turn coil, or loop, is shown, and when unbiased, the magnetic field lies at right angles to the loop. This is a time-varying field, and the arrows shown will reverse in direction periodically at the frequency of the input signal. Such a field can be resolved into two rotating steady fields, as shown in Fig. 6.10. This is also shown in Fig. 6.21(b), where the steady fields are represented by counterrotating bar magnets. These are identified by the labels C and CC. Figure 6.21(c) shows the positions of the rotating magnets as they pass through the loop. This is the instant at which the induced EMF in the loop is at a peak, and the magnetic field goes through zero. It will be recalled that self-induced EMF is proportional to rate of change of flux linkages, and hence leads the magnetic flux by a time phase of 90°. Both magnets (steady fields) rotate at the same rate, which is equal to the input angular frequency. The input circuit is resonant at this frequency and hence the input voltage and current are in phase. The inductance of the loop is determined by the unbiased permeability of the ferrite.

A magnetic bias may now be applied to the ferrite such that the clockwise rotating field experiences a permeability μ_+, and the counterclockwise rotating field a permeability μ_-. Operation is below the cutoff

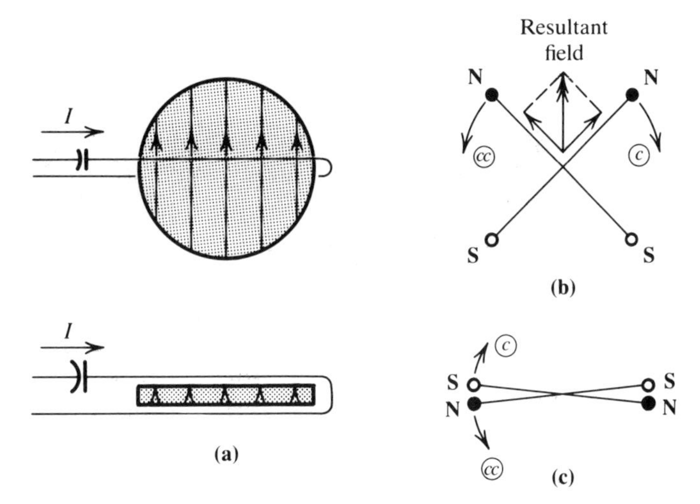

FIGURE 6.21. (a) Ferrite disk forming part of the inductance of a series resonant circuit. The disk is unmagnetized. (b) Time-varying magnetic field in the disk represented by two counterrotating bar magnets. (c) Magnet positions for peak induced EMF.

point, ω_{oc}/ω of Fig. 6.9(c), for which μ_+ is smaller than the intrinsic permeability, and μ_- is larger. From the circuit point of view, this means that the equivalent self-inductance for the clockwise mode is decreased, or, for the circuit to resonate with the rotation, the frequency would have to be increased. Similarly, for the counterclockwise mode, the frequency would have to be decreased.

At some frequency intermediate between these resonant limits, the induced voltage from the clockwise rotating field will lead the current in time phase (since the actual frequency is lower than the resonant value), while that from the counterclockwise rotating field will lag the current in time phase. This is equivalent to advancing the clockwise rotating magnet by an amount equal to the angle of lead, since one complete rotation is equivalent to one cycle in the time domain. Similarly, the counterclockwise magnet is retarded by an amount equal to the angle of lag. It should be kept in mind that the rotational velocities are both equal to the input angular frequency. The input frequency can be adjusted such that the magnitude of phase lead is equal to the magnitude of phase lag and thus the equivalent circuit appears purely resistive, similar to the resonant circuit. For the Y-junction circulator, the angle of lead or lag is made equal to 30°. This means that the clockwise rotating field must undergo a 30° advance and the counterclockwise rotating field a 30° retardation in space. Since retardation for counterclockwise rotation is equivalent to an advance in the clockwise direction, both fields are displaced 30° in a clockwise direction. This is illustrated in Fig. 6.22(a) for the instant of peak induced

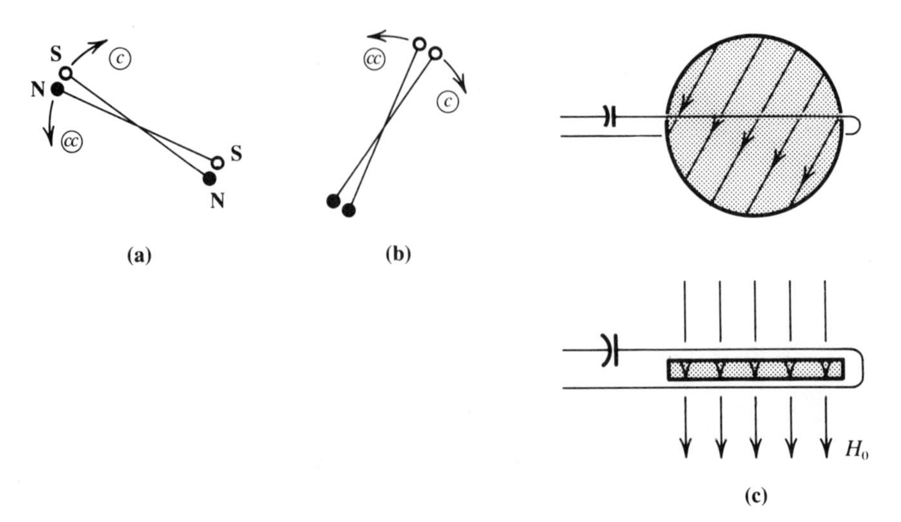

FIGURE 6.22. (a) Rotating fields of Fig. 6.14.2(a) each displaced by 30°; (b) rotating fields aligned for peak instantaneous value of time-varying field; (c) biased ferrite disk, showing polarization shift of magnetic field.

EMF [corresponding to that shown in Fig. 6.21(c)]. The position of the steady fields, when adding to give a peak in the time-varying field, is shown in Fig. 6.22(b), and the corresponding peak field is shown in Fig. 6.22(c). This shows the polarization line of the resultant time-varying field, which is seen to be shifted by 30° from its position in Fig. 6.21(a).

By placing two other resonant loops as shown in Fig. 6.23, it is seen that circuit 1, which generates the field, is coupled to circuit 2, while circuit

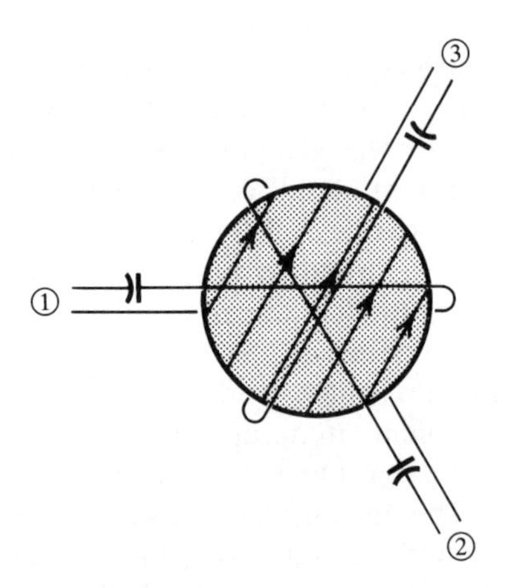

FIGURE 6.23. Lumped-component Y-junction circulator.

3 is not coupled, since the magnetic field is always parallel to the circuit 3 loop. By symmetry, it follows that if port 2 is the input, coupling will be to port 3, and port 1 will be isolated, while if port 3 is the input, coupling will be to port 1, and port 2 will be isolated.

Lumped component circulators are useful at frequencies below about 1 GHz. For higher frequencies, both stripline and waveguide versions are used. The principles of operation in each case are similar to the lumped component version, although the actual field distributions are more complicated. The construction of a stripline version of the Y-junction coupler is shown in Fig. 6.24(a), and of a waveguide version in Fig. 6.24(b).

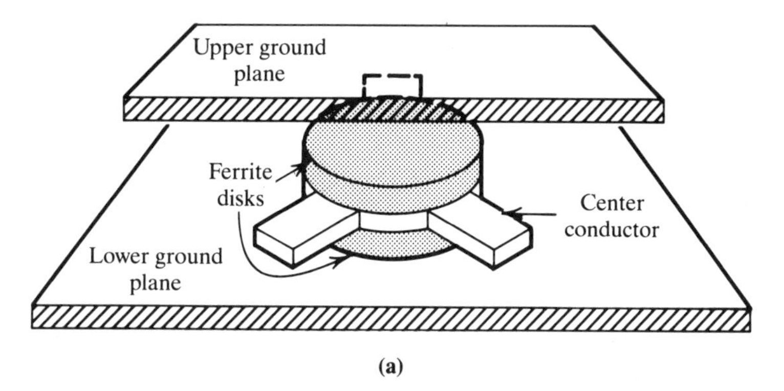

(a)

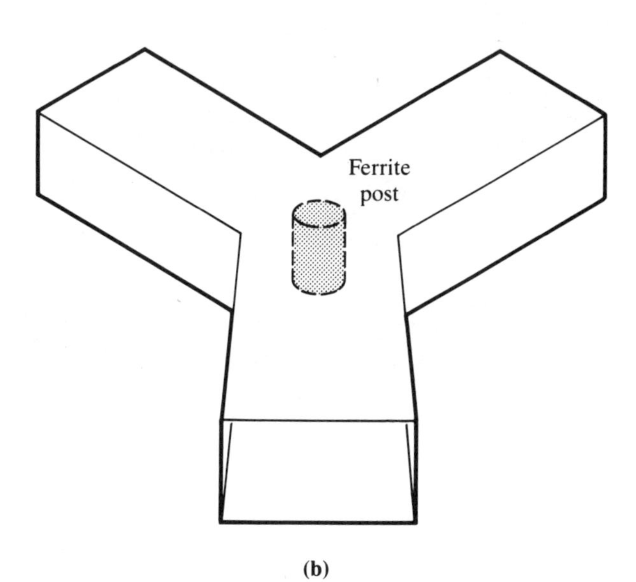

(b)

FIGURE 6.24. Y-junction circulators: (a) stripline construction; (b) waveguide construction.

6.15 FILTERS AND RESONATORS

Ferrimagnetic resonance is used as the basis for filters employing ferrites. Losses in ferrites can be reduced by removing surface roughness, and highly polished spheres of yttrium iron garnet (YIG) exhibit very high Q-factors, up to values of 10,000 (Q-factor is defined in Section 6.5). These spheres are physically small, of the order of a few millimeters in diameter, and the bulk of the size of YIG devices lies in the magnet assemblies required for the biasing fields, as shown in Fig. 6.25. As shown by Eq. (6.11), the resonant frequency of spherical specimens is independent of the saturation magnetization. The resonant frequency is determined by Eq. (6.11), which when anisotropy (see Section 6.4) is taken into account becomes

$$\omega_0 = |\gamma|(H_0 - H^a) \tag{6.37}$$

H^a is given by Eq. (6.14) for which it will be recalled that with k_1 negative, H assists H_0.

FIGURE 6.25. Several tunable YIG devices. The electromagnets, whose connections can be seen on the front of the housings, determine the size of the devices. (*Courtesy of Philips Tech. Rev., Vol. 32, 1971.*)

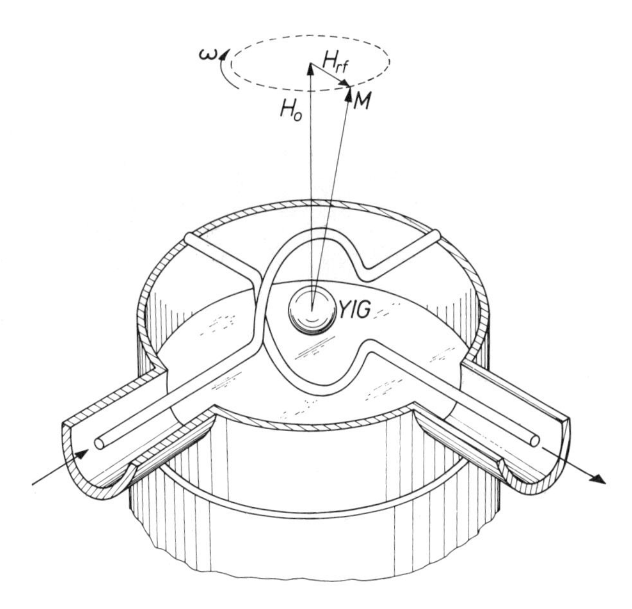

FIGURE 6.26. Bandpass filter with YIG resonator. Input and output are coupled to the YIG sphere by two orthogonal half-loops. The sphere is magnetized by a static external field H_o; if an RF magnetic field H_{rf} is coupled in, the resulting magnetization M precesses about the direction of H_o because of the gyroscopic property of electron spins in the material. The precession introduces magnetic field components that can be coupled out by the output semiloop. The precession angular frequency ω increases linearily with the magnetic field strength H_o, which is adjusted to tune the filter to the frequency required. Other frequencies do not excite the precession and are therefore not transmitted by the filter. (*Courtesy of Philips Tech. Rev., Vol. 32, 1971.*)

Electrical tuning of the resonators can be accomplished over a range of about 10:1, through variation of the biasing field.

Figure 6.26 shows the principle of one type of bandpass YIG filter. The input microwave signal alters the resultant magnetic field at resonance in a similar manner to that described for the junction circulator, and as a result, the magnetic field also couples with the output loop. The coupling is sharply pronounced at resonance, that is, when the input frequency is equal to the ferrimagnetic resonant frequency as given by Eq. (6.37). Figure 6.27 shows the constructional details of a two-stage bandpass filter designed on the principle outlined in Fig. 6.26, and Fig. 6.28 shows the frequency response of this filter.

Many other types of filters and resonators are possible, some of which are described in the paper by Roschmann (1971).

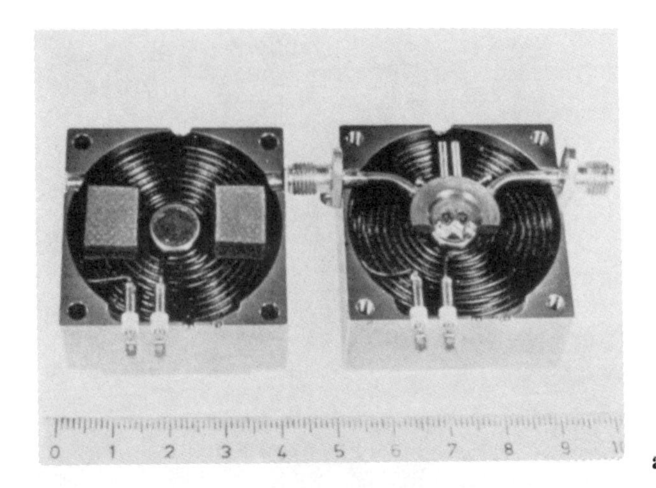

a

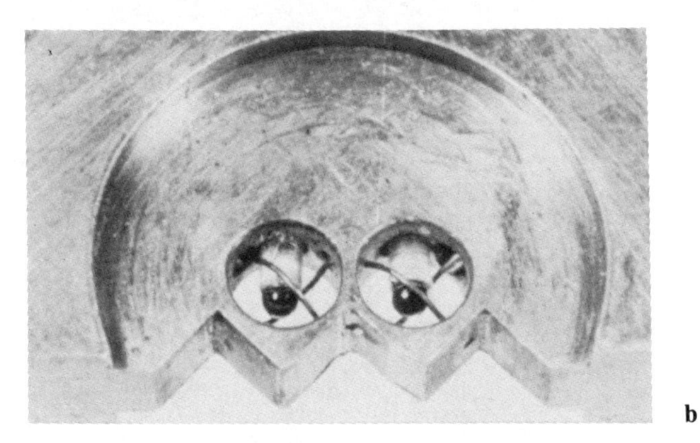

b

FIGURE 6.27. (a) Two-stage bandpass filter designed on the principle outlined in Fig. 6.26. The upper half of the electromagnet has been taken off and is shown on the left. This filter can be tuned from 1 to 20 GHz, the 3-dB bandwidth increasing from 20 MHz to 45 MHz. (b) The two YIG resonators with coupling loops. (*Courtesy of Philips Tech. Rev., Vol. 32, 1971.*)

6.16 FERRITE SUBSTRATES

Substrates for microwave integrated circuits may be constructed from ferrite material, the circuits being deposited in the form of microstrip, as discussed in Chapter 5. These substrates are sufficiently dense that they can be polished to an optical finish, which greatly reduces propagation losses. The magnetic and dielectric losses in the ferrite are also low, and the substrates are strong and rugged mechanically.

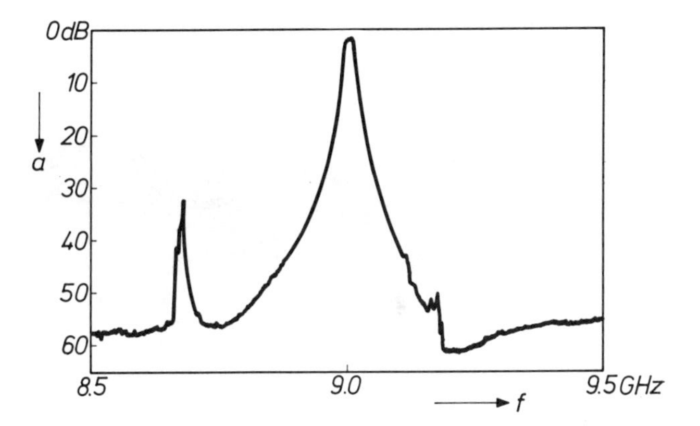

FIGURE 6.28. Attenuation *a* of the filter shown in Fig. 6.27 as a function of frequency. The filter is tuned to 9 GHz. Apart from a spurious response due to a higher-order magnetostatic resonance, the rejection is greater than 50 dB outside the passband. (*Courtesy of Philips Tech. Rev., Vol. 32, 1971.*)

Ferrite composition can be changed to produce high-quality substrates which do not have ferrimagnetic properties. These are useful in themselves as microstrip substrates, and in addition, ferrimagnetic regions can be sintered in, yielding composite substrates. Thus composite circuits can be fabricated on these using common deposition and photoetching processes for the magnetic and nonmagnetic components.

Figure 6.29(a) shows the constructional features of a junction circulator built on a ferrite substrate. The principle of operation is the same as that for the Y-junction circulator described in Section 6.14. Figure 6.29(b) shows the insertion loss and the voltage standing-wave ratio as a function of frequency about the operating frequency of 16 GHz. Circuits incorporating isolators, phase shifters, and so on, are also possible, and some of these are described in the paper by Lemke and Schilz (1971).

6.17 PROBLEMS

1. The gyromagnetic resonant frequency for a specimen of ferrite was measured as 1.0 GHz, and the *g*-factor is known to be equal to 2. Calculate the magnetic field strength at which the measurement was made.
2. A ferrite specimen is placed in the air gap between the poles of a permanent magnet. The magnetic flux density of the magnet is known to be $0.15T$. The *g*-factor for the ferrite material is 2.19. Calculate the gyromagnetic resonant frequency under these conditions.

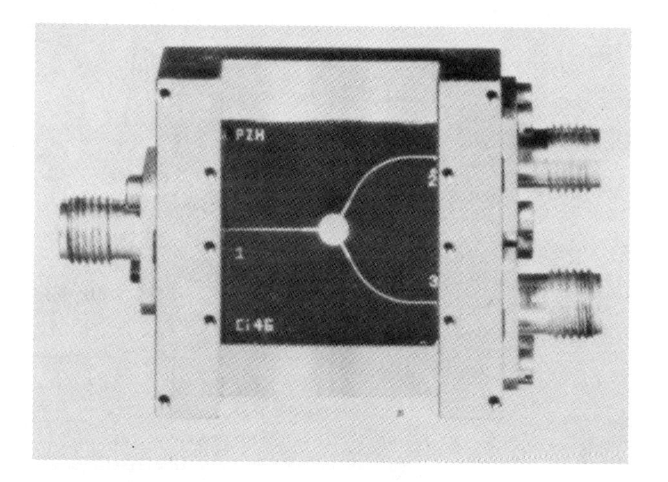

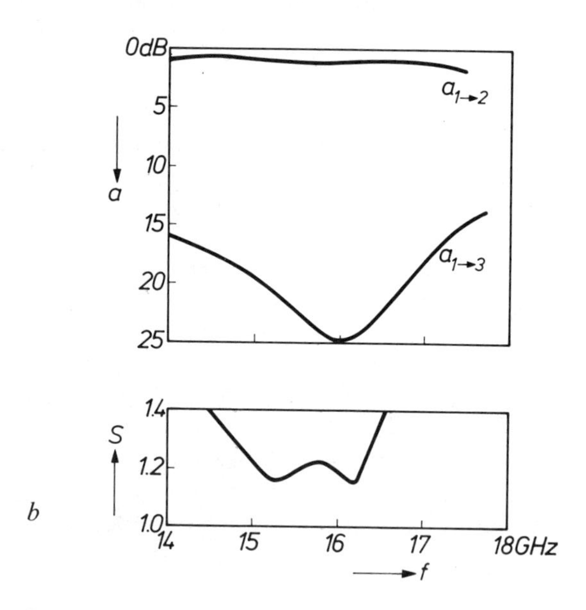

b

FIGURE 6.29. (a) Circulator for 16 GHz. Three microstrip lines are coupled to a resonator in the form of a circular disk. The circuit is mounted on a ferrimagnetic ferrite substrate, which is magnetized in a direction perpendicular to the surface by a permanent magnet. Energy transfer is only possible from input 1 to 2, from 2 to 3, and from 3 to 1. (b) Insertion loss ($a_{1\to 2}$), isolation ($a_{1\to 3}$), and voltage standing-wave ratio (S) as a function of the frequency f. (*Courtesy of Philips Tech. Rev. vol 32, 1971.*)

3. For a ferrite specimen, the g-factor is 2.01 and the saturation magnetization is 111.4 kA/m. Determine the resonant frequency for the saturation magnetization.
4. An external magnetic field of 300 kA/m is applied along the z-axis of a ferrite, for which the g-factor is 2, and the saturation magnetization is 150 kA/m. Calculate the internal magnetic field and the gyromagnetic resonant frequency for each of the following specimen shapes: (a) sphere; (b) thin disk, the plane of which is perpendicular to the z-axis; (c) thin circular rod, the axis of symmetry being the z-axis.
5. The resonant line width for a ferrite specimen is measured as 10 kA/m. (a) If the g-factor is equal to 2, calculate the -3-dB bandwidth. (b) If the effective internal field for the measurement is 100 kA/m, calculate the Q-factor.
6. Calculate the gyromagnetic ratio, the resonant frequency for saturation magnetization, the anisotropy field strength, and the -3-dB bandwidth for each of the ferrites listed in Table 6.1.
7. Convert (a) 3000 G and (b) 70 Oe to the corresponding SI units. Convert (c) 0.25 T and (d) 340 kA/m to the corresponding CGSEM units.
8. From the condition that $\mu_{r+} = 0$, derive Eq. (6.26).
9. From the condition that μ_{re} crosses the abscissa at an infinite discontinuity [Fig. 6.9(a)], derive Eq. (6.27).
10. From the condition that μ_{r+} crosses the axis at an infinite discontinuity [Fig. 6.9(a)], derive Eq. (6.28).
11. A ferrite material has a saturation magnetization of 150 kA/m and a g-factor of 2. If an effective internal field of 60 kA/m is established for the ferrite, determine the relative permeabilities μ_{r+}, μ_{r-}, and μ_{re} for an input angular frequency of 160 Grad/s. The ferrite medium may be assumed infinite in extent.
12. The magnetic field and microwave frequency applied to a certain ferrite are such that $\mu_{r+} = 0.9$ and $\mu_{r-} = 1.071$. The relative permittivity of the ferrite is 9. Calculate the phase velocity, the phase-shift coefficient, and the wavelength in the ferrite relative to their free-space values, for each of the three relative permeability values, μ_{r+}, μ_{r-}, and μ_{re} for a microwave frequency of 20 GHz.
13. A rod of the ferrite material specified in Problem 12 is used to produce a Faraday rotation of 45°, in a manner similar to that described in Section 6.10. If the frequency of the applied signal is 20 GHz, calculate the length of rod required.
14. A rod of the ferrite material specified in Problem 12 is used in the construction of a coaxial phase shifter of the type shown in Fig. 6.14. The length of the rod is 20 cm and the initial relative permeability of

the ferrite is 0.65. Calculate the phase shift produced when the applied magnetic field is 8000 A/m and the frequency of the microwave signal is 15 GHz. Assume g = 2.

15. State the purpose of an isolator and explain the prinicples of operation of one type of ferrite isolator. Describe one application of an isolator.

16. State the purpose of a circulator and explain the principles of operation of one type of circulator. Describe one application of a circulator.

17. Explain the principles of operation of a ferrite resonator. A sphere of the YIG material listed in Table 6.1 is used in the construction of a resonator as described in Section 6.15. Calculate the resonant frequency for an applied field H_0 = 250 kA/m.

REFERENCES

CLARRICOATS, P. J. B., 1961. *Microwave Ferrites*. London: Chapman & Hall Ltd.

FOX, A. G., S. E. MILLER, and M. T. WEISS, 1955. "Behavior and applications of ferrites in the microwave region," *Bell Syst. Tech. J.*, Vol. 34.

LEMKE, M., and W. SCHILZ, 1971. "Microwave integrated circuits on a ferrite substrate," *Philips Tech. Rev.*, Vol. 32, Nos. 9/10/11/12, pp. 315–321.

ROBERTS, J., 1960. *High Frequency Applications of Ferrites*. London: English Universities Press.

ROSCHMANN, P., 1971. "YIG filters," *Philips Tech. Rev.*, Vol. 32, Nos. 9/10/11/12, pp. 322–327.

SAY, M. G., (ed.), 1954. *Magnetic Alloys and Ferrites*. George Newnes Ltd.

VON AULOCK, WILHELM H., and CLIFFORD E. FAY, 1968. *Linear Ferrite Devices for Microwave Applications*. New York: Academic Press, Inc.

FURTHER READING

BUTTON, KENNETH J., 1958. "Theoretical analysis of the operation of the field displacement ferrite isolator," *IRE Trans. Microwave Theory Tech.*, July, pp. 303–308.

FAY, C. E., and R. L. COMSTOCK, 1965. "Operation of the ferrite junction circulator," *IEEE Trans. Microwave Theory Tech.*, Vol. MTT-13, No. 15.

IKUSHIMA, ICHIRO, and MINORU MAEDA, 1974. "A temperature stabilized broad-band lumped element circulator," *IEEE Trans. Microwave Theory Tech.*, Vol. MTT-22, No. 12, pp. 1220–1225.

LAX, BENJAMIN, KENNETH J. BUTTON, and LAURA M. ROTH, 1954. "Ferrite phase shifters in rectangular wave guide," *J. Appl. Phys.*, Vol. 25, No. 11.

LAX, BENJAMIN, and KENNETH J. BUTTON, 1962. *Microwave Ferrites and Ferrimagnetics*. New York: McGraw-Hill Book Company.

MATTHAEI, GEORGE L., LEO YOUNG, and E. M. T. JONES, 1973. *Microwave Filters, Impedance Matching Networks, and Coupling Structures*. New York: McGraw-Hill Book Company.

seven
Microwave Diodes and Diode Circuits

7.1 INTRODUCTION

The diode is a two-electrode device which has a nonlinear current–voltage characteristic. The most familiar diode is probably the rectifying diode, widely used at low frequencies for power rectification and signal detection. At microwaves, use is made of a number of special properties, leading to a wide range of applications, including rectification, oscillation, amplification, and harmonic generation. These are described in this chapter.

7.2 THE VARACTOR DIODE

The name *varactor* is derived from *var*iable re*actor*, and in the case of the varactor diode, the variable reactance is obtained through the voltage–capacitance variation of a *p-n* junction. When a *p-n* junction is formed, carriers initially diffuse across the junction, giving rise to a depletion region, that is, one deplete of mobile charges. The electric field set up by the exposed donor and acceptor centers establishes an equilibrium conditions that limits the diffusion of carriers. The depletion region is sketched in Fig. 7.1(a). Short-circuiting the ends of the *p-n* diode as shown does not result in an external current flow from the internal field because the contact potential between the external *p* and *n* contacts exactly balances the internal potential. Applying a reverse bias V_R, as shown in Fig. 7.1(b), widens the depletion region and therefore increases the exposed charge. This is similar to the action of a capacitor, in that the positive side of the battery is

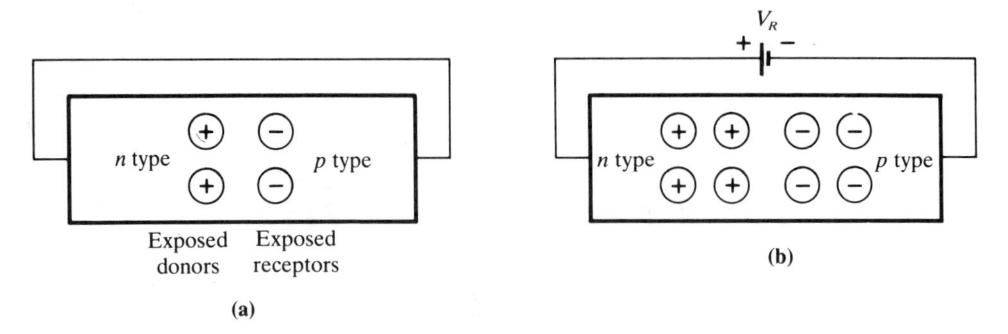

FIGURE 7.1. Diode depletion region: (a) with zero bias; (b) with reverse bias.

connected to the positively charged side of the depletion region, and this charge increases as the magnitude of the reverse bias is increased. Of course, an equal and opposite charge is maintained at all times on the negative side. Note that the *n*-material forms the positive side of the system, and the *p*-material the negative side. From the point of view of the external circuit, however, the battery appears to supply extra charge as the magnitude of V_R is increased.

Figure 7.2(a) shows how charge and voltage are related in an ordinary, or what is referred to as a linear, capacitor. Here, charge is proportional to voltage, the constant of proportionality being the capacitance C, the equation being

$$Q = CV \tag{7.1}$$

It will be seen that capacitance is given by the slope of the Q/V line, being 5 pF for the example shown in Fig. 7.2(a). On the same graph is shown the charge–voltage curve for a *p-n* junction diode operated under reverse-bias conditions. The charge here is the change relative to the equilibrium conditions shown in Fig. 7.1(a); that is, the charge for zero applied voltage is zero. Here, too, the slope of the $Q–V$ curve gives the capacitance, but this will vary from point to point along the curve, and therefore the capacitance must be defined as

$$\frac{dQ}{dV} = C(V) \tag{7.2}$$

Figure 7.2(b) shows how the capacitance varies with the voltage across the junction. The theoretical equation for the capacitance–voltage variation is

$$C(V) = \frac{C_{j0}}{(1 - V/\phi)^{\gamma}} \tag{7.3}$$

Here ϕ is the junction potential approximately equal to $\frac{1}{2}$ volt. V is the voltage across the junction and this will be a negative number for reverse bias, or $V = -V_R$. The capacitance at zero voltage is denoted by C_{j0}. The index γ depends on the type of junction. For an abrupt junction, in which the doping densities are uniform up to the junction and then change abruptly from p-type to n-type, the index $\gamma = \frac{1}{2}$. For a linear graded junction,

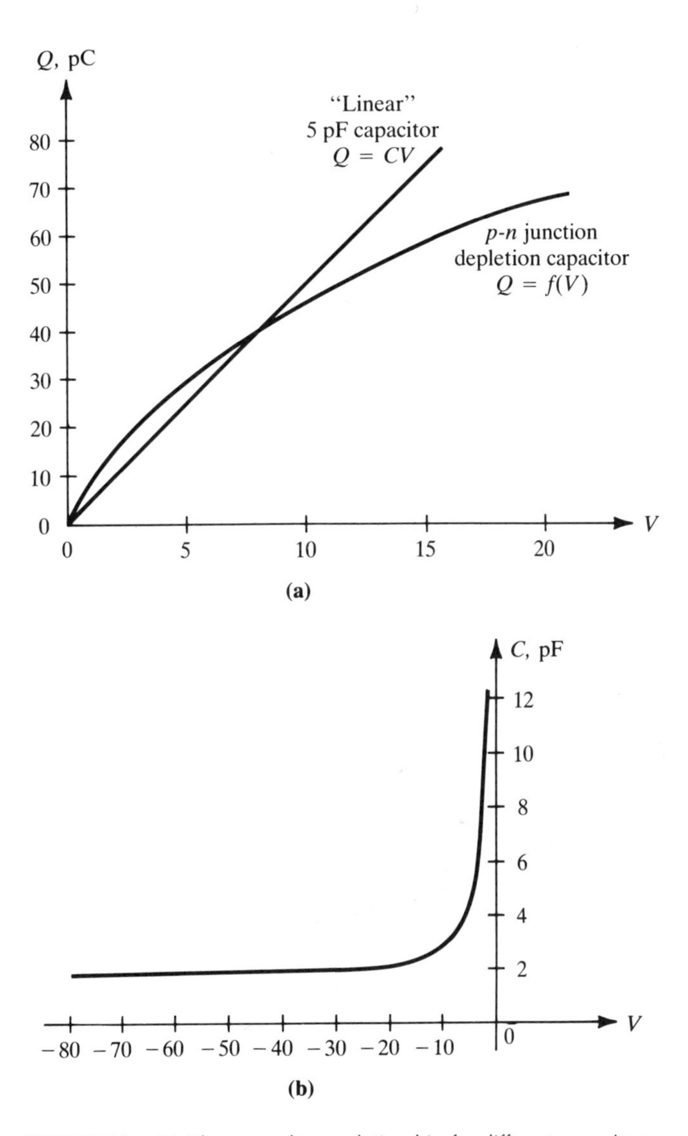

FIGURE 7.2. (a) Charge–voltage relationship for different capacitors; (b) capacitance–voltage relationship for a reverse-biased p-n junction.

in which the densities taper to zero at the junction in a linear manner with distance, $\gamma = \frac{1}{3}$.

Figure 7.3 shows how a periodic voltage may be used to "pump" the diode. The diode is biased to some suitable point on the reverse characteristic, and a sinusoidal signal, termed the pumping signal, superimposed on this. The diode voltage is

$$V = -(V_{\text{bias}} + V_p \sin \omega_p t) \tag{7.4}$$

Since the voltage is now a function of time, the capacitance will also be a function of time and is given by

$$C_p(t) = \frac{C_{j0}}{\left(1 + \dfrac{V_{\text{bias}} + V_p \sin \omega_p t}{\phi}\right)^{\gamma}} \tag{7.5}$$

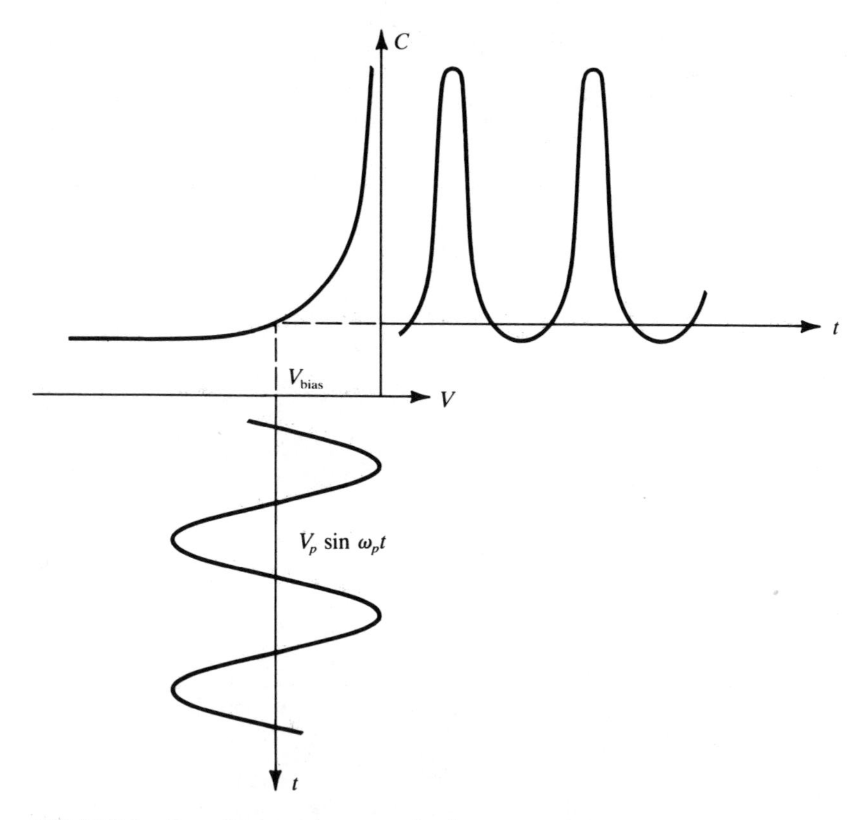

FIGURE 7.3. Capacitance–voltage variation for a pumped diode.

The subscript p is to show that the variation is a result of the pumping voltage. Equation (7.5) describes the rather peaky capacitance–time variation shown in Fig. 7.3. This periodic capacitance variation is made use of in a number of microwave varactor applications to be described. It proves to be more convenient to work in terms of the reciprocal of capacitance, or elastance S; thus

$$S_p(t) = \frac{1}{C_p(t)} \tag{7.6}$$

Thus the equation for elastance is

$$S_p(t) = S_{j0}\left(1 + \frac{V_{\text{bias}} + V_p \sin \omega_p t}{\phi}\right)^{\gamma} \tag{7.7}$$

where $S_{j0} = 1/C_{j0}$.

Equation (7.7) can be expanded in a harmonic series as

$$S_p(t) = S_0 + S_1 \sin \omega_p t + S_2 \sin 2\omega_p t + \cdots \tag{7.8}$$

Here S_0 is the average value of the elastance, S_1 is the peak of the fundamental component, S_2 the peak of the second harmonic component, and so on. In practice, not all components need be considered together because circuits are arranged to tune to particular harmonics and to reject the others. In the parametric amplifier application described in the following section, only the S_0 term and the first harmonic term need be considered.

The equivalent circuit for the varactor diode, including the package, is shown in Fig. 7.4(a). The variable capacitance is shown as $C(t)$, and R_s in series with this is the combined series resistance of the p and n material on either side of the junction. This resistance also varies with applied voltage but the variation is swamped out by the $C(t)$ component and can usually be ignored. Stray capacitance will exist in the vicinity of the junction and this is represented by C_s. The series inductance of the leads is represented by L_s, and C_p is the capacitance of the package. Most of these parasitic elements can be tuned out by external circuitry and the equivalent circuit then reduces to that shown in Fig. 7.4(b). The resistance R_s cannot be removed, and it sets a limit on the device performance by the noise it introduces.

7.3 THE PARAMETRIC AMPLIFIER

The varactor diode may be arranged to present a negative resistance at a desired signal frequency, and the negative resistance in turn may be incorporated in a circuit to provide signal amplification. To show how the

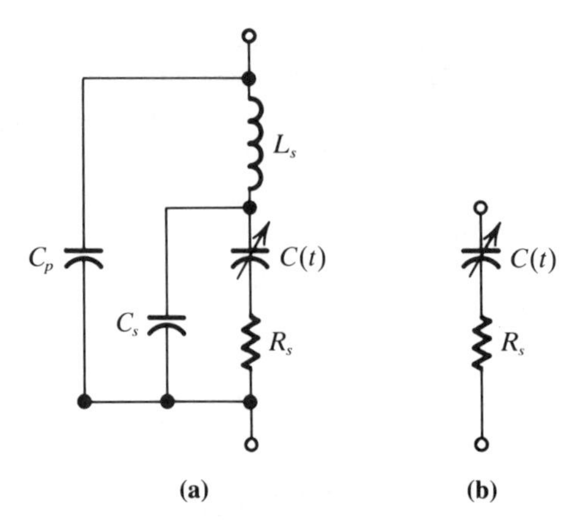

FIGURE 7.4. (a) Equivalent circuit for the varactor diode; (b) simplified equivalent circuit.

negative resistance comes about, the simplified circuit of Fig. 7.5(a) will be analyzed. The diode is being pumped at the pumping frequency f_p, and for clarity the resonant pumping circuit is not shown. An input resonance loop consisting of C_1, $R_1 + R_s$, L_1, and S_0 is formed, where S_0 is the mean value of the varactor elastance. The L_1, C_1, and R_1 components are part of the circuit external to the diode, L_1 and C_1 being used for tuning the loop to resonance at f_s. Resistance R_1 is an unavoidable part of the external circuit, and this is kept as small as possible to minimize losses and noise generation. A second resonant loop is formed consisting of L_2, $R_2 + R_s + R_{iL}$, C_2 and S_0. This loop is resonant at the difference frequency $f_i = f_p - f_s$, known as the idler frequency for the reason explained in Section 7.5. As with the input loop, the components L_2, C_2 and R_2 are external to the diode, and resistance R_2 is kept as small as possible to reduce losses and noise. In some situations it is necessary to use an external load resistance with the idler circuit and this is shown as R_{iL}. It is assumed that the parasitic elements shown in Fig. 7.4(a) are absorbed in the external tuning elements so that the diode can be represented by the simplified equivalent circuit of Fig. 7.4(b).

Here, only the small-signal amplifying action will be considered. The only voltages to the diode are the small-signal voltage to be amplified, and the large-signal pumping voltage. The nonlinear capacitance shown in Fig. 7.3 results in small-signal currents at harmonics of the input frequency, and sum and difference components of the pumping and input frequencies. In order, therefore, to select the desired frequencies, a bandpass filter is included which allows only current at f_s to circulate in the input loop, and a bandpass filter is included which allows only current at f_i to circulate in the idler loop. Filtering is also included in the pumping circuit to restrict the current to the frequency component at f_p.

Because the tuned circuits involve only the first harmonic of the pumping signal, only the first two terms of the trigonometric expansion for $S_p(t)$ need be used, so that

$$S_p(t) = S_0 + S_1 \sin \omega_p t \tag{7.9}$$

It may be assumed that only the signal current flows in the input loop. This current is given by

$$i_s = I_{sm} \sin \omega_s t \tag{7.10}$$

Similarly, in the idler loop, the only current is

$$i_i = I_{im} \sin \omega_i t \tag{7.11}$$

The current flowing in the diode (apart from the pumping current) is

$$i = i_s + i_i \tag{7.12}$$

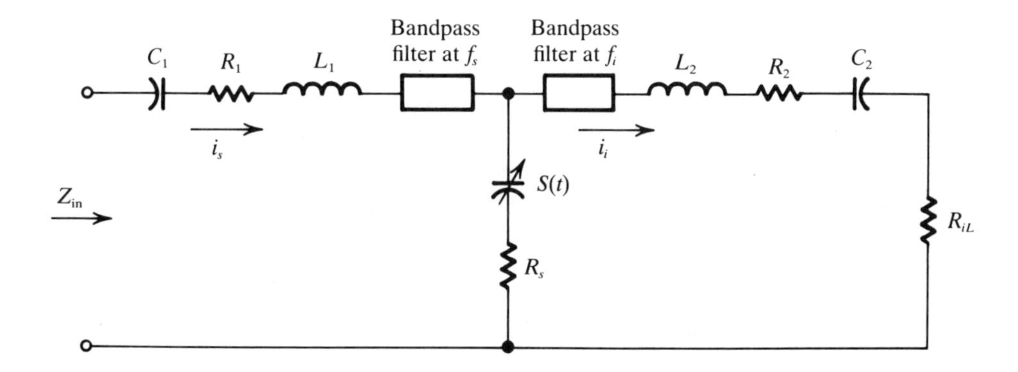

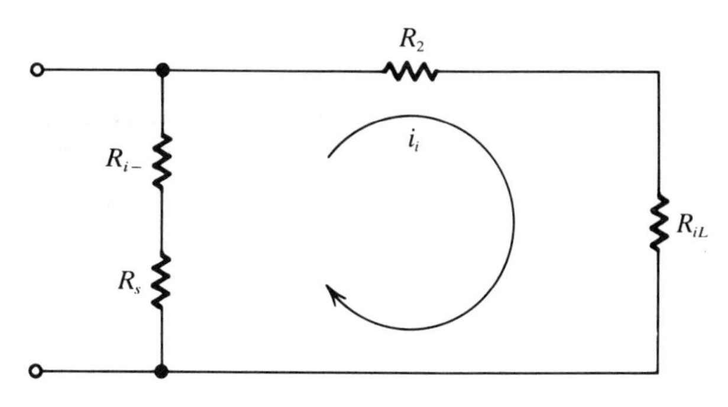

FIGURE 7.5. (a) Parametric amplifier equivalent circuit; (b) "idler" loop.

It is assumed that this is a small-signal current, much smaller than the pumping current, and therefore it does not contribute to the pumping of the diode. It is required to find the signal component of voltage across the diode and to do this, use is made of the relationship $v_c = q/C = qS$, where q is the charge on the capacitor and v_c the voltage across the capacitor. In this case the elastance term is given by Eq. (7.9) and the expression for charge must be obtained by integrating the current, which using Eqs. (7.10), (7.11) and (7.12) gives

$$\begin{aligned} q &= \int i\,dt \\ &= -\left(\frac{I_{sm}}{\omega_s}\cos\omega_s t + \frac{I_{im}}{\omega_i}\cos\omega_i t\right) \end{aligned} \tag{7.13}$$

Therefore, the voltage across the variable capacitive element is

$$v_c = -(S_0 + S_1 \sin\omega_p t)\left(\frac{I_{sm}}{\omega_s}\cos\omega_s t + \frac{I_{im}}{\omega_i}\cos\omega_i t\right) \tag{7.14}$$

This contains the product term $\sin\omega_p t \cos\omega_i t$, which can be expressed, by means of trigonometric identities, as $0.5[\sin(\omega_p + \omega_i)t + \sin(\omega_p - \omega_i)t]$. Because there is no resonant circuit at the sum frequency, the $(\omega_p + \omega_i)$ term can be ignored. The difference frequency $(\omega_p - \omega_i)$ is equal to the signal frequency ω_s and therefore this product term can be replaced by $0.5(\sin\omega_s t)$. Thus the voltage component across the elastance, at the signal frequency, is

$$v_{cs} = -\frac{S_1 I_{im}}{2\omega_i}\sin\omega_s t \tag{7.15}$$

This can be expressed as

$$v_{cs} = -\frac{S_1 I_{im}}{2\omega_i I_{sm}} I_{sm}\sin\omega_s t \tag{7.16}$$

This shows that the voltage component is in antiphase to the current, and therefore the pumped elastance effectively has a negative resistance to the signal frequency. The negative resistance is given by

$$R_- = -\frac{S_1 I_{im}}{2\omega_i I_{sm}} \tag{7.17}$$

This negative resistance is the key to the parametric amplifying action. The pumping action results in the capacitance supplying a signal current from the stored charge, the current being 180° out of phase with the signal voltage. The negative resistance is seen to need the idler current I_{im}. Without this, the negative resistance would be zero.

A negative-resistance component also appears in the idler circuit as a result of the cross-product term $\sin \omega_p t \cos \omega_s t$, and a similar analysis yields, for the negative resistance in the idler loop,

$$R_{i-} = -\frac{S_1 I_{sm}}{2\omega_s I_{im}} \tag{7.18}$$

This equation enables the current ratio to be eliminated from Eq. (7.17). Since there is no externally impressed voltage in the idler loop, then from the circuit of Fig. 7.5(b),

$$i_i(R_s + R_{i-} + R_2 + R_{iL}) = 0 \tag{7.19}$$

Substituting for R_{i-} from Eq. (7.18) and rearranging yields

$$\frac{I_{im}}{I_{sm}} = \frac{S_1}{2\omega_s(R_s + R_2 + R_{iL})} \tag{7.20}$$

This in turn can be substituted in Eq. (7.17) to give

$$R_- = -\frac{S_1^2}{4\omega_i\omega_s(R_s + R_2 + R_{iL})} \tag{7.21}$$

Dynamic Q-factors for the diode are defined as

$$Q_i = \frac{S_1}{2\omega_i R_s} \tag{7.22}$$

$$Q_s = \frac{S_1}{2\omega_s R_s} \tag{7.23}$$

In terms of these Q-factors, which are measurable parameters, the negative resistance component becomes

$$R_- = -Q_i Q_s \frac{R_s^2}{R_s + R_2 + R_{iL}} \tag{7.24}$$

Before taking this analysis any further it is worth looking at some numerical values. As already mentioned, all resistor values are kept as small as possible to minimize losses and reduce noise. Resistance R_s is part of the diode and is the major resistive component. It will be assumed that resistors R_1, R_2, and R_{iL} are all equal to zero. Typical Q values are $Q_i = 2$ and $Q_s = 10$. This gives a negative rsistance of $R_- = -(2 \times 10)R_s = -20R_s$.

The total input resistance of the resonant input loop is

$$R_{\text{in}} = R_s + R_1 + R_- \tag{7.25}$$

The overall resistance must be negative to achieve gain and this requires that $|R_-| > R_s + R_1$. Note that with the external resistors R_1, R_2, and R_{iL} equal to zero, the input resistance becomes

$$R_{\text{in}} = R_s(1 - Q_s Q_i) \tag{7.26}$$

One of the most common forms of amplifier circuit is where the negative resistance is used to load one port of a circulator as shown in Fig. 7.6. This is known as a circulator-type amplifier. The available power from the source, P_{av}, circulates around to the negative resistance port. Here the negative resistance results in a reflection coefficient greater than unity, so that the reflected power is greater than the incident power. The reflected power then circulates around to the next port, where it is absorbed by the load.

The voltage reflection coefficient is given by

$$\Gamma = \frac{r_{\text{in}} - 1}{r_{\text{in}} + 1} \tag{7.27}$$

where $r_{\text{in}} = R_{\text{in}}/Z_0$. Z_0 is the source resistance (assumed purely resistive). This is simply the reflection coefficient as discussed in Section 2.5. The power reflected from the negative resistance port is $|\Gamma|^2 P_{\text{av}}$, and this is the power that reaches the load. The power gain is therefore

$$G = |\Gamma|^2 \tag{7.28}$$

EXAMPLE 7.1

Calculate the power gain for a circulator-type parametric amplifier for which $Q_i = 3$, $Q_s = 10$, $R_s/Z_0 = 0.08$, $R_{iL} = 0$, and R_1 and R_2 may be assumed equal to zero.

SOLUTION From Eq. (7.26),

$$\begin{aligned} r_{\text{in}} &= \frac{R_s}{Z_0}(1 - Q_s Q_i) \\ &= 0.08(1 - 30) \\ &= -2.32 \end{aligned}$$

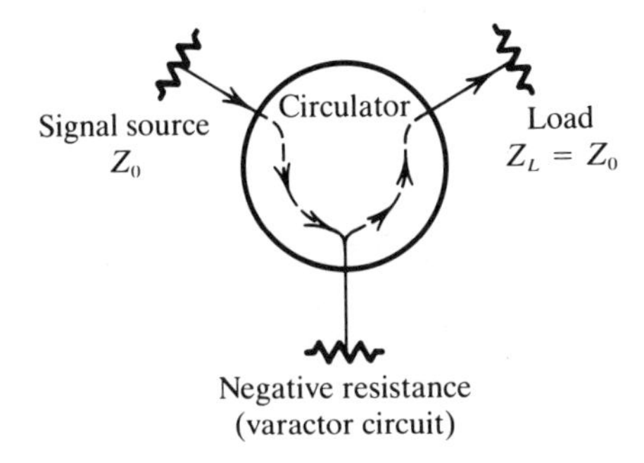

FIGURE 7.6. Negative resistance circulator-type amplifier.

From Eq. (7.27),

$$\Gamma = \frac{-2.32 - 1}{-2.32 + 1}$$

$$= 2.515$$

$$\therefore G = |2.515|^2 = 6.33$$

In decibels this is 10 log 6.33 = **8.01 dB**.

The parametric amplifier is a very low noise amplifier, which is the reason it is used at microwaves. To reduce noise further, the amplifier may be cooled, and Fig. 7.7(a) shows the block diagram of a 4-GHz two-stage amplifier. The input signal, at 4.17 GHz, is fed into port 1 of the first circulator. The diode is connected to port 2, and the pump frequency is 23 GHz. The output from port 3 is fed to the second stage, which is similar to the first except that it is not cooled. The first stage is cooled to liquid-nitrogen temperature, as shown in Fig. 7.7(a), to maintain a low-noise temperature. A cross-sectional view of one of the amplifier stages is shown in Fig. 7.7(b).

7.4 SMALL-SIGNAL UP-CONVERTER

As shown in Section 7.3, the input-signal circuit introduces a negative resistance into the idler circuit. This reduces the effective resistance of the idler circuit, and therefore a greater current can flow than would otherwise be the case. Physically, power is transferred from the pumping source to the idler circuit, at a frequency $f_i = f_p - f_s$. By taking power from the

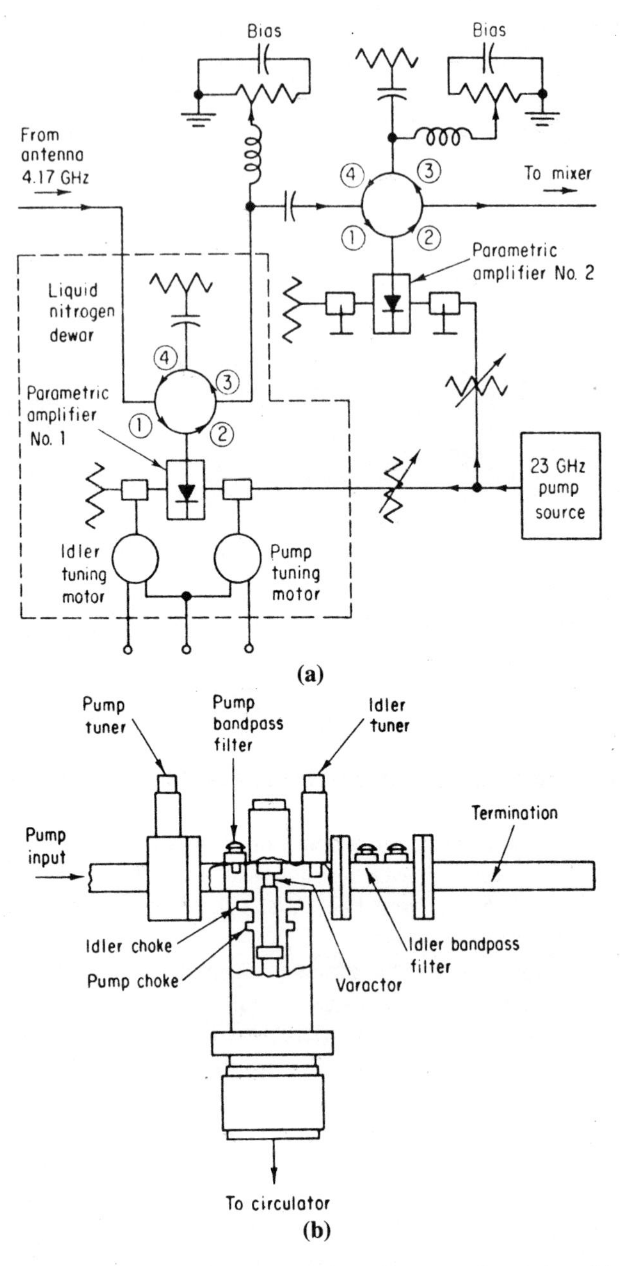

FIGURE 7.7. (a) Block diagram of 4-GHz parametric receiver; (b) cross-sectional view of 4-GHz parametric amplifier. (*From M. Uenohara and J. W. Gewartowski, "Varactor Applications," in Microwave Semiconductor Devices and Their Circuit Applications, ed. H. A. Watson, (New York: McGraw-Hill, © 1969); with permission.*)

idler circuit (in which case it is, of course, no longer an "idler" circuit, but the notation f_i will be retained), it is possible to achieve a power gain, that is, the power into the idler load R_{iL} is greater than the power supplied from the signal source. The increase in power to the load comes from the pumping source. Since $f_i = f_p - f_s$, f_i in this case is known as a lower sideband, and when f_i is greater than f_s the amplifier is termed a lower-sideband up-converter. Such a converter is capable of power gain with low noise. Thus the arrangement shown in Fig. 7.8 may be used to achieve front-end gain with a low-noise figure. It will be seen that the input signal at 961 MHz is up-converted to 10,769 MHz, and this is strictly to take advantage of the low-noise amplification provided by the up-conversion process. The up-converter is immediately followed by a down-converter, which converts the signal to a 30-MHz intermediate frequency, but the noise figure is determined almost entirely by the up-converter. It is also possible to tune the idler circuit to the frequency $f_p + f_s$, in which case the circuit is known as an upper-sideband up-converter. The detailed operation of this circuit differs somewhat from that of the lower-sideband circuit. Both circuits are used in practice, but it is found that the upper-sideband up-converter is useful only at input frequencies below about 100 MHz.

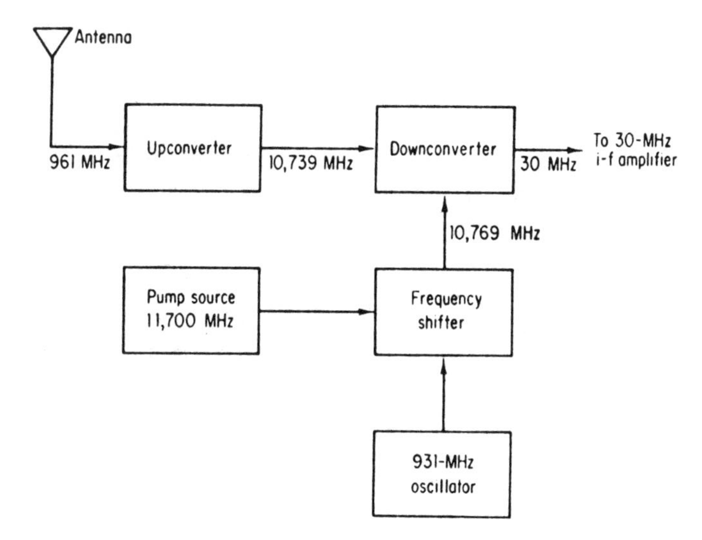

FIGURE 7.8. Receiver front end with 1.6-dB noise figure. (*From M. Uenohara and J. W. Gewartowski, "Varactor Applications," in Microwave Semiconductor Devices and Their Circuit Applications, ed. H. A. Watson (New York: McGraw-Hill, © 1969); with permission.*)

7.5 FREQUENCY MULTIPLIERS

The varactor diode makes an almost ideal frequency multiplier. Because the nonlinearity that generates the harmonic currents is capacitive, very little power loss occurs and the multiplier circuit operates at high efficiency. In practice, efficiencies of the order of 80% can be achieved, as there are losses in the diode resistance R_s as well as in the parasitic resistances associated with the external tuning circuits. The usefulness of frequency multipliers at microwave frequencies arises because highly stable, lower-frequency signal sources are readily available, and the varactor multiplier enables the frequency to be increased at high efficiency without adding excessive noise.

The actual mixing process in the diode is complicated. The input signal pumps the diode, and in the process the second harmonic is generated. This second harmonic then mixes with the input frequency to produce the third harmonic. The third harmonic in turn can mix with the fundamental to produce the fourth harmonic, and so on. Thus the process is one of frequency doubling and mixing, and provision must be made to allow the second harmonic current to flow as well as the fundamental component and the desired output component. As an example of the terminology used, a 1–2–3 tripler means that three resonant circuits are used, one at the fundamental, one at the second harmonic, and one at the third harmonic, which is the output. Because the second harmonic circuit is used only as an intermediate stage between fundamental and desired output, it is termed an *idler stage*. As a second example, a 1–2–3–4 quadrupler would have two idler circuits, one at the second harmonic and one at the third harmonic.

Figure 7.9 shows how varactor multipliers may be used to generate a microwave output from a low-frequency input. The crystal controlled generator is at 125 MHz. This is fed through a buffer amplifier to the input of a transistor doubler, which in turn feeds another transistor doubler. The output from this transistor stage, at 500 MHz, is fed to a varactor tripler, which increases the frequency to 1500 MHz, and this is fed to a varactor quadrupler, which produces the final output of 6000 MHz. It will be seen that the transfer efficiency of the quadrupler is 2.5/6.3 = 39.68% and the transfer efficiency of the tripler is 6.3/8.6 = 73.25%. The transistor doublers actually provide signal gain, but of course they require a dc input power. As shown, the total dc input is 34 W for the complete circuit and therefore the conversion efficiency is 2.5/34 = 7.4%.

Figure 7.10 shows in cross section the constructional features of a 1–2–3 tripler. The input signal at 4 GHz is fed from a coaxial cable through the sidewall of a rectangular waveguide to the varactor diode. The inner conductor of the cable is continued to form a tuning stub, which together with a tuning screw resonates at the second harmonic and therefore forms

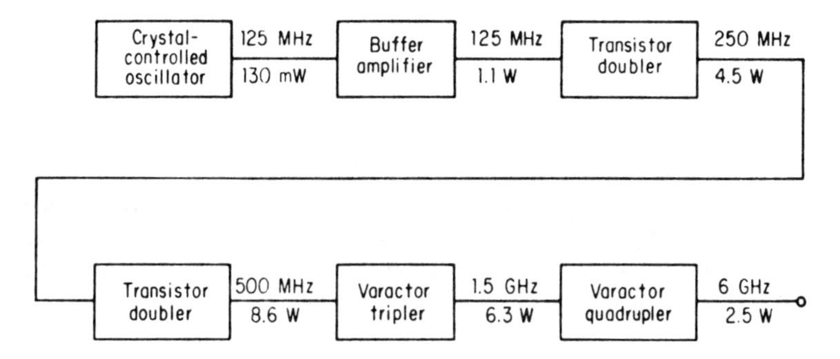

FIGURE 7.9. Six-gigahertz solid-state source. (*From M. Uenohara and J. W. Gewartowski, "Varactor Applications," in Microwave Semiconductor Devices and Their Circuit Applications, ed. H. A. Watson, (New York: McGraw-Hill, © 1969); with permission.*)

the idler circuit. The waveguide TE_{10} mode cutoff frequency is 9.486 GHz; therefore, neither the fundamental at 4 GHz nor the idler frequency at 8 GHz can propagate down the guide. The 12-GHz harmonic is propagated and is available at the output port of the guide. The other end of the guide is terminated in a movable short circuit which allows for fine tuning of the

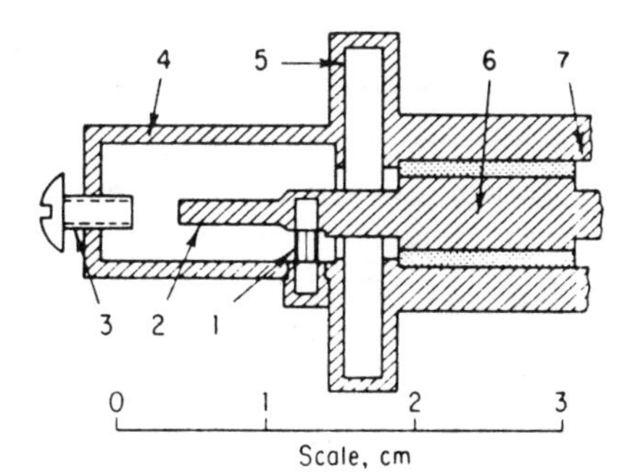

FIGURE 7.10. Sectional view of a 4- to 12-GHz tripler. 1, Varactor diode; 2, transverse stub resonant at 8 GHz; 3, idler tuning screw; 4, 12-GHz output waveguide, 1.58 × 0.79 mm; 5, 12-GHz radial rejection choke on input coaxial line; 6, 4-GHz matching transformer, supported by dielectric sleeve; 7, 50-Ω input coaxial line. (*From M. Uenohara and J. W. Gewartowski, "Varactor Applications," in Microwave Semiconductor Devices and Their Circuit Applications, ed. H. A. Watson (New York: McGraw-Hill, © 1969); with permission.*)

output. A 12-GHz radial choke is placed around the coaxial input to prevent the 12-GHz signal from being fed back along the cable. The input section of the coaxial cable at the waveguide is designed as a matching transformer section to provide matching to the 4-GHz source.

The *step recovery diode* (also known as a *snapback diode*) is a special type of varactor diode used for frequency multiplication. In the step recovery diode, the capacitance under forward bias is utilized, whereas with the varactor diodes described previously, the reverse bias capacitance is the important element. Under forward bias minority carriers are injected into the depletion region. The step recovery diode is designed so that minimum recombination of minority carriers occur, and the carriers remain close enough to the junction so that they are withdrawn when the pump voltage swings into reverse bias. The signal voltage is also the pump voltage in this application. While the stored charge is being withdrawn a comparatively large reverse current flows, but this suddenly ceases when all the charge is withdrawn. This current transient contains harmonics of the input signal, and the output may be tuned to the desired harmonic.

The terms *step recovery* and *snapback* refer to the withdrawal action of the stored charge. Step recovery diode circuits are simpler than the normal varactor circuits since no idler circuits are required, and they can provide high-frequency multiplication factors. The power conversion efficiency of the step recovery diode tends to be less than that of the normal varactor diode. A Ku-band, ×8 multiplier constructed in a waveguide assembly is shown in Fig. 7.11(a), and a simplified schematic in Fig. 7.11(b). The input, at 2 GHz, is fed to the diode through a coaxial input. The length of the lower section of coaxial line can be adjusted by means of the spacers shown, for maximum power output. The movable waveguide short circuit is for matching purposes, and a tapered section of waveguide couples the diode to the output filter. This utilizes two $\lambda/2$ resonator sections which can be fine-tuned by means of the tuning screws shown, and the filter has a bandwidth of 500 MHz at a center frequency of 16 GHz.

7.6 THE PIN DIODE

The letters P-I-N stand for *p*-type, *i*ntrinsic, and *n*-type material and describe a special kind of diode constructed in silicon. Schematically, such a diode may be represented as shown in Fig. 7.12(a). At dc and low frequencies, the intrinsic region has little effect, and the diode behaves as a normal *p-n* junction diode. Under forward bias, a relatively large current flows. Mobile charges from the *p*- and *n*-regions are injected into the intrinsic layer, and this becomes highly conductive. Under reverse bias, the intrinsic layer becomes part of the depletion region, which reduces current flow to practically zero. If a low-frequency signal is applied, such

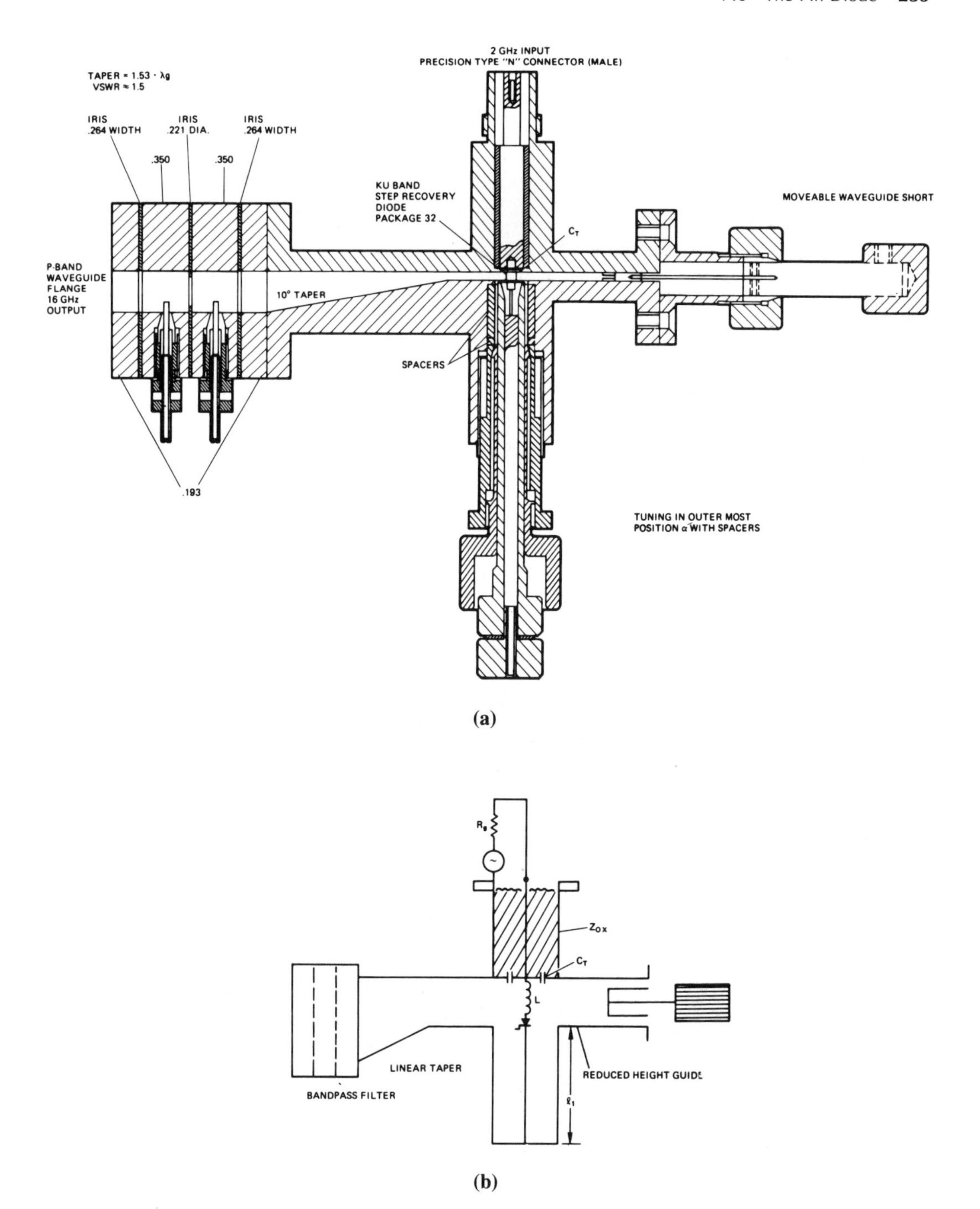

FIGURE 7.11. (a) X8, Ku-band waveguide multiplier; (b) schematic diagram, Ku-band, X8 multiplier. (*Courtesy of Hewlett-Packard, AN-928.*)

(a)

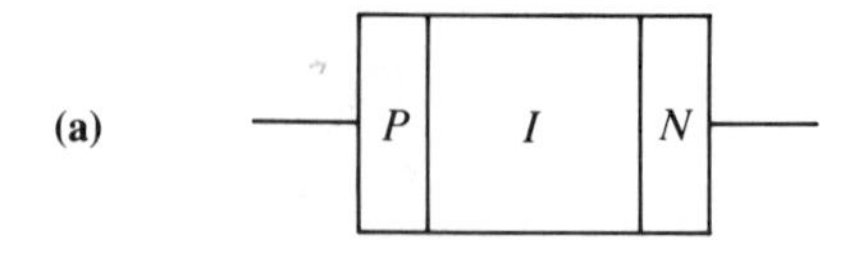

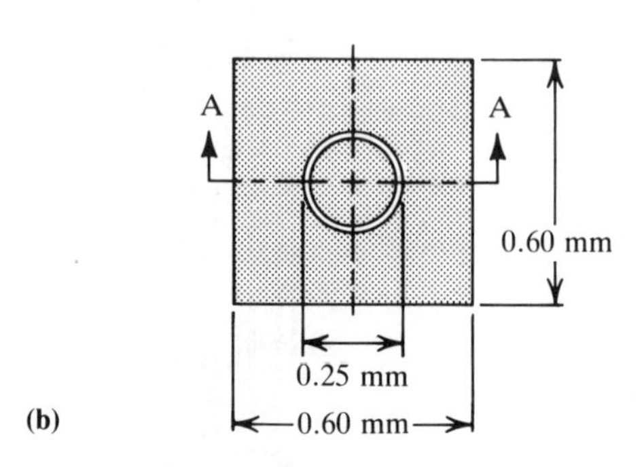

(b)

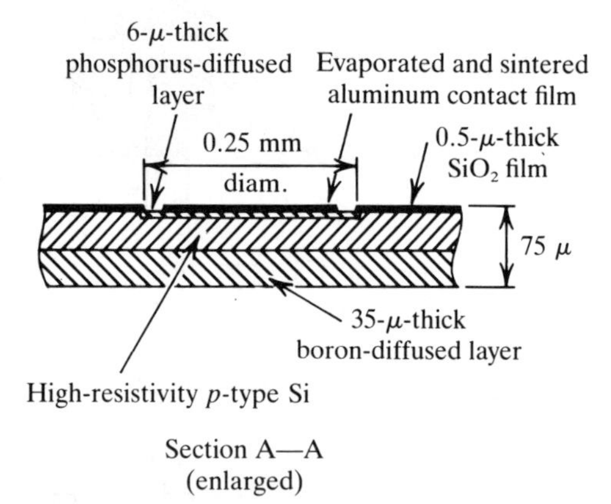

FIGURE 7.12. (a) The PIN diode; (b) constructional details. [*Part (b) from H. M. Olson, "p-i-n Diodes," in Microwave Semiconductor Devices and Their Circuit Applications, ed. H. A. Watson, (New York: McGraw-Hill, © 1969); with permission.*]

that the periodic time of the signal is much greater than the time taken for the charge to "fill and empty" the intrinsic layer, the low-frequency signal will be rectified in the normal way. Furthermore, if a low-frequency small signal is superimposed on dc forward bias, it will modulate the conductivity of the intrinsic layer.

A high-frequency signal, such that the periodic time of the signal is very much less than the "filling and emptying" time of the intrinsic layer, will have a negligible effect on the intrinsic layer. This is because of the very short distance traveled in the intrinsic layer by mobile charges under the influence of the high-frequency signal. A useful measure of the separation between low- and high-frequency modes of operation is provided by the recombination lifetime, τ. This is a measure of the time it takes the intrinsic layer to return to its high-resistivity intrinsic state once forward bias has been removed. Values for τ are usually specified on data sheets, and may range from a nanosecond to several microseconds. A frequency may be defined as

$$f_\tau = \frac{1}{2\pi\tau} \tag{7.29}$$

At frequencies well below f_τ, the diode behaves as a normal *p-n* junction diode, and at frequencies well above f_τ, the diode behaves as a simple linear network the parameters of which are determined by the bias conditions. This is the key to understanding the usefulness of the PIN diode. The bias may be switched from forward to reverse, changing the impedance seen at microwaves from a very low to a very high value. Also, the bias may be varied by a low-frequency signal, and this will result in modulation of the microwave signal. The constructional features of a typical planar PIN diode are shown in Fig. 7.12(b). In a practical diode, the intrinsic layer will receive some doping from either the *n* or the *p* doping processes. If it is slightly *p*-type it is referred to as a π-region, and slightly *n*-type, as a ν-region (ν is the Greek lowercase letter nu).

The equivalent circuit for a PIN diode and package, under three different bias conditions, is shown in Fig. 7.13. In all cases, the constant elements are L_s, the lead inductance; C_p, the package capacitance; C_f, the fringing capacitance around the edge of the junction to case; and R_s, the series resistance of the diode, mainly in the *p* and *n* contact regions. Typical values, obtained from the chapter by H. M. Olson in Watson (1969), are shown.

Under a 50-V reverse bias, the circuit is as shown in Fig. 7.13(a). Here any residual carriers in the intrinsic region are swept out by the reverse bias, and the intrinsic layer is absorbed in the depletion region, which is represented by the comparatively high resistor R_j in parallel with the depletion capacitance C_j. At zero bias [Fig. 7.13(b)], the intrinsic layer will be only partially swept out; that is, it will be either a π-region or a ν-region, as already mentioned. The effect of this can be represented by the parallel combination of R_i and C_i as shown. Note that this parallel branch is in series with the depletion-region equivalent circuit of $R_j // C_j$. The total reactance of these two branches is practically constant, independent of bias. This means that the PIN diode presents an almost constant capacitance under reverse bias, in contrast to the varactor diode, which presents a bias-

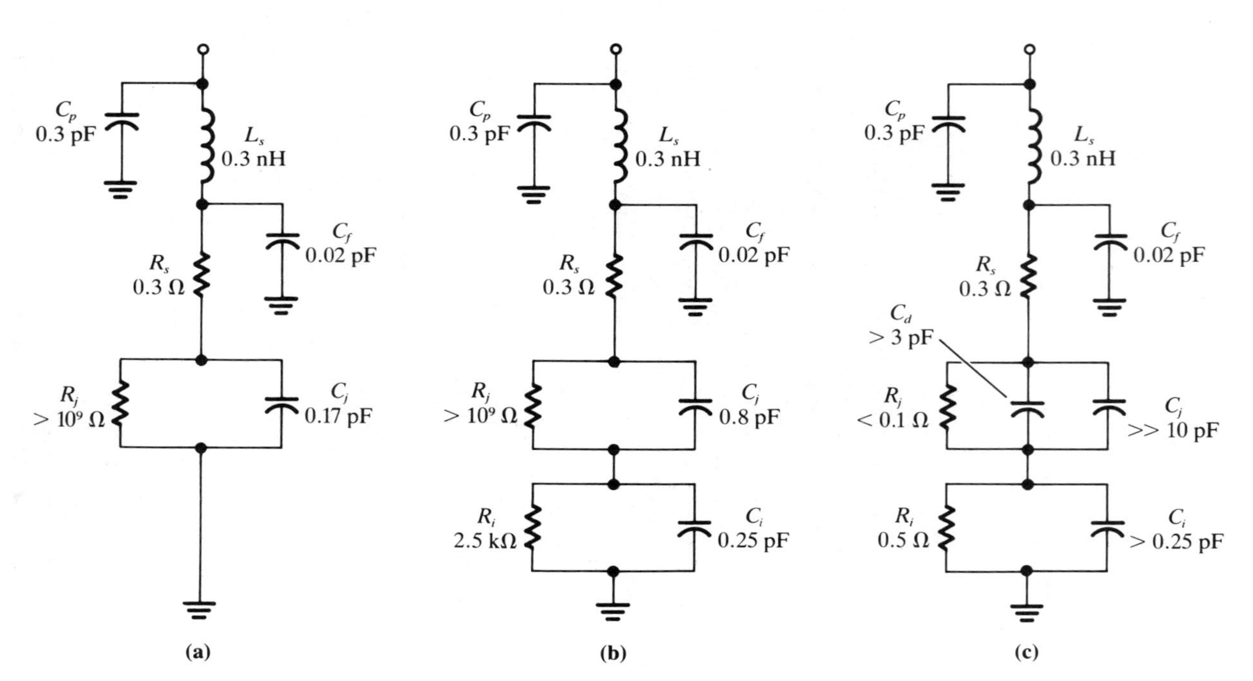

FIGURE 7.13. Equivalent circuit for a PIN diode: (a) reverse bias; (b) zero bias; (c) forward bias.

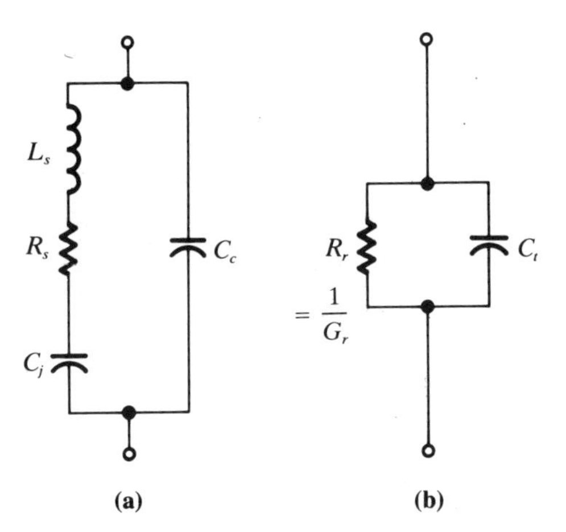

FIGURE 7.14. Simplified equivalent circuits for the reverse-biased PIN diode: (a) first approximation; (b) second approximation.

dependent capacitance. Under forward bias [Fig. 7.13(c)] the resistance of the depletion region is reduced to almost zero, as is normal for a *p-n* junction, and the intrinsic layer resistance is also reduced to a low value because of the presence of the injected carriers. In addition, a diffusion capacitance C_d, occurs under forward bias, which is in parallel with the depletion capacitance, which itself is increased considerably. Typical values are shown in Fig. 7.13 for these elements.

From the point of view of circuit applications, the equivalent circuit can be considerably simplified. The capacitances C_p and C_f may be combined without much error into the one-case capacitance C_c. Under reverse bias the resistance R_j is very large. Since it is in parallel with capacitance C_j, it may be ignored. The diode is normally operated with sufficient bias to sweep out the intrinsic layer. Taking all these factors into account, the reverse-bias equivalent circuit of Fig. 7.13 reduces to that shown in Fig. 7.14(a). The operating microwave frequency is normally chosen to be much lower than the series resonant frequency of the L_s, C_j branch, and therefore the inductive reactance of L_s can be neglected. The admittance of the circuit of Fig. 7.14(a) is therefore

$$\begin{aligned} Y &\simeq \frac{1}{R_s + 1/j\omega C_j} + j\omega C_c \\ &= \frac{j\omega C_j}{1 + j\omega C_j R_s} + j\omega C_c \\ &= \frac{j\omega C_j(1 - j\omega C_j R_s)}{1 + (\omega C_j R_s)^2} + j\omega C_c \end{aligned} \tag{7.30}$$

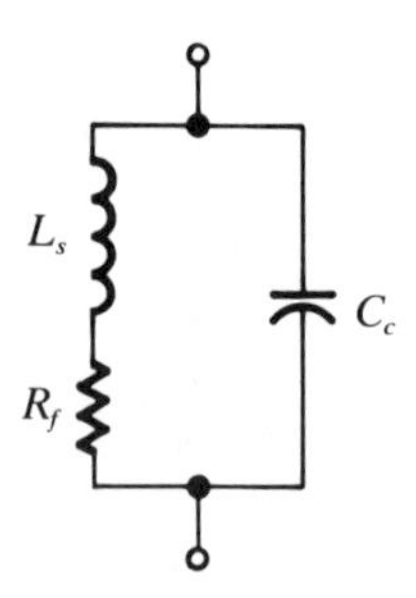

FIGURE 7.15. Simplified equivalent circuit for the PIN diode under forward bias.

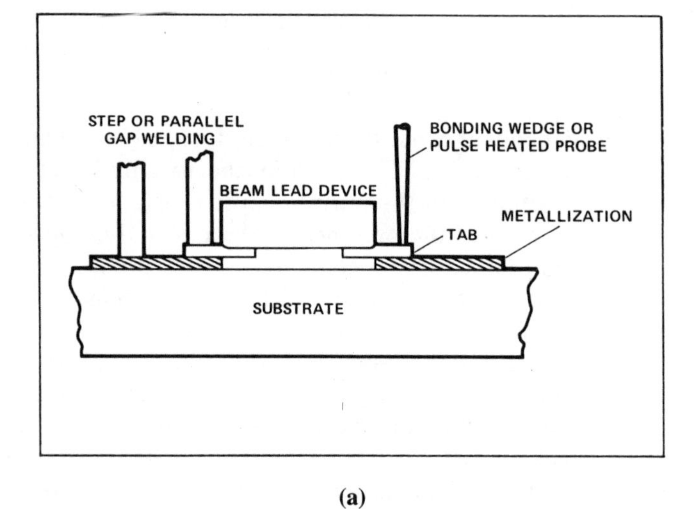

(a)

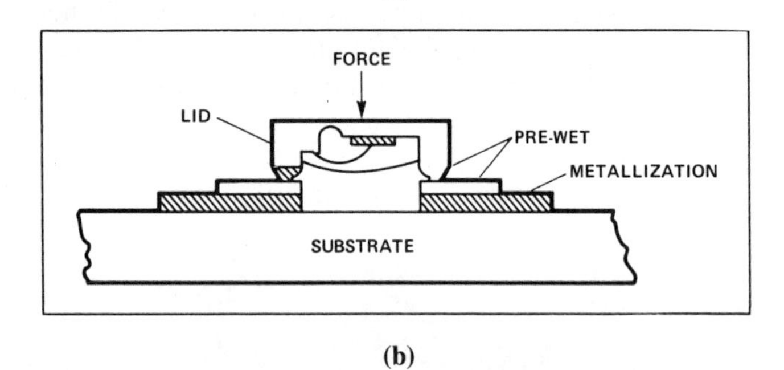

(b)

FIGURE 7.16. (a) LID mounting is the easiest type to perform in hybrid microcircuitry. It can generally be accomplished with a single solder operation. (b) Beam lead attachment requires thermocompression bonding to the substrate metallization. (*Courtesy of Hewlett-Packard, AN-940.*)

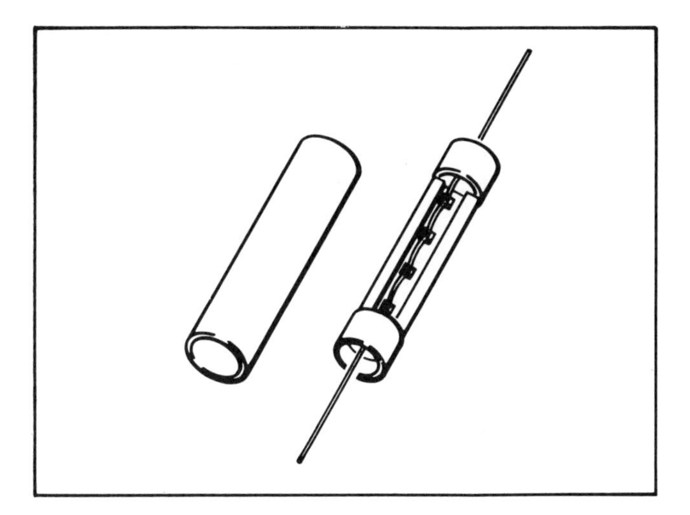

FIGURE 7.17. Cylindrical module construction. (*Courtesy of Hewlett-Packard, AN-932.*)

In practice, R_s is made as small as possible to reduce losses and the further approximation, $(\omega C_j R_s)^2 << 1$ is valid. Therefore, the admittance becomes

$$\begin{aligned} Y &\simeq j\omega C_j(1 - j\omega C_j R_s) + j\omega C_c \\ &= G_r + j\omega C_t \end{aligned} \tag{7.31}$$

where $C_t = C_j + C_c$ and $G_r = \omega^2 C_j^2 R_s$. Thus the simplified equivalent circuit for the PIN diode under reverse bias is as shown in Fig. 7.14(b).

Under forward bias the shunting effect of the resistors completely swamp the capacitive elements and the equivalent circuit reduces to that of Fig. 7.15. Here R_f is the total effective forward resistance, which includes R_s, R_j, and R_i.

The characteristics of the diode depend greatly on the type of construction and packaging. Figure 7.16(a) shows one arrangement in which the diode is mounted in a carrier which is then inverted and soldered into place. This is referred to as a *leadless inverted device* (LID). Figure 7.16(b) shows the beam-lead mounting arrangement. Diode modules are also available in which a number of diodes are mounted in a common housing. The housing is carefully designed to present a constant characteristic impedance. Figure 7.17 shows a 50-Ω coaxial module and Fig. 7.18 shows how this may be fitted into a rectangular waveguide. Rectangular modules are also available as shown in Fig. 7.18. The rectangular modules are particularly suitable for stripline, as they are designed to fit into recesses in the stripline assemblies. Figure 7.19 shows a SPDT switch module mounted in microstrip.

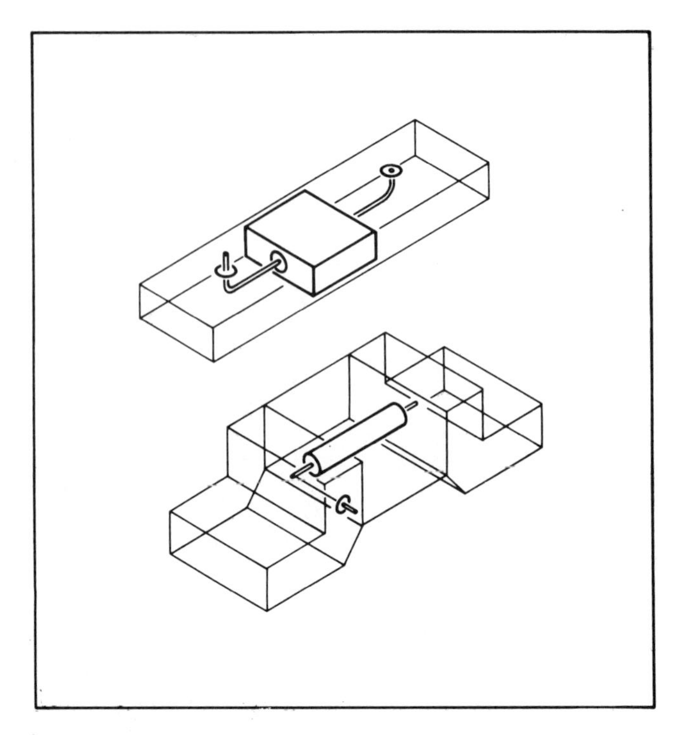

FIGURE 7.18. Waveguide switches using modules. (*Courtesy of Hewlett-Packard, AN-932.*)

7.7 PIN-DIODE REFLECTIVE ATTENUATORS AND SWITCHES

The PIN diode can be inserted between a signal source and a load to provide controlled attenuation. With transmission lines the diode may be connected either in series or in parallel, whereas with waveguides it has to be connected in parallel. When connected in parallel, the diode presents a high impedance in the reverse-bias state and allows the signal to pass to the load with little attenuation. With forward bias, the diode impedance in parallel with the line is low, and most of the signal energy is reflected back along the line to be absorbed by the source. The diode itself does not dissipate much power, and thus this relatively low powered device can control the flow of much larger amounts of power through its effect on the reflection coefficient. A similar situation arises with the series connection except that in the forward-bias or low-impedance state, most of the signal power is allowed to flow on to the load, and in the reverse bias or high-impedance state, the power flow is reflected back to the source. To illustrate the action further, the parallel connection will be analyzed in detail.

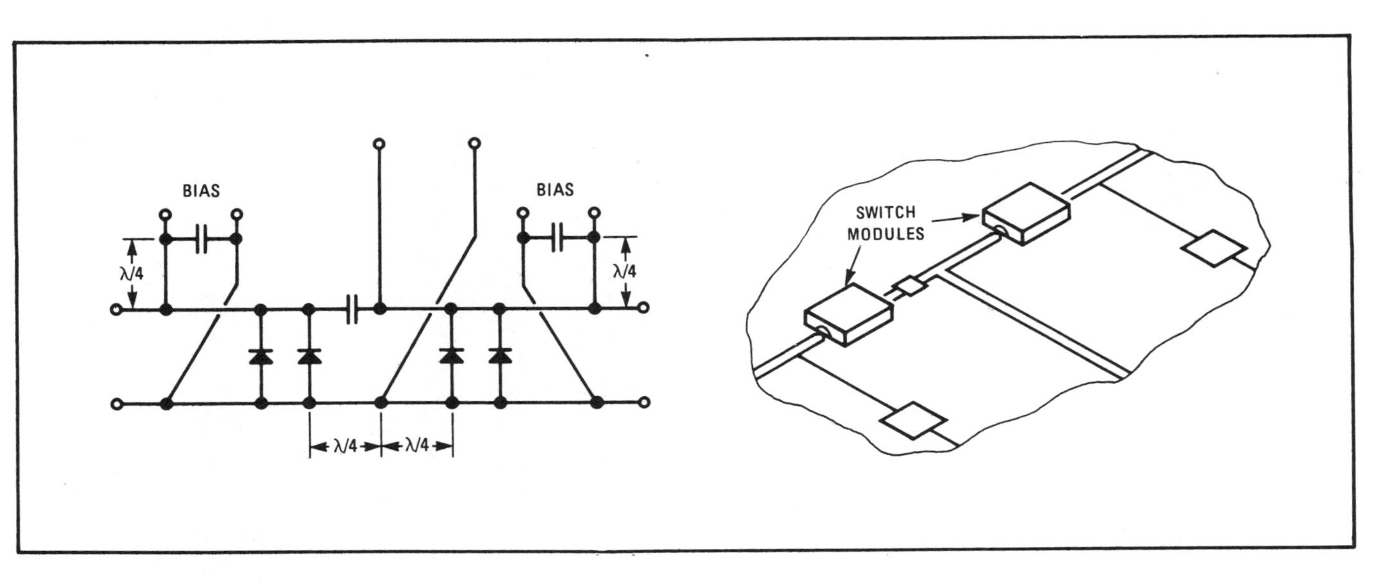

FIGURE 7.19. Shunt-mode SPDT switch. (*Courtesy of Hewlett-Packard, AN-932.*)

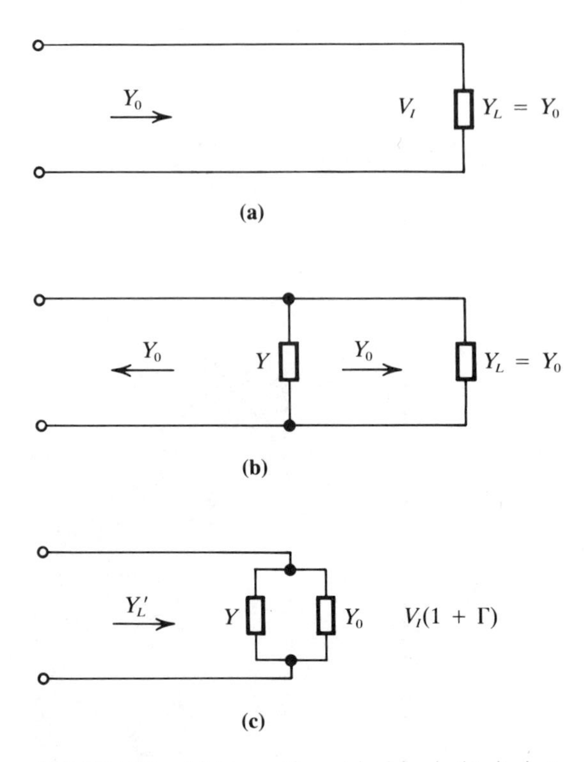

FIGURE 7.20. (a) Line with matched load; (b) diode in parallel with line; (c) equivalent circuit of part (b).

Figure 7.20(a) shows a transmission line with a matched load, where Y_0 is the characteristic admittance of the line (assumed purely resistive) and Y_L is the admittance of the load. The power delivered to the load is

$$\begin{aligned} P_{L0} &= |V_I|^2 Y_L \\ &= |V_I|^2 Y_0 \end{aligned} \tag{7.32}$$

The latter equation may be written since the load matches the line. Figure 7.20(b) shows the line with the diode connected in parallel. Use of admittances instead of impedances facilitates the analysis of parallel-connected circuits. Because the line to the right of the diode is matched, the circuit can be simplified to that shown at Fig. 7.20(c). The load voltage will be as given by Eq. (1.56) and is

$$V_L = V_I(1 + \Gamma) \tag{7.33}$$

Hence the power delivered to the load becomes

$$\begin{aligned} P_L &= |V_L|^2 Y_0 \\ &= |V_I|^2 |1 + \Gamma|^2 Y_0 \end{aligned} \tag{7.34}$$

It is assumed that the incident voltage V_I is not attenuated over the length of the line between diode and load.

The insertion loss is the ratio of the power delivered to the load with the diode in circuit to that delivered when the diode is out of circuit. In decibels the insertion loss is

$$\mathrm{IL(dB)} = -10 \log \frac{P_L}{P_{L0}} \tag{7.35}$$

The negative sign is used so that the loss in decibels appears as a positive number. Substituting for P_{L0} and P_L gives

$$\mathrm{IL(dB)} = -10 \log |1 + \Gamma|^2 \tag{7.36}$$

From Fig. 7.20(c), the effective load admittance seen by the line is

$$Y_L' = Y + Y_0 \tag{7.37}$$

This normalized to Y_0 becomes

$$y_L' = y + 1 \tag{7.38}$$

From Eq. (3.13) the voltage reflection coefficient in terms of normalized admittance is

$$\begin{aligned} \Gamma &= \frac{1 - y_L'}{1 + y_L'} \\ &= -\frac{y}{2 + y} \end{aligned} \tag{7.39}$$

Hence

$$1 + \Gamma = \frac{1}{1 + y/2} \tag{7.40}$$

Substituting this back into Eq. (7.36) gives for the insertion loss in decibels:

$$\mathrm{IL(dB)} = 10 \log \left| 1 + \frac{y}{2} \right|^2 \tag{7.41}$$

The two specific conditions for the diode can now be evaluated.

Low insertion loss (reverse bias). Equation (7.31) shows that the admittance of the diode under reverse bias is of the form $Y = G_r + jB$, and normalized, this becomes

$$y = g_r + jb \tag{7.42}$$

Hence, using this in Eq. (7.41), the insertion loss in decibels becomes

$$\text{IL(dB)} = 10 \log \left[\left(1 + \frac{g_r}{2}\right)^2 + \left(\frac{b}{2}\right)^2 \right] \tag{7.43}$$

In this case the insertion loss should be as low as possible, and this is achieved by making the susceptance b equal to zero. To do this, an inductance L_p must be added in parallel with C_t of the diode for parallel resonance, as shown in Fig. 7.21(a). The insertion loss for the resonant condition becomes

$$\text{IL(dB)} = 20 \log \left(1 + \frac{g_r}{2}\right) \tag{7.44}$$

High insertion loss (forward bias). Under forward bias the diode

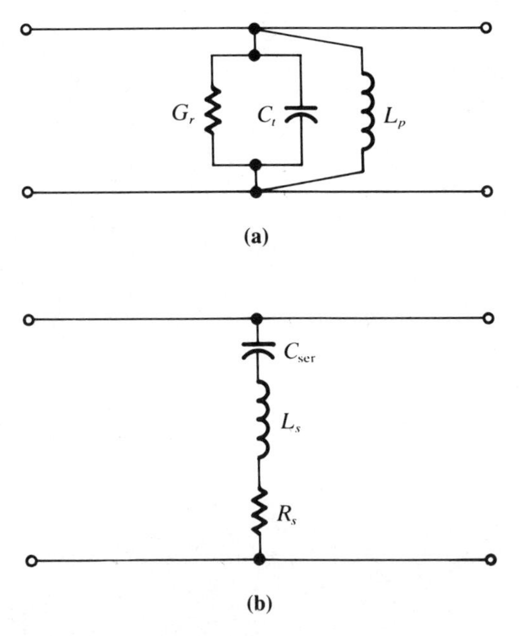

FIGURE 7.21. (a) Addition of inductance L_p to resonate with C_t; (b) addition of capacitance C_{ser} to resonate with L_s.

presents an impedance of the form $Z = R_f + jX$, where X is mainly the reactance of the lead inductance L_s, as shown in Fig. 7.21(b). This assumes that the microwave frequency is well below the parallel resonance of the $L_s C_c$ combination. The impedance normalized to Z_0 is $r_f + jx$, and therefore the normalized admittance is

$$y = \frac{1}{r_f + jx} = \frac{r_f}{r_f^2 + x^2} - j\frac{x}{r_f^2 + x^2} \tag{7.45}$$

In Eq. (7.43), g_r can be replaced with $r_f/(r_f^2 + x^2)$ and b with $-x/(r_f^2 + x^2)$.The insertion loss under these conditions is referred to as the isolation, and this becomes

$$(\text{Isolation})\,\text{dB} = 10\log\left[\left(1 + \frac{r_f}{2(r_f^2 + x^2)}\right)^2 + \left(\frac{x}{2(r_f^2 + x^2)}\right)^2\right] \tag{7.46}$$

This expression is a maximum when $x = 0$, which requires that the series lead inductance be tuned to series resonance by addition of a capacitor as shown in Fig. 7.21(b). The insertion loss then becomes:

$$\text{isolation(dB)} = 20\log\left(1 + \frac{1}{2r_f}\right) \tag{7.47}$$

Figure 7.22 shows how both conditions may be combined. In practice the series capacitor C_{ser} is large and has a negligible effect on the reverse-bias resonance. It also acts as a dc blocking capacitor, which prevents the bias from entering the transmission line.

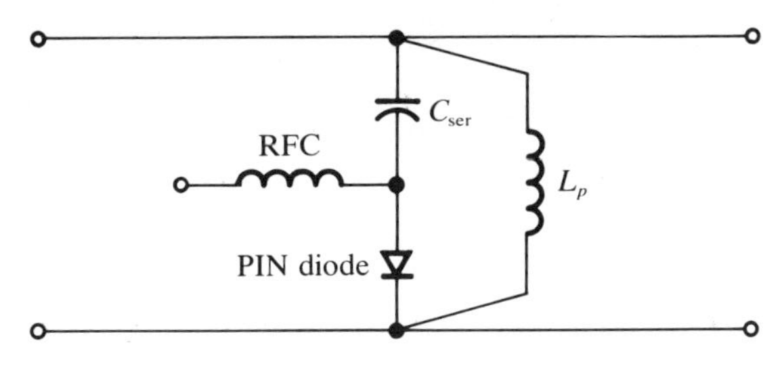

FIGURE 7.22. Combined circuit for L_p and C_{ser}.

EXAMPLE 7.2

A PIN diode has the following characteristics: $L_s = 0.3$ nH, $C_j = 0.5$ pF, $C_c = 0.3$ pF, $R_s = 1.5\ \Omega$, and $R_f = 1\ \Omega$. The diode is to be used as a parallel-connected attenuator in a 50-Ω line system at a frequency of 1.5 GHz. Calculate the values of L_p and C_{ser} needed, the insertion loss, and the isolation.

SOLUTION

$$C_{ser} = \frac{1}{(2\pi f)^2 L_s}$$

$$= \frac{1}{(2\pi \times 1.5 \times 10^9)^2 \times 0.3 \times 10^{-9}}$$

$$= \mathbf{37.53\ pF}$$

$$L_p = \frac{1}{(2\pi f)^2 C_b}$$

$$= \frac{1}{(2\pi \times 1.5 \times 10^9)^2 \times 0.8 \times 10^{-12}}$$

$$= \mathbf{14.07\ nH}$$

From Eq. (7.31),

$$G_r = (2\pi \times 1.5 \times 10^9 \times 0.5 \times 10^{-12})^2 \times 1.5$$

$$= 33.31\ \mu S$$

The normalized value is

$$g_r = 50 G_r$$

$$= 1.665 \times 10^{-3}$$

From Eq. (7.44),

$$\text{IL(dB)} = 20 \log \left(1 + \frac{0.001665}{2}\right)$$

$$= \mathbf{0.007\ dB}$$

Also, $r_f = 1/50 = 0.02$, and from Eq. (7.47),

$$\text{isolation(dB)} = 20 \log \left(1 + \frac{1}{0.04}\right)$$

$$= \mathbf{28.3\ dB}$$

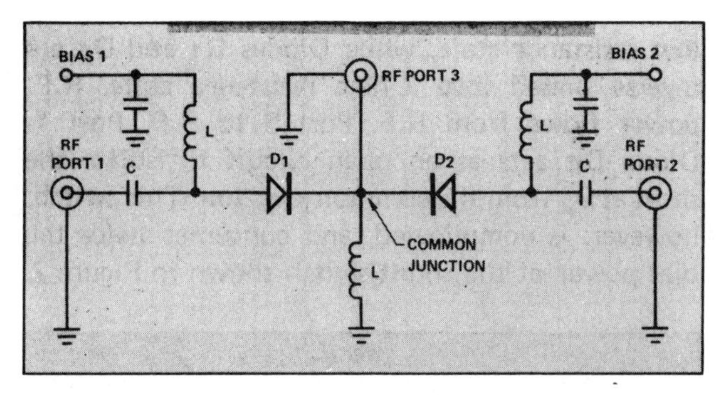

FIGURE 7.23. Series PIN SPDT switch. (*Courtesy of Hewlett-Packard, AN-957-1.*)

Figures 7.23 through 7.25 show how PIN diodes may be used in a variety of switching arrangements. Here SPDT stands for single-pole double-throw, and Fig. 7.23 shows a series PIN SPDT switch. The *LC* filters in the bias lines prevent the microwave signal from entering the bias lines, and capacitors *C* block the dc from the RF ports 1 and 2. The inductance *L* connected to the common diode point provides a dc return for the bias but presents a high impedance to the microwave signal. When diode D_1 is forward biased, diode D_2 is reverse biased and therefore RF port 1 is connected to RF port 3 while RF port 2 is isolated. Reversing the bias, switches port 2 to port 3 and isolates port 1. This particular circuit arrangement is well suited to beam lead diodes (see Fig. 7.16) in microstrip construction, but presents some difficulties in stripline construction with packaged diodes. Also, the parasitic capacitances tend to give poor isolation at the higher microwave frequencies, the isolation showing a 6-dB rolloff with frequency.

Figure 7.24 shows a shunt PIN diode SPDT switch which is relatively easy to construct in stripline with packaged diodes. By transforming a short

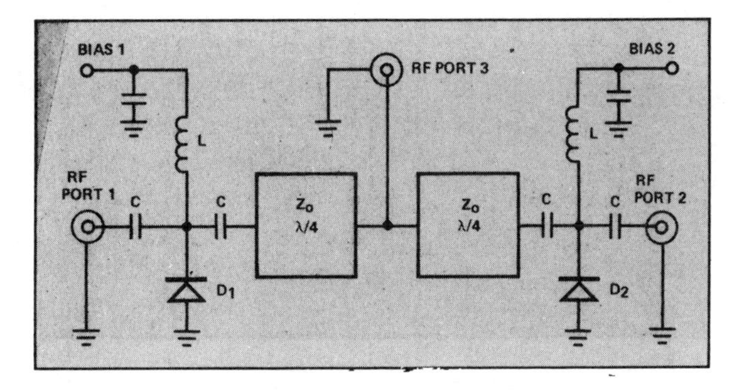

FIGURE 7.24. Shunt PIN SPDT switch. (*Courtesy of Hewlett-Packard, AN-957-1.*)

circuit to an open circuit, the λ/4 transformers eliminate any reactive loading on RF port 3 from whichever diode is forward biased. For example, when diode D_1 is forward biased, diode D_2 will be reverse biased. Therefore, port 2 will be connected to port 3, any microwave signal from port 1 will be reflected back to port 1 by the low-impedance diode D_1, and this is effectively isolated from port 3 by the λ/4 transformer. The main disadvantage of this circuit is the bandwidth restriction placed by the λ/4 transformers.

The series-shunt arrangement shown in Fig. 7.25 eliminates the need for the λ/4 transformers of Fig. 7.24. When diode D_1 is forward biased, diode D_3 is reverse biased and therefore isolates the low-impedance D_1 from RF port 3. Diode D_4 provides a similar function when D_2 is forward biased. Other arrangements are described in *Hewlett-Packard Application Note 957-1*.

7.8 ABSORPTIVE ATTENUATOR

In some situations the source may not be able to handle reflected power and then an absorptive switch is required. A number of arrangements are possible, one being shown in Fig. 7.26. Here a circulator is used, and when the diode is in the off, or high-impedance condition, all the power is absorbed in the matched impedance Z_0 at the second port. As forward dc bias is applied through the input port, the diode progressively shunts the Z_0 impedance, producing a mismatch at the second port. The reflected power travels around the circulator to the output port. The amount of reflected power depends on the level of dc bias, and the attenuation is controlled in this manner. Note that the RF input and output feed into constant impedances.

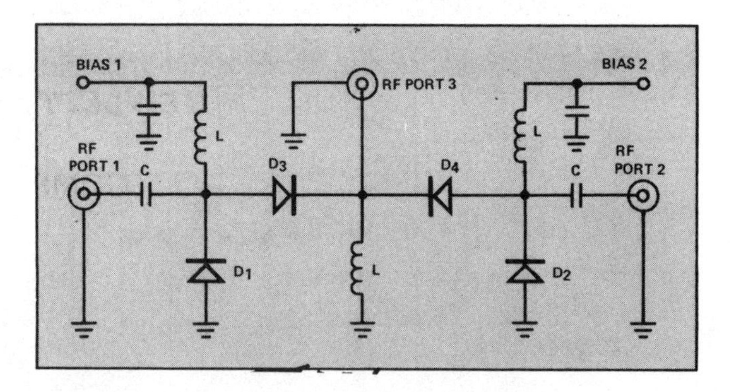

FIGURE 7.25. Series/shunt PIN switch. (*Courtesy of Hewlett-Packard, AN-957-1.*)

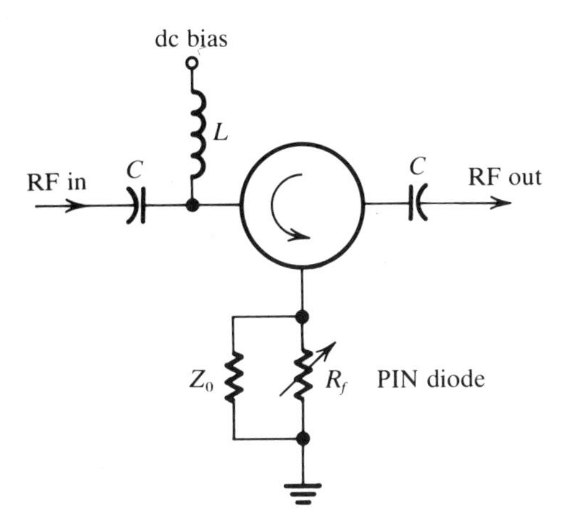

FIGURE 7.26. Schematic for an absorptive attenuator.

7.9 PIN-DIODE PHASE SHIFTERS

PIN diodes may be used to switch reactive elements into a circuit, in this way producing a phase shift. Phase shifters are used for a variety of purposes at microwaves. For example they are widely used in phased array antennas for radar, where the pointing direction of an antenna beam may be shifted by altering the phase of the signal to successive elements in the antenna array. Reflective-type phase shifters utilize the change in the phase angle of the reflection coefficient which can be produced by altering the load on a transmission line. Figure 7.27(a) shows the basic arrangement. The PIN diode acts as a switch which alters the length of a short-circuited length of line. Of importance is the differential phase shift introduced by the additional length d, and this is

$$\Delta\phi = -2\beta d \tag{7.48}$$

where β is the phase-shift coefficient $= 2\pi/\lambda$. The negative sign shows that the extra length introduces a phase lag, and the factor 2 appears because the signal has to traverse forward and back along length d. One method of utilizing this differential phase shift of the reflected signal is shown in Fig. 7.27(b). The RF signal is fed into port 1 of a hybrid ring, and the output is taken from port 3. Shunt PIN diode switches are shown by the $\times$ on the arms from ports 2 and 4. When open, these switches connect in short-circuited lines of length d on each arm. An additional $\lambda/4$ length is included in arm 2. The signal here must traverse forward and back along this, equivalent to $\lambda/2$, and therefore an additional shift of 180° occurs in

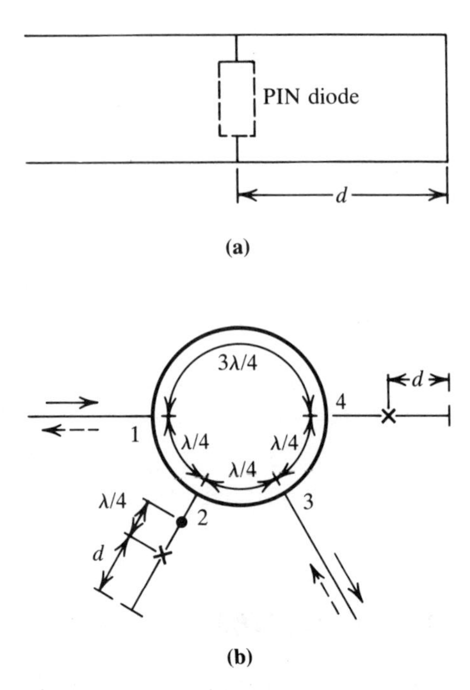

FIGURE 7.27. (a) Basic reflective phase-shift element; (b) phase shifter utilizing a hybrid ring and the basic element.

arm 2. It will be recalled from Section 4.15 that no output appears at port 3 of the hybrid ring when ports 2 and 4 are terminated in identical impedances and the input is to port 1. However, with the additional $\lambda/4$ length in arm 2, the reflected signal at port 2 is 180° out of phase with the reflected signal at port 4. This exactly compensates for the 180° phase shift introduced by the hybrid ring, and therefore the signals at port 3 are in phase. With both diodes forward biased, a given reference signal appears at port 3. When the diodes are reverse biased, they open-circuit and the extra length d is introduced in both sections. This results in the differential phase shift $\Delta\phi$, as given by Eq. (7.48), appearing at port 3.

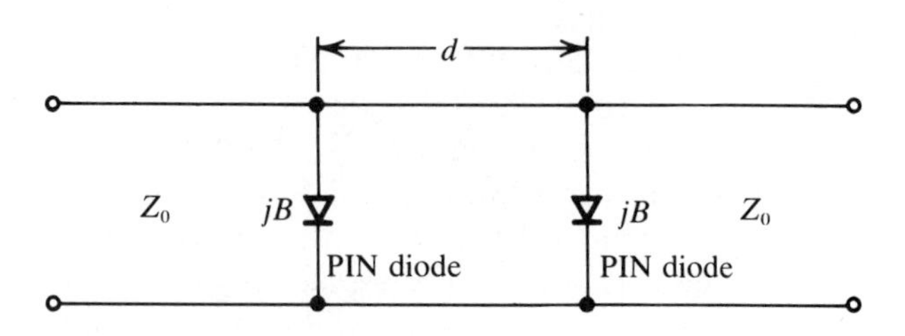

FIGURE 7.28. In-line phase shifter.

Phase shift may also be introduced by switching a suitably loaded section into the main line, as shown in Fig. 7.28. The susceptance jB is determined by the diodes, and will change from capacitive to inductive as the diodes are switched from reverse- to forward-bias states. The characteristic impedance and length of the intermediate section of line must be chosen to avoid reflections on the main line.

Figure 7.29 shows the layout for a 3-GHz phase shifter built on the principles already described. In practice it is found that the reflective type is more suited for phase shifts of 90° and 180°, while the variable-delay-line type is better suited for small phase angles, here 22.5° and 45°. For the phase shifter shown in Fig. 7.29, a 4-bit control signal can switch the phase through 0 to 360° in steps of 22.5°. Figure 7.30 shows the details of the diode mounting.

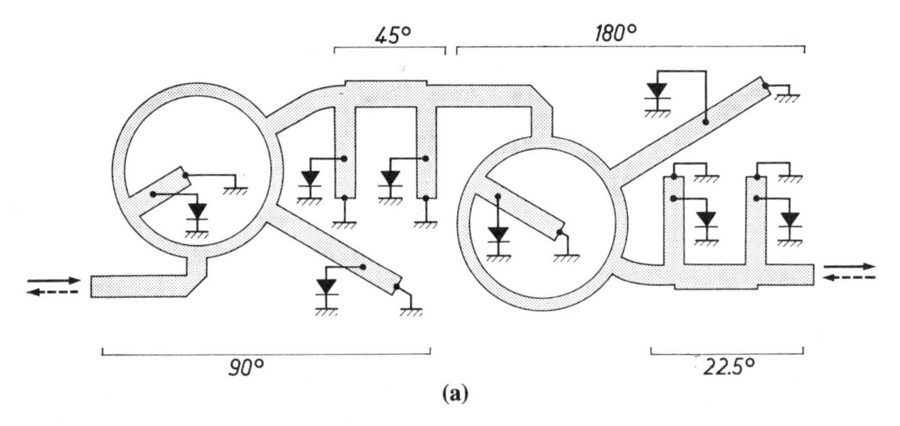

(a)

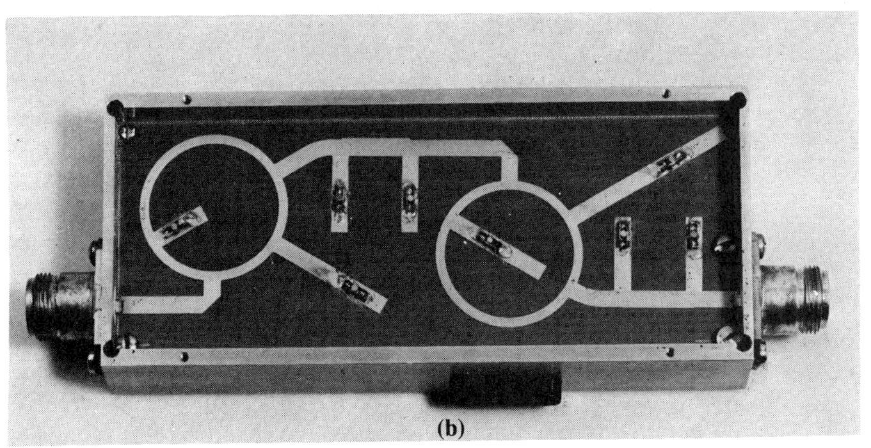

(b)

FIGURE 7.29. Experimental phase shifter for 3 GHz. (a) Diagram of the phase shifter. The locations of switching diodes and short circuits are indicated. The extent to which the phase can be changed in various parts of the circuit is also indicated. (b) The phase shifter. (*Courtesy of Philips Tech. Rev., Vol. 32, 1971.*)

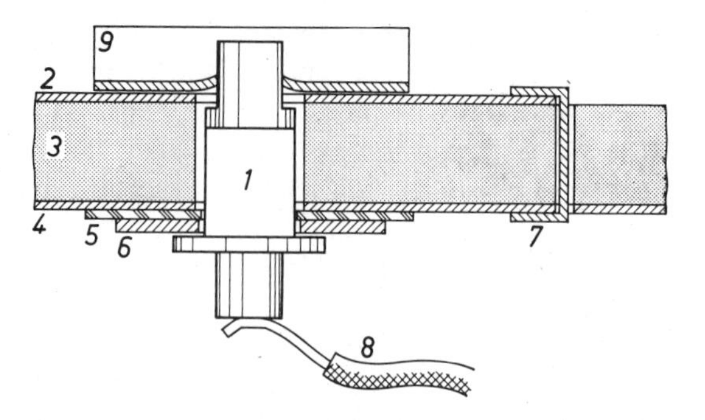

FIGURE 7.30. Switching diode mounted in a microstrip circuit. 1, encapsulated diode; 2, microstrip line; 3, dielectric; 4, ground plane. The insulating sheet 5 and the metal plate 6 form a capacitor, whose function is to ensure that the short-circuited termination 7 of the microstrip does not short circuit the diode control voltage. 8, supply lead for the control voltage; 9, mounting bracket. (*Courtesy of Philips Tech. Rev., 1971.*)

7.10 THE SQUARE-LAW DIODE DETECTOR

A microwave signal may be detected using a simple diode circuit. When microwave power is fed into the circuit, the nonlinearity of the diode results in an increase in diode current, and this increase is a measure of the input power. As will be shown shortly, it is a square-law term in the voltage–current characteristic of the diode which is most significant in low-level detection, hence the description "square-law detector." These detectors are also referred to as "low-level detectors" because the input signal level is sufficiently low to restrict operation to the square-law region of the characteristic. Low-level detectors are widely used in such applications as short-range radar receivers, beacon receivers, missile guidance receivers, and power-monitoring detectors. Signal levels in such applications are in the range −60 to −30 dBm. Figure 7.31 shows the circuit for a video receiver that utilizes a low-level detector. The term "video" is used because the output is usually a wideband pulse which requires similar amplifier circuitry to that required by video signals. The RF filter and matching transformer at the input are for the purpose of matching the signal source to the diode, to provide maximum power transfer. The input radio frequency choke, RFC_1, acts as an open circuit to the microwave signal while providing a dc path for bias current and also a short-circuit return for the video output current. Capacitor C_b acts as a short circuit to the microwave signal, allowing the microwave signal to be applied across the diode, and at the same time, this capacitor presents essentially an open circuit to the video signal so that the video signal voltage is developed across the load resistor R_L. Capacitor C_A is the input capacitance of the video amplifier. Bias to the diode is applied through the filter RFC_2, C_c, which prevents the video signal (and any stray microwave signal) from entering the bias line. Before analyzing the detector action, a brief description of the diode types used as microwave detectors will be given.

Schottky barrier diodes and point contact diodes are used for microwave detection and mixing. These diodes differ from the normal *p-n* junction diode in that their operation depends on the nonlinear characteristic of a metal–semiconductor junction. When for example, aluminum is welded onto *n*-type silicon, electrons diffuse from the silicon into the aluminum. This creates a potential barrier and as a result the junction exhibits rectifying action. In the manufacture of such junctions, the semiconductor surface must be extremely clean and free of defects. Silicon and gallium arsenide are the two most common semiconductors used and a number of different manufacturing methods are employed. Schottky barrier diodes are made by vacuum evaporation of metal onto the semiconductor surface, and in one type, known as the mesh diode, a large number of very small contact areas are first formed, as shown in Fig. 7.32(a). These contacts are then probed in a random manner, using a tungsten whisker, until a junction exhibiting the desired diode characteristics is found. A good manufacturing yield is achieved by this method. However, the mesh diode is limited in operation to frequencies below about 7 GHz because of the relatively large contact area and hence large capacitance. The contact area can be reduced by etching a contact window through a silicon oxide mask in the case of silicon diodes, and depositing the metal through this.

Another method used to reduce contact area is to make a direct metallic point contact with the semiconductor surface as shown in Fig. 7.32(b), such diodes being known as point contact diodes. The theory of the point contact diode is not as well understood as that of the Schottky

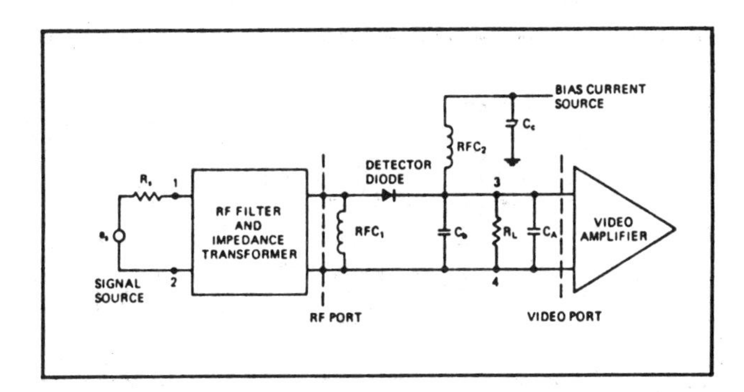

FIGURE 7.31. Typical video receiver. (*Courtesy of Hewlett-Packard, AN-923.*)

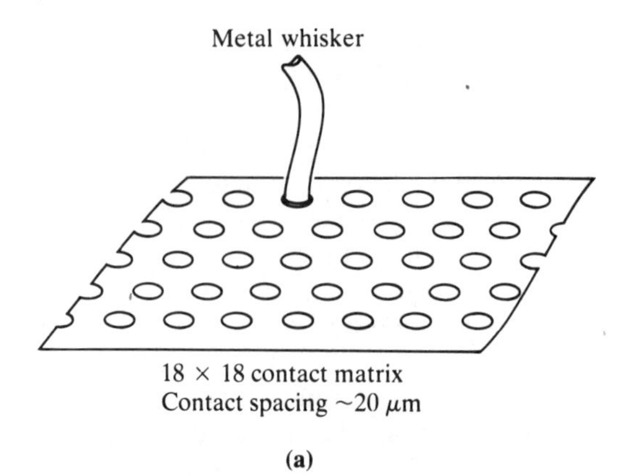

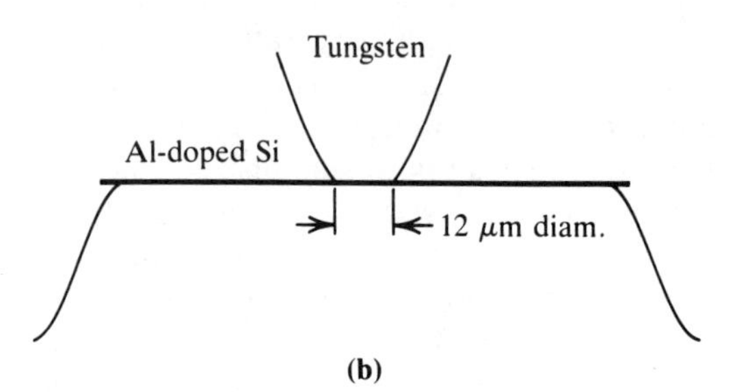

FIGURE 7.32. (a) Detail of a mesh diode; (b) detail of a point contact diode.

barrier diode, but manufacturing techniques are very well established and such diodes are widely used in noncritical microwave applications.

Diodes may be encapsulated as discrete devices, and are available as such in a variety of standard packages. They are also readily integrated into circuits, as shown in Fig. 7.33. The main advantage of the metal–semiconductor type of diode for microwave applications is that it responds very fast to changes in bias from forward to reverse. This is partly a result of the small physical size and hence small capacitance, but more important, forward conduction is by means of majority carriers, and therefore there is no delay associated with minority-carrier withdrawl, which is a limiting factor with *p-n* junctions.

The diode equivalent circuit, shown in Fig. 7.34, for Schottky and point contact diodes, is similar to that described for the PIN diode and can be developed from Fig. 7.13(c). L_s is the lead inductance, C_c the case capacitance, R_s the series lead resistance, and C_j and R_j the effective capacitance and resistance for the junction proper. Both of these components are functions of the diode voltage. For the low-level detectors described

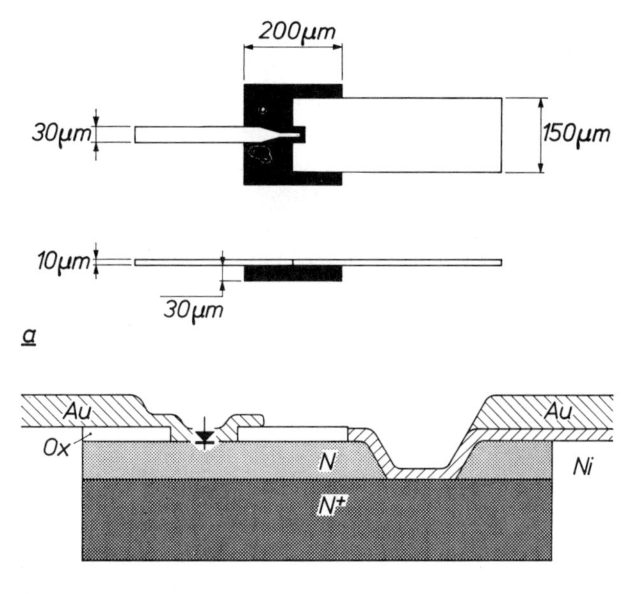

FIGURE 7.33. Schottky barrier diode with connections projecting outside the edge of the chip ("beam leads"). (a) Plan and side view. The diode is directly connected to the strip conductors by the flat leads. (b) Enlarged cross section. Ox, silicon oxide; Au, gold; Ni, nickel contact electrode (sometimes made of titanium or palladium). A diode symbol indicates the polarity of the diode at the position of the rectifying contact. (*Courtesy of Philips Tech. Rev., 1971.*)

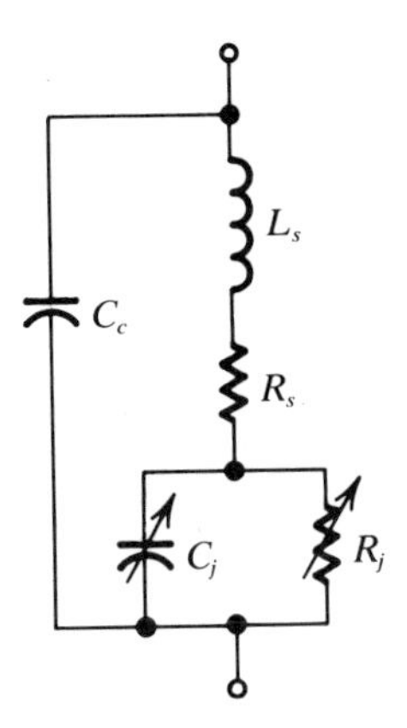

FIGURE 7.34. Small-signal equivalent circuit for a Schottky barrier diode detector.

below, the small-signal values for all components are evaluated at the bias point; that is, no large-signal variations take place. The diode resistance can be adjusted by means of the bias to a value close to that required for impedance matching to the input line, and a matching network can be used to tune out L_s and C_c, as shown in Fig. 7.31.

The diode detection action depends on the voltage dependence of R_j, the junction resistance, and for this reason the diode is referred to as a *varistor*. The voltage–current relationship for R_j is given by

$$I = I_s(e^{V/nV_T} - 1) \tag{7.49}$$

Here I_s is the diode saturation current, V_T is the thermal voltage, equal to 26 mV at room temperature, and n is a constant, typically 1.1 for Schottky barrier diodes and 1.4 for point contact diodes. I is the current through R_j and V is the voltage across R_j. The junction slope resistance R_j is an important circuit parameter, and as shown in Fig. 7.35, it is determined from the slope of the I/V curve around the bias point. Hence differentiating Eq. (7.49) gives for $G_j = 1/R_j$:

$$G_j = \frac{I_0 + I_s}{nV_T} \tag{7.50}$$

For all practical detector applications, $I_0 >> I_s$, and so

$$G_j = \frac{I_0}{nV_T} \tag{7.51}$$

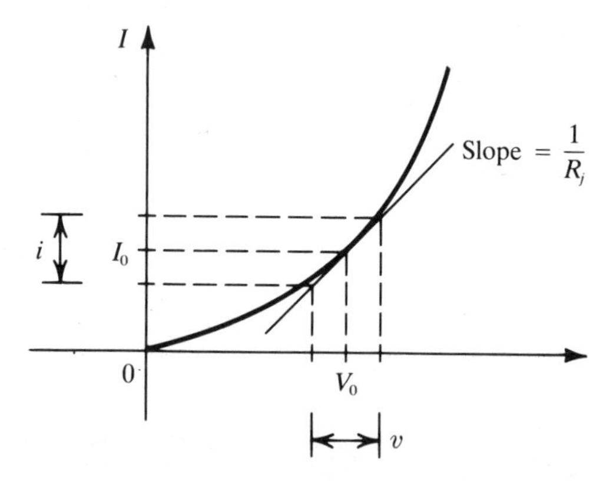

FIGURE 7.35. Forward-bias operating point for the low-level detector.

For the low-level detector described next, Eq. (7.49) can be approximated by the first two terms of a power series expansion, and for this reason the low-level detector is also referred to as a square-law detector.

7.10.1 The Low-Level Detector

A low-level detector is one in which the input signal is small, such that operation takes place over the curved portion of the diode I/V characteristic. The I/V characteristic is sketched in Fig. 7.35 for forward bias. Operation is around some fixed bias point denoted by V_0 and I_0, and the small-signal quantities are shown as i and v. Now the I/V curve may be expanded about the operating point I_0/V_0 by means of Taylor's theorem. In the present context this expansion may be written in the form

$$i = G_j v + \frac{G_j}{2nV_T} v^2 + \cdots \tag{7.52}$$

Here, $v = V - V_0$ is the small-signal (microwave) variation about V_0, and $i = I - I_0$ is the corresponding small-signal current variation. These quantities are shown in Fig. 7.35. To illustrate the detection process, let the input signal voltage v be a sinusoidal microwave carrier given by

$$v = V_{cp} \sin \omega_c t \tag{7.53}$$

Here V_{cp} is the peak carrier voltage, and ω_c is the carrier angular frequency. Making use of the trignometric identity $\sin^2 A = 0.5(1 - \cos 2A)$ for v^2 in Eq. (7.52) results in

$$i = G_j V_{cp} \sin \omega_c t + \frac{G_j V_{cp}^2}{4nV_T} (1 - \cos 2\omega_c t)$$

and arranging this in order of terms yields

$$i = \frac{G_j V_{cp}^2}{4nV_R} + G_j V_{cp} \sin \omega_c t - \frac{G_j V_{cp}^2}{2nV_T} \cos 2\, \omega_c t \tag{7.54}$$

The first term is a steady or dc component which is seen to arise as a result of the microwave input. Also, it is proportional to V_{cp} squared, hence the name "square-law detector." If the microwave carrier is now ON/OFF-modulated by a pulsed waveform, the diode direct current will follow this waveform, and in this way the carrier is demodulated.

The microwave input power is proportional to V_{cp} squared, and therefore the increase in the diode direct current is proportional to the microwave input power. Denoting the available input power (i.e., the power obtained under matched conditions) by P_{av} and the corresponding increase in diode direct current by Δi, the current sensitivity, denoted by the symbol β, is defined as

$$\beta \triangleq \frac{\Delta i}{P_{av}} \tag{7.55}$$

Current sensitivity (sometimes referred to as current responsivity) is dependent on the bias current through its effect on G_j, and typical values range from 1 to 15 μA/μW. Another useful figure of merit for the diode is the voltage sensitivity, which is the ratio of the video output EMF to microwave power input. The equivalent video source resistance of the diode is simply

$$R_v = R_s + R_j \tag{7.56}$$

The source EMF will be $R_v \cdot \Delta i$, and hence the voltage sensitivity, denoted by the symbol γ, is

$$\gamma = \frac{R_v \, \Delta i}{P_{av}} = R_v\beta \tag{7.57}$$

Values for both the video resistance and the voltage sensitivity are usually specified on diode data sheets. It will also be seen that, knowing the voltage sensitivity, the video source EMF can be determined for a given input power P as

$$E_v = \gamma P \tag{7.58}$$

EXAMPLE 7.3

The video resistance for the Hewlett-Packard 5082 family of Schottky diodes is specified as typically 1400 Ω, and the voltage sensitivity as 6.6 mV/μW. Determine (a) the current sensitivity, and (b) the output EMF when the input power is -45 dBm.

SOLUTION (a) From Eq. (7.57),

$$\beta = \frac{6.6 \times 10^{-3}}{1400} \text{ A/}\mu\text{W}$$

$$= \mathbf{4.71} \ \mu\text{A/}\mu\text{W}$$

(b) The input power of −40 dBm expressed in microwatts is 0.1 μW, and hence from Eq. (7.58), $E = 6.6 \times 0.1 =$ **0.66 mV.**

The component values for the matching network and the frequency bandwidth of the overall circuit are determined largely by L_s and C_c. The series resistance R_s and the junction capacitance C_j limit the sensitivity of the detector circuit by reducing the signal power reaching the active component R_j. The power loss resulting from these two components may be found as follows. Assuming that L_s and C_c are taken care of in the matching network, the diode equivalent circuit reduces to that shown in Fig. 7.36. The rms signal current input is denoted by I, and the rms signal voltage across R_j by V. The total power into the circuit is seen to be

$$P_{\text{tot}} = |I|^2 R_s + \frac{|V|^2}{R_j} \tag{7.59}$$

The power input to R_j alone is

$$P = \frac{|V|^2}{R_j} \tag{7.60}$$

Hence the ratio of useful power input (i.e., that reaching R_j) to total power input is

$$\frac{P}{P_{\text{tot}}} = \frac{1}{1 + (|I|^2/|V|^2)\, R_s R_j} \tag{7.61}$$

But $|I|/|V| = |Y|$, where Y is the admittance of $R_j//C_j$:

$$Y = G_j + j\omega C_j \tag{7.62}$$

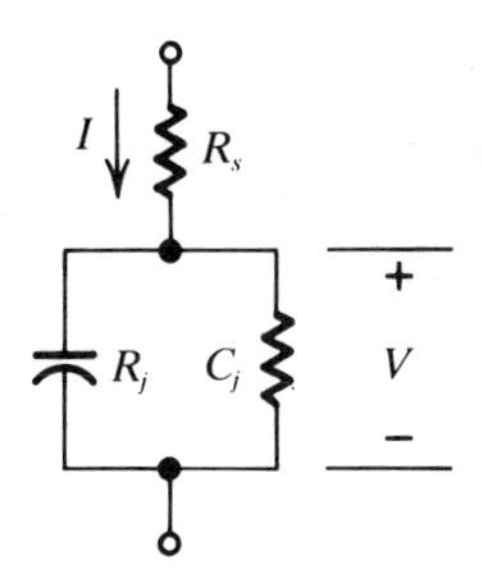

FIGURE 7.36. Equivalent circuit for the low-level detector.

Hence $|I|^2/|V|^2 = G_j^2 + (\omega C_j)^2$. Substituting this in Eq. (7.61) and rearranging gives

$$\frac{P}{P_{\text{tot}}} = \frac{1}{1 + R_s G_j + \omega^2 C_j^2 R_s R_j} \tag{7.63}$$

This loss expressed in decibels is

$$\begin{aligned} \text{loss(dB)} &= -10 \log \frac{P}{P_{\text{tot}}} \\ &= 10 \log (1 + R_s G_j + \omega^2 C_j^2 R_s R_j) \end{aligned} \tag{7.64}$$

EXAMPLE 7.4

Calculate the power loss as a percentage, and in decibels, for a low-level point contact detector. The bias for the detector is adjusted to make $R_j = R_s = 10\ \Omega$ and $C_j = 0.3$ pF. The frequency of operation is 4 GHz.

SOLUTION

$$\begin{aligned} \frac{P}{P_{\text{tot}}} &= \frac{1}{1 + 1 + (2\pi \times 4 \times 10^9 \times 0.3 \times 10^{-12})^2 \times 10 \times 10} \\ &= \frac{1}{2.0057} \end{aligned}$$

In decibels the loss is

$$\text{loss (dB)} = 10 \log 2.0057 = 3.02 \text{ dB}$$

The percentage loss is

$$\left(1 - \frac{1}{2.0057}\right) \times 100\% = \mathbf{50.14\%}$$

A method widely used for measuring and specifying the sensitivity of a low-level detector is known as the *tangential sensitivity* (TSS) *method*. In this method, the noise output of the detector circuit is observed on an oscilloscope, as shown in Fig. 7.37(a). When the microwave carrier is ON/OFF-modulated, the resultant shift in direct current during the carrier ON periods will "lift" the noise as shown. The carrier power is adjusted until the bottom of the noise wave during ON periods is level (or tangent to) the top of the noise waveform during OFF periods. The microwave input

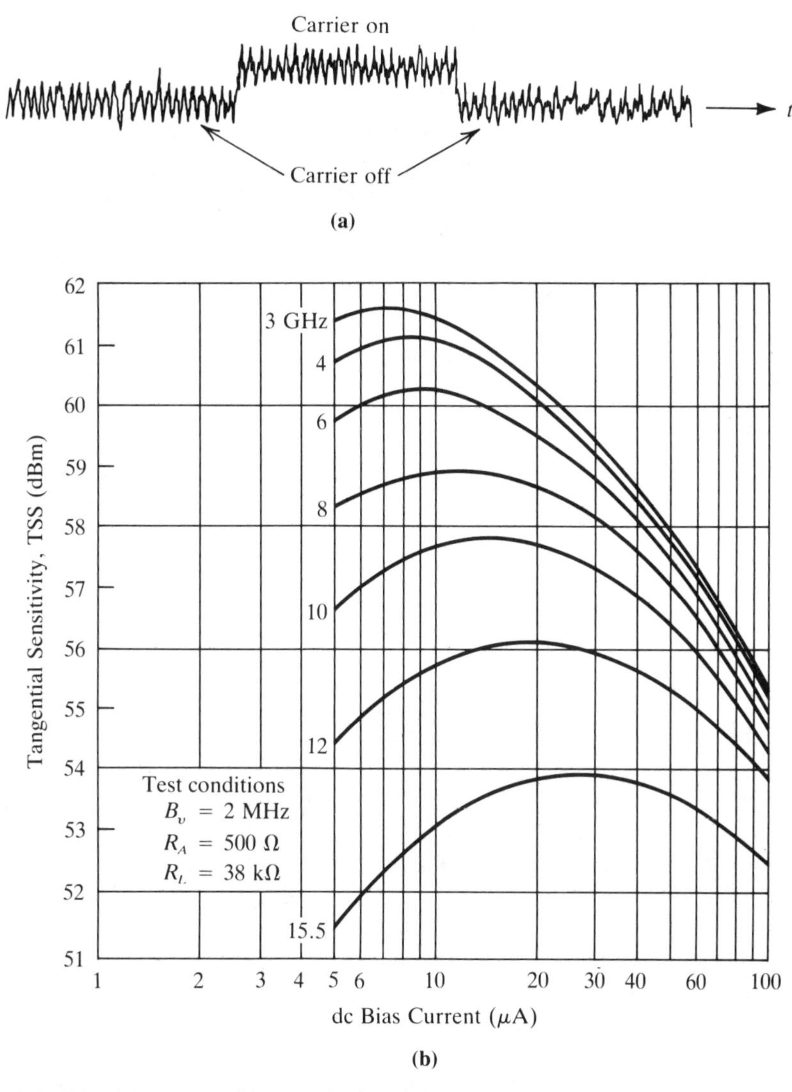

FIGURE 7.37. (a) Oscilloscope display of the tangential sensitivity criterion; (b) tangential sensitivity (TSS) values for HP-5082-2755 diode. Video bandwith B_V; video load impedance R_L; equivalent noise resistance of video amplifier R_A. [*Part (b) courtesy of Hewlett-Packard, AN-923.*]

power required to meet this condition is then thc tangential sensitivity. The value is usually specified in dBm and typically may be in the range -60 to -30 dBm. The tangential sensitivity depends on many factors, which must also be specified, the most important of these being the microwave frequency, the video bandwidth, the video amplifier noise figure, the diode bias current, and the test mount parameters. Figure 7.37(b) shows some measured values of TSS which are typical of the Hewlett-Packard diode 5082-2755.

The method is subjective in that the tangential condition cannot be precisely defined and depends on the judgment of the observer, but in practice reasonably consistent results are obtained. Some manufacturers relate the tangential sensitivity to an output signal-to-noise ratio, for example, in *Application Note 923,* Hewlett-Packard specifies the tangential sensitivity as the lowest input power level required to produce an 8-dB output signal-to-noise ratio.

7.11 DIODE MIXER CIRCUITS

The simple diode detector described in the preceding section is limited in sensitivity. An increase in sensitivity, in the range 35 to 40 dB, can be achieved by using the superheterodyne principle. Here, the incoming microwave signal is first converted to a lower frequency through a mixer stage. This lower frequency, termed the intermediate frequency (IF), is then amplified before being passed onto the detector stage, in this case a high-level detector. The general principles of the superhet and high-level detectors are described in Roddy and Coolen (1984). The frequency converter or mixer stage has to operate at the incoming microwave frequency, and therefore Schottky barrier diodes and point contact diodes, similar to those used for low-level detectors, are employed. The mixing process occurs mainly through the varistor action of the diode. The nonlinear capacitance, C_j, and the lead inductance, L_s, also have a significant effect on the mixing process. The mixer may be used to detect the Doppler frequency shift in the case of Doppler radar (see Section 12.6), in which case the output frequency may be as low as a few hundred hertz. In other situations it is advantageous to use a mixer to convert the microwave carrier to a relatively high intermediate frequency, typically 70 MHz, before detection. Greater amplification of the signal can be achieved at the intermediate frequency.

A mixer circuit therefore has three ports: one for signal input, one for local oscillator input, and one for intermediate frequency output. These are usually denoted as RF, LO, and IF, or sometimes just by the letters *R*, *L*, and *I*.

The microwave signal frequency may be lower than the local oscillator frequency, in which case the mixer is referred to as a lower-sideband mixer, or it may be higher, in which case it is an upper-sideband mixer. Upper-sideband operation will be described here for purposes of illustration. In practice, the local oscillator level is always large compared to the microwave signal level. As a result of the relatively large local oscillator input, a significant second harmonic component, at frequency $2f_{osc}$, is produced in the mixer and this plays an important part in the mixing process. The important frequency components at the output of the upper-sideband mixer are

$$\text{IF} = f_s - f_{\text{osc}} \tag{7.65}$$

$$f_{\text{sum}} = f_s + f_{\text{osc}} \tag{7.66}$$

$$f_{\text{image}} = 2f_{\text{osc}} - f_s \tag{7.67}$$

The latter component, f_{image}, is known as the image because it appears as the mirror image of the signal frequency about the oscillator frequency. This follows since $2f_{\text{osc}} - f_s = f_{\text{osc}} - (f_s - f_{\text{osc}}) = f_{\text{osc}} - \text{IF}$, while from Eq. (7.65), $f_s = f_{\text{osc}} + \text{IF}$. These frequency components are shown in Fig. 7.38. It is important to realize that the image is internally generated in the mixer, and as such its energy can be recovered. Means of doing so will be described shortly. Students familiar with the basic superheterodyne receiver will know that an external image can occur, it being possible for an external signal to be present at $f_{\text{osc}} - \text{IF}$, which is the image frequency for the upper-sideband mixer. Except where a very broadband input circuit is required, the selectivity of the input signal circuit is usually sufficient to reject any external image signal, and the most important image component is that generated internally. The image can be used to enhance circuit performance, the circuit then being known as an image recovery mixer. In fact, the energy at the image and at the sum frequency can be recovered and made to contribute to the useful power output at IF.

Figure 7.39(a) shows a mixer diode assembly in coaxial line construction, and Fig. 7.39(b) shows the equivalent circuit. This is a broadband mixer, useful up to frequencies of about 3 GHz. The particular mixer illustrated is used in the General Radio slotted line-measuring equipment, which uses an intermediate frequency of 30 MHz. No image or sum frequency recovery is employed.

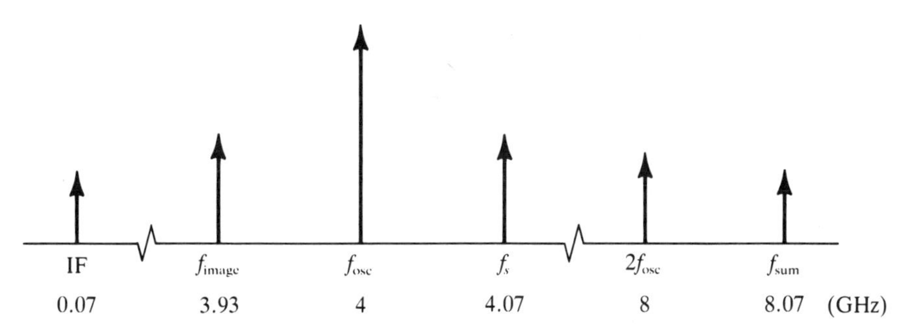

FIGURE 7.38. Mixer frequency components.

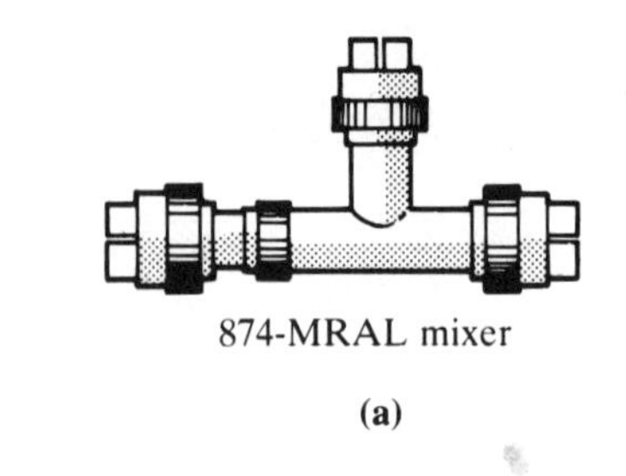

(a)

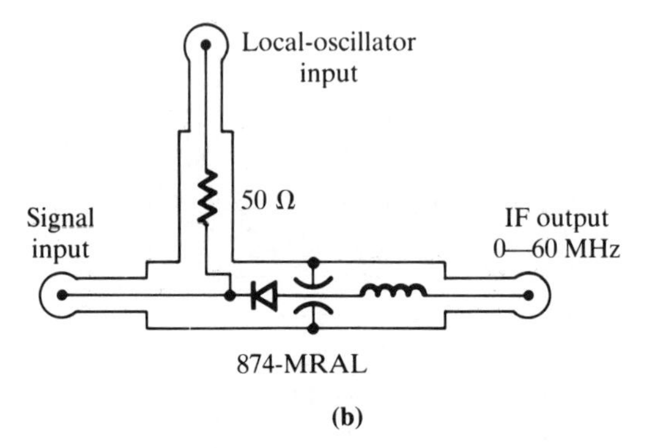

(b)

FIGURE 7.39. Single diode mixer in a coaxial assembly: (a) mixer unit; (b) equivalent circuit. (*Courtesy of GenRad, Inc., West Concord, Mass.*)

Figure 7.40(a) shows in block schematic form how image recovery may be introduced. The image generated in the mixer diode is transmitted along the line to the signal filter. The filter is arranged to reflect the image back to the mixer, where it mixes with the oscillator frequency f_{osc} to produce an IF:

$$f_{osc} - (2f_{osc} - f_s) = \text{IF} \tag{7.68}$$

The length l_1 is adjusted such that this new IF is in phase with the primary IF and thus the IF output is increased.

Figure 7.40(b) shows a block schematic diagram for a circuit in which recovery of both image and sum frequency energies takes place. The image recovery works as for Fig. 7.40(a). In addition, the low-pass filter in the signal line passes f_{osc}, f_s, and f_{image} and reflects back to the mixer the f_{sum} frequency generated in the mixer. There the f_{sum} component mixes with the second harmonic of the oscillator to produce an IF:

$$(f_s + f_{osc}) - 2f_{osc} = \text{IF} \tag{7.69}$$

The phase of the f_{sum} component reflected back to the mixer is adjusted through length l_2, so that the IF signals are in phase. Mixers in which the image is not recovered are sometimes termed *resistively terminated image mixers* because the image energy is dissipated in resistance in the circuit, and mixers in which the image is recovered are sometimes termed *reactively terminated image mixers*.

Single diode mixers are useful where signals are relatively large and noise is not a problem. For most communications applications mixers are of the balanced type. Single-balanced mixers use two diodes, and double-balanced mixers use four diodes. Figure 7.41 shows the schematic diagram for a single-balanced mixer and Fig. 7.42 shows the construction of the mixer in microstrip, and the equivalent circuit. Referring to Fig. 7.41, the mixer utilizes a hybrid ring as described in Section 4.15. The RF is applied to port 1, and the LO to port 3, and these ports are decoupled as described in Section 4.15. The RF signal splits at the input and recombines at port 2 in phase. The RF signal also recombines at port 4 in phase, but notice that the RF signal at port 2 leads that at port 4 by 180°. The LO signals also add at port 2 and at port 4. These ports are in phase with each other

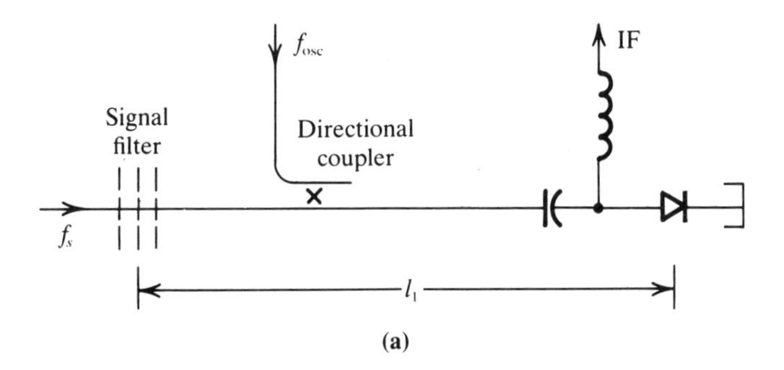

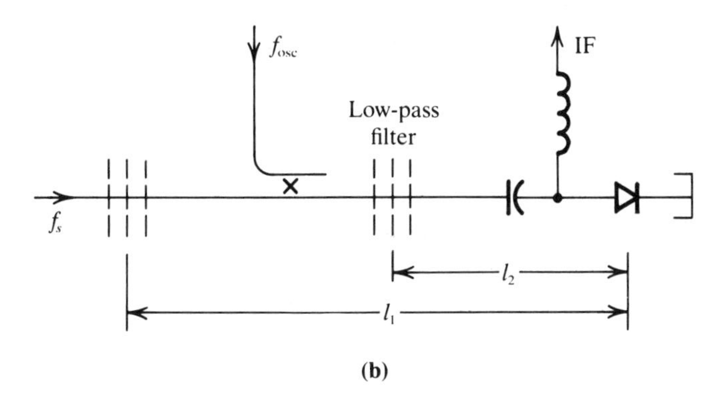

FIGURE 7.40. Schematic diagrams for (a) image recovery mixing and (b) image and sum-frequency recovery mixing.

at the LO frequency. It is assumed that the RF and LO are close enough in frequency to allow the hybrid ring to function reasonably well for both signals. The 180° phase shift at RF between ports 2 and 4 is transferred to the IF output, and therefore a large IF current flows in the inductors connected across the diode outputs. The IF is taken from the inductor center tap as shown. The 180° phase relationship is important for cancellation of noise arising in the oscillator. Any oscillator noise, close in frequency to the RF signal, will arrive at ports 2 and 4 in phase. Since the circuit is arranged for 180° phased signals to add at IF, the noise signals will cancel. Filters F_1 are to reduce the amount of RF reaching the output, and filters F_2 are to reduce the oscillator second harmonic reaching the diodes. A waveguide version of the single-balanced mixer is described in Section 4.14.

Figure 7.43(a) shows a broadband, double-balanced mixer, and Fig. 7.43(b) the schematic diagram for the mixer. With the double-balanced mixer, all ports are isolated from each other. This simplifies the filtering requirements at the output port and reduces the amount of oscillator signal energy reaching the RF port. Oscillator signal energy that reaches the RF port tends to be radiated and is a potential source of interference. Double-balanced mixers also produce less harmonic distortion than do single-balanced mixers.

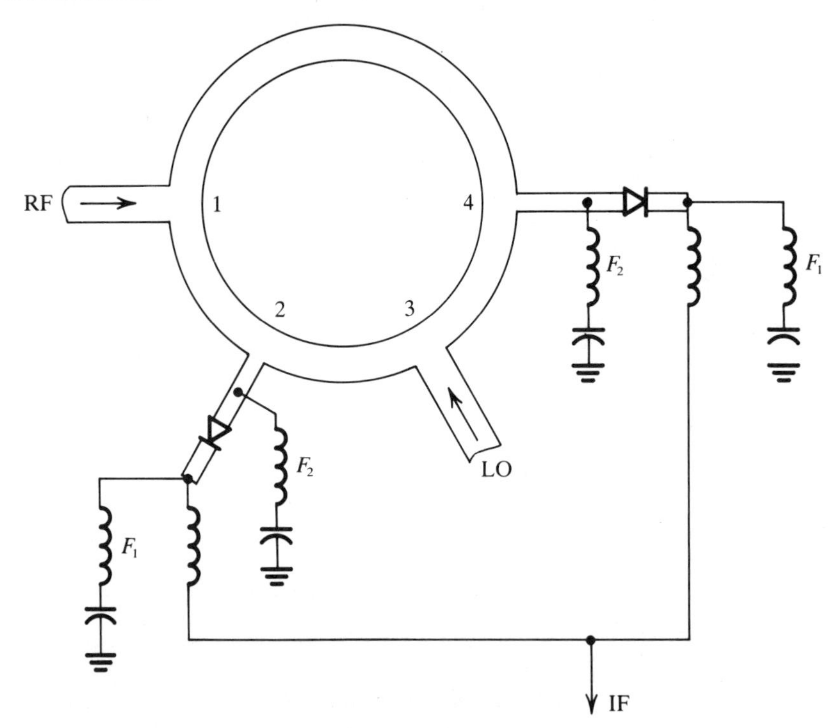

FIGURE 7.41. Schematic diagram for a single balanced mixer.

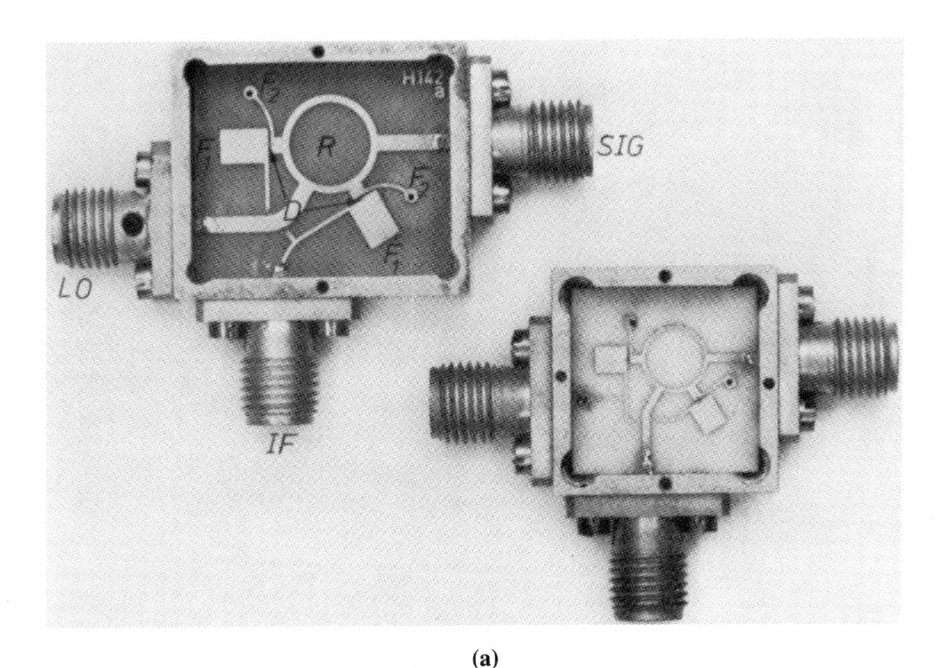

(a)

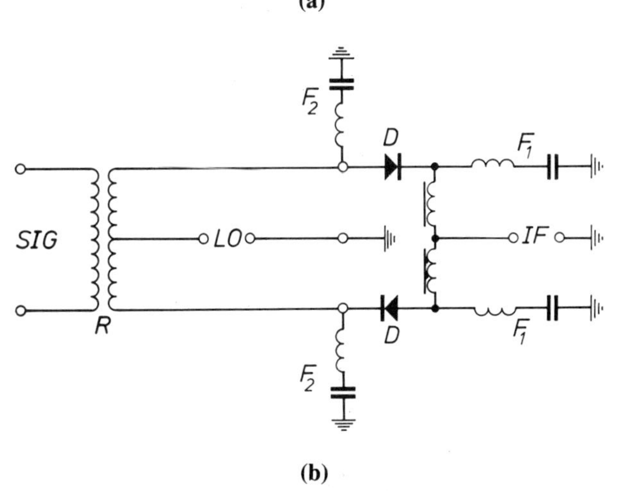

(b)

FIGURE 7.42. (a) Balanced mixer for 12 GHz: left, on quartz glass; right, on aluminum oxide. R, hybrid ring; D, Schottky barrier diodes; F_1, bandstop filters for 12 GHz. F_2, bandstop filters for the second harmonic of the oscillator frequency; LO, oscillator input; SIG, aerial input; IF, intermediate-frequency output. (b) Low-frequency equivalent of the mixer of part (a). (*Courtesy of Philips Tech. Rev., Vol. 32, 1971.*)

When the oscillator signal is large, it may be considered as a switching signal which results in pairs of diodes open-circuiting or short-circuiting as shown in Fig. 7.44. Thus the output at the IF port is

$$v = p(t) \sin \omega_s t \tag{7.70}$$

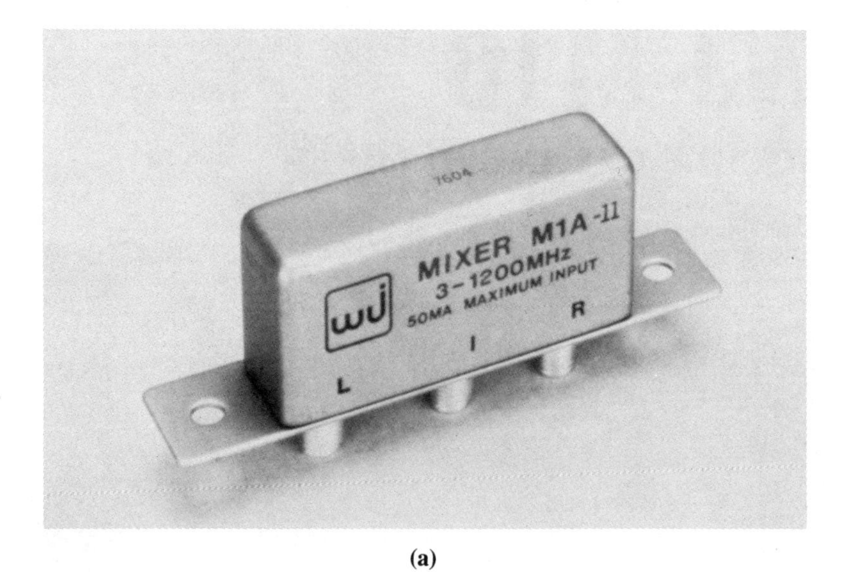

(a)

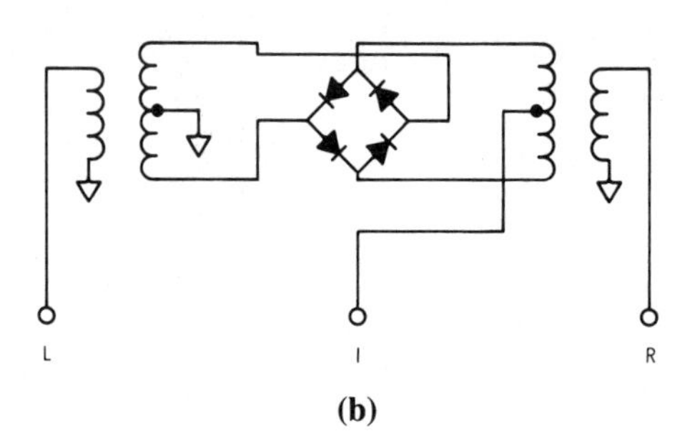

(b)

FIGURE 7.43. (a) Double-balanced broadband mixer; (b) circuit schematic. (*Courtesy of Watkins-Johnson Co., Palo Alto, Calif.*)

where $p(t)$ is a square wave that switches between ± 1, at the oscillator frequency f_{osc}. Now, $p(t)$ can be expanded by means of a Fourier series as

$$p(t) = \frac{4}{\pi} (\sin \omega_{osc} t + \sin 3\omega_{osc} t + \cdots) \tag{7.71}$$

Thus the product term $(\sin \omega_{osc} t)(\sin \omega_s t)$ occurs in Eq. (7.70) and this in turn can be expanded as

$$(\sin \omega_{osc} t)(\sin \omega_s t) = \tfrac{1}{2}[\cos (\omega_s - \omega_{osc})t - \cos(\omega_s + \omega_{osc})t] \tag{7.72}$$

The first term on the right-hand side is at frequency $(\omega_s - \omega_{osc})$, which is the desired IF. The other terms are removed by filtering.

Certain terms are used to specify mixer performance, the most important of these being the following:

Conversion loss: This is the ratio, expressed in decibels, of output power at IF to input power at the signal frequency. The conversion loss is normally specified for a 50-Ω system and at a local oscillator input of +7 dBm for low-level mixers. A typical value for conversion loss is 6 dB.

Noise figure: This is the difference in decibels between the input signal-to-noise ratio and the output signal-to-noise ratio, both expressed in decibels. Thus it gives a measure of how the mixer degrades the signal-to-noise performance. In measuring the noise figure of a mixer, the noise of the following IF amplifier stage is not included. The lower limit to IF frequency is usually specified as 400 kHz, so that $1/f$ noise

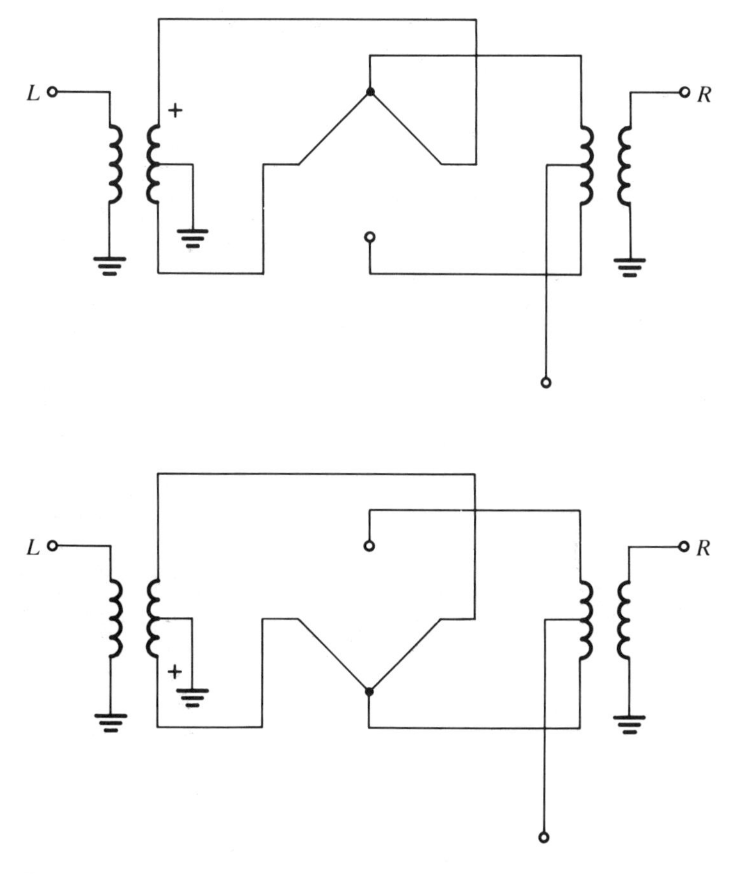

FIGURE 7.44. Switching action in the double-balanced mixer.

is also excluded. For a mixer, the decibel number for the noise figure is approximately equal to that for the conversion loss.

Isolation This is a measure of the leakage between ports. Because the local oscillator drive is much larger than the RF input, the LO-IF, and LO-RF leakage figures are the most significant. Isolation levels vary from about 25 dB to greater than 70 dB. The higher the isolation value, the better.

Conversion compression For a mixer, the IF power output is directly proportional to RF input up to a certain level. Above this level, the output increases less rapidly and the output is said to be compressed. The 1-dB compression point is specified by the input at which the output drops to 1 dB below the projected linear level, as illustrated in Fig. 7.45.

Dynamic range This is the input signal range over which the mixer can be used without serious degradation of performance. The lower limit is set by the noise figure and the required signal-to-noise ratio, and the upper limit is set by the compression point.

Third-order intercept point When two RF signals, f_{s1} and f_{s2}, are present at the input, an IF given by $(2f_{s2} - f_{s1}) \pm f_{osc}$ can be generated by the third-order term in the expansion given by Eq. (7.52). The theoretical line relating input to output has a slope of 3, as shown in Fig. 7.45, compared to a slope of 1 for the fundamental line. The input level of the second RF signal is assumed equal to that of the first. The third-order intercept point is where the fundamental line and third-order line intercept, as shown in Fig. 7.45. Although this

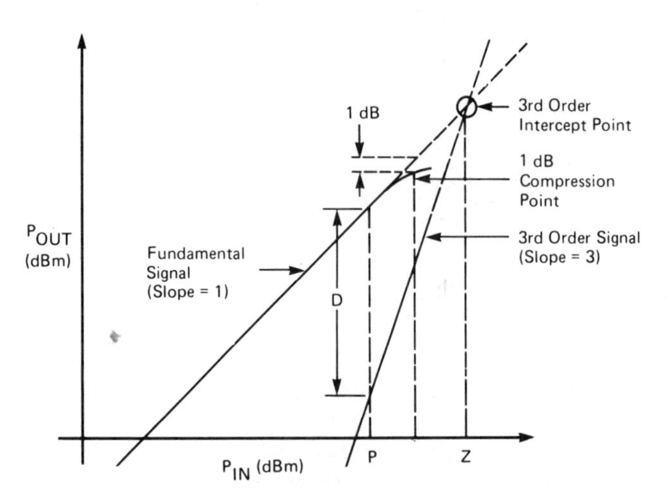

FIGURE 7.45. Input–output characteristic of a double-balanced mixer. *(Courtesy of Aertech Co., Catalog 0279.)*

point cannot be measured directly, it can be calculated from measurements of P and D (Fig. 7.45). It is left as an exercise for the student to show that from the geometry of the figure, $Z(\text{dBm}) = P + D/2$. The higher the intercept point, the better the third-order suppression. The input frequencies and terminating impedances must be specified together with the third-order intercept point.

7.12 THE IMPATT DIODE

Under certain reverse-bias conditions a *p-n* junction exhibits a negative ac resistance. This negative resistance can be used to sustain oscillations in a resonant circuit, and it can also be used to provide amplification in a reflection-type amplifier similar to that described in Section 7.3. The name *Impatt* is derived from the physical mechanisms involved, which are, *imp*act ionization, *a*valanching, and *t*ransit *t*ime drift.

Figure 7.46 shows the reverse-bias characteristic of a *p-n* junction. Breakdown occurs at some voltage V_B as a result of avalanche multiplication of hole and electrons. At voltage V_B electrons and holes making up the reverse current I_s gain sufficient kinetic energy to ionize further atoms on impact. The process is cumulative and the carrier density increases very rapidly, as shown by the steep increase in the reverse current magnitude in Fig. 7.46. As mentioned, the process is given the descriptive name "impact avalanching." In practice, to prevent a diode from avalanching to destruction, a constant-current bias source is used to maintain the average current at some safe value I_0, as shown in Fig. 7.46.

When the diode, biased very close to breakdown, is connected into a sharply resonant circuit, any noise voltage spike will be sufficient to excite a resonant component of voltage across the circuit. This voltage will alternately swing the diode into and out of the avalanching condition. Throughout, the diode remains reverse biased and carriers generated by the avalanching process drift in the depletion region toward the end contacts. The time taken for the carriers to drift from the junction to the end contacts is the *transit time*. The diode is designed so that carriers are drifting with the steady component of field and against the alternating component when this changes direction. Also, the drift time is short enough that the carriers are removed from the drift region before the voltage reverses phase again. The alternating field therefore takes energy from the carriers. In effect, the dc bias source supplies energy to the ac field and in this way the oscillatory voltage builds up across the diode. Another way of looking at this is that the diode presents a negative ac resistance, the required phase delay between voltage and current being the sum of transit time delay and a quarter-period delay which occurs between the current peak and the voltage peak in the avalanching process. Note that the latter delay is not

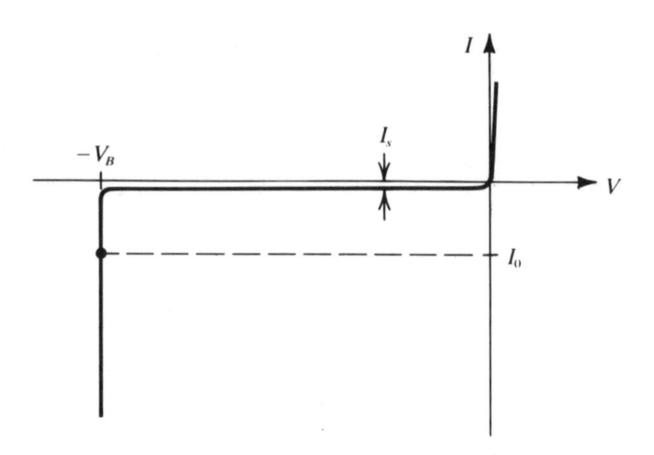

FIGURE 7.46. Avalanche characteristic for a reverse-biased *p-n* junction.

apparent in Fig. 7.46, which is for steady-state conditions. Figure 7.47(a) illustrates the drift process for a $p+ \; n \; n+$ structure. Avalanching occurs in a narrow region close to the $p+ \; n$ junction. Holes generated by the avalanche process are immediately absorbed in the $p+$ contact region. Electron bunches drift through the depletion region and are eventually absorbed by the $n+$ contact region. This type of diode is referred to as a *single-drift Impatt diode*.

The width of the drift region is determined by the required transit time. A detailed analysis [see, e.g., de Nobel, and Vlaardingerbroek (1971)] shows that the transit time should be equal to $0.37T$, where T is the periodic time of the microwave signal. The reverse-bias field is always such that the drift velocity is at its maximum, which for silicon is about 10^7 cm/s. Thus for a width d cm,

$$\tau = 0.37T = \frac{d}{v} = d10^{-7}$$

Therefore,

$$d = 0.37 \times T \times 10^7 \tag{7.73}$$

EXAMPLE 7.5

Calculate the layer thickness required for a 5-GHz Impatt diode.

SOLUTION

$$T = \frac{1}{f} = 0.2 \text{ ns}$$

From Eq. (7.73),

$$d = 0.37 \times 0.2 \times 10^{-9} \times 10^{7}$$
$$= 0.00074 \text{ cm}$$
$$= \mathbf{7.4\ \mu m}$$

A *double-drift diode* is illustrated in Fig. 7.47(b). The double-drift device is a $p+$ p-n $n+$ structure. Avalanching occurs at the p-n junction, holes generated by avalanching drift in the p region to the $p+$ contact and the electrons drift in the n region to the $n+$ contact. The double-drift device is capable of providing greater power output at better efficiency

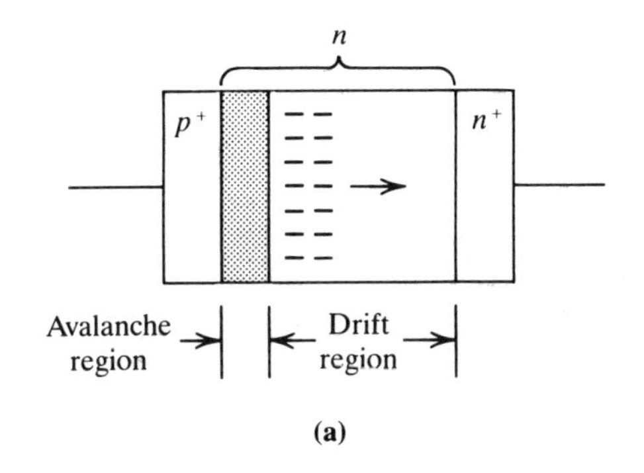

(a)

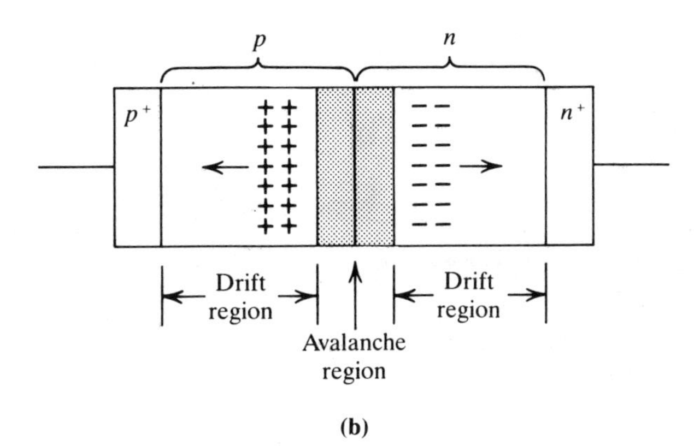

(b)

FIGURE 7.47. (a) Single-drift IMPATT diode; (b) double-drift IMPATT diode.

and lower noise than the single-drift device (see, e.g., *Hewlett-Packard Application Notes 961 and 962*).

Figure 7.48 shows the constructional and mounting details for one form of single-drift diode. Figure 7.48(a) shows the chip arrangement, and Fig. 7.48(b) shows one form of packaging. Figure 7.48(c) shows the diode mounted in a coaxial structure, and Fig. 7.48(d) shows it mounted in a microstrip line. Figure 7.49 shows a simplified equivalent circuit which takes into account the diode chip, the package reactances, and the load. The diode chip consists of a resistor R_D in series with a capacitor C_j, which is the capacitance of the junction at the breakdown voltage V_B. The resistance R_D consists of two components in series, R_s, the series lead resistance, and the negative resistance representing the Impatt action. The overall value of R_D is a negative number (typical values being shown in Fig. 7.50). When the diode chip is packaged, the package lead inductance and capacitance must be taken into account, and these can be modeled by the simple filter section shown in Fig. 7.49. Since the diode operates in a

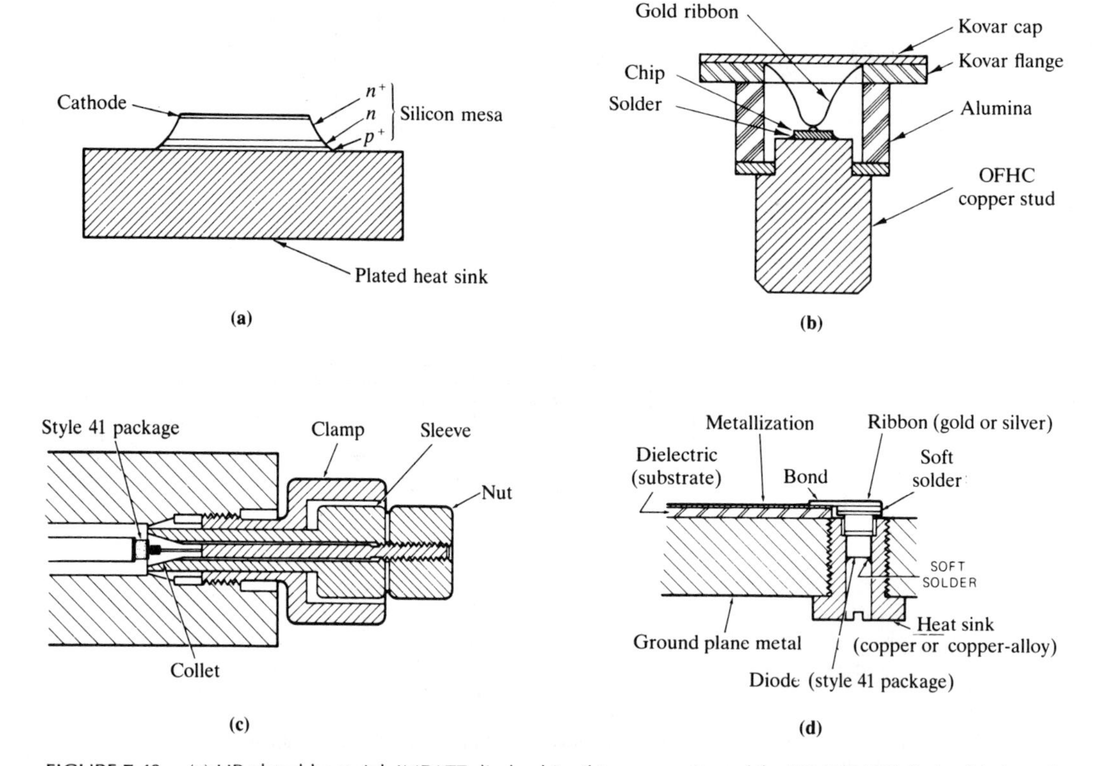

FIGURE 7.48. (a) HP plated-heat-sink IMPATT diode chip; (b) cross section of the HP IMPATT diode chip in style 41 package; (c) coaxial end mount; (d) microstrip mounting. (*Courtesy of Hewlett-Packard, AN-935.*)

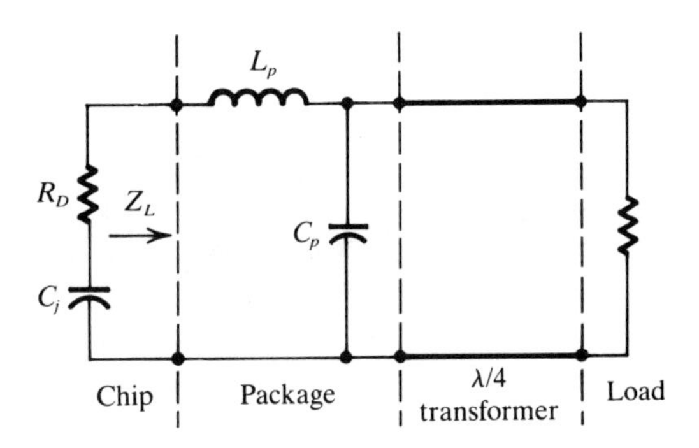

FIGURE 7.49. Simplified equivalent circuit for IMPATT diode and package.

resonant mode, the overall reactance of the circuit must be tuned to zero. This is often achieved by varying the geometry of the diode mount, and hence L_p (see Fig. 7.51). In addition to tuning the total reactance to zero, the total resistance must also equal zero, which in effect is saying that the power extracted by the ac field from the dc field exactly compensates for the power dissipated in real positive resistance of the circuit. Since the magnitude of R_D is low, a matching section is required to match this to a 50-Ω load, which is the usual value. A one-quarter-wavelength matching transformer may be used as shown in Fig. 7.51, which also shows how the package inductance can be varied by varying the recess depth of the diode in the line. Where broadband operation is required a multisection transformer may be used as described in Section 1.14.

Denoting the chip impedance by

$$Z_D = R_D - \frac{j}{\omega C_j}$$

$$= -|R_D| - \frac{j}{\omega C_j} \tag{7.74}$$

the required load impedance, including the case parasitics, is

$$Z_L = -Z_D$$

or

$$R_L + jX_L = |R_D| + \frac{j}{\omega C_j} \tag{7.75}$$

The real part of the load, as seen by the chip, must equal the magnitude of the dynamic resistance of the chip, and the imaginary part must equal

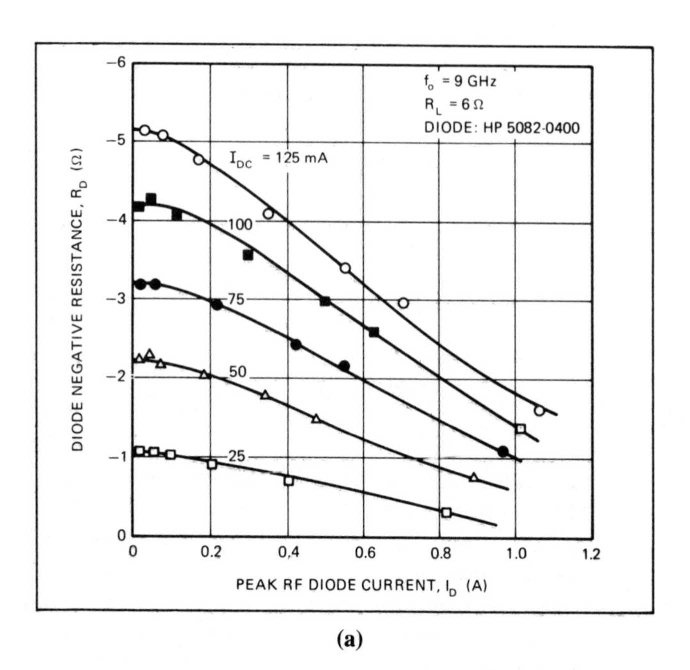

(a)

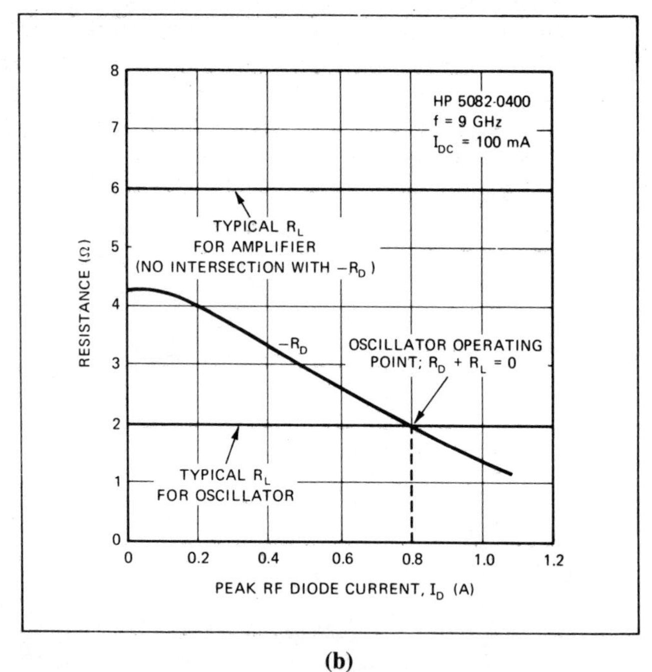

(b)

FIGURE 7.50. (a) R_D vs. I_D for an HP-5082-0400 diode at 9 GHz; (b) variation of R_D (a negative number) with the RF current amplitude I_D. The operating point is determined by the condition $R_L = -R_D$. (*Courtesy of Hewlett-Packard, AN-935.*)

the magnitude of the C_j reactance. As shown in the Example 7.6, for low-Q circuits the case inductance is the major part of the load reactance.

The value of R_D varies with both the bias current and the signal current in the manner shown in Fig. 7.50(a). For a given bias current and load, oscillator operation will stabilize at the point where R_D is equal to

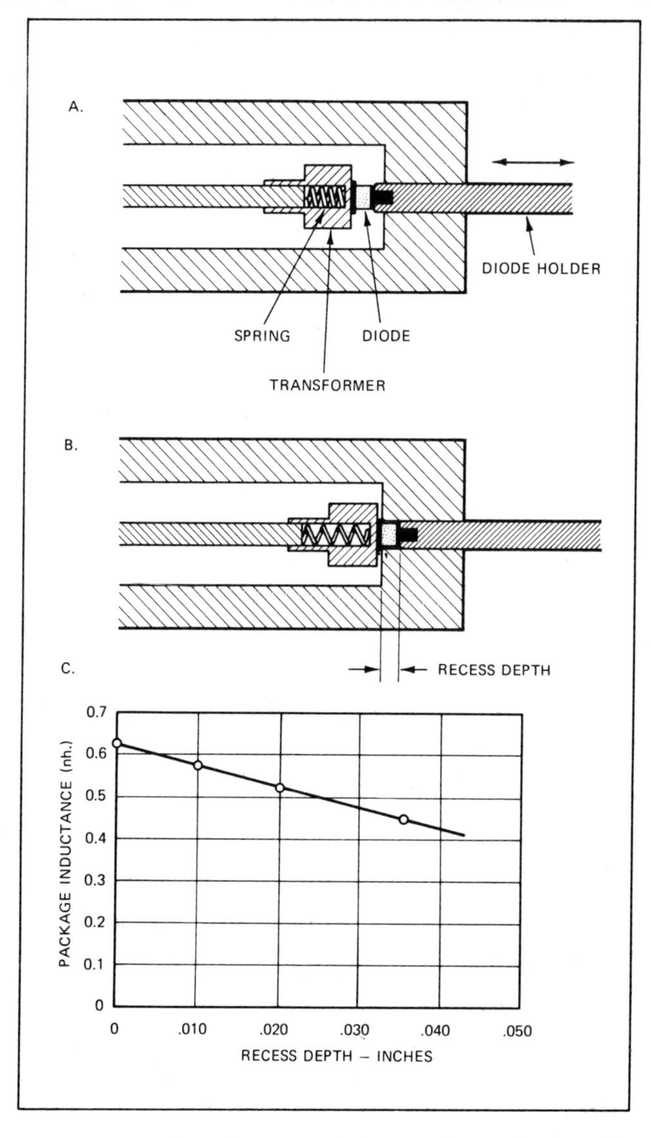

FIGURE 7.51. Effect of recessing diode on package inductance; the data are for HP style 41 package. (*Courtesy of Hewlett-Packard, AN-935.*)

R_L, as shown in Fig. 7.50(b). From this curve the peak *RF* current can be determined and hence the load power. Also, knowing the bias current and the breakdown voltage, the average input power can be calculated and hence the efficiency.

EXAMPLE 7.6

Typical values for an Impatt diode, Hewlett-Packard 5082-0400, are nominal frequency 10 GHz, $R_D = -2\ \Omega$, $C_j = 0.52$ pF at the breakdown bias of 78 V, $L_p = 0.6$ nH, and $C_p = 0.3$ pF. The bias current is 75 mA. For oscillator operation a resistive load $R_L = 2\ \Omega$ is connected across C_p. Determine (a) the resonant frequency, (b) the average power output, (c) the average power input, and (d) the conversion efficiency.

SOLUTION

(a) The effective load impedance is

$$Z_l = j\omega L_p + R_L \| C_p$$

$$= j\omega L_p + \frac{R_L}{1 + j\omega C_p R_L}$$

At the nominal operating frequency,

$$\omega C_p R_L = 2\pi \times 10^{10} \times 0.3 \times 10^{-12} \times 2$$

$$= 0.0377$$

Thus $(\omega C_p R_L)^2 << 1$ and the effective load impedance is, to a close approximation,

$$Z_L \simeq R_L + j\omega L_p$$

Circuit resonance will therefore be determined by C_j and L_p in series:

$$f = \frac{1}{2\pi\sqrt{L_p C_j}} = \frac{1}{2\pi\sqrt{0.6 \times 10^{-9} \times 0.52 \times 10^{-12}}}$$

$$= \mathbf{9.01\ GHz}$$

(b) From Fig. 7.50, at $R_L = 2\ \Omega$ and $I_0 = 75$ mA, the peak RF current is 0.6 A. Hence the power output is

$$P_L = \tfrac{1}{2} I_0^2 \mathrm{R}_L$$

$$= \tfrac{1}{2} \times 0.6^2 \times 2$$

$$= \mathbf{0.36\ W}$$

(c) The average input power is that taken from the dc source and is

$$P_{dc} = V_B \times I_0$$
$$= 78 \times 0.075$$
$$= \mathbf{5.85\ W}$$

(d) The conversion efficiency is

$$\eta = \frac{P_L}{P_{dc}}$$
$$= \frac{0.36}{5.85} \times 100\%$$
$$= \mathbf{6.15\%}$$

The circuits described so far are low-Q circuits in that they use the intrinsic diode and package reactances to tune to resonance. The circuit for a high-Q oscillator is shown Fig. 7.52(a). Here the waveguide section to the left forms a high-Q resonator, which may be tuned by means of the tuning screws shown. The diode is coupled to the resonator through the magnetic field surrounding the coaxial center conductor. This conductor is placed at a magnetic field maximum in the resonator. A coaxial line matching section is used to match the diode into the circuit. Oscillator output is obtained from the round coupling iris shown on the right. Bias to the diode is supplied through the coaxial line from the BNC connector. A wedge-shaped 50-Ω resistive load is incorporated in this part of the line. This resistance dissipates negligible oscillator power but is required for circuit stability. Figure 7.52(b) shows a photograph of the oscillator.

Impatt amplifiers use the same basic circuit arrangement as oscillators, but it is necessary that R_L be greater than the magnitude of R_D, as shown in Fig. 7.50(b). Amplification is achieved by using the negative resistance in a circulator-type amplifier as described for the varactor circuit (Fig. 7.6). With the amplifier tuned to resonance, the power gain at center frequency is given by Eq. (7.28), which in the present instance becomes

$$G = \left| \frac{R_D - R_L}{R_D + R_L} \right|^2 \tag{7.76}$$

Figure 7.53 shows how two such amplifiers may be connected in cascade. The gain of the first stage is approximately 9 dB. Because of the large drive into the second stage, its R_D value is reduced and the gain is reduced to 3 dB, giving an overall gain of 12 dB for the two-stage amplifier.

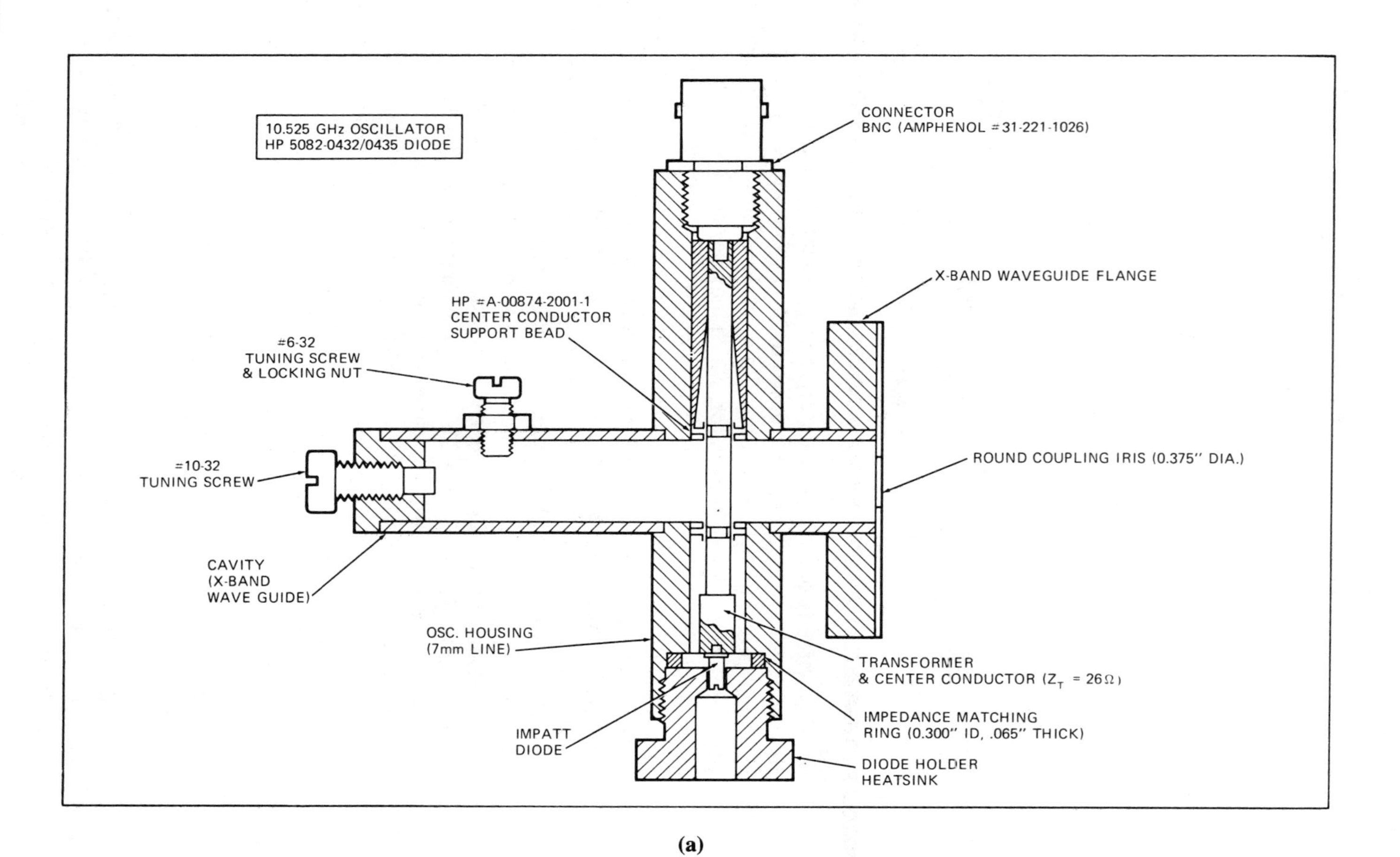

(a)

FIGURE 7.52. Coupled coax-waveguide IMPATT oscillator: (a) schematic; (b) photograph. (*Courtesy of Hewlett-Packard, AN-935.*)

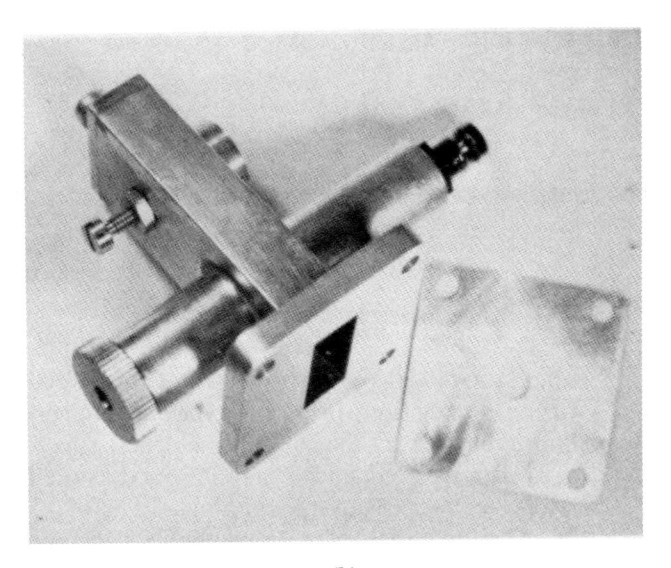

(b)

FIGURE 7.52. Continued.

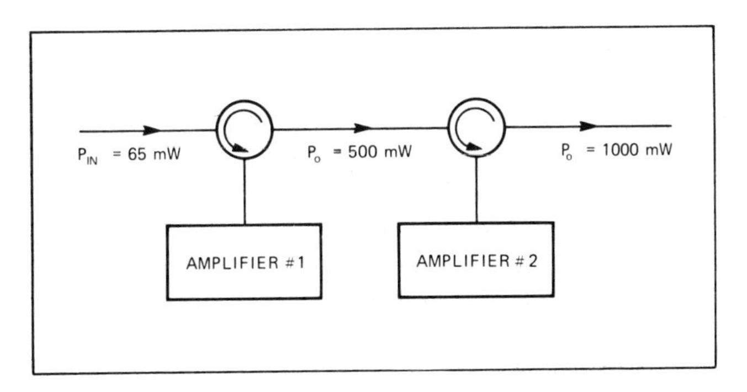

FIGURE 7.53. Cascaded IMPATT amplifiers.

EXAMPLE 7.7

A small-signal Impatt amplifier operates with an R_D of $-4\ \Omega$ and a load resistance of $5\ \Omega$. Calculate the power gain at the resonant frequency.

SOLUTION

$$G = \left|\frac{-4 - 5}{-4 + 5}\right|^2$$

$$= 81$$

In decibels this is

$$G(\text{dB}) = 10 \log 81$$
$$= \mathbf{19.1\ dB}$$

7.13 THE GUNN DIODE

The Gunn diode is named after J. B. Gunn, who in 1963 reported observing current fluctuations in certain types of semiconductors when the applied direct voltage exceeded a threshold level. These current fluctuations occur at microwave frequencies, and are used as the basis for microwave oscillators [see, e.g., Robson and Mahrous (1965)].

The material most widely used for the construction of Gunn diodes is the compound semiconductor gallium arsenide (GaAs). The diode itself consists of a thin section of GaAs sandwiched between two ohmic contacts, as shown in Fig. 7.54(a). Thus the Gunn diode differs from most other diodes in that it does not require a rectifying junction. When the applied voltage exceeds the threshold level, electrons are transferred from the main conduction band in the GaAs to a subconduction band, where their effective mass appears to increase. These heavier electrons bunch together to form an electric field domain. The electric field across the domain is greater than the average field, and because the applied voltage remains constant, the electric field across the rest of the GaAs decreases below the threshold level, thus preventing the formation of further domains. All the conduction-band electrons, including the domain bunch, drift across the GaAs at the same velocity, and the presence of the bunch of heavier electrons has the effect of reducing the drift velocity. The current, which is proportional to velocity, also decreases while the domain is present. Once the domain has passed from the GaAs into the end contact, the current returns to its higher level, but immediately, the electric field increases above threshold value and a new domain is formed.

Figure 7.54(a) shows in cross section the construction of a Gunn diode *(G.E. App. Note TPD-6104)*. The active n-layer is 10 μm thick for a frequency of oscillation of 10 GHz. The intrinsic frequency of the diode is determined by the drift velocity v and diode length L as

$$f = \frac{v}{L} \tag{7.77}$$

For GaAs $v \simeq 10^7$ cm/s, independent of applied voltage (see e.g. Robson and Mahrous 1965).

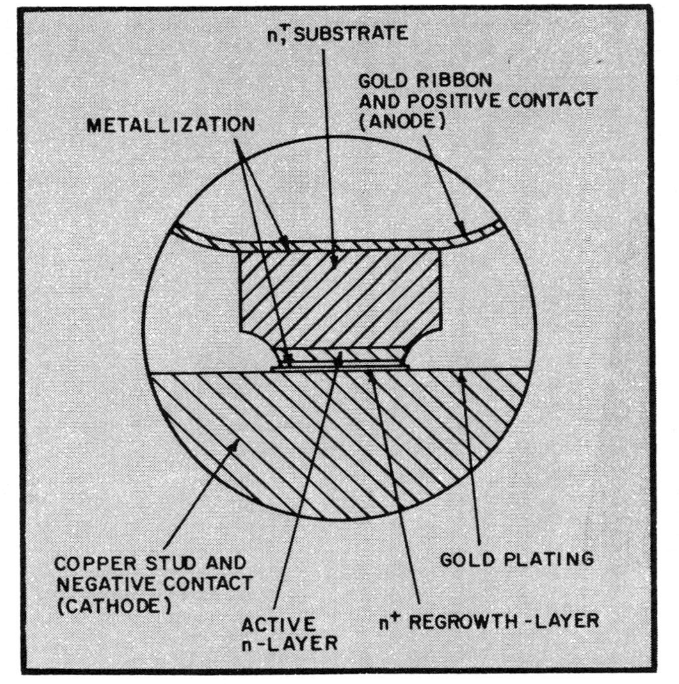

(a)

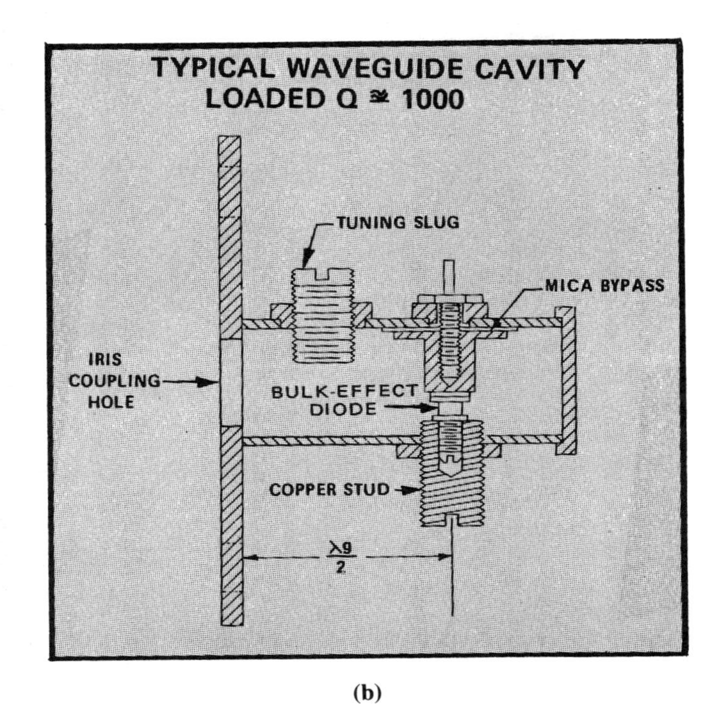

(b)

FIGURE 7.54. (a) Constructional details of the Gunn diode; (b) cutaway sketch of an X-band Gunn diode oscillator. (*Courtesy of G.E., Application Note TPD-6104.*)

Considerable heat is generated in the diode, and because GaAs itself is a poor conductor of heat, it must be well bonded onto a heat sink, which is the copper stud in Fig. 7.54(a).

The diode may be mounted in a cavity resonator as shown in Fig. 7.54(b). Normally, the cavity will be tuned to resonate at the intrinsic frequency of the diode. By reducing the resonant frequency of the cavity, the combined alternating and direct electric field appearing across the diode can be adjusted to delay onset of the threshold level, and in this way the output frequency is reduced to that of the cavity. The cavity may also be tuned to a higher frequency. In this case, the combined field drops below the value need to sustain the domain for certain periods in the cycle. The domain is quenched before it reaches the electrode, and therefore the transit time is shortened or the frequency is increased. Figure 7.54(b) is a cutaway sketch of an X-band oscillator.

The Gunn diode also exhibits negative resistance and can therefore be used in a circulator-type amplifier of the type shown in Fig. 7.6.

7.14 THE TUNNEL DIODE

The tunnel diode is a *p-n* junction diode in which both sides are very heavily doped. In a normal *p-n* junction, an energy barrier is formed by the exposed donor and acceptor sites in the depletion region, and this energy barrier balances carrier diffusion. Forward bias applied to such a junction has the effect of lowering the barrier, permitting current flow. When the doping levels are increased by a factor of 1000 or more on both sides, as is done in the tunnel diode, the width of the depletion region becomes very thin. In addition, an overlap occurs between the conduction-band level on the *n*-side and the valence-band level on the *p*-side. Under slight forward bias, conduction-band electrons can "tunnel" through the narrow depletion region to the same available energy levels on the *p*-side, giving rise to a forward current which at first increases with increase in voltage as shown in Fig. 7.55. This current is considerably larger than the normal diffusion current that occurs for a *p-n* junction under forward bias. As the forward bias is increased further, the conduction-band electron energies are eventually raised above the available energy states in the valence band and appear level with the forbidden energy band. Direct tunneling between energy bands cannot now occur and the band-to-band tunneling current decreases. It is this decrease in current with increase in applied voltage, shown as region A to B in Fig. 7.55, which gives the tunnel diode its negative resistance characteristic and makes it a useful amplifier and oscillator device. A second tunneling mechanism occurs which causes the current to increase again after point B. Electrons in the conduction band can drop to lower energy levels provided by impurities such as impurity atoms and defects within the semiconductor crystal structure. At the higher voltages, these electrons can then tunnel from the impurity states to the valence band. The resulting current increases with forward bias. At the same time the normal injection current for the *p-n* junction also becomes significant, and the current–voltage curve has a positive slope, as shown in Fig. 7.55. It should be noted that the tunneling current consists of majority carriers, for example, electrons tunneling from the *n*-side to the *p*-side, whereas the normal *p-n* injection current consists of minority carriers, for example, holes being injected from the *p*-side into the *n*-side. Majority carrier current responds very much faster to voltage changes than minority carrier current, and hence operation in the tunneling region is ideally suited to microwaves.

Under reverse bias there is considerable overlap between conduction and valence bands, so that carriers can tunnel through to available energy states on opposite sides, and a large reverse bias current flows, as shown in Fig. 7.55.

The curve through the origin is nonlinear and at relatively small values of reverse bias a large current flows. The diode is used in this region as a passive frequency converter, when it is referred to as a backward diode. The negative-resistance portion of the characteristic, *A* to *B* in Fig. 7.55, is used in tunnel diode amplifiers and oscillators. The Intelsat communications satellite series have relied on the tunnel diode amplifier in the spacecraft to provide low-noise front-end amplification at 6 GHz and computer studies have shown that tunnel diode amplifiers (TDAs) are suitable

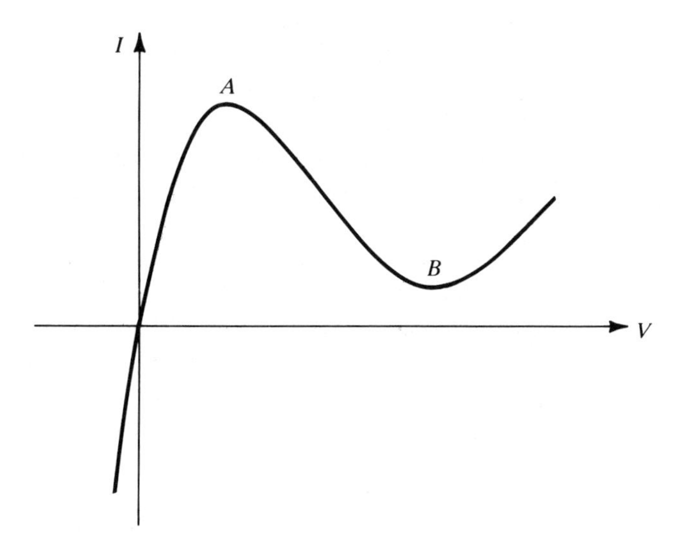

FIGURE 7.55. Current–voltage characteristic for a tunnel diode.

for the 14-GHz band [see Mott (1979) and Watson (1969)]. These amplifiers are of the circulator type shown schematically in Fig. 7.6.

A number of different semiconductor materials have been used in the construction of tunnel diodes, the choice of material largely being dependent on the application (e.g., oscillator, amplifier, or backward diode). Silicon is not a good material, and the preferred materials at present seem to be germanium (Ge) and gallium arsenide (GaAs).

7.15 PROBLEMS

1. Plot the capacitance $C(V)$ as a function of voltage V in the range $-10 \leq V \leq 0$, for an abrupt *p-n* junction diode. Assume that $\phi = \frac{1}{2}$ and $C_{j0} = 1$ pF.
2. Plot the charge–voltage curve for the diode in Problem 1. Find C when $V = -5$ V.
3. Repeat Problem 1 for a linear graded junction.
4. Repeat Problem 2 for a linear graded junction.
5. A sinusoidal signal having an angular frequency of 10^{10} rad/s and a peak value 5 V is superimposed on a reverse bias of 5 V, applied to the *p-n* junction of Problem 1. Plot the capacitance and elastance as functions of time over one complete cycle.
6. For a parametric diode amplifier, the Q-factors are $Q_i = 2.5$, $Q_s = 12$, and $R_s = 4\ \Omega$. Assuming that $R_L = R_1 = R_2 = 0$, calculate (a) the negative resistance component and (b) the input resistance.
7. Given that the characteristic impedance of the amplifier of Problem 6 is 50 Ω, calculate (a) the reflection coefficient and (b) the power gain.
8. Explain the principles of operation of an up-converter amplifier.
9. Explain what is meant by the notation 1-2-3 tripler when applied to varactor frequency multipliers. Given that the input frequency to such a multiplier is 3 GHz, determine the idler frequency and the output frequency.
10. Explain what is meant by a step-recovery diode and how this may be used as a frequency multiplier.
11. The recombination time for a PIN diode is 20 ns. Determine the frequency above which the equivalent circuit for the diode is linear.
12. Explain the approximations involved in reducing the equivalent circuit of Fig. 7.13(a) to that of Fig. 7.14(b).
13. For a PIN diode equivalent circuit $C_c = 0.32$ pF, $C_j = 0.15$ pF, and $R_s = 0.25\ \Omega$. The series inductance L_s may be assumed negligible. Show that $(\omega C_j R_s)^2 << 1$, and calculate the admittance of the diode at a frequency of 3 GHz.

14. For a PIN diode under forward bias, $R_f = 2\ \Omega$, $L_s = 0.3$ nH, and $C_c = 0.3$ pF. Calculate the diode admittance at a frequency of 2 GHz.

15. A PIN diode operated between reverse and forward bias as a reflective attenuator has $C_c = 0.25$ pF, $C_j = 0.15$ pF, $R_s = 0.25\ \Omega$, $R_f = 2\ \Omega$, and $L_s = 0.3$ nH. Calculate (a) the insertion loss, in decibels, under reverse-bias conditions; (b) the value of the parallel inductance needed to minimize the insertion loss; and (c) the minimum insertion loss. The frequency of operation is 4 GHz and $Z_o = 50\ \Omega$.

16. For the diode in Problem 15, calculate (a) the isolation, in decibels, achieved under forward bias; (b) the value of series capacitance needed to maximize the isolation; and (c) the maximum isolation.

17. Explain the difference between PIN diode reflective and absorptive attenuators, and state when the absorptive type would be used in preference to the reflective type.

18. Explain what is meant by differential phase shift. Describe one type of phase shifter which utilizes PIN diode switching of differential phase shift.

19. Draw the circuit for a square-law diode detector and explain its operation. State the range of signal level over which the circuit operates.

20. Draw the equivalent circuit for a Schottky barrier diode. Briefly explain the significance of each component of the equivalent circuit.

21. Explain the principles of operation of a low-level detector and state why this is also referred to as a square-law detector.

22. The dc through a point contact diode is 1 mA. Determine the slope conductance and slope resistance, assuming that $n = 1.4$ and $V_T = 26$ mV.

23. A sinusoidal microwave signal of peak voltage 50 mV is applied to the diode of Problem 22. Calculate the resulting dc component.

24. Define and explain what is meant by current sensitivity as applied to low-level detectors. The current sensitivity for a detector is specified as 10 A/W. Calculate the current change produced by an input power of -40 dBm.

25. Define and explain what is meant by voltage sensitivity as applied to low-level detectors. The video resistance for the detector of Problem 24 is 1200 Ω. Calculate the voltage sensitivity.

26. Explain what is meant by the power loss of a low-level detector. A low-level detector is adjusted so that $R_j = R_s = 8\ \Omega$ and $C_j = 0.35$ pF. Calculate the power loss in decibels when the frequency of operation is 6 GHz.

27. Explain what is meant by tangential sensitivity and state briefly the main factors that affect the tangential sensitivity measurement. Two diodes measured under identical conditions yield tangential sensitivity figures of -35 dBm and -40 dBm. Which diode is the most sensitive?

28. In a diode mixer circuit the signal frequency is 4030 MHz and the oscillator frequency is 3960 MHz. Determine the sum frequency, the image frequency, and the IF.

29. Explain what is meant by an image recovery mixer. Draw the schematic for an image recovery mixer and describe briefly the principle of operation.

30. Explain briefly what is meant by (a) a resistively terminated image mixer and (b) a reactively terminated image mixer. List the principal advantages and disadvantages of each type.

31. Draw the schematics for, and explain the principles of operation of, (a) a single-balanced and (b) a double-balanced mixer.

32. Define and explain the term *conversion loss* as applied to mixer circuits.

33. Define and explain the term *noise factor* as applied to mixer circuits. The conversion loss of a mixer is 6 dB. What is the approximate value of noise figure?

34. Explain what is meant by isolation in a mixer circuit and why high isolation is desirable.

35. With reference to an input-output curve, explain what is meant by conversion compression and by the 1-dB compression point.

36. Explain what is meant by the dynamic range of a mixer and the factors that determine the limits of the range.

37. Explain what is meant by the third-order intercept point. With reference to Fig. 7.45, show that the intercept Z on the input axis is given by $P + D/2$.

38. Explain the principles of operation of an Impatt diode oscillator. Describe briefly the differences between single-drift and double-drift Impatt diodes.

39. Draw the equivalent circuit for an Impatt diode and its package and briefly explain the significance of each component.

40. Equivalent-circuit values for an Impatt diode oscillator are $R_D = -2.2\ \Omega$, $C_j = 0.5$ pF at a breakdown voltage of 80 V, $L_p = 0.55$ nH, and $C_p = 0.3$ pF. The bias current is 75 mA. The oscillator operates into a resistive load of 2.2 Ω connected across C_p. Determine (a) the resonant frequency, (b) the average power output, (c) the average power input, and (d) the conversion efficiency. The curves of Fig. 7.50 apply to the diode.

41. Explain how an Impatt diode may be used to provide signal amplification. A small-signal Impatt diode amplifier has an R_D of $-3.5\ \Omega$ and is operated into a load of 6 Ω. Calculate the power gain in decibels at the resonant frequency.

42. Describe briefly how negative resistance occurs in a Gunn diode. A Gunn diode constructed in gallium arsenide has a drift length of 10 μm. Determine the intrinsic frequency of the diode oscillations.

43. Explain what is meant by the tunneling effect in a tunnel diode and how this gives rise to negative resistance. How could such a diode be used as (a) a passive frequency converter, (b) an amplifier, and (c) an oscillator?

REFERENCES

G.E. Application Note TPD-6104. Bulk Effect Diodes and Solid State Microwave Circuit Modules.

Hewlett-Packard Application Note 923, Hot Carrier Diode Video Detectors.

Hewlett-Packard Application Note 957-1, Broadbanding the Shunt PIN Diode SPDT Switch.

Hewlett-Packard Application Note 961, Silicon Double-Drift IMPATT Diodes for Pulse Applications.

Hewlett-Packard Application Note 962, Silicon Double-Drift IMPATTs for High Power CW Applications.

MOTT, RICHARD C., 1979. "Tunnel diodes keep pace with Ku-band satellites," *Microwaves*, February.

ROBSON, P. N., and S. M. MAHROUS, 1965. "Some aspects of Gunn effect oscillators," *Radio Electron. Eng.*, December.

RODDY, D., and JOHN COOLEN, 1984. *Electronic Communications*, 3rd ed., Reston, Va.: Reston Publishing Company, Inc.

WATSON, H. A. (ed.), 1969. *Microwave Semiconductor Devices and Their Circuit Applications*. New York: McGraw-Hill Book Company.

eight
Transistor Amplifiers

8.1 INTRODUCTION

Microwave transistor amplifiers can be broadly classified into three groups: (1) small-signal, low-noise amplifiers; (2) small-signal, linear power amplifiers; and (3) large-signal power amplifiers. The circuit configurations for the first two are similar, but with different design objectives. In the first case the noise figure is optimized, while in the second, the linear gain is optimized. These two aspects of amplifier design are discussed in this chapter. Power amplifiers required to deliver relatively large amounts of power under nonlinear operating conditions (such as class C) are not covered.

The two types of transistors most widely used at present are the silicon bipolar transistor (BJT), and the gallium arsenide field-effect transistor (GaAsFET). The bipolar transistor provides useful amplification up to about 6 GHz, while the GaAS FET is used in the higher bands, up to about 18 GHz.

8.2 MICROWAVE BIPOLAR JUNCTION TRANSISTORS

Microwave bipolar transistors are constructed in planar form; that is, the emitter and base electrodes are diffused in from the top surface of the silicon, and the collector connection is made by way of the substrate. The cross section of a planar transistor is shown in Fig. 8.1

The input capacitance of the transistor is directly proportional to the

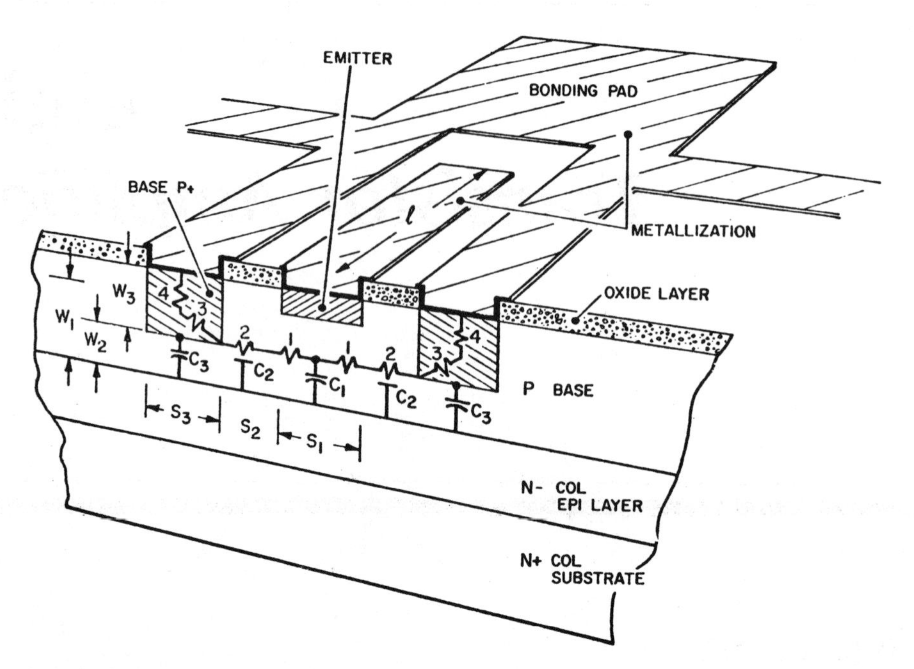

FIGURE 8.1. Cross section of a planar transistor showing r_b' and capacitance; 1, 2, and 3 are parts of r' and 4 is R_c (contact resistance). (*From Harry F. Cooke, Microwave Transistors—Theory and Design, Proc. IEEE, Vol. 59, Aug. 1971; © IEEE 1971, with permission.*)

emitter area, and hence in order to get good high-frequency amplification, the area has to be kept small. At the same time, the current-carrying capacity is proportional to emitter periphery, and the base resistance, which should be as small as possible, is inversely proportional to emitter length. To meet these requirements, a long, narrow emitter is required in principle. In practice this is achieved by forming a number of short emitter sites, which are then interconnected by means of a top metallization layer. Figure 8.2 shows the three principal types of geometry used. With the interdigitated type [Fig. 8.2(a)], the emitter and base stripes are interleaved as shown. In the overlay type [Fig. 8.2(b)], the emitter sites are interconnected by a metallization layer which overlays the base sites. In the mesh type [Fig. 8.2(c)], a high periphery-to-area ratio is achieved through use of smaller base sites, which are then interconnected by a metallization layer. The overlay and mesh types are more widely used for VHF-UHF power applications, while the interdigitated type is preferred for S-band and C-band small signal applications (Cooke 1971).

Ballast resistors are included in power transistors to equalize the current flow to the base-emitter junctions formed at the various sites. Overlay and mesh transistors use diffused resistors which are integral with

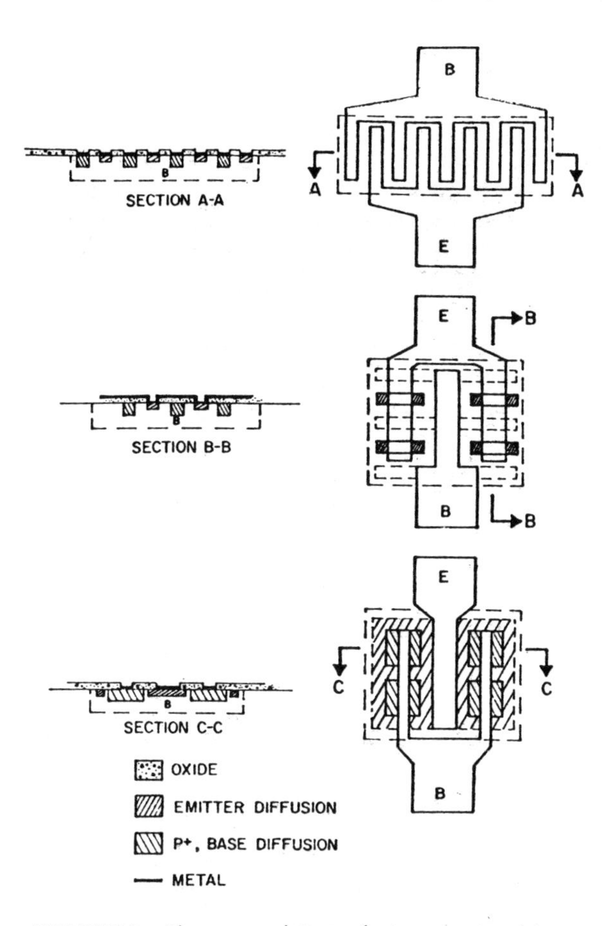

FIGURE 8.2. Three general types of microwave transistor geometry. (*From Harry F. Cooke, Microwave Transistors—Theory and Design, Proc. IEEE, Vol. 59, Aug. 1971; © IEEE 1971, with permission.*)

the emitter (or base) electrodes. Thin-film resistors are generally used in series with the emitter fingers in the interdigitated type, and thin-film resistors have also been used with the mesh type as an alternative to diffused resistors. Without current equalization two undesirable and potentially damaging effects can occur. One effect is associated with the temperature at the center of the transistor, which is generally higher than that at the edge. This can result in greater current flow through the center section, which in turn causes further heating at the center. The effect can be cumulative unless current limiting is employed. The other undesirable effect, known as secondary breakdown, occurs where there is unequal channeling of the current. The unequal current flow can result in high-current-density spots forming, which in turn can lead to burnout in the base region.

8.3 THE GaAs FET

A cross section of a gallium arsenide field-effect transistor is shown in Fig. 8.3. Gallium arsenide is a semi-insulating material, and the conducting channel is formed in the *n*-type epitaxial layer grown on the top surface. Ohmic source and drain contacts are formed, one method being to alloy in gold-germanium contacts. The gate metal is deposited directly on top of the epitaxial channel and forms a Schottky barrier contact. Application of a negative bias between gate and source increases the depletion width of the barrier, and this reduces the width of the conducting channel between source and drain. In this way, a small-signal voltage applied between gate and source modulates the source–drain current. Figure 8.4 shows a scanning electron microscope photograph of a transistor structure.

The GaAs FET is also referred as a MESFET, this being an acronym for Metal Semiconductor Field-Effect Transistor. For power applications,

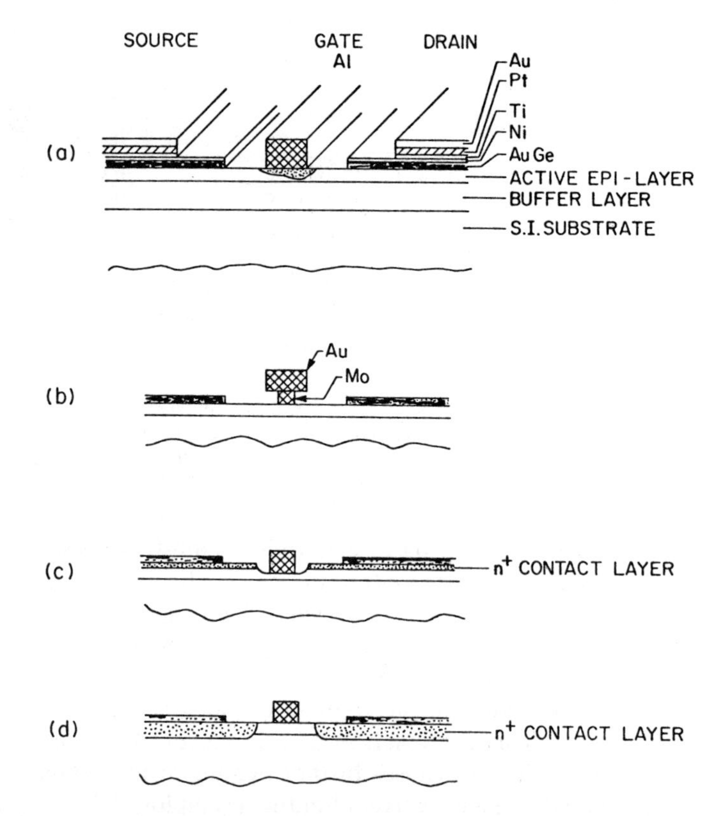

FIGURE 8.3. Fundamental structures of low-noise GaAs FETs: (a) flat-type FET; (b) improved gate structure; (c) recessed structure; (d) FET with a selective n^+ contact layer. [*From F. Hasegawa, "Low Noise GaAs FETs," in GaAs FET Principles and Technology, ed. J. DiLorenzo and D. Khandelwal (Dedham, Mass.: Artech House, 1982); with permission.*]

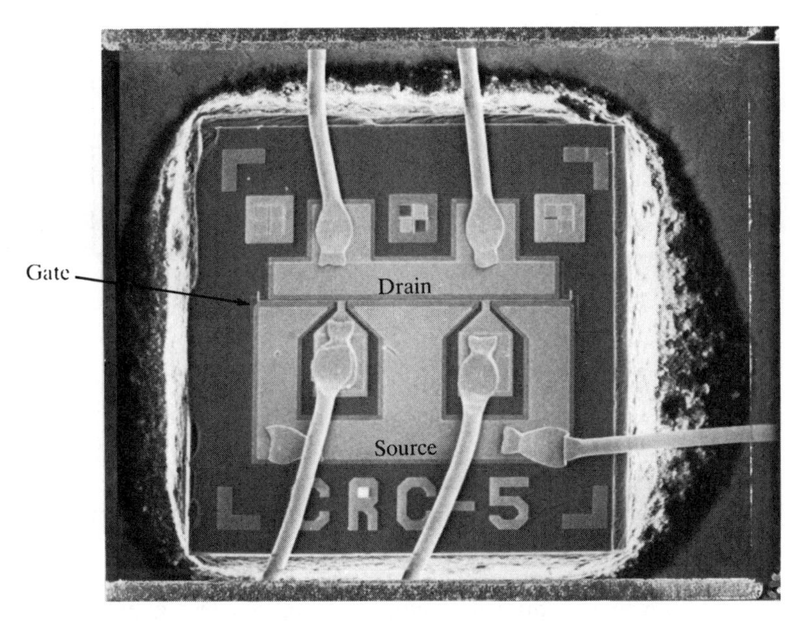

FIGURE 8.4. GaAs FET. The gate width is 1 μm. Note the double-gate connection used to reduce resistance. (*Courtesy of John Noad, Communications Research Centre, Department of Communications, Ottawa, Canada.*)

source and drain can be interdigitated as discussed for the silicon BJT. An interdigitated structure is illustrated in the data sheet for the AT-8141 in Section 8.14. A number of other types of field-effect transistors, including silicon types, are described by Liechti (1976).

8.4 THE UNILATERAL SMALL-SIGNAL AMPLIFIER

In this section the transistor amplifier will be treated as a linear two-port. In this way, quite general expressions can be developed for gain and noise performance, without reference to the type of transistor used or circuit configuration. Once the significant gain and noise parameters have been identified, these will be examined for particular circuit conditions in later sections.

S-parameters are used to characterize microwave transistors and amplifier circuits both for small-signal and large-signal conditions. Here only the small-signal conditions are considered, this generally being required for linear operation. *S*-parameters are described in detail in Chapter 2 and the transducer power gain of a linear two-port is given by Eq. (2.55). This is identical to the power gain of the amplifier considered as a linear two-port. Figure 2.6 may be redrawn as shown in Fig. 8.5, in which the source may, for example, be a receiving antenna, and the load the input of a

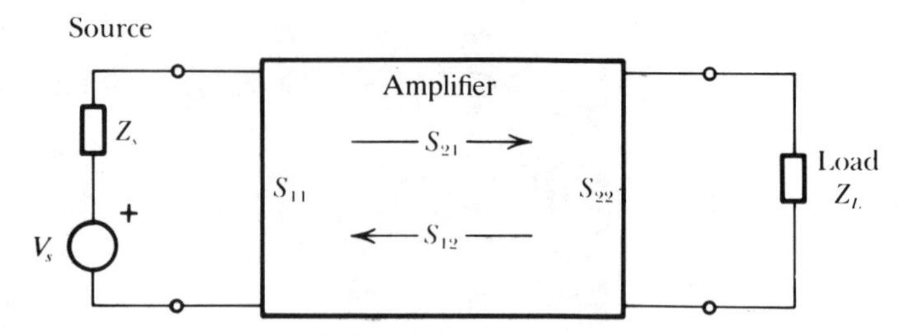

FIGURE 8.5. Microwave amplifier as a linear two-port.

mixer stage. The function of the amplifier is to deliver small-signal power P_L to the load, which is an amplified version of the available source power P_{av}, and the transducer power gain, as defined by Eq. (2.47), is P_L/P_{av}.

In a unilateral amplifier the signal is transmitted in one direction only, from input to output, there being no reverse transmission of signal from output to input. Unilateral operation is of course the desired mode, but in practice some leakage of signal from output to input may take place, which complicates amplifier design. The general principles of operation are more easily explained with reference to the unilateral amplifier, and this approach is followed here.

For an amplifier to be unilateral, the reverse transmission coefficient S_{12} (see Table 2.1) becomes zero:

$$S_{12} = 0 \tag{8.1}$$

When this condition is substituted in Eq. (2.55), and after some rearrangement of terms, the unilateral transducer gain is obtained as

$$G_{\mathrm{Tu}} = \frac{1 - \rho_s^2}{|1 - S_{11}\Gamma_s|^2} |S_{21}|^2 \frac{1 - \rho_L^2}{|1 - S_{22}\Gamma_L|^2} \tag{8.2}$$

The right-hand side of this equation has been grouped into three distinct terms, the first depending only on the source side coefficients Γ_s and S_{11}, the second depending only on the forward transmission coefficient S_{21}, and the third depending only on the load side coefficients Γ_L and S_{22}. Thus three distinct gain terms can be identified:

$$G_S = \frac{1 - \rho_s^2}{|1 - S_{11}\Gamma_s|^2} \tag{8.3}$$

$$G_0 = |S_{21}|^2 \tag{8.4}$$

$$G_L = \frac{1 - \rho_L^2}{|1 - S_{22}\Gamma_L|^2} \tag{8.5}$$

TABLE 8.1
Typical S-Parameters for Hewlett-Packard GaAs FET (HFET-2202)[a]

	S_{11}		S_{21}			S_{12}			S_{22}	
Frequency (GHz)	*Mag.*	*Ang.*	*(dB)*	*Mag.*	*Ang.*	*(dB)*	*Mag.*	*Ang.*	*Mag.*	*Ang.*
2.0	0.96	−51	8.72	2.73	134	−34.4	0.02	61	0.79	−23
3.0	0.89	−75	8.27	2.59	113	−31.7	0.03	49	0.76	−35
3.5	0.87	−87	8.07	2.53	103	−31.1	0.03	43	0.76	−41
3.6	0.86	−89	8.04	2.52	101	−30.8	0.03	43	0.75	−43
3.7	0.86	−92	8.01	2.52	99	−30.7	0.03	43	0.75	−44
3.8	0.85	−94	7.92	2.49	97	−30.5	0.03	41	0.74	−46
3.9	0.84	−97	7.89	2.48	94	−30.5	0.03	40	0.74	−47
4.0	0.84	−99	7.85	2.47	93	−30.5	0.03	40	0.74	−48
4.1	0.84	−102	7.83	2.46	91	−30.5	0.03	38	0.74	−49
4.2	0.84	−104	7.74	2.44	88	−30.2	0.03	38	0.74	−50
4.3	0.83	−107	7.69	2.43	86	−29.9	0.03	36	0.74	−51
4.4	0.82	−109	7.69	2.43	84	−29.9	0.03	36	0.73	−52
4.5	0.82	−111	7.64	2.41	82	−29.9	0.03	35	0.73	−54
5.0	0.80	−122	7.35	2.34	72	−29.4	0.03	31	0.72	−61
6.0	0.76	−143	6.90	2.21	52	−29.1	0.04	29	0.71	−75
7.0	0.72	−164	6.21	2.04	33	−28.4	0.04	27	0.68	−91
8.0	0.68	178	5.51	1.88	15	−27.1	0.04	27	0.67	−107
9.0	0.68	159	4.91	1.76	−3	−25.7	0.05	25	0.65	−124
10.0	0.67	143	4.36	1.65	−20	−23.9	0.06	20	0.66	−142
11.0	0.65	128	3.83	1.55	−37	−21.6	0.08	4	0.67	−160
12.0	0.62	112	3.24	1.45	−55	−21.1	0.09	1	0.68	−176

[a] $R_0 = 50\ \Omega$; high-gain bias: $V_{DS} = 3.5$ V, $V_{GS} = 0$ V; S_{11}, input reflection coefficient for $a_2 = 0$; S_{21}, forward transmission coefficient for $a_2 = 0$; S_{12}, reverse transmission coefficient for $a_1 = 0$; S_{22}, output reflection coefficient for $a_1 = 0$.

In terms of these three expressions, the overall transducer gain is

$$G_{\text{tu}} = G_S G_0 G_L \tag{8.6}$$

Some typical S-parameters are shown in Table 8.1. It will be seen from these that the magnitude of S_{12} is small compared to the other values, and in the following examples, S_{12} will be assumed equal to zero.

EXAMPLE 8.1

Transistor HFET-2202 is used in an amplifier circuit at 4 GHz, for which the S-parameters given in Table 8.1 apply. Calculate the power gain if the source and load impedances are each 50 Ω purely resistive.

SOLUTION From Table 8.1, at 4 GHz:

$$|S_{12}| = 0.03$$

$$|S_{21}| = 2.47$$

Assume that S_{12} is negligible so that unilateral theory applies; then

$$G_0 = |2.47|^2$$
$$= 7.85 \text{ dB}$$

From Eq. (2.19),

$$\Gamma_s = \frac{50 - 50}{50 + 50} = 0$$

Hence from Eq. (8.3),

$$G_s = 1$$
$$= 0 \text{ dB}$$

Similarly, $\Gamma_L = 0$ and therefore $G_L = 0$ dB. Hence

$$G_T = 0 + 7.85 + 0$$
$$= 7.85 \text{ dB}$$

EXAMPLE 8.2

The amplifier in Example 8.1 is used with a 100-Ω source and a 500-Ω load, both purely resistive. Calculate the power gain.

SOLUTION As before, $G_0 = 7.85$ dB.
From Eq. (2.19),

$$\Gamma_s = \frac{100 - 50}{100 + 50}$$
$$= \frac{1}{3}$$

From Table 8.1,

$$S_{11} = 0.84 \angle -99^\circ$$

Therefore,

$$1 - \Gamma_s S_{11} = 1 - \frac{1}{3} \times 0.84 \angle -99^\circ$$
$$= 1 - 0.28 \angle -99^\circ$$
$$= 1 - 0.28 \cos 99^\circ + j0.28 \sin 99^\circ$$
$$= 1.044 + j0.277$$

Then

$$|1 - \Gamma_s S_{11}|^2 = 1.166$$

and from Eq. (8.3),

$$G_s = \frac{1 - (\frac{1}{3})^2}{1.166}$$
$$= 0.762$$
$$= -1.179 \text{ dB}$$
$$\Gamma_L = \frac{500 - 50}{500 + 50}$$
$$= 0.818$$

From Table 8.1, $S_{22} = 0.74 \angle -48°$. Therefore,

$$1 - \Gamma_L S_{22} = 1 - 0.818 \times 0.74 \angle -48°$$
$$= 1 - 0.605 \cos 48° + j0.605 \sin 48°$$
$$= 0.595 + j0.450$$

Then

$$|1 - \Gamma_L S_{22}|^2 = 0.556$$

and

$$G_L = \frac{1 - (0.818)^2}{0.556}$$
$$= 0.595$$
$$= -2.25 \text{ dB}$$

Hence

$$G_T = -1.1 + 7.85 - 2.25$$
$$= 4.42 \text{ dB} \quad \text{or} \quad \text{a power gain ratio of } 2.77 : 1$$

From these two examples it will be seen that the external loading at both input and output has a marked effect on the gain, through G_s and G_L. The question then naturally arises: How can G_s and G_L be maximized?

8.5 MAXIMUM UNILATERAL POWER GAIN

It will be seen that the equations for G_s and G_L are similar, the general form being

$$G = \frac{1 - \rho^2}{|1 - \Gamma S|^2} \tag{8.7}$$

The problem then is: Given a value for S, what value of Γ maximizes G? Since both S and Γ are complex in general, two conditions must be sought. Let

$$S = |S| \angle\theta \qquad \text{and} \qquad \Gamma = \rho \angle\phi$$

Then

$$\Gamma S = \rho |S| \angle\theta + \phi$$

and the denominator for G is

$$\begin{aligned} |1 - \Gamma S|^2 &= [1 - \rho |S| \cos(\theta + \phi)]^2 + [\rho |S| \sin(\theta + \phi)]^2 \\ &= 1 + \rho^2 |S|^2 - 2\rho |S| \cos(\theta + \phi) \end{aligned}$$

Considering ϕ first, it will be seen that $|1 - \Gamma S|^2$ is a minimum, and hence G a maximum, when $\cos(\theta + \phi) = 1$, or

$$\theta = -\phi \tag{8.8}$$

With this condition in effect, Eq. (8.7) for G becomes

$$\begin{aligned} G &= \frac{1 - \rho^2}{1 + \rho^2 |S|^2 - 2\rho |S|} \\ &= \frac{1 - \rho^2}{[1 - \rho |S|]^2} \end{aligned}$$

The condition for maximum can be found by differentiating G with respect to ρ and equating to zero, yielding the result

$$\rho = |S| \tag{8.9}$$

Combining Eq. (8.9) and (8.8), it is seen that to maximize G,

$$\Gamma = S^* \tag{8.10}$$

Specifically, this means that for maximum G_s,

$$\Gamma_s = S_{11}^* \tag{8.11}$$

and for maximum G_L,

$$\Gamma_L = S_{22}^* \tag{8.12}$$

Equation (8.11) effectively states that the input impedance must conjugately match the source, a result that could have been deduced from the maximum power transfer theorem. Equation (8.12) effectively states that the load reflection coefficient must conjugately match the output reflection coefficient, but in this case it is not readily apparent what impedance match is involved, since the output impedance of the amplifier has not been defined.

With conditions (8.11) and (8.12) inserted in the gain equations (8.3) and (8.5), the maximum gain equations become

$$G_{s\max} = \frac{1}{1 - |S_{11}|^2} \tag{8.13}$$

$$G_{L\max} = \frac{1}{1 - |S_{22}|^2} \tag{8.14}$$

EXAMPLE 8.3

For the amplifier in Example 8.1, calculate the maximum power gain, and the source and load impedances required to realize this.

SOLUTION As before,

$$S_{12} = 0 \text{ (assumed)}$$

$$S_{21} = 2.47$$

Therefore,

$$G_0 = (2.47)^2$$

$$= 7.85 \text{ dB}$$

From Eq. (8.13),

$$G_{s\max} = \frac{1}{1 - |0.84|^2}$$

$$= 3.397$$

$$= 5.31 \text{ dB}$$

From Eq. (8.14),

$$G_{L\max} = \frac{1}{1 - |0.74|^2}$$
$$= 2.21$$
$$= 3.44 \text{ dB}$$

Therefore,

$$G_{T\max} = 5.31 + 7.85 + 3.44$$
$$= 16.6 \text{ dB}$$

The conditions required for maximum gain are Eq. (8.11) and (8.12):

$$\Gamma_s = S_{11}^*$$
$$= 0.84 \angle 99^\circ$$

and

$$\Gamma_L = S_{22}^*$$
$$= 0.74 \angle 48^\circ$$

Hence, from Eq. (2.18),

$$\frac{Z_s}{50} = \frac{1 + 0.84 \angle 99^\circ}{1 - 0.84 \angle 99^\circ}$$
$$= \frac{1 + 0.84 \cos 99^\circ + j0.84 \sin 99^\circ}{1 - 0.84 \cos 99^\circ - j0.84 \sin 99^\circ}$$
$$= 0.1495 + j0.843$$

Hence

$$Z_s = 50(0.1495 + j0.843)$$
$$= 7.48 + j42.2\ \Omega \text{ (or } 42.8 \angle 79.9^\circ\ \Omega)$$

Also

$$\frac{Z_L}{50} = \frac{1 + \Gamma_L}{1 - \Gamma_L}$$

$$= \frac{1 + 0.74 \angle 48^\circ}{1 - 0.74 \angle 48^\circ}$$

$$= \frac{1 + 0.74 \cos 48^\circ + j0.74 \sin 48^\circ}{1 - 0.74 \cos 48^\circ - j0.74 \sin 48^\circ}$$

$$= 1.897 + j1.974$$

Hence

$$Z_L = 50(1.897 + j1.974)$$

$$= 94.86 + j98.68\ \Omega \text{ (or } 136.9 \angle 46.13^\circ\ \Omega)$$

8.6 INPUT AND OUTPUT NETWORKS

Source and load impedances must have specific values as dictated by Eqs. (8.11) and (8.12), for maximum gain. There are other situations where impedance transformation may be desirable, for example in minimizing the noise figure, which will be discussed later. In general, there is the requirement for source and load matching networks as shown in Fig. 8.6. This illustrates a common situation in microwave amplifier design where the amplifier must work between a 50-Ω source and a 50-Ω load.

The matching networks are usually lossless (negligible resistance), so, in general, *LC* components are used. At microwaves these are most easily realized using transmission line sections as described in Section 1.15. Design options are quite flexible, since length of line, characteristic impedance, and use of open- and short-circuited stubs are variables available to the designer. Microstrip and stripline are the two most common forms of transmission line encountered in circuit applications.

Because it is simpler to connect transmission-line elements in parallel rather than in series, it becomes desirable to work in terms of admittances rather than impedances. In general, admittance y is equal to $1/z$, and in particular, the characteristic admittance is $Y_0 = 1/Z_0$. Thus, in terms of normalized values,

$$y = \frac{Y}{Y_0} = \frac{1}{z} \tag{8.15}$$

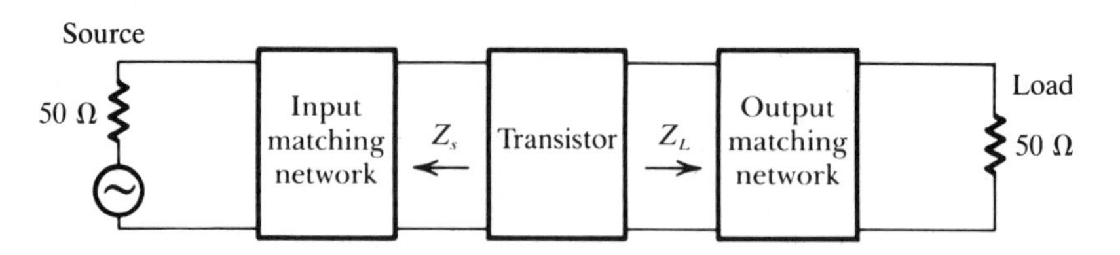

FIGURE 8.6. Three main circuit blocks of a microwave amplifier: the input matching network, the transistor gain block, and the output matching network.

Equation (2.18) may be rearranged as

$$\Gamma = \frac{1 - y}{1 + y} \tag{8.16}$$

The two examples that follow illustrate the design principles for matching networks.

EXAMPLE 8.4

Transistor 2N-3563 (Fairchild) has $S_{11} = 0.5 + j0.1$, normalized to 50 Ω at 1 GHz (see *Hewlett-Packard Application Note 77-1*). Assuming that the reverse transmission coefficient is zero, design an input-matching network to maximize the gain term G_s for a source impedance of 50 $\angle 0$ Ω.

SOLUTION The line layout of Fig. 8.7(a) will be tried, in which the line elements have a characteristic impedance of 50 Ω, and the line lengths ℓ_1 and ℓ_2 are the design variables. From Eq. (8.11),

$$\Gamma_s = S_{11}^*$$
$$= 0.5 - j0.1$$

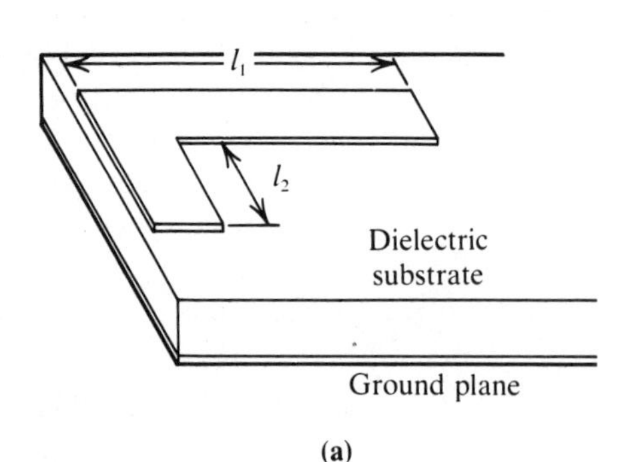

(a)

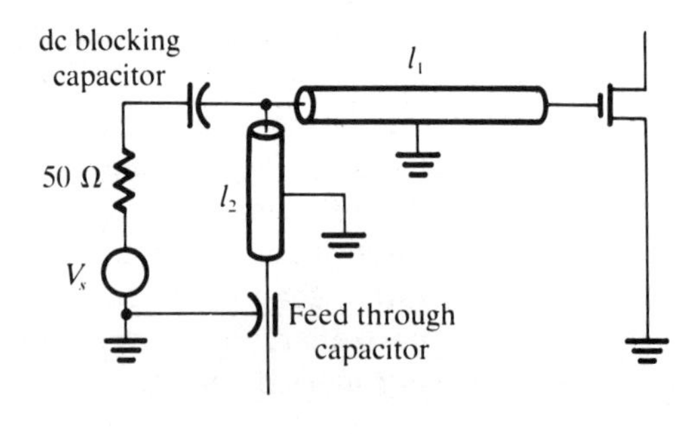

(c)

FIGURE 8.7. Example 8.4.

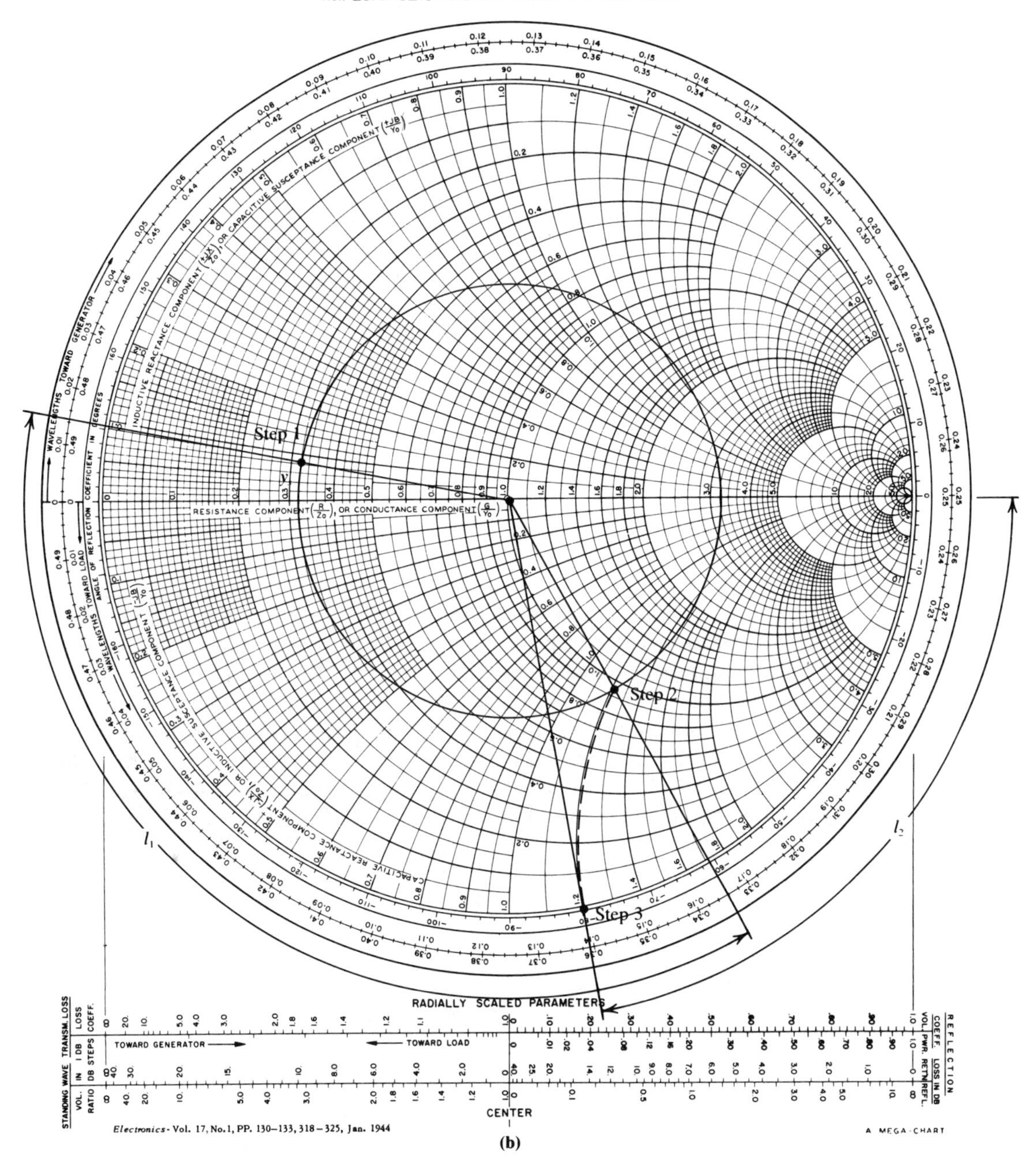

(b)

FIGURE 8.7. continued

From Eq. (8.16), the normalized value of the required source admittance is

$$y_s = \frac{1 - (0.5 + j0.1)}{1 + (0.5 - j0.1)}$$

$$= 0.327 + j0.089$$

Step 1: Locate y_s on the Smith chart [Fig 8.7(b)].

Step 2: Move a distance ℓ_1 toward the load (note that the 50 Ω source impedance is the "load" in this context), until admittance point $1 + jb$ is reached. The 50-Ω source impedance provides the required conductance component of 1, and a stub must be added in parallel to provide the susceptance component of jb. From the chart

$$\ell_1 = 0.184\lambda$$

$$jb = -j1.2$$

Step 3: From position $-j1.2$ on the susceptance circle, move distance ℓ_2 toward the load. In this instance, the "load" will be either a short circuit or an open circuit, depending on whether the infinite admittance or zero admittance point is reached first. From the chart it is seen that the infinite admittance point is reached first, when $\ell_2 = 0.11\lambda$.

This completes this aspect of the design. A feedthrough capacitor would be used to provide the short circuit on ℓ_2, and this would also provide a base bias input point for the BJT, as described later. A dc isolating capacitor would be required at the input, and the circuit would appear as shown in Fig. 8.7(c).

The physical lengths of the lines would depend on the dielectric constant of the substrate, given that the frequency is 1 GHz. A commonly used material is polyolefin, for which $\varepsilon_r = 2.32$. Hence

$$\lambda = \frac{3 \times 10^8}{10^9 \sqrt{2.32}} \times 100 \text{ cm}$$

$$= 19.7 \text{ cm}$$

Therefore,

$$\ell_1 = 0.184 \times 19.7 = \mathbf{3.62\ cm}$$

$$\ell_2 = 0.11 \times 19.7 = \mathbf{2.17\ cm}$$

EXAMPLE 8.5

For the HFET 2202 (see Table 8.1), $S_{11} = 0.84 \angle -99°$ at 4 GHz, normalized to 50 Ω. Design an input-matching network to maximize the gain term G_s for a source impedance of $50 \angle 0$ Ω, assuming that the reverse transmission coefficient is zero.

SOLUTION In this approach, Fig. 8.8(a), fixed lengths of line will be used, the characteristic impedance being the design variable. Because Z_0 is different for each line, it is best to work directly in admittance values rather than normalized values. From Eq. (8.11),

$$\Gamma_s = S_{11}^*$$
$$= 0.84 \angle 99°$$

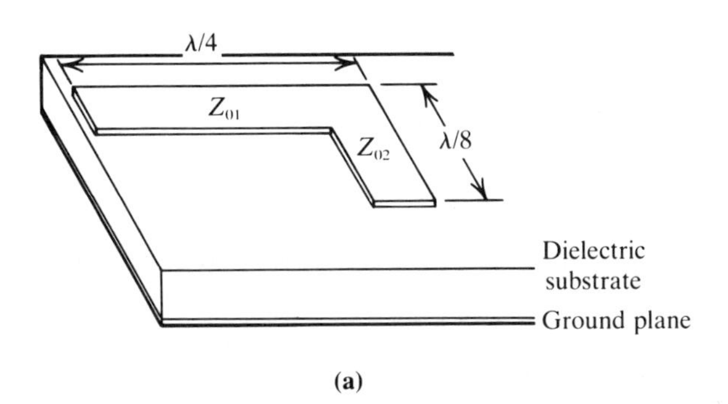

(a)

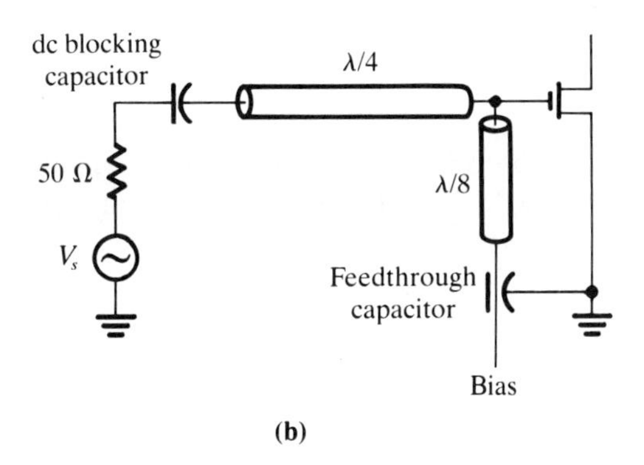

(b)

FIGURE 8.8. Example 8.5.

From Eq. (8.16):

$$Y_s = Y_0 \frac{1 - \Gamma_s}{1 + \Gamma_s}$$

$$= \frac{1}{50} \left(\frac{1 - 0.84 \cos 99 - j0.84 \sin 99}{1 + 0.84 \cos 99 + j0.84 \sin 99} \right)$$

$$= \frac{0.204 - j1.15}{50}$$

$$= 4.08 - j23 \text{ mS}$$

Step 1: The $\lambda/8$ stub must provide the susceptance component, $-j23$ mS. This is inductive susceptance, hence using Eq. (1.102),

$$Z_{sc} = jZ_0 \tan \beta\ell$$

But for $\ell = \frac{1}{8}\lambda$, $\tan \beta\ell = 1$, and in this case $Z_0 = Z_{02}$; hence,

$$Z_{sc} = jZ_{02}$$

or

$$Y_{sc} = \frac{-j}{Z_{02}}$$

or

$$\frac{-j23}{10^3} = \frac{-j}{Z_{02}}$$

Therefore,

$$Z_{02} = \mathbf{43.48\ \Omega}$$

Step 2: The $\frac{1}{4}\lambda$ line must transform the input (source) admittance of $\frac{1}{50}$ S to 4.08 mS. Applying Eq. (1.93) yields

$$Z_s' = \frac{Z_{01}^2}{50}$$

Therefore,

$$\frac{10^3}{4.08} = \frac{Z_{01}^2}{50}$$

so

$$Z_{01} = \sqrt{\frac{10^3 \times 50}{4.08}}$$
$$= \mathbf{111\ \Omega}$$

8.7 CONSTANT-GAIN CIRCLES

The gain factors G_S and G_L can be plotted on a Smith chart, which provides a graphical means of determining reflection coefficient for a given gain value. As will be shown later, the noise factor can be shown on the chart in a similar manner, and the input reflection coefficient can be selected graphically for a desired gain–noise factor combination.

Because the equations for G_S and G_L are similar, the general form, as given by Eq. (8.7) will be used. This can be rearranged as

$$|1 - \Gamma S|^2 = \frac{1-\rho^2}{G} \tag{8.17}$$

It will be recalled that when the input circuit is being considered, Γ stands for Γ_{11} and S for S_{11}, while for the output circuit, Γ stands for Γ_{22} and S for S_{22}. Both Γ and S will be complex in general, so that $\Gamma = \rho \angle\phi$ and $S = |S| \angle\theta$.

Expansion of Eq. (8.17) in terms of these complex values gives

$$1 + (\rho|S|)^2 - 2\rho S \cos(\phi + \theta) = \frac{1 - \rho^2}{G} \tag{8.18}$$

Collecting the ρ^2 terms yields

$$1 + \rho^2\left(|S| + \frac{1}{G}\right) - 2\rho|S| \cos(\phi + \theta) = \frac{1}{G} \tag{8.19}$$

Dividing through by the coefficient of ρ^2 and rearranging, we have

$$\rho^2 - 2\rho \frac{G|S|}{G|S|^2 + 1} \cos(\phi + \theta) = \frac{1 - G}{G|S|^2 + 1} \tag{8.20}$$

As described in Chapter 3, the reflection coefficients $\rho\angle\phi$ and $|S|\angle\theta$ may be represented on a Smith chart as shown in Fig. 8.9. The complex conjugate of S, which is $S^* = |S|\angle{-\theta}$, is also shown. Applying the cosine rule to the triangle shown in Fig. 8.9 gives

$$\rho^2 + C^2 - 2\rho C \cos A = R^2 \tag{8.21}$$

C is the distance from the center of the small circle to the center of the Smith chart, and R is the radius of the small circle. The objective now is to relate the gain expression. Eq. (8.20), to Eq. (8.21). From the figure it will be seen that angle $A = (\phi - (360 - \theta)$, and therefore $\cos A = \cos(\phi + \theta)$. Equation (8.21) may therefore be written as

$$\rho^2 - 2\rho C \cos (\phi + \theta) = R^2 - C^2 \tag{8.22}$$

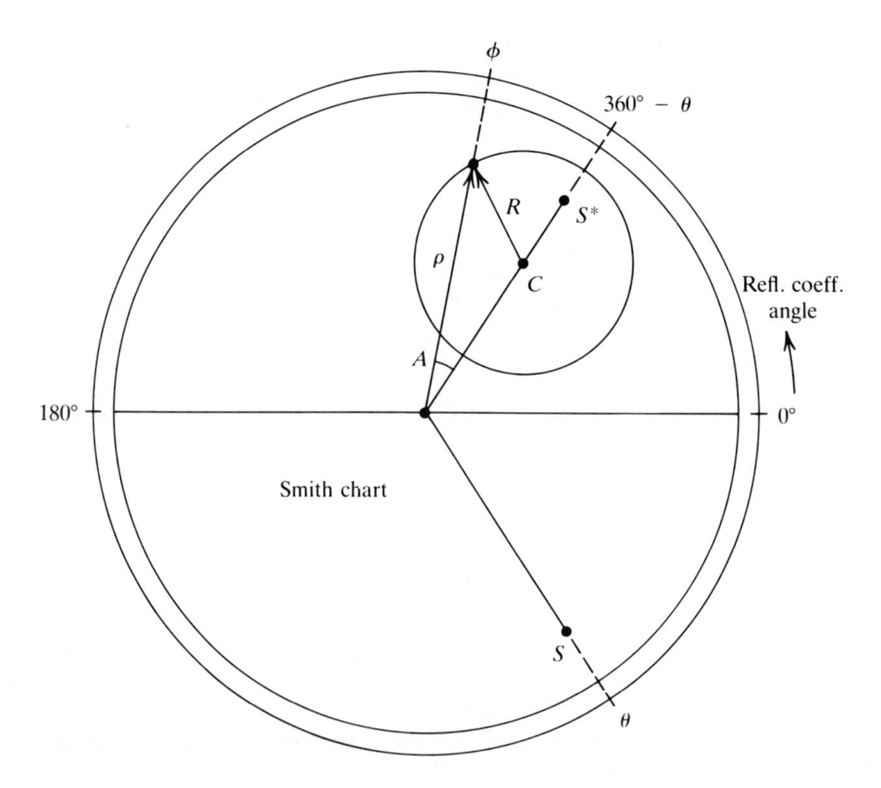

FIGURE 8.9. Construction of a constant-gain circle on the Smith chart.

Comparing Eqs. (8.20) and (8.22), the coefficients of the cosine term may be equated to yield

$$C = \frac{G|S|}{G|S|^2 + 1} \tag{8.23}$$

Equating terms independent of ρ yields

$$R^2 - C^2 = \frac{1 - G}{G|S|^2 + 1}$$

from which

$$R = \sqrt{C^2 + \frac{(1-G)\,C}{|S|\,G}} \tag{8.24}$$

Equations (8.23) and (8.24) are the desired relationships. They give the center and radius of the circle in terms of gain G and $|S|$, the circle being centered on the S^* line. The reflection coefficient is the independent variable. The length of the line joining the center of the Smith chart to any point on the circumference of the circle gives the magnitude ρ for that point, and the projection of the line onto the angle scale around the chart gives the angle φ.

Before presenting worked examples to illustrate the construction of the gain circles, some general notes are in order. Although Eq. (8.24) could be developed to show R in terms of G and $|S|$ only, it is easier to use it in the form shown. The procedure is first to determine C using Eq. (8.23), and then R using Eq. (8.24). Various publications show C and R in their independent forms, and usually in terms of a normalized gain factor (see, e.g., *Hewlett-Packard Application Note 77–1*). These are equivalent to the equations given here but they require more arithmetic operations. This in turn can lead to considerable round-off error because of the difference terms of nearly equal quantities, such as $(1 - |S|^2)$.

EXAMPLE 8.6

From Table 8.1 the input reflection coefficient S_{11} at $f = 2$ GHz is given as $0.96 \angle -51°$. Draw the constant-gain circle for $G_S = +2$ dB. From this determine the source reflection coefficient that has the minimum phase angle and which satisfies the gain requirements.

SOLUTION

It is first necessary to convert the decibel value to a power ratio: 2 dB = 1.585 : 1 power ratio. From Eq. (8.23),

$$C = \frac{0.96 \times 1.585}{1.585 \times 0.96^2 + 1}$$

$$= \mathbf{0.618}$$

From Eq. (8.24),

$$R = \sqrt{0.618^2 + \frac{(1 - 1.585) \times 0.618}{0.96 \times 1.585}}$$

$$= \mathbf{0.380}$$

Locate the S^* line at $+51°$ on the Smith chart, as shown in Fig. 8.10. On this, mark off the center point at $C = 0.618$ and draw the circle of radius $R = 0.380$. Recall from Chapter 3 that the maximum radius of the Smith chart corresponds to a reflection coefficient of unity, and that the reflection coefficient is a voltage ratio. The voltage ratio scale is normally shown along the bottom of the chart, as shown in Fig. 8.10, and this can be used to scale C and R.

The reflection coefficient with the minimum phase angle is determined by the lower tangent to the circle, as shown in Fig. 8.10, and from the chart scales this is $0.495 \angle 13.9°$.

Note that although G must be used as a power ratio in the calculations, the circle can be labeled with its decibel value.

EXAMPLE 8.7

For a certain transistor, $S_{22} = 0.8 \angle -90°$. Draw the constant-gain circles for $G_L = +3$ dB, 0 dB, and -3 dB.

SOLUTION

First convert the decibel gain values to power ratios: $+3$ dB = 2 : 1, 0 dB = 1 : 1, and -3 dB = 0.5:1. Applying Eqs. (8.23) and (8.24), we have:

$$C = \frac{0.8 \times 2}{2 \times 0.8^2 + 1} = \mathbf{0.702}$$

$$R = \sqrt{0.702^2 + \frac{(1 - 2) \times 0.702}{0.8 \times 2}} = \mathbf{0.232}$$

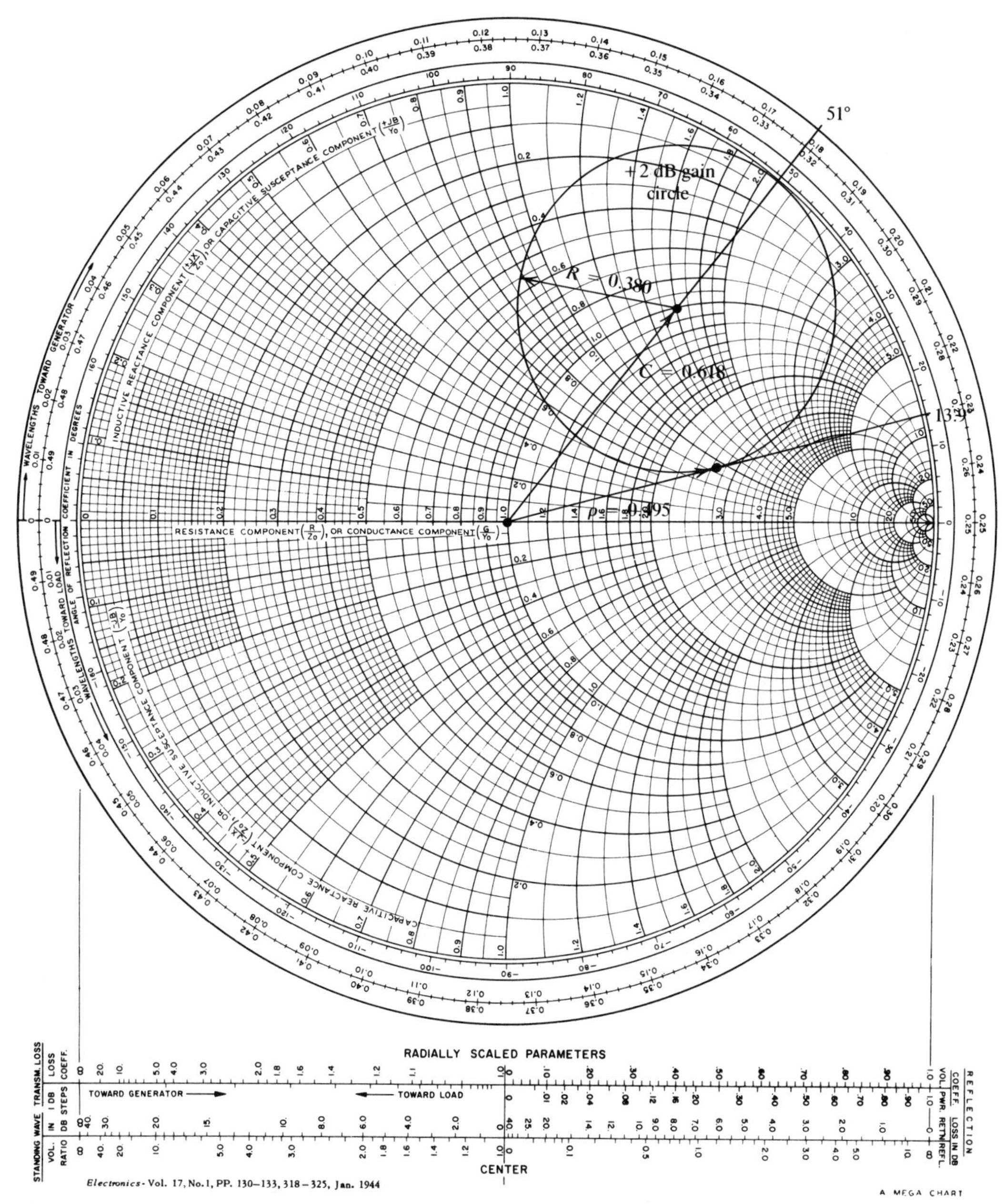

FIGURE 8.10. Construction for Example 8.6

For the 0-dB circle:

$$C = \frac{0.8 \times 1}{1 \times 0.8^2 + 1} = \mathbf{0.489}$$

$$R = \sqrt{0.489^2 + \frac{(1 - 1) \times 0.489}{0.8 \times 1}} = \mathbf{0.489}$$

For the -3-dB circle:

$$C = \frac{0.8 \times 0.5}{0.5 \times 0.8^2 + 1} = \mathbf{0.303}$$

$$R = \sqrt{0.303^2 + \frac{(1 - 0.5) \times 0.303}{0.8 \times 0.5}} = \mathbf{0.686}$$

Locate the S^* line at $+90°$, and along this, center the circles as shown in Fig. 8.11.

From this example it is seen that the 0-dB-gain circle passes through the center of the Smith chart, positive-dB-gain circles are smaller, and negative-dB-gain circles are larger than the 0-dB-gain circle. These results can also be deduced from Eq. (8.24) for $G = 1$, $G > 1$, and $G < 1$.

Also from Eq. (8.23), when G is equal to its maximum value given by $1/(1 - |S|^2)$ [see Eqs. (8.13) and (8.14)], C takes on a maximum value:

$$C_{\max} = |S| \tag{8.25}$$

Substituting this value for C in Eq. (8.24) shows that $R = 0$. Thus the maximum-gain circle degenerates into the point S^*.

To repeat, only the gain factor G_S and G_L from the total-gain expression [Eq. (8.6)] are shown on the Smith chart.

8.8 AVAILABLE POWER GAIN

The available power gain of an amplifier is the ratio of available output power to available input power. In Chapter 2 the transducer power gain is defined as the ratio of load power to available input power. The transducer power gain becomes the available power gain when the load is conjugately matched to the amplifier output.

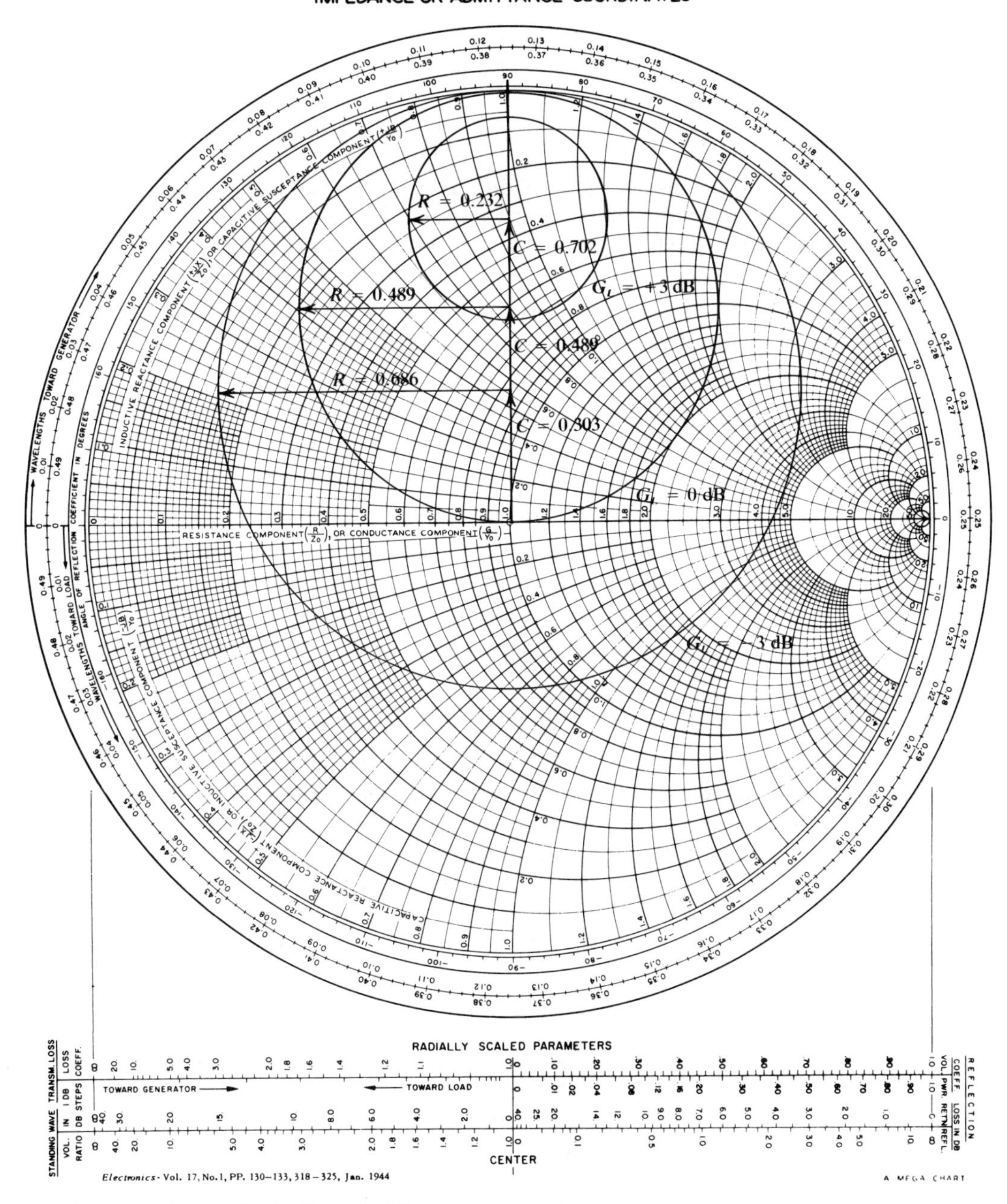

FIGURE 8.11. Construction for Example 8.7.

For the unilateral amplifier, matching of the output in this way results in G_L becoming a maximum, as shown by Eq. (8.14). Under this condition, Eq. (8.6) for unilateral power gain becomes, for available power gain,

$$G_{\mathrm{Au}} = G_S G_0 G_{L\max} \tag{8.26}$$

It will be seen that the available power gain does not depend on the load reflection coefficient, but it does depend on the source reflection coefficient through G_S. Thus, although available power is used in the definition, the input does not have to be conjugately matched for the available power gain to be used. Because G_0 and $G_{L\max}$ are fixed by the transistor parameters, only the G_S gain factor need be plotted on the Smith chart when this method is used.

When the input is conjugately matched, the available power gain becomes a maximum given by

$$G_{\mathrm{Au,max}} = G_{S\max} G_0 G_{L\max} \tag{8.27}$$

where $G_{S\max}$ is given by Eq. (8.13). Available power gain is used in the definition of noise factor, discussed in Section 8.9.

8.9 TRANSISTOR NOISE FACTOR

Random variations of charge movement within a transistor give rise to noise currents and voltages which degrade the available signal-to-noise ratio at the input to an amplifier. There are a number of different mechanisms by which noise is generated within a transitor [see, e.g., Roddy and Coolen (1984)], but from the applications point of view, the noise performance of an amplifier may be characterized by its noise factor. For a source at room temperature connected to the input of an amplifier, the noise factor of the combination is defined as

$$\begin{aligned} F &= \frac{\text{available signal-to-noise power ratio at input}}{\text{available signal-to-noise power ratio at output}} \\ &= \frac{P_{ss}}{P_{ns}} \times \frac{P_{no}}{P_{so}} \end{aligned} \tag{8.28}$$

Available signal powers are denoted by P, the first subscript referring either to signal (s) or noise (n), and the second subscript either to output, (o) or source (s).

As defined in Section 8.8, the available power gain is

$$G_A = \frac{P_{so}}{P_{ss}} \tag{8.29}$$

The noise factor can therefore be written as

$$F = \frac{P_{no}}{G_A P_{ns}} \tag{8.30}$$

The available noise power from a source at temperature T kelvin is kTB, where k is Boltzmann's constant, equal to 1.38×10^{-23} J/K and B is the bandwidth in hertz over which the noise is spread. Denoting room temperature by T_0, the noise factor becomes

$$F = \frac{P_{no}}{G_A k T_0 B} \tag{8.31}$$

Alternatively, the output noise may be written as

$$P_{\text{no}} = F G_A k T_0 B \tag{8.32}$$

If the amplifier were noiseless, the noise output would simply be $G_A k T_0 B$, that is, the available input noise power multiplied by the available power gain. The noise output is further increased by the factor F for the practical amplifier that contributes noise.

Noise factor is a specified parameter for a transistor and it usually varies with frequency. When the value is specified for one frequency, it is known as the *spot noise factor*. In some situations, a value averaged over a range of frequencies is specified, this being known as the *average noise factor*. In this chapter only the spot noise factor will be considered.

Noise factor, as defined above, is a multiplying factor, and basically it is a power ratio. The noise figure is simply the noise factor expressed in decibels; thus

$$F(\text{dB}) = 10 \log F \tag{8.33}$$

Specification sheets for transistors usually show the noise figure. The measurement of the noise factor is discussed in Section 14.10.

8.9.1 Amplifier Input Noise in Terms of F

The total noise at the input of the amplifier can be considered as made up of two components, amount kT_0B contributed by the source, and some amount contributed by the amplifier. Since Eq. (8.32) gives the total noise at the output, the total noise at the input is found by dividing the output noise by G_A. The source contribution may be subtracted from this to give the amplifier noise referred to the input as

$$\begin{aligned} P_{nc} &= \frac{P_{no}}{G_A} - kT_0B \\ &= FkT_0B - kT_0B \\ &= (F - 1)\,kT_0B \end{aligned} \tag{8.34}$$

8.9.2 Amplifiers in Cascade

Consider first the situation where two amplifiers are connected in cascade, as shown in Fig. 8.12. It is desired to find the overall noise factor for the combination, and by extension of Eq. (8.31) this is

$$F = \frac{P_{no2}}{G_{A1}G_{A2}kT_0B} \tag{8.35}$$

Now the noise output from amplifier 1 is $G_{A1}F_1kT_0B$, and this is fed into amplifier 2, where by Eq. (8.34) the already existing input noise is $(F_2 - 1)kT_0B$. The total input noise to amplifier 2 is the sum of these two components, and the output noise is G_{A2} times this input:

$$P_{no2} = G_{A2}[G_{A1}F_1kT_0B + (F_2 - 1)kT_0B] \tag{8.36}$$

Substituting this in Eq. (8.35) and simplifying gives for the overall noise factor,

$$F = F_1 + \frac{F_2 - 1}{G_{A1}} \tag{8.37}$$

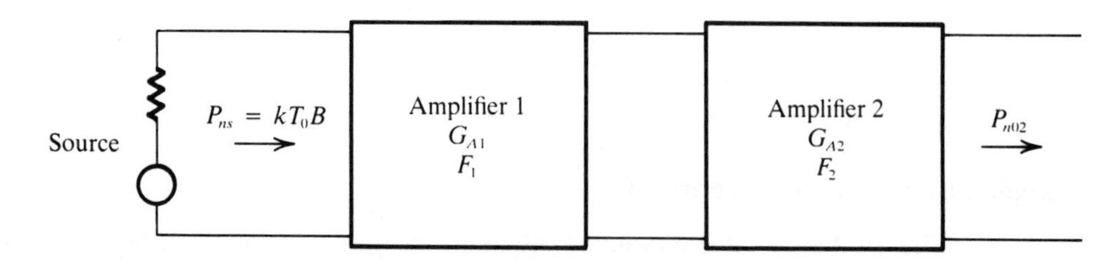

FIGURE 8.12. The noise factor of two amplifiers in cascade is $F = F_1 + (F_2 - 1)/G_{A1}$.

The argument is easily extended for additional amplifiers to give

$$F = F_1 + \frac{F_2 - 1}{G_{A1}} + \frac{F_3 - 1}{G_{A2}G_{A1}} + \cdots \tag{8.38}$$

This is known as *Friis's formula.*

8.10 NOISE CIRCLES

The noise factor of an amplifier is a function of the input reflection coefficient Γ_s. The reflection coefficient can be adjusted to give a minimum noise factor $F_{\min}$, and both values, the minimum noise factor and the reflection coefficient $\Gamma_{\min}$, required for this are specified noise parameters for a transistor. In addition to these, the noise in the transistor itself can be represented by an equivalent noise resistance R_n, which is also a specified parameter. Keeping in mind that $\Gamma_{\min}$ is complex in general, and therefore consists of two parts $\rho_{\min} \angle \phi_{\min}$, the four noise parameters which must be specified to characterize the noise of a transistor completely are

$$F_{\min}, \quad R_n, \quad \rho_{\min}, \quad \text{and} \quad \phi_{\min} \tag{8.39}$$

Constant-noise-factor circles may be constructed in a manner similar to that for constant-gain circles. The noise factor as a function of source reflection coefficient may be written as [see Fukui (1981) and *Hewlett-Packard Application Bulletin 17*]

$$F = F_{\min} + 4r_n \frac{|\Gamma_s - \Gamma_{\min}|^2}{(1 - \rho_s^2)|1 + \Gamma_{\min}|^2} \tag{8.40}$$

where $r_n = R_n/Z_0$ is the noise resistance normalized to a characteristic impedance (usually 50 Ω) for the system.

Equation (8.40) can be rearranged as

$$|\Gamma_s - \Gamma_{\min}|^2 = \frac{(F - F_{\min})|1 + \Gamma_{\min}|^2 (1 - \rho_s^2)}{4r_n}$$

$$= N_i(1 - \rho_s^2) \tag{8.41}$$

where

$$N_i = \frac{(F - F_{\min})|1 + \Gamma_{\min}|^2}{4r_n}$$

The left-hand side can be expanded as

$$\rho_s^2 + \rho_{\min}^2 - 2\rho_s\rho_{\min} \cos(\phi_{\min} - \phi_s) = N_i(1 - \rho_s^2) \tag{8.42}$$

Following the procedure used for the gain circles, Eq. (8.42) can be rearranged with ρ_s terms on one side as

$$\rho_s^2 - 2\rho_s \frac{\rho_{\min}}{1 + N_i} \cos(\phi_{\min} - \phi_s) = \frac{N_i - \rho_{\min}^2}{1 + N_i} \tag{8.43}$$

Figure 8.13 shows a construction similar to Figure 8.9. The angle A in this case is $A = \phi_s - \phi_{\min}$. Applying the cosine rule to the triangle shown gives

$$\rho_s^2 + C^2 - 2\rho_s C \cos(\phi_s - \phi_{\min}) = R^2 \tag{8.44}$$

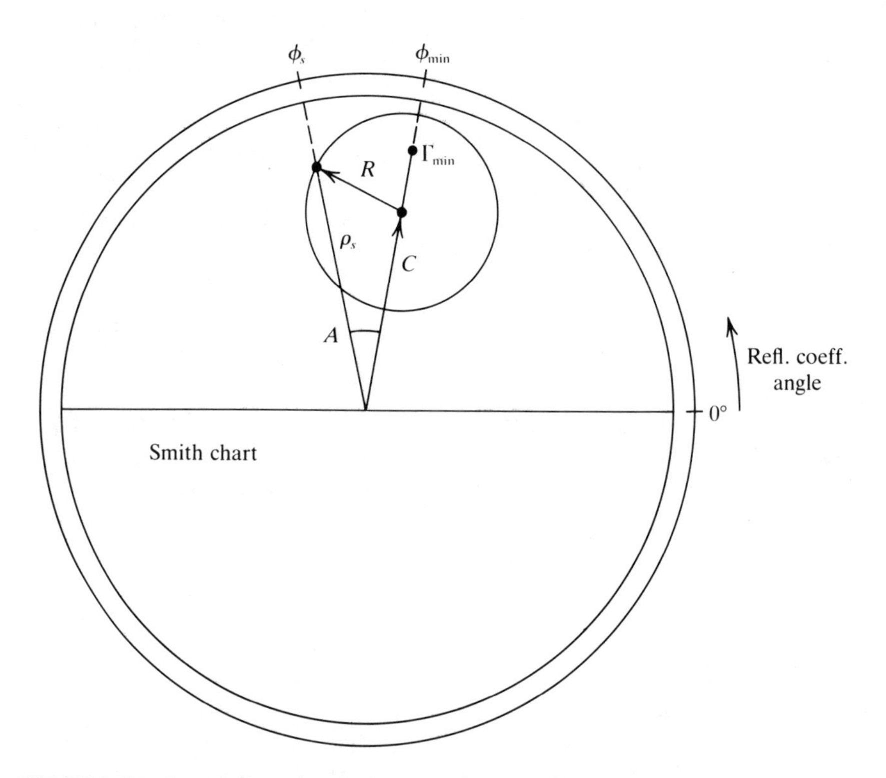

FIGURE 8.13. Construction of a constant-noise factor circle on the Smith chart.

Since cos ($\phi_s - \phi_{min}$) = cos ($\phi_{min} - \phi_s$), the coefficients of the cosine term in Eqs. (8.43) and (8.44) can be equated to yield

$$C = \frac{\rho_{min}}{1 + N_i} \tag{8.45}$$

Comparing the terms independent of ρ_s yields

$$R^2 - C^2 = \frac{N_i - \rho_{min}^2}{1 + N_i}$$

from which

$$R = \sqrt{C^2 + \frac{N_i - \rho_{min}^2}{1 + N_i}} \tag{8.46}$$

These equations for R and C enable circles, each representing a constant-noise factor, to be drawn on the Smith chart. This is illustrated in the following example.

EXAMPLE 8.8

Parameters specified for transistor HFET-1101 are (see *Hewlett-Packard Application Bulletin 19*):

F (GHz)	Γ_{min}	R_n (Ω)	F_{min} (dB)
4	0.618 $\angle 98°$	23.14	1.60
12	0.660 $\angle -87°$	49.10	4.50

Draw (a) the 3-dB noise circle for the 4-GHz values, and (b) the 6-dB noise circle for the 12-GHz values. (c) State why a 3-dB noise circle cannot be drawn at 12 Ghz.

SOLUTION

(a) At 4 GHz, F(dB) = 3 dB, therefore F = 2:1. F_{min}(dB) = 1.6 dB, therefore F_{min} = 1.45:1.

$$N_i = \frac{(2 - 1.45)\,[(1 + 0.618 \cos 98)^2 + (0.618 \sin 98)^2] \times 50}{4 \times 23.14}$$

$$= 0.359$$

Therefore,

$$C = \frac{0.618}{1.359} = 0.455$$

$$R = \sqrt{0.455^2 + \frac{0.359 - 0.618^2}{1.359}} = \mathbf{0.436}$$

This circle is shown in Fig. 8.14.

(b) At 12 GHz, $F(\text{dB}) = 6$ dB; therefore $F = 3.98{:}1$. $F_{\text{min}}(\text{dB}) = 4.5$ dB; therefore, $F_{\text{min}} = 2.82{:}1$.

$$N_i = \frac{(3.98 - 2.82) \times [(1 + 0.66\cos(-87))^2 + (0.66\sin(-87))^2] \times 50}{4 \times 49.1}$$

$$= 0.445$$

Therefore,

$$C = \frac{0.66}{1.445} = \mathbf{0.457}$$

$$R = \sqrt{0.457^2 + \frac{0.445 - 0.66^2}{1.445}} = \mathbf{0.464}$$

This circle is also shown on Fig. 8.14.

(c) A 3-dB-noise-figure circle cannot be shown in this case because F_{min} is 4.58 dB, which is greater than 3 dB.

The next example illustrates the use of both noise and gain circles on the same Smith chart.

EXAMPLE 8.9

Data for the HXTR-6101 silicon bipolar transistor at 4 GHz are

$S_{11} = 0.522 \angle 169°$ $S_{21} = 1.681 \angle 26°$ $S_{22} = 0.839 \angle -67°$

$F_{\text{min}} = 2.5$ dB $\Gamma_{\text{min}} = 0.475 \angle 166°$ $R_n = 3.5\ \Omega$

Draw the $F = 3$ dB noise figure circle and the $G_s = 0$ dB gain circle.

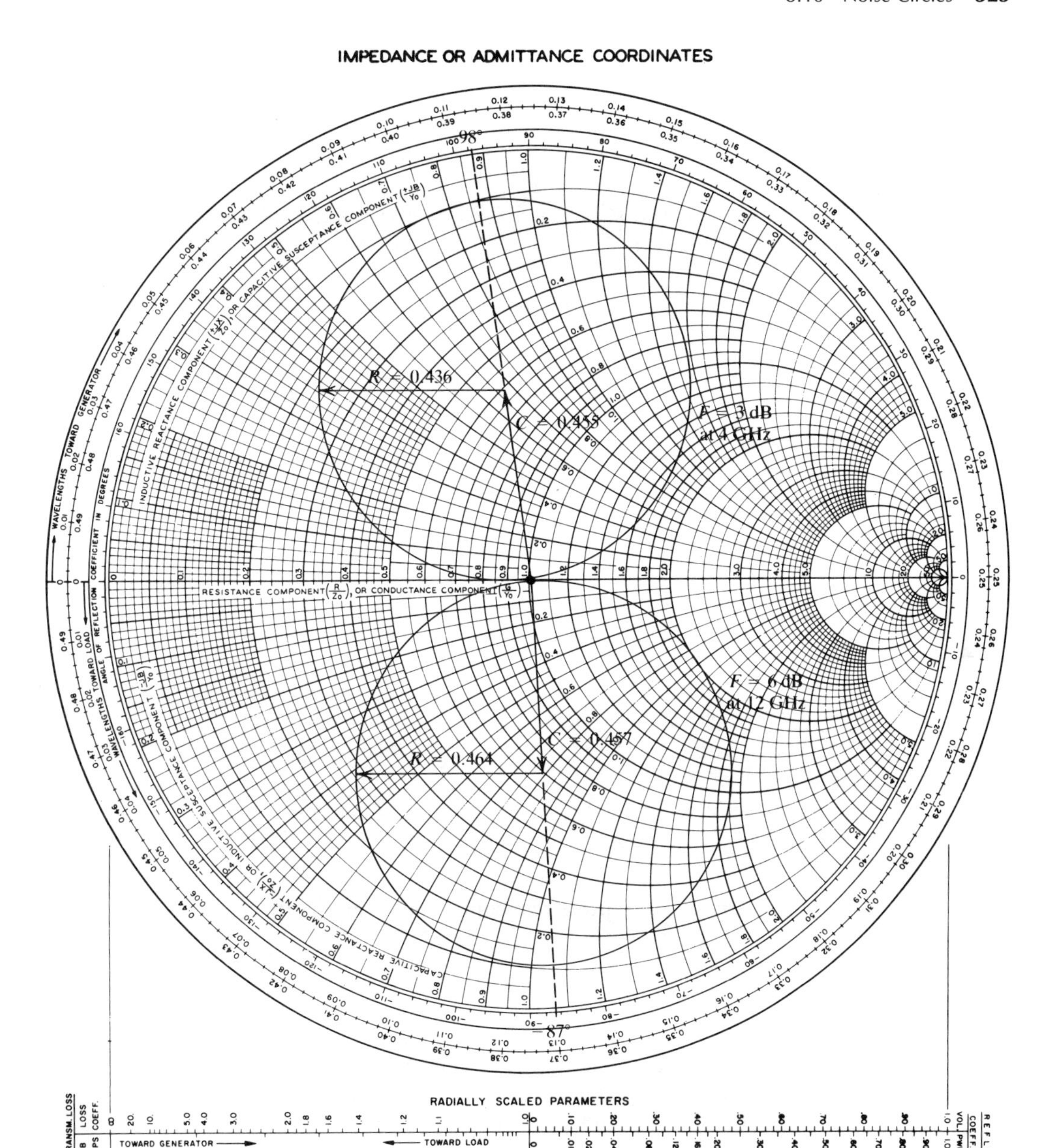

FIGURE 8.14. Construction for Example 8.8.

From these determine the source reflection coefficient, and hence the source impedance required to realize both values. Assume that the characteristic impedance of the system is 50 Ω and that $S_{12} = 0$. What is the maximum available gain under these conditions?

SOLUTION For the noise circle F_{min} as a noise factor is 1.78:1, and the required noise factor is 2:1. Hence

$$N_i = \frac{(2-1.78)\times[(1+0.475\cos 166)^2+(0.475\sin 166)^2]\times 50}{4\times 3.5}$$

$$= \mathbf{0.239}$$

Therefore,

$$C = \frac{0.475}{1.239} = 0.383$$

$$R = \sqrt{0.383^2 + \frac{0.239 - 0.475^2}{1.239}} = \mathbf{0.397}$$

The center of the noise circle is located at 0.383 along the Γ_{min} line, as shown in Fig. 8.15. This noise circle passes very close to the center of the Smith chart.

For the gain circle, $G_s = 0$ dB gives a power ratio of 1 : 1 and therefore

$$C = \frac{0.522}{0.522^2 + 1} = \mathbf{0.410}$$

$$R = \mathbf{0.410}$$

The center of this circle is located on the S_{11}^* line as shown in Fig. 8.14, and the circle does pass through the center of the Smith chart.

Where the two circles intersect gives the common value of Γ_s required. Two values satisfy this condition: $\Gamma_s \simeq 0$ (intersect at the center of the Smith chart) and the point z. The $\Gamma_s = 0$ would naturally be chosen, and thus the source impedance required is 50 Ω.

As a point of interest, if point z were chosen, the normalized impedance could be read directly from the Smith chart as

$$z = 0.13 + j0.075$$

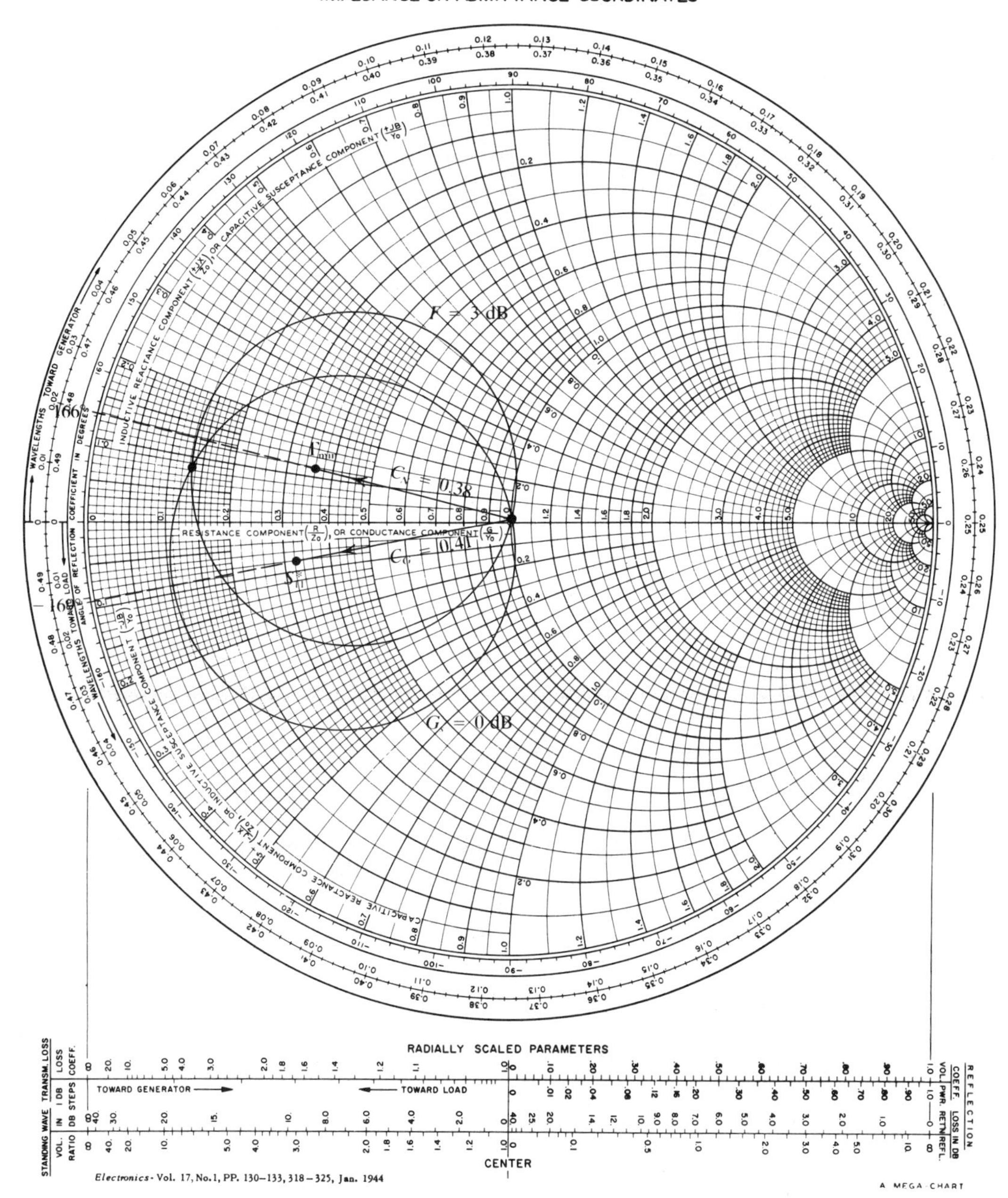

FIGURE 8.15. Construction for Example 8.9.

Therefore,

$$Z_s = 50z = 6.5 + j3.75\ \Omega$$

Thus for an actual source impedance of 50 Ω, a matching network would have to be designed as described in Section 8.6. The available power gain for these conditions is found as follows:

$$G_s = 0\ \text{dB}$$

From Eq. (8.4),

$$G_0 = 20 \log 1.681 = 4.51\ \text{dB}$$

From Eq. (8.14),

$$G_{L\max} = 10 \log (1 - 0.839) = 5.29\ \text{dB}$$

From Eq. (8.26),

$$G_{Au} = 0 + 4.51 + 5.29 = \mathbf{9.8\ dB}$$

8.11 EQUIVALENT NOISE TEMPERATURE OF AN AMPLIFIER

As shown in Section 8.9.1, the noise referred to the input of an amplifier can be represented by the quantity $(F - 1)kT_0B$. An alternative way of representing this noise is as an available noise power of amount kT_eB, where T_e is an equivalent noise temperature. Thus

$$kT_eB = (F - 1)kT_0B$$

Therefore,

$$T_e = (F - 1)T_0 \tag{8.47}$$

It must be realized that T_e is not a physical temperature; it is a mathematical representation of amplifier noise. For example, amplifiers (and converters) used with satellite antennas for home reception usually have noise specified by T_e, a typical value being 100 K (or −173°C). In noise calculations for low-noise amplifiers, equivalent noise temperature is often a more convenient quantity to use than noise factor, and as seen from Eq. (8.47), noise temperature is known if noise factor is known, and vice versa.

Friis's formula [Eq. (8.38)] can be expressed in terms of noise temperature. Subtracting 1 from both sides of Eq. (8.38) gives

$$F - 1 = F_1 - 1 + \frac{F_2 - 1}{G_{A1}} + \frac{F_3 - 1}{G_{A1}\, G_{A2}} + \cdots \qquad (8.48)$$

Substituting for the various $(F_i - 1)$ from Eq. (8.47) yields

$$\frac{T_e}{T_0} = \frac{T_{e1}}{T_0} + \frac{T_{e2}/T_0}{G_{A1}} + \frac{T_{e3}/T_0}{G_{A1}\, G_{A2}} + \cdots$$

or

$$T_e = T_{e1} + \frac{T_{e2}}{G_{A1}} + \frac{T_{e3}}{G_{A1}\, G_{A2}} + \cdots \qquad (8.49)$$

Noise circles may also be shown for T_e rather than F, the most direct way being simply to calculate the noise temperature equivalent for each noise factor. Alternatively, the substitution of Eq. (8.47) in the expression for N_i, shown in Eq. (8.41) gives

$$N_i = \frac{(T_e - T_{e\min})\,|1 + \Gamma_{\min}|^2}{4r_n T_0} \qquad (8.50)$$

Equations (8.45) and (8.46) can now be used to draw the constant-noise-temperature circles directly.

8.12 DATA SHEETS

To illustrate some of the foregoing points, data sheets are included for three different transistors. The Avantek AT-2845 transistor is a bipolar type suitable for small-signal amplification up to 4 GHz. It will be seen that typical constant-gain and noise-figure circles are provided as part of the data sheet. The Dexcel 2502a series of GaAs FETs illustrate the transistor in chip form. Noise parameters are listed in the data tables. The Hewlett-Packard HFET-2202 is also a GaAs FET, shown mounted in a HPAC-100A package. Again, typical noise parameters are provided in the data tables. In all cases, the small-signal S-parameters are specified.

AT-2845

General Purpose
Bipolar Transistor

FEATURES

- **3.2 dB Noise Figure at 2 GHz**
- **11.0 dB Gain at NF**
- **Hermetic 100 Mil Microstrip Package**
- **Gold Metal System**
- **Phosphorous Emitter**

DESCRIPTION

The Avantek AT-2845 is designed for low noise figure, high gain small signal amplification at frequencies up to 4 GHz. It is a particularly cost effective choice for amplifiers in the 500 MHz through 2500 MHz frequency range where low noise figure, high gain and wide dynamic range are required. This transistor is widely used in tuned front-end and signal procesing amplifiers in radar, telemetry and point-to-point communications receivers as well as in wideband amplifiers for instrumentation and EW applications.

This transistor features an etchless gold metal system that produces conductive films of 1 μm thickness and extremely uniform coverage. A dielectric layer protects the transistor chip from scratching or contamination before they are packaged.

The 100 Mil metal/ceramic package is easy to install in conventional printed circuits or hybrid thin or thick film circuits and will withstand handling, soldering and welding processes. Each package is filled with a dry, inert atmosphere and hermetically sealed to assure long-term protection from humidity and corrosive gases.

OUTLINE DRAWING 100 MIL PACKAGE

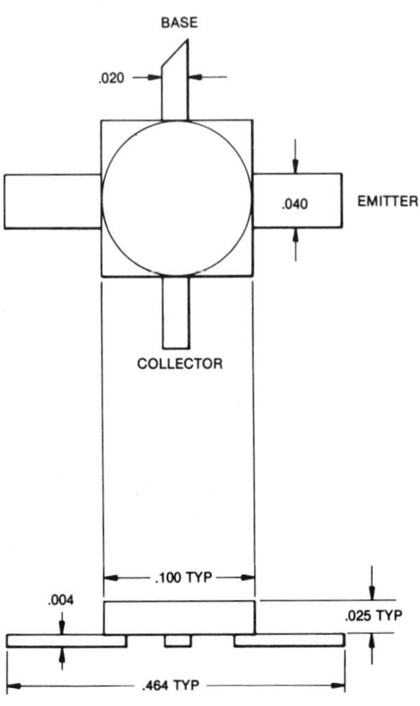

COMMON EMITTER OPERATING CHARACTERISTICS ($T_A = 25°C$)

Parameter	Symbol	Test Cond $V_{CE}I_C$	Freq GHz	Min	AT-2845 Typ	Max
Spot Noise Figure	NF_{opt}	10V 5 mA	1		2.3 dB	
Spot Noise Figure	NF_{opt}	10V 5 mA	2		3.2 dB	3.5 dB
Spot Noise Figure	NF_{opt}	10V 5 mA	4		5.5 dB	
Gain at Optimum Noise Figure	G_{NF}	10V 5 mA	1		14.0 dB	
Gain at Optimum Noise Figure	G_{NF}	10V 5 mA	2		11.0 dB	
Gain at Optimum Noise Figure	G_{NF}	10V 5 mA	4		7.0 dB	
Max Available Power Gain	G_{max}	10V 15 mA	2		15.0 dB	
Max Available Power Gain	G_{max}	10V 15 mA	4		9.0 dB	

AT-2845

MAXIMUM RATINGS ($T_A = 25°C$)

Parameter	Symbol	Limit
Reverse Emitter Base Voltage	V_{EB}	3.0V
Reverse Collector Base Voltage	V_{CB}	20.0V
Open Base Collector-Emitter Voltage	V_{CEO}	12.0V
Collector Current	I_C	50 mA
Continuous Dissipation	P_T (T_{case} = 25°C)	400 mW
Junction Temperature	T_j	200°C
Storage Temperature Range	T_{STG}	− 65 to 200°C
Thermal Resistance	θ_{jc}	300°C/watt

POWER DERATING CURVE

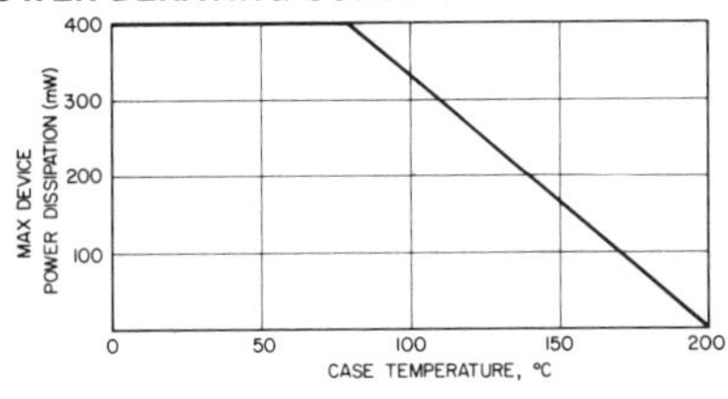

ELECTRICAL CHARACTERISTICS ($T_A = 25°C$)

Parameter	Symbol	Test Conditions	Freq.	Min	Typ	Max
Collector-Base Breakdown	$V_{(BR)CBO}$	$I_C = 10\ \mu A$		20V		
Emitter-Base Breakdown	$V_{(BR)EBO}$	$I_E = 10\ \mu A$		3.0V		
Collector-Emitter Breakdown	$V_{(BR)CEO}$	$I_C = 100\ \mu A$		12V		
Collector Cutoff Current	I_{CBO}	$V_{CB} = 10V$				20 nA
Forward Current Transfer Ratio	h_{FE}	$V_{CE} = 10V$, $I_C = 15$ mA		20	75	
Collector-Base Capacitance	C_{cb}	$V_{CB} = 10V$, $I_E = 0$				0.5 pF

TYPICAL PERFORMANCE CURVES ($T_A = 25°C$)

Maximum Available Gain, $|S_{21E}|^2$ vs. Frequency, $V_{CE} = 10V$ $I_C = 15$ mA

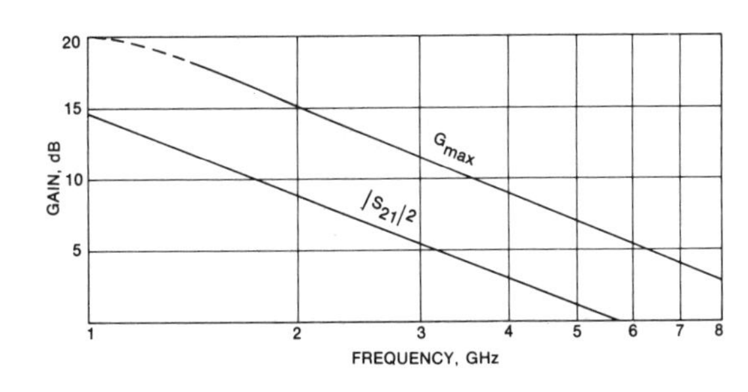

Maximum Available Gain, $|S_{21E}|^2$ vs. Collector Current, F = 2 GHz, $V_{CE} = 10V$

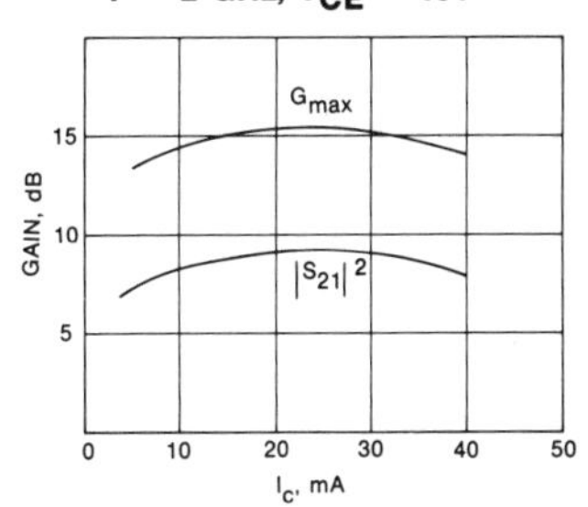

AT-2845

Spot Noise Figure vs. Frequency
V_{CE} = 10V, I_C = 5mA

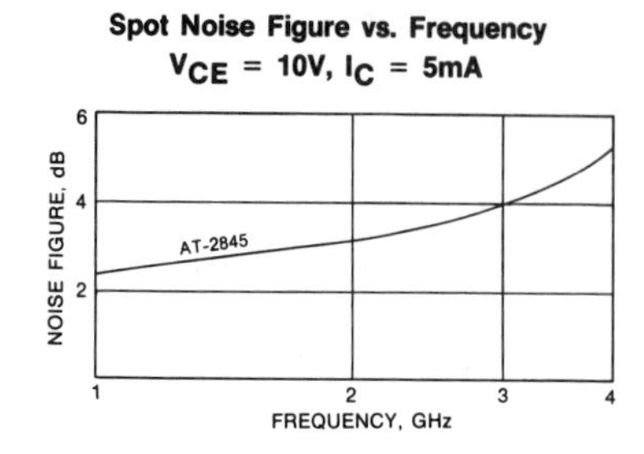

Spot Noise Figure vs. Collector Current
F = 2 GHz, V_{CE} = 10V

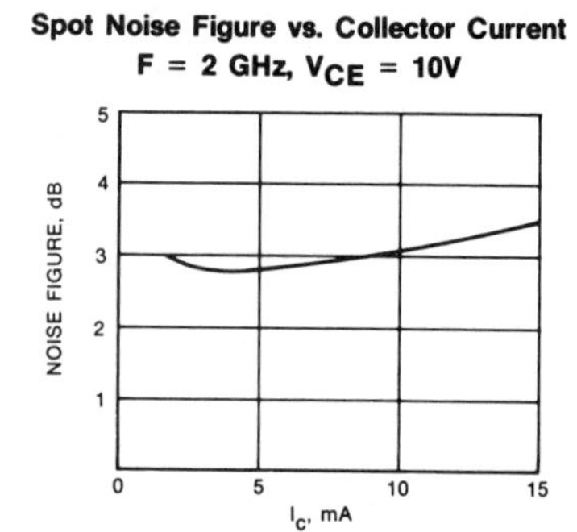

TYPICAL CONTOURS OF CONSTANT GAIN AND NOISE FIGURE, AT-2645

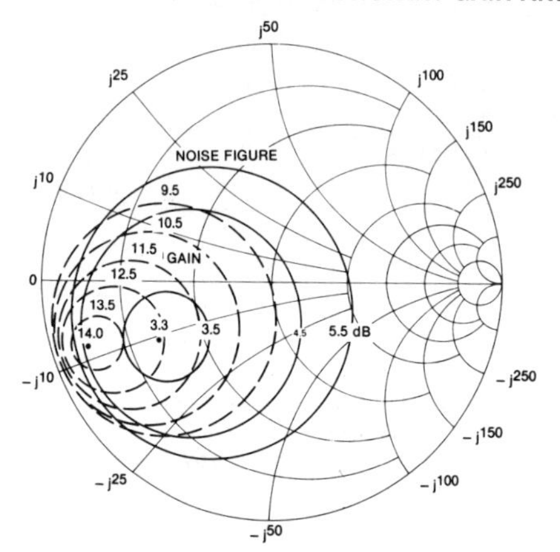

Frequency = 2 GHz, 10V 5 mA

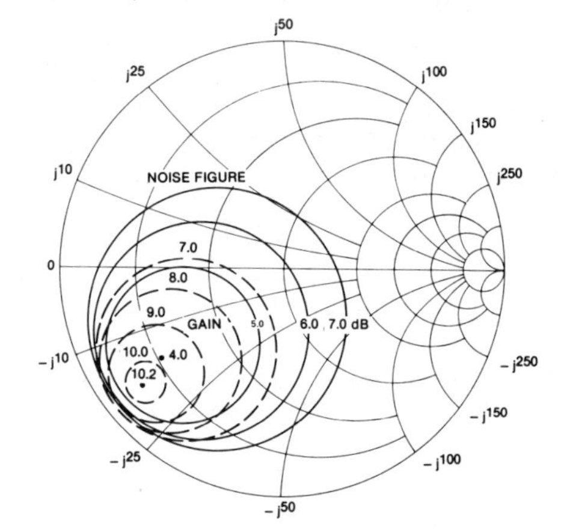

Frequency = 3 GHz, 10V 5 mA

AT-2845

TYPICAL SCATTERING PARAMETERS

Bias = 10.00 Volts, 5.00 mA

S—MAGN AND ANGLES

FREQ	11		21		12		22	
500.00	.679	−117.8	7.663	107.6	.045	39.1	.710	−28.9
1000.00	.650	−159.4	4.390	80.7	.054	28.3	.603	−35.9
1500.00	.654	177.3	3.031	61.9	.059	26.4	.571	−44.7
2000.00	.655	162.1	2.286	45.8	.064	23.7	.569	−54.6
2500.00	.662	148.4	1.853	31.1	.070	23.3	.561	−66.3
3000.00	.674	136.1	1.544	16.6	.078	21.8	.565	−80.0
3500.00	.688	124.5	1.329	2.4	.085	21.2	.577	−93.8
4000.00	.711	114.2	1.167	−10.1	.094	21.1	.580	−103.3
4500.00	.721	104.1	1.036	−24.7	.105	18.0	.596	−118.7
5000.00	.736	94.5	.915	−36.3	.119	13.9	.616	−133.9
5500.00	.732	85.0	.796	−47.8	.132	9.7	.631	−149.2
6000.00	.784	75.3	.747	−58.1	.153	5.3	.697	−163.6

Bias = 10.00 Volts, 15.00 mA

S—MAGN AND ANGLES

FREQ	11		21		12		22	
500.00	.617	−148.6	10.509	96.1	.029	41.3	.551	−30.3
1000.00	.626	−177.5	5.544	74.6	.039	42.2	.479	−34.2
1500.00	.645	165.4	3.755	58.3	.049	42.2	.464	−42.8
2000.00	.651	153.7	2.803	44.0	.058	40.7	.465	−52.2
2500.00	.662	142.1	2.265	30.1	.068	38.1	.462	−64.0
3000.00	.680	131.2	1.876	16.8	.079	34.4	.467	−77.6
3500.00	.696	120.4	1.611	3.3	.090	31.2	.479	−91.8
4000.00	.721	110.8	1.413	−8.7	.100	29.2	.482	−101.4
4500.00	.733	101.0	1.255	−21.7	.112	24.4	.502	−116.8
5000.00	.748	91.8	1.117	−34.1	.127	18.9	.523	−132.4
5500.00	.745	82.6	.977	−45.4	.141	13.6	.544	−147.6
6000.00	.801	73.0	.926	−55.8	.164	8.5	.609	−161.9

CHIP CODE M14

Low Noise, High Gain Microwave GaAs FETs DXL 2502A–CHIP DXL 2502A–CR DXL 2502A–CR–RES

DEXCEL INC.
2285C MARTIN AVE.
SANTA CLARA, CAL 95050
TEL (408) 244-9833
TWX 910-338-0180

GENERAL DESCRIPTION

The DXL2502A-CHIP is a high gain, low noise GaAs FET designed for small signal amplifier applications through Ku-band. It employs Dexcel's proven excellence in GaAs FET material technology and microwave device fabrication.

The DXL2502A-CR chip carrier is provided when no chip handling facilities are available. This package features excellent heat sinking, minimum parasitics, and wide bandwidths. The chip carrier is directly compatible with standard 25 mil thick alumina microstripline.

The DXL2502A-CR-RES includes a bypassed source resistor which can be set for any drain current for specific amplifier applications. The gate must be DC common for this bias configuration.

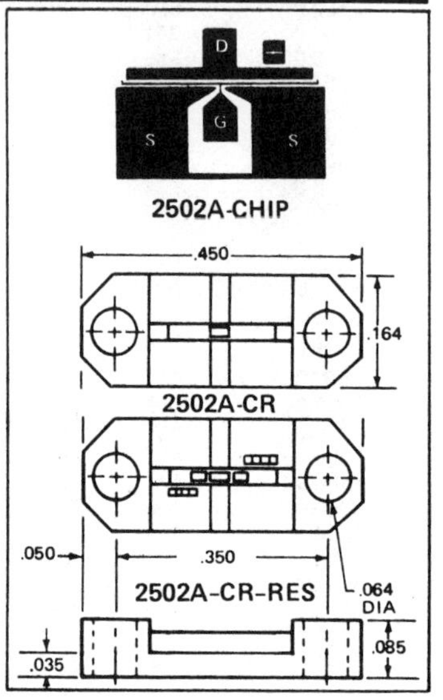

FEATURES

- RUGGED DEVICE (GOLD METALLIZATION)
- WIDE BANDWIDTH
- HIGH GAIN
- CONSISTENT PERFORMANCE
- NO DRIFT
- LOW DISTORTION
- LOW NOISE

ELECTRICAL CHARACTERISTICS

SYMBOL	PARAMETER	MIN	TYP	MAX	UNITS	TEST CONDITIONS
NF_{opt}	Optimum Noise Figure		1.8	2.4	dB	V_{ds}=3.5V; I_{ds}=10 mA; f= 4 GHz
			2.5	3.0	dB	V_{ds}=3.5V; I_{ds}=10 mA; f= 8 GHz
			3.5	4.0	dB	V_{ds}=3.5V; I_{ds}=10 mA; f=12 GHz
G_A	Available Gain at Optimum Noise Figure	11.0	13.0		dB	V_{ds}=3.5V; I_{ds}=10 mA; f= 4 GHz
		7.0	9.0		dB	V_{ds}=3.5V; I_{ds}=10 mA; f= 8 GHz
		4.5	6.0		dB	V_{ds}=3.5V; I_{ds}=10 mA; f=12 GHz
I_{dss}	Saturated Drain Current	25	40	70	mA	V_{ds}=3V; V_{gs}=0V
G_M	Transconductance	17	20		mmho	V_{ds}=3V; V_{gs}=0V

ABSOLUTE MAXIMUM RATINGS

SYMBOL	PARAMETER	MIN	TYP	MAX	UNITS	TEST CONDITIONS
V_{ds}	Drain-Source Voltage			8	volts	at $0.5I_{dss}$
I_{gf}	Gate Forward Current			50	mA	$V_{gs} > 0V$
P_T	Total Power Dissipation			0.5	W	note 1
T_{ch}	Channel Temperature			175	°C	
T_{stg}	Storage Temperature	-65		125	°C	

Notes: 1. Derate maximum power dissipation at 5.33 mW/°C above 25°C.

OCT 1977

DC PARAMETERS

SYMBOL	PARAMETER	TYP	UNITS	TEST CONDITIONS
V_P	Pinch Off Voltage	3	volts	$V_{ds} = 3$ V; $I_{ds} = 1$ mA
θ_T	Thermal Resistance	90	°C/W	
BV_{gd}	Breakdown Voltage, Gate to Drain, Source Open	8	volts	$I_g = 10\ \mu A$
BV_{gs}	Breakdown Voltage, Gate to Source, Drain Open	8	volts	$I_g = 10\ \mu A$

TYPICAL SMALL SIGNAL S-PARAMETERS

$V_{ds} = 3.5V$; $I_{ds} = 10$ mA

f (GHz)	S11		S21		S12		S22	
4.0	0.87	− 44	1.85	136	0.030	73	0.81	−12
6.0	0.74	− 65	1.59	113	0.030	70	0.77	−18
8.0	0.67	− 81	1.53	107	0.031	87	0.78	−21
10.0	0.57	−102	1.34	84	0.032	100	0.77	−28
12.0	0.51	−129	1.32	63	0.031	97	0.77	−39
15.0	0.48	−161	0.99	35	0.028	90	0.78	−51
18.0	0.41	178	0.98	9	0.029	78	0.91	−57

$V_{ds} = 3.5V$; $I_{ds} = 0.9I_{dss}$

f (GHz)	S11		S21		S12		S22	
4.0	0.86	− 48	2.17	135	0.017	82	0.84	−10
6.0	0.72	− 71	1.84	111	0.015	98	0.81	−16
8.0	0.66	− 89	1.74	105	0.023	134	0.82	−19
10.0	0.57	−111	1.51	81	0.034	145	0.83	−27
12.0	0.52	−139	1.47	60	0.039	140	0.75	−40
15.0	0.50	−171	1.10	31	0.053	134	0.86	−47
18.0	0.44	167	1.23	3	0.041	122	0.99	−52

TYPICAL NOISE PARAMETERS

$V_{ds} = 3.5V$; $I_{ds} = 10$ mA

f (GHz)	NF_{opt} (dB)	Γ_{opt}		$r_n = \frac{R_n}{50}$
4.0	1.8	0.62	45	0.60
8.0	2.5	0.42	79	0.50
12.0	3.5	0.42	125	0.36

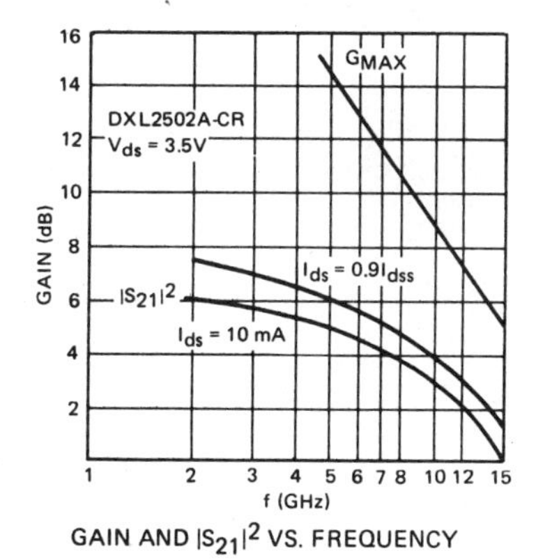

GAIN AND $|S_{21}|^2$ VS. FREQUENCY

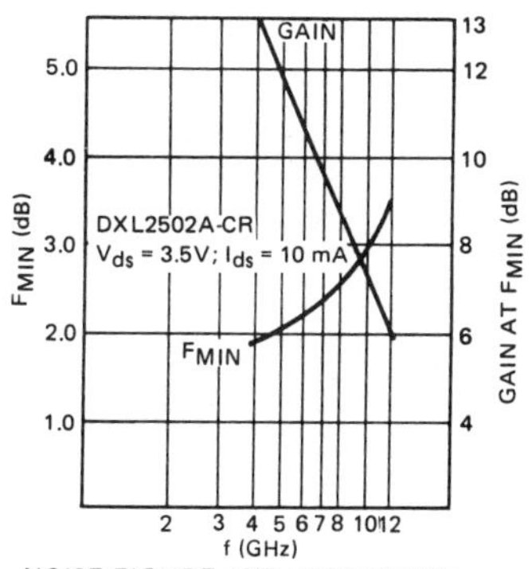

NOISE FIGURE AND ASSOCIATED GAIN VS. FREQUENCY

LOW NOISE MICROWAVE GaAs FET

HFET-2202

TENTATIVE DATA JULY 1979

LOW NOISE FIGURE
1.1 dB Typical NF at 4 GHz, 1.4 dB Maximum
1.9 dB Typical NF at 8 GHz

HIGH ASSOCIATED GAIN
13.6 dB Typical G_a at 4 GHz, 12.0 dB Minimum
9.6 dB Typical at 8 GHz

HIGH OUTPUT POWER
14.5 dBm Linear Power at 4 GHz

CHARACTERIZED TO 12 GHz

RUGGED HERMETIC PACKAGE

0.5 MICROMETER GATE

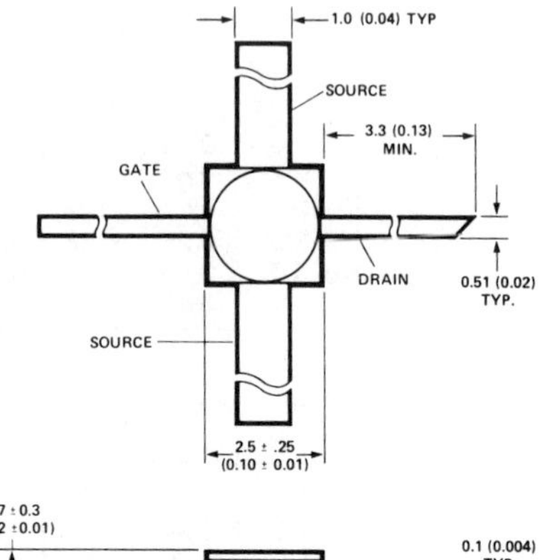

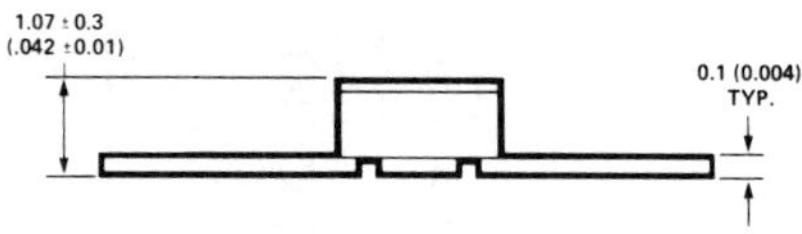

DIMENSIONS IN MILLIMETERS (INCHES).

HPAC-100A Package Outline

The HFET-2202 is a gallium arsenide Schottky gate field effect transistor. It features a rugged, hermetic package that is designed for consistent operation over the frequency range of 2 GHz to 12 GHz. The device's superior noise and gain performance, coupled with its wide dynamic range capability, make it ideally suited for such applications as land and satellite communications and radar.

The HFET-2202 is packaged in the HPAC-100A. The part is capable of meeting the environmental requirements of MIL-S-19500 and the test requirements of MIL-STD-750/883.

Symbol	Parameters and Test Conditions		Units	Min.	Typ.	Max.
I_{DSS}	Saturated Drain Current, V_{DS} = 3.5V, V_{GS} = 0V		mA	25	45	90
V_{GSP}	Pinch Off Voltage, V_{DS} = 3.5V, I_{DS} <500 μA		V	-0.5	-2.0	-4.0
g_m	Transconductance, V_{DS} = 3.5V, ΔV_{GS} = 0V to -0.5V		mmho	20	32	45
$G_{a(max)}$	Maximum Available Gain V_{DS} = 3.5V, V_{GS} = 0V	f = 6 GHz 8 GHz	dB		16.0 13.0	
F_{MIN}	Minimum Noise Figure V_{DS} = 3.5V, I_{DS} = 15% I_{DSS} (Typ. 7.5 mA)	f = 4 GHz 6 GHz 8 GHz	dB		1.1 1.4 1.9	1.4
G_a	Associated Gain at N.F. Bias	f = 4 GHz 6 GHz 8 GHz	dB	12.0	13.6 11.3 9.6	
P_{1dB}	Power at 1 dB Gain Compression	f = 4 GHz	dBm		14.5	
G_{1dB}	Associated 1 dB Compressed Gain V_{DS} = 4.0V, I_{DS} = 50% I_{DSS} (0 dBm Input Matching, Tuned for Maximum Output)	f = 4 GHz	dB		13.3	

Symbol	Parameter	Values
V_{DS}	Drain to Source Voltage, $-4\ V \leq V_{GS} \leq 0V$	4V
V_{GS}[2]	Gate to Source Voltage $4V \geq V_{DS} \geq 0V$	-4V
T_{CH}[3]	Maximum Channel Temperature	125°C
T_{STG}	Storage Temperature	-65°C to +125°C

Notes:
1. Operation of this device in excess of any one of these conditions is likely to result in a reduction in device mean time between failure (MTBF) to below the design goal of 1×10^7 hours at $T_{CH} = 125°C$ (assumed Activation Energy = 1.6 eV).
2. Maximum continuous forward gate current should not exceed 1.5 mA.
3. Θ_{jc} — Thermal resistance, channel to case = 260°C/W.

Symbol	Parameter	Limits
V_{DS}	Drain to Source Voltage $-4V \leq V_{GS} \leq 0V$	10V
V_{GS}[2]	Gate to Source Voltage $4V \geq V_{DS} \geq 0V$	-6V
T_{CH}	Maximum Channel Temperature	300°C
$T_{STG(max)}$	Maximum Storage Temperature	250°C

Lead Soldering Temperature[3] 250°C for 10 sec. each lead.

Notes:
1. Operation in excess of any one of these conditions may result in permanent damage to this device.
2. Maximum forward gate current should not exceed 2 mA.
3. See Handling and Use Precautions (page 3).

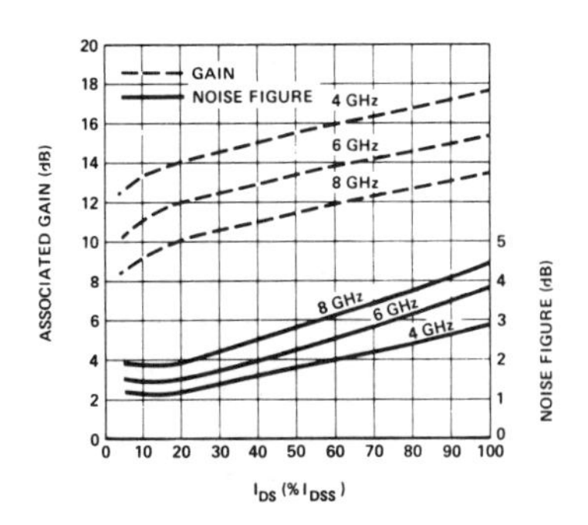

Figure 1. Typical Associated Gain and Noise Figure (F_{MIN}) vs. I_{DS} as a percentage of I_{DSS} when tuned for minimum noise figure. Frequency from 4 GHz to 8 GHz, $V_{DS} = 3.5V$.

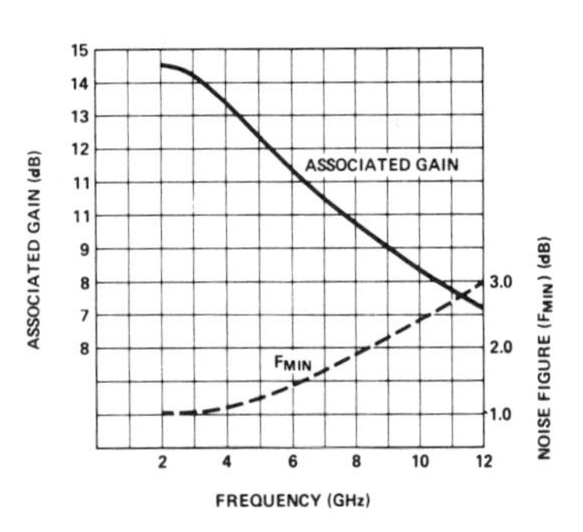

Figure 2. Typical Noise Figure (F_{MIN}) and Associated Gain vs. Frequency. $V_{DS} = 3.5V$, $I_{DS} = 15\%\ I_{DSS}$.

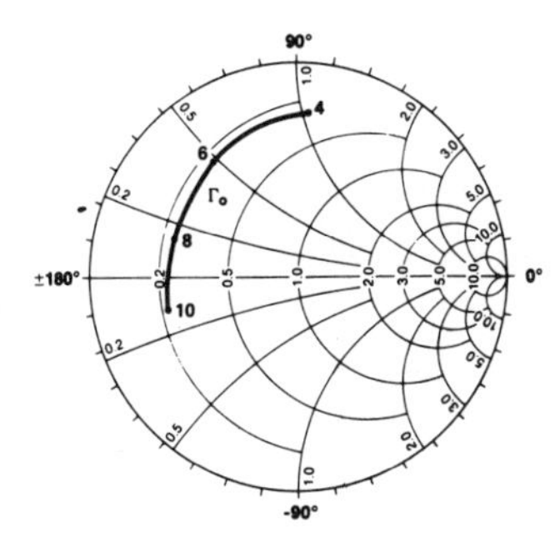

Figure 3. Typical Γ_o in the 4 to 10 GHz range for $V_{DS} = 3.5V$, $I_{DS} = 15\%\ I_{DSS}$. (Γ_o = Input Match for Minimum Noise).

TABLE I. HFET-2202 Typical Noise Parameters[1]

Frequency	Γ_o	$\Gamma_L = [S'_{22}]^*$	F_{MIN} (dB)	R_N (Ω)
4 GHz	.75 ∠ 86°	.74 ∠ 70°	1.1	45
6 GHz	.63 ∠ 119°	.65 ∠ 93°	1.4	21
8 GHz	.62 ∠ 161°	.64 ∠ 124°	1.9	7
10 GHz	.62 ∠ -168°	.65 ∠ 159°	2.5	4

Note:
1. Optimum Input Reflection Coefficient (Γ_o), Output Match for Minimum Noise (Γ_L), Associated Noise Figure (F_{MIN}) and Noise Resistance (R_N) at $V_{DS} = 3.5V$, $I_{DS} = 15\%\ I_{DSS}$.

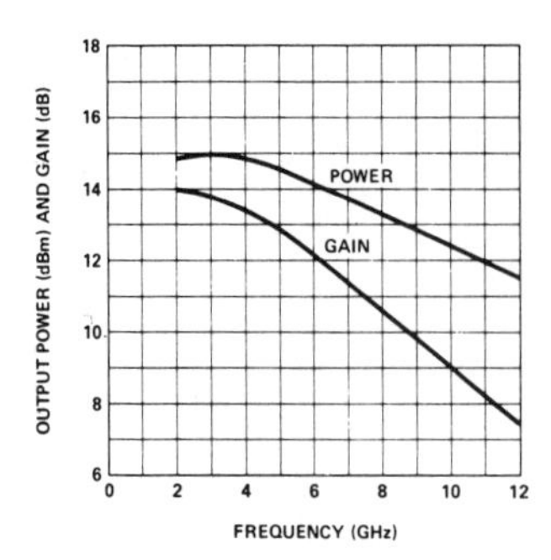

Figure 4. Typical P_{1dB} Linear Power and Associated 1dB Compressed Gain vs. Frequency at V_{DS} = 4.0V, I_{DS} = 50% I_{DSS}.

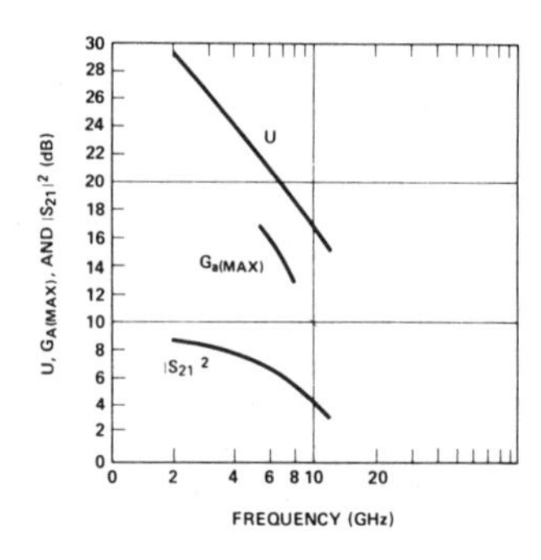

Figure 5. Mason's Gain (U), $G_{a(max)}$ and $|S_{21}|^2$ vs. Frequency, V_{DS} = 3.5V, V_{GS} = 0.0V.

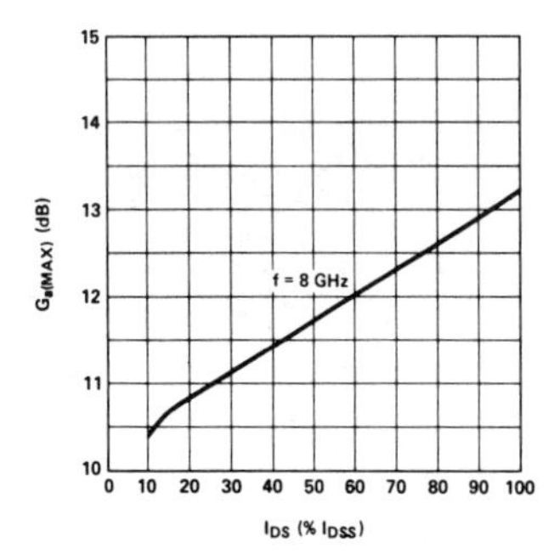

Figure 6. Typical $G_{a(max)}$ vs. I_{DS} as a percentage of I_{DSS}. Frequency = 8 GHz. V_{DS} = 3.5V.

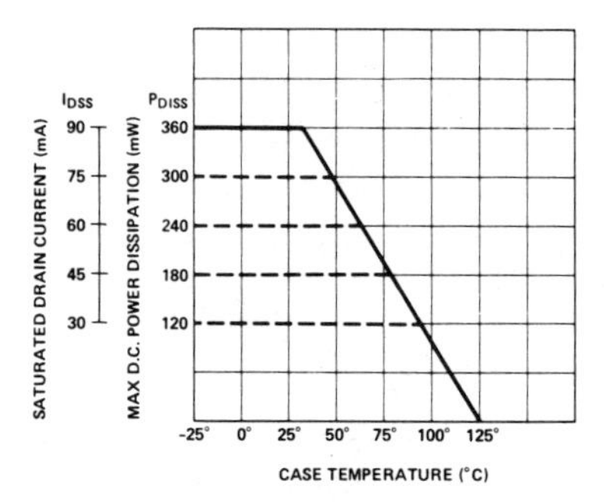

Figure 7. P_{DISS} vs. Temperature, Power Derating Curve at V_{DS} = 4V. Maximum power dissipation is a function of device I_{DSS}. Begin derating at P_{DISS} corresponding to individual device I_{DSS}, following a horizontal line until it intersects with the solid diagonal line.

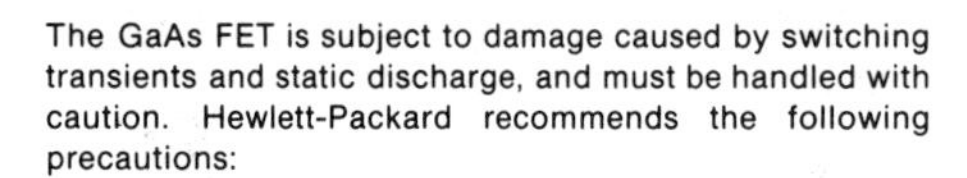

The GaAs FET is subject to damage caused by switching transients and static discharge, and must be handled with caution. Hewlett-Packard recommends the following precautions:

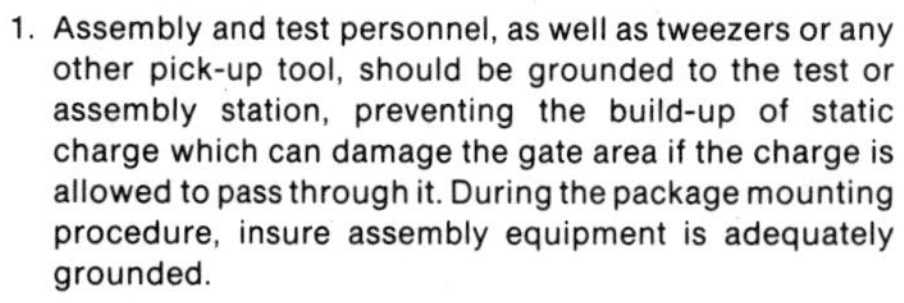

1. Assembly and test personnel, as well as tweezers or any other pick-up tool, should be grounded to the test or assembly station, preventing the build-up of static charge which can damage the gate area if the charge is allowed to pass through it. During the package mounting procedure, insure assembly equipment is adequately grounded.

   Static discharge during handling, testing, and assembly can induce increased reverse gate leakage of a resistive nature.

   To prevent the buildup of static charge on the package during storage, the device should be held in a conductive medium (e.g., metal container, conductive foam).

2. Spurious pulses generated by test equipment (i.e. contact bounce during switching, induced voltage in the leads, etc.) must be eliminated. Avoid turning instrument power on and off, or switching between instrument ranges when bias is applied to the device.
3. Inductive pickup from large transformers, switching power supplies, inductive ovens, etc., must also be eliminated. Use shielded signal and power cables.
4. Assembly equipment (i.e., soldering irons) must be adequately grounded.
5. Application of bias. When applying bias to the FET, first apply the gate voltage, then the drain voltage. When removing bias, remove the gate voltage last.

MINIMUM NOISE FIGURE BIAS V_{DS} = 3.5V, I_{DS} = 15% I_{DSS}

	S_{11}			S_{21}			S_{12}		S_{22}	
Freq. (GHz)	**Mag.**	**Ang.**	**(dB)**	**Mag.**	**Ang.**	**(dB)**	**Mag.**	**Ang.**	**Mag.**	**Ang.**
2.0	0.98	-44	5.23	1.83	138	-29.6	0.03	58	0.80	-24
3.0	0.93	-66	5.02	1.78	118	-26.6	0.05	43	0.77	-37
3.5	0.91	-77	4.94	1.77	108	-25.5	0.05	36	0.76	-44
3.6	0.91	-79	4.96	1.77	106	-25.2	0.06	35	0.75	-46
3.7	0.90	-81	4.95	1.77	104	-25.2	0.06	34	0.75	-47
3.8	0.90	-83	4.88	1.75	102	-25.0	0.06	32	0.74	-49
3.9	0.89	-85	4.87	1.75	100	-24.7	0.06	30	0.74	-50
4.0	0.89	-88	4.88	1.75	98	-24.6	0.06	29	0.74	-52
4.1	0.89	-91	4.88	1.75	96	-24.6	0.06	28	0.74	-52
4.2	0.89	-92	4.80	1.74	94	-24.4	0.06	27	0.73	-53
4.3	0.88	-95	4.78	1.73	91	-24.3	0.06	25	0.73	-55
4.4	0.88	-97	4.81	1.74	89	-24.2	0.06	24	0.73	-56
4.5	0.87	-99	4.77	1.73	88	-24.1	0.06	23	0.73	-58
5.0	0.86	-109	4.63	1.71	77	-23.5	0.07	15	0.71	-65
6.0	0.83	-130	4.40	1.66	57	-23.1	0.07	2	0.69	-81
7.0	0.78	-150	3.87	1.56	37	-23.1	0.07	-9	0.66	-97
8.0	0.74	-169	3.32	1.47	19	-23.2	0.07	-18	0.63	-114
9.0	0.73	174	2.80	1.38	1	-23.6	0.07	-24	0.61	-132
10.0	0.72	157	2.38	1.32	-16	-23.6	0.07	-29	0.61	-151
11.0	0.71	142	1.93	1.25	-34	-22.2	0.08	-34	0.62	-168
12.0	0.67	126	1.44	1.18	-52	-23.0	0.07	-38	0.62	175

HIGH GAIN BIAS V_{DS} = 3.5V, V_{GS} = 0V

	S_{11}			S_{21}			S_{12}		S_{22}	
Freq. (GHz)	**Mag.**	**Ang.**	**(dB)**	**Mag.**	**Ang.**	**(dB)**	**Mag.**	**Ang.**	**Mag.**	**Ang.**
2.0	0.96	-51	8.72	2.73	134	-34.4	0.02	61	0.79	-23
3.0	0.89	-75	8.27	2.59	113	-31.7	0.03	49	0.76	-35
3.5	0.87	-87	8.07	2.53	103	-31.1	0.03	43	0.76	-41
3.6	0.86	-89	8.04	2.52	101	-30.8	0.03	43	0.75	-43
3.7	0.86	-92	8.01	2.52	99	-30.7	0.03	43	0.75	-44
3.8	0.85	-94	7.92	2.49	97	-30.5	0.03	41	0.74	-46
3.9	0.84	-97	7.89	2.48	94	-30.5	0.03	40	0.74	-47
4.0	0.84	-99	7.85	2.47	93	-30.5	0.03	40	0.74	-48
4.1	0.84	-102	7.83	2.46	91	-30.5	0.03	38	0.74	-49
4.2	0.84	-104	7.74	2.44	88	-30.2	0.03	38	0.74	-50
4.3	0.83	-107	7.69	2.43	86	-29.9	0.03	36	0.74	-51
4.4	0.82	-109	7.69	2.43	84	-29.9	0.03	36	0.73	-52
4.5	0.82	-111	7.64	2.41	82	-29.9	0.03	35	0.73	-54
5.0	0.80	-122	7.35	2.34	72	-29.4	0.03	31	0.72	-61
6.0	0.76	-143	6.90	2.21	52	-29.1	0.04	29	0.71	-75
7.0	0.72	-164	6.21	2.04	33	-28.4	0.04	27	0.68	-91
8.0	0.68	178	5.51	1.88	15	-27.1	0.04	27	0.67	-107
9.0	0.68	159	4.91	1.76	-3	-25.7	0.05	25	0.65	-124
10.0	0.67	143	4.36	1.65	-20	-23.9	0.06	20	0.66	-142
11.0	0.65	128	3.83	1.55	-37	-21.6	0.08	4	0.67	-160
12.0	0.62	112	3.24	1.45	-55	-21.1	0.09	1	0.68	-176

LINEAR POWER BIAS V_{DS} = 4.0V, I_{DS} = 50% I_{DSS}

	S_{11}			S_{21}			S_{12}		S_{22}	
Freq. (GHz)	**Mag.**	**Ang.**	**(dB)**	**Mag.**	**Ang.**	**(dB)**	**Mag.**	**Ang.**	**Mag.**	**Ang.**
2.0	0.97	-47	7.78	2.45	136	-32.8	0.02	61	0.78	-23
3.0	0.91	-71	7.42	2.35	115	-30.2	0.03	47	0.75	-35
3.5	0.88	-82	7.27	2.31	105	-29.1	0.04	41	0.74	-42
3.6	0.88	-85	7.26	2.31	103	-28.9	0.04	40	0.73	-44
3.7	0.87	-87	7.23	2.30	101	-28.9	0.04	39	0.73	-45
3.8	0.87	-89	7.16	2.28	99	-28.6	0.04	37	0.73	-47
3.9	0.86	-92	7.13	2.27	97	-28.6	0.04	36	0.73	-48
4.0	0.86	-94	7.12	2.27	95	-28.4	0.04	36	0.72	-49
4.1	0.86	-96	7.10	2.26	93	-28.4	0.04	35	0.72	-49
4.2	0.86	-99	7.02	2.24	91	-28.2	0.04	34	0.72	-51
4.3	0.85	-101	6.98	2.23	89	-28.0	0.04	33	0.72	-52
4.4	0.85	-103	6.98	2.23	87	-28.0	0.04	32	0.72	-53
4.5	0.84	-105	6.94	2.22	85	-28.0	0.04	31	0.72	-55
5.0	0.83	-116	6.72	2.17	75	-27.5	0.04	24	0.70	-62
6.0	0.79	-137	6.34	2.08	55	-27.3	0.04	19	0.68	-76
7.0	0.74	-158	5.72	1.93	35	-26.9	0.05	14	0.66	-92
8.0	0.70	-176	5.08	1.79	17	-26.4	0.05	11	0.64	-108
9.0	0.69	166	4.52	1.68	-1	-25.7	0.05	10	0.62	-126
10.0	0.68	150	4.02	1.59	-18	-24.3	0.06	7	0.63	-144
11.0	0.67	135	3.53	1.50	-35	-22.3	0.08	-6	0.64	-161
12.0	0.63	119	2.96	1.41	-53	-21.9	0.08	-9	0.65	-178

8.13 STABILITY

Stability refers to an amplifier's ability to provide gain without going into oscillation. An amplifier is said to be unconditionally stable if the real part of its input impedance remains positive when the load impedance is changed arbitrarily, and also if the real part of its output impedance remains positive when the source impedance is changed arbitrarily. The only condition on the changes of load and source impedances is that their real parts must remain positive. The way in which load and source impedances influence the input and output impedances is shown in Eqs. (2.58) and (2.60).

Kurokawa (1965) has shown that the necessary and sufficient conditions for an amplifier to be unconditionally stable are

$$|S_{12} S_{21}| < 1 - |S_{11}|^2 \tag{8.51}$$

$$|S_{12} S_{21}| < 1 - |S_{22}|^2 \tag{8.52}$$

$$2|S_{12} S_{21}| < 1 + |S_{12} S_{21} - S_{11} S_{22}|^2 - |S_{11}|^2 - |S_{22}|^2 \tag{8.53}$$

The stability factor k, obtained from the last condition, is

$$k = \frac{1 + |S_{12} S_{21} - S_{11} S_{22}|^2 - |S_{11}|^2 - |S_{22}|^2}{2|S_{12} S_{21}|} \tag{8.54}$$

It will be seen that the last condition for unconditional stability requires that $k > 1$. When $k < 1$, the amplifier is conditionally stable, that is, it may oscillate under some conditions. The value of k for given bias conditions is often stated on a specification sheet. For example, in *Hewlett-Packard Application Note 970*, $k = 1.504$ for transistor HFET-1101.

The maximum available power gain for the unconditionally stable amplifier, for $S_{12} S_{21} \neq 0$, is given by

$$G_{A\max} = \frac{|S_{21}|}{|S_{12}|}(k - \sqrt{k^2 - 1}) \tag{8.55}$$

For the unilateral amplifier $S_{12} = 0$. It is left as an exercise for the reader (Problem 41) to show that for the unilateral amplifier, Eq. (8.55) for maximum available power gain reduces to Eq. (8.27), and that condition (8.53) is redundant (Problem 42). Unconditional stability for the unilateral amplifier therefore requires that

$$|S_{11}| < 1 \tag{8.56}$$

and

$$|S_{22}| < 1 \tag{8.57}$$

Another way of looking at these requirements for unconditional stability of the unilateral amplifier is that the gain factors $G_{S\max}$ and $G_{L\max}$ both remain finite and positive. Application of Eqs. (8.56) and (8.57) to Eqs. (8.13) and (8.14) shows this to be the case.

8.14 LINEAR POWER AMPLIFIERS

Where high power output is required, the bias point must be selected accordingly. This is shown, for example, in the data sheet for the HFET-2202, in Section 8.12, where the S-parameters are tabulated for three different bias conditions: minimum noise figure, high gain, and linear power amplification.

When power output in dBm is plotted as a function of power input, also in dBm, a straight line results over part of the range, as shown in Fig. 8.16. (P_{out}) dBm − (P_{in}) dBm gives the power gain. At relatively high levels, the curve departs from linearity, and a measure of this departure is the 1-dB compression point. This is the point where the actual output/input curve drops by 1 dB below the extrapolated straight line. This is similar to the 1-dB conversion compression point shown in Fig. 7.45.

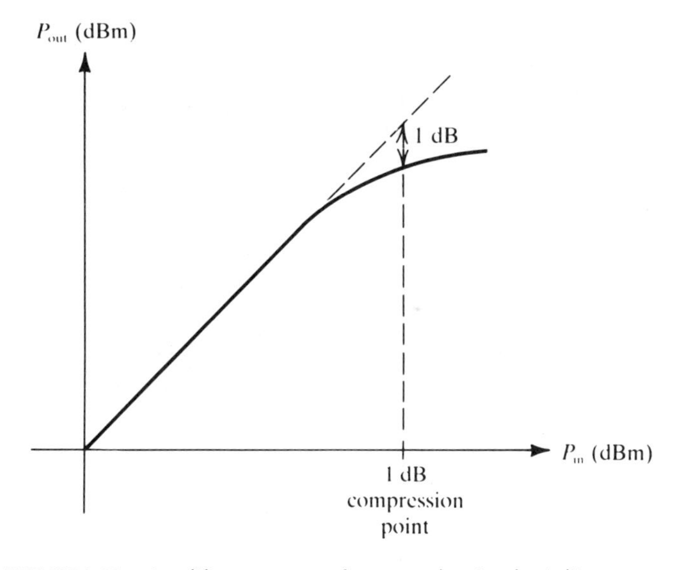

FIGURE 8.16. Amplifier power transfer curve, showing the 1-dB compression point.

Apart from biasing a transistor to operate in a linear power mode, transistors are specifically designed for this application. The Avantek AT-8141 transistor is a 1-W device, and the chip illustrated on the data sheet shows how high power rating is achieved through multiple-device geometry. The transistor is available in chip and packaged forms.

A data sheet is also shown for the Hewlett-Packard HXTR-3002 linear power transistor chip. This is a bipolar device, and the chip illustrated on the data sheet shows that a similar multiple-emitter construction is used to increase the power ratings.

8.15 BIASING

As mentioned in Section 8.14, the bias point must be selected for the particular application in mind. This is illustrated in Fig. 8.17, where typical bias points for the three different conditions—low noise, high gain, and high power—are shown. Also shown are the five basic circuits used to apply bias (Vendelin 1978). With the bipolar supply (meaning two polarities of supply and not to be confused with bipolar transistor), shown in panel (1a) of Fig. 8.17, the source can be directly grounded, and therefore maximum gain achieved. The disadvantage of this arrangement is the susceptibility to damage if the drain supply is connected first. The precaution must always be taken of connecting the gate supply first and disconnecting it last. A similar precaution applies to the other circuits that utilize a separate gate bias supply. The circuits shown in panels (1b) and (1c) of Fig. 8.17 show how different polarities of supply can be applied. In both cases an RF choke is required in the source lead, and the source connected to ground through a bypass capacitor. This arrangement can lead to gain instability. The circuits shown in panels (1d) and (1e) of Fig. 8.17 show a source resistance used to generate self-bias. This provides automatic bias protection (since the level of bias is directly proportional to the drain–source current). As well as the decoupling problem, similar to that encountered with the source choke arrangement, another disadvantage is that dc power is dissipated in the resistor.

Figure 8.18 shows a practical arrangement for connecting bias supplies into a circuit (*Hewlett-Packard Application Note 973*). The dc bias lines are $\lambda/4$ in length and therefore transform the low impedances presented at the ground bar end to high impedances at the amplifier connection points.

With bipolar transistors, the emitter is usually grounded directly in discrete component designs, to prevent instability. Instability may occur even where a good bypass capacitor is provided for the emitter resistor at the operating frequency. The problem is that at lower frequencies the bypass impedance increases, and the effective S_{11}-parameter can increase

AT-8141

One Watt, 2-10 GHz
Gallium arsenide FET Chip
December, 1980

FEATURES

- +30 dBm Min. Power Output at 4 GHz
- 8 dB Min. Associated Gain (G_P)
- Up to 35% Power Added Efficiency
- Gold-Based Metallization
- Optimized Power Epitaxy and Doping
- Suitable for Broadband Applications
- Optimum Thermal and Electrical Design
- Large Bonding Pads

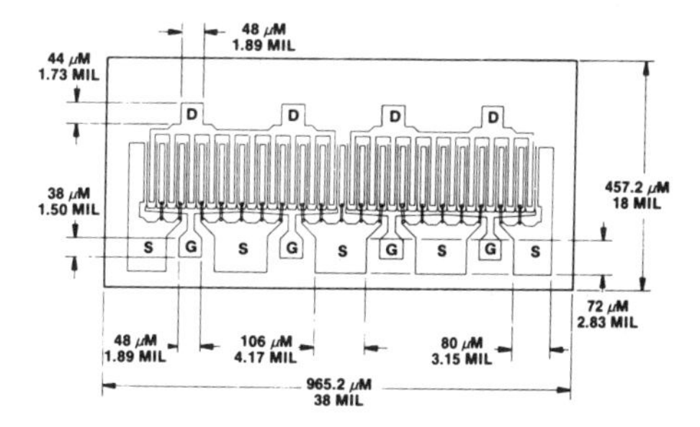

DESCRIPTION

The Avantek® AT-8141 is an unpackaged gallium arsenide Schottky-gate field effect transistor chip designed for medium power, linear amplification in the 2 to 10 GHz frequency range. This rugged, reliable device is suitable for a wide variety of applications such as communications and radar equipment operating in the space, airborne, military and commercial environments.

This GaAs FET chip (also available in packaged form as the AT-8140) has a four-cell, 5-millimeter gate periphery structure with airbridge interconnects between source pads. All metal surfaces are gold plated for ease of bonding and die attach. Large bonding pads facilitate bonding into hybrid integrated circuits.

ELECTRICAL SPECIFICATIONS, $T_A = 25°C$

Symbol	Parameters and Test Conditions	Freq.	Units	Min.*	Typ.	Max.
$P_{O(-1\ dB)}$	Output Power at 1 dB Gain Compression, $V_{DS} = 9V$, $I_{DS} = 50\%\ I_{DSS}$, Tuned for maximum output power at input level producing 1 dB gain compression.	4 GHz 8 GHz	dBm	30.0	31.5 30.0	
G_P	Associated Small Signal Gain, $V_{DS} = 9V$, $I_{DS} = 50\%\ I_{DSS}$,	4 GHz 8 GHz	dB	8.0	9.0 6.0	
G_{max}	Maximum Available Gain, $V_{DS} = 9V$, $I_{DS} = 50\%\ I_{DSS}$	8 GHz	dB		8.0	
g_m	Transconductance: $V_{DS} = 3V$, $I_{DS} = I_{DSS}$		mmho		450	
I_{DSS}	Saturated Drain Current: $V_{DS} = 3V$, $V_{gs} = 0$		mA		1000	
V_P	Pinchoff Voltage: $V_{DS} = 3V$, $I_{DS} = 5.0$ mA		V		-3.0	

*(Measured in Avantek's 100 MIL FET flange package.)

AT-8141

MAXIMUM RATINGS (T_A = 25°C)

Parameter	Symbol	Recommended max. for continuous operation	Absolute max.
Drain-Source Voltage	V_{DS}	+9V	+14V
Gate-Source Voltage	V_{GS}	−5V	−7V
Drain Current	I_{DS}	500 mA	I_{DSS}
Continuous Dissipation	P_T (T_{case} = 25°C)	4.5 W	7.0 W
Channel Temperature	T_{ch}	150°C	300°C
Storage Temperature	T_{stg}	−65° to +150°C	250°C

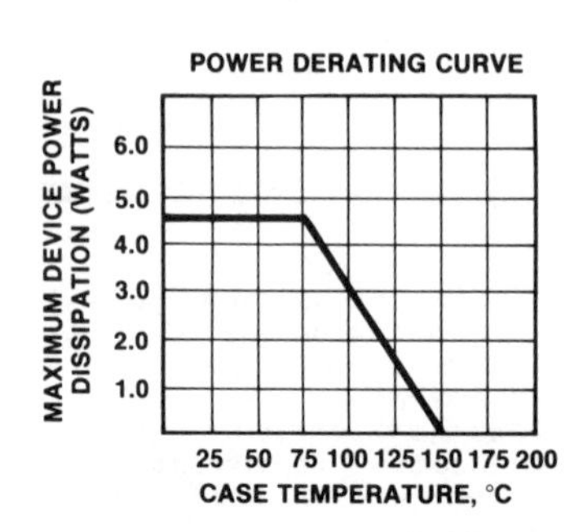

Thermal Resistance	Θ_{jc}	18°C/W

Note: The indicated thermal resistance (θ_{jc}) and power derating curve apply to this GaAs FET chip when installed in the Avantek 100-mil FET flange package. The actual thermal characteristics of this unpackaged device will depend on the type and quality of bond and characteristics of the surface to which it is bonded.

TYPICAL PERFORMANCE CURVES (T_A=25°C)

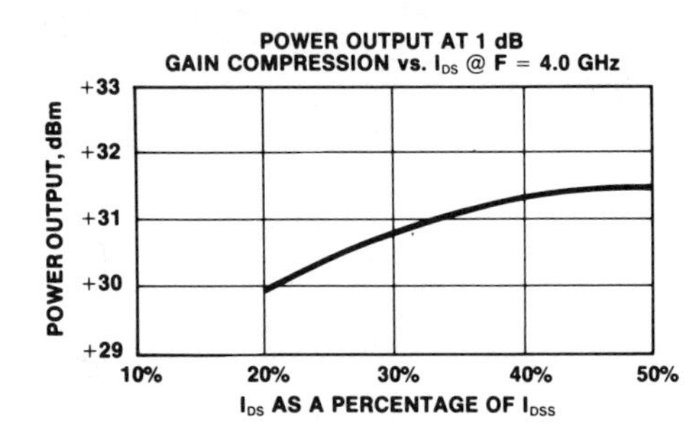

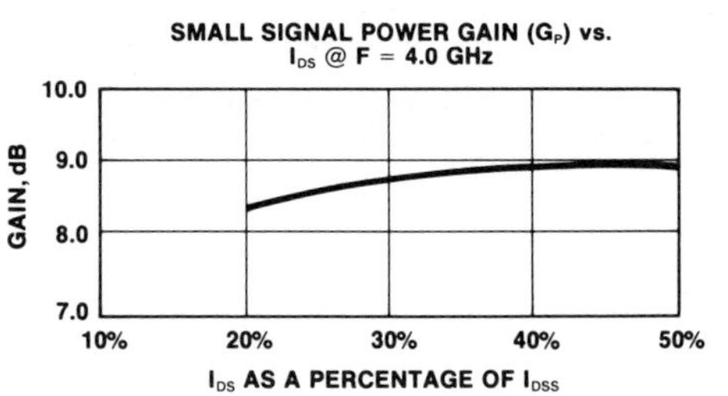

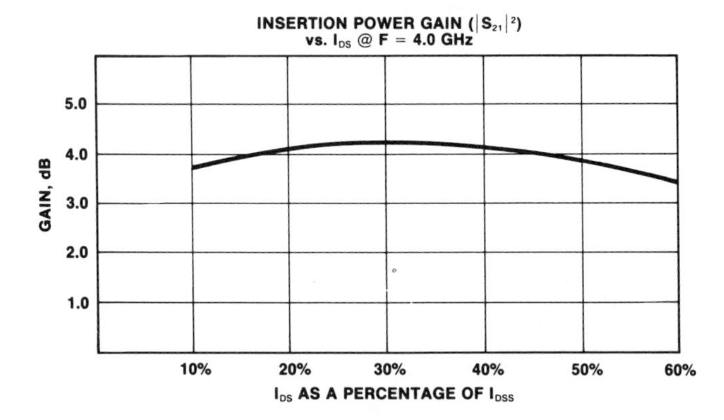

AT-8141

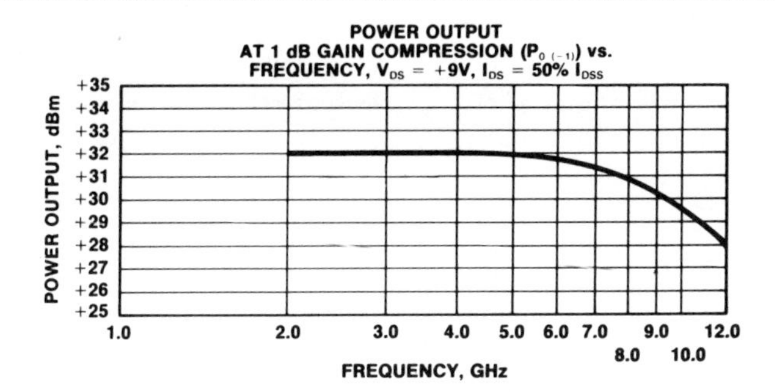
POWER OUTPUT
AT 1 dB GAIN COMPRESSION ($P_{0(-1)}$) vs.
FREQUENCY, V_{DS} = +9V, I_{DS} = 50% I_{DSS}
POWER OUTPUT, dBm
+35
+34
+33
+32
+31
+30
+29
+28
+27
+26
+25
1.0
2.0
3.0
4.0
5.0
6.0
7.0
8.0
9.0
10.0
12.0
FREQUENCY, GHz

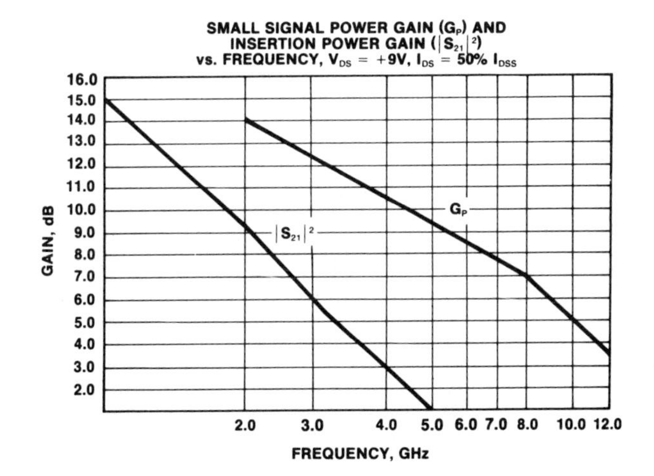
SMALL SIGNAL POWER GAIN (G_P) AND
INSERTION POWER GAIN ($|S_{21}|^2$)
vs. FREQUENCY, V_{DS} = +9V, I_{DS} = 50% I_{DSS}
GAIN, dB
16.0
15.0
14.0
13.0
12.0
11.0
10.0
9.0
8.0
7.0
6.0
5.0
4.0
3.0
2.0
G_P
$|S_{21}|^2$
2.0
3.0
4.0
5.0
6.0
7.0
8.0
10.0
12.0
FREQUENCY, GHz

AT-8141

TYPICAL SCATTERING PARAMETERS, COMMON SOURCE*

S — MAGN AND ANGLES: AT-8141 BIAS = 9.00 VOLTS, 425.00 MA

FREQ.	11		21		12		22	
1000.00	.879	−91.2	5.563	101.3	.028	40.9	.227	−52.3
2000.00	.841	−136.0	2.933	83.1	.034	49.4	.182	−83.2
3000.00	.846	−157.9	1.891	73.9	.040	61.2	.207	−97.3
4000.00	.858	−168.5	1.440	66.7	.047	72.4	.228	−106.6
5000.00	.851	−174.8	1.128	58.8	.056	80.8	.254	−111.8
6000.00	.882	−179.6	.924	52.5	.065	84.0	.284	−115.6
7000.00	.843	−179.6	.764	49.9	.074	90.7	.326	−123.9
8000.00	.838	172.4	.694	45.8	.086	92.6	.347	−127.6
9000.00	.856	160.1	.619	42.0	.097	94.0	.383	−137.9
10000.00	.835	145.0	.577	36.6	.111	95.6	.399	−148.4
11000.00	.778	129.2	.530	30.9	.129	92.3	.433	−160.0
12000.00	.845	116.5	.494	21.4	.137	84.8	.480	−177.7
13000.00	.796	108.6	.403	15.0	.136	81.6	.524	157.6

*(S-parameters include bond wire inductance and are measured in Avantek's standard 50 ohm test carrier.)

LINEAR POWER TRANSISTOR CHIP

HXTR-3002

TECHNICAL DATA APRIL 1980

Features

HIGH P_{1dB} LINEAR POWER
22 dBm Typical at 1 GHz

HIGH P_{1dB} GAIN
18.0 dB Typical at 1 GHz

HIGH S_{21E} GAIN
16.5 dB Typical at 500 MHz

WIDE DYNAMIC RANGE

LARGE GOLD BONDING PADS

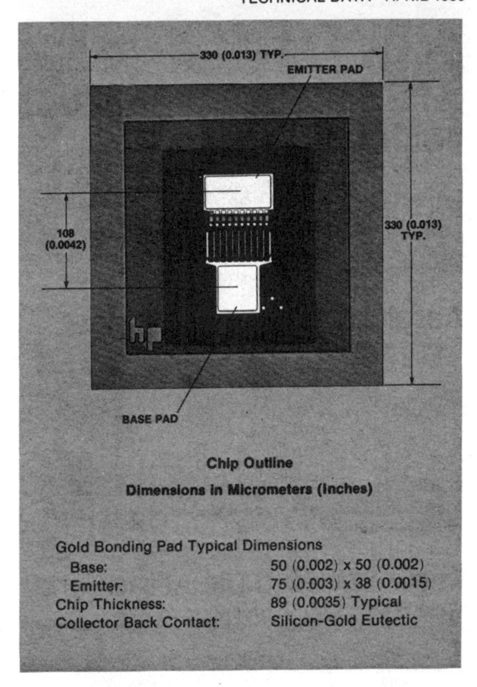

Chip Outline
Dimensions in Micrometers (Inches)

Gold Bonding Pad Typical Dimensions	
Base:	50 (0.002) x 50 (0.002)
Emitter:	75 (0.003) x 38 (0.0015)
Chip Thickness:	89 (0.0035) Typical
Collector Back Contact:	Silicon-Gold Eutectic

Description/Applications

The HXTR-3002 is an NPN bipolar transistor chip intended for use in hybrid applications requiring superior UHF and microwave performance. Use of ion implantation and self-alignment techniques in its manufacture produce uniform devices requiring little or no individual circuit adjustment. The HXTR-3002 features a Ti/Pt/Au metallization system, a dielectric scratch protection over its active area and Ta_2N ballast resistors for ruggedness. Its large gold bonding pads facilitate use of 25 μm (1 mil) gold bond wires frequently used in hybrid applications.

The superior power, gain and dynamic range performance of the HXTR-3002 commend it for RF and IF use in broad and narrow band commercial and military communications, radar, and ECM hybrid applications.

Electrical Specifications at $T_A = 25°C$

Symbol	Parameters and Test Conditions		Test MIL-STD-750	Units	Min.	Typ.	Max.
BV_{CBO}	Collector-Base Breakdown Voltage at $I_C = 3$ mA		3001.1*	V	40		
BV_{CEO}	Collector-Emitter Breakdown Voltage at $I_C = 15$ mA		3011.1*	V	24		
BV_{EBO}	Emitter-Base Breakdown Voltage at $I_B = 30$ μA		3026.1*	V	3.3		
I_{EBO}	Emitter-Base Leakage Current at $V_{EB} = 2$ V		3061.1*	μA			2
I_{CES}	Collector-Emitter Leakage Current at $V_{CE} = 32$ V		3041.1**	nA			200
I_{CBO}	Collector-Base Leakage Current at $V_{CB} = 20$ V		3036.1**	nA			100
h_{FE}	Forward Current Transfer Ratio at $V_{CE} = 18$ V, $I_C = 30$ mA		3076.1*		15	40	75
P_{1dB}	Power Output at 1 dB Gain Compression	f = 1 GHz		dBm		22	
G_{1dB}	Associated 1 dB Compressed Gain	f = 1 GHz		dB		18	
S_{21E}	Transducer Gain	f = 500 MHz 1 GHz		dB		16.5 13.6	
η	Power-Added Efficiency at 1 dB Compression	f = 1 GHz		%		29	
	Test Conditions: $V_{CE} = 18$ V, $I_C = 30$ mA, $\theta_{JA} = 210°$ C/W						

*300 μs wide pulse measurement at $\leq 2\%$ duty cycle.
**Measured under low ambient light conditions.

Recommended Maximum Continuous Operating Conditions[1]

Symbol	Parameter	Value
V_{CBO}	Collector to Base Voltage	40V
V_{CEO}	Collector to Emitter Voltage	22V
V_{EBO}	Emitter to Base Voltage	3.3V
I_C	DC Collector Current	50 mA
P_T	Total Device Dissipation[2]	700 mW
T_J	Junction Temperature	200°C
T_{STG}	Storage Temperature	-65°C to +200°C

Notes:

1. Operation of this device in excess of any one of these conditions is likely to result in a reduction in device mean time between failure (MTBF) to below the design goal of 1×10^7 hours at $T_J = 175°C$ (assumed Activation Energy = 1.5 eV).
2. Power dissipation derating should include a Θ_{JB} (Junction-to-Back contact thermal resistance) of 125°C/W.

   Total Θ_{JA} (Junction-to-Ambient) will be dependent upon the heat sinking provided in the individual application.

Absolute Maximum Ratings*

Symbol	Parameter	Limit
V_{CBO}	Collector to Base Voltage	45V
V_{CEO}	Collector to Emitter Voltage	27V
V_{EBO}	Emitter to Base Voltage	4.0V
I_C	DC Collector Current	100 mA
P_T	Total Device Dissipation	1.4W
T_J	Junction Temperature	300°C
$T_{STG(MAX)}$	Maximum Storage Temperature	300°C

*Operation in excess of any one of these conditions may result in permanent damage to this device.

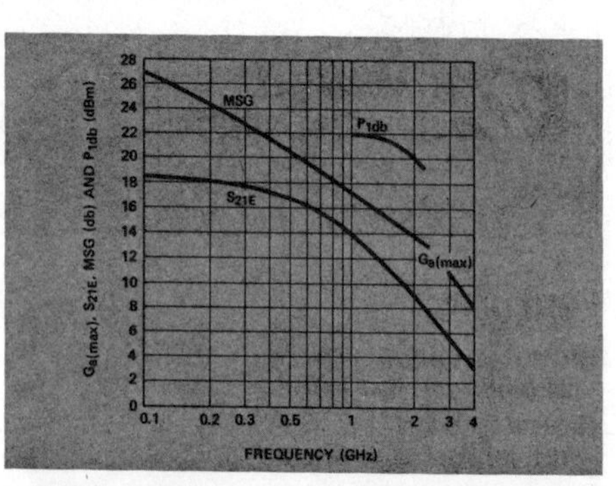

Figure 1. Typical $G_{a(max)}$, S_{21E}, Maximum Stable Gain (MSG) and Power Output at 1 dB Gain Compression (P_{1dB}) vs. Frequency, V_{CE} = 18V, I_C = 30 mA.

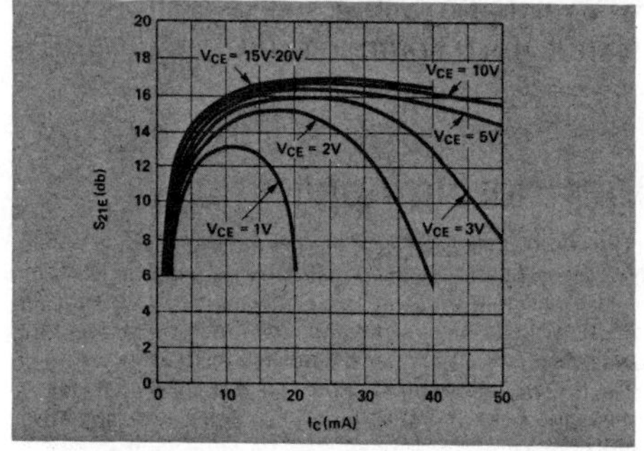

Figure 2. Typical S_{21E} vs. Current at 500 MHz

Recommended Die Attach and Bonding Procedures

Eutectic Die Attach at a stage temperature of 410 ± 10°C under an N_2 ambient. Chip should be lightly scrubbed using a tweezer and eutectic should flow within five seconds.

Thermocompression Wire Bond at a stage temperature of 310 ± 10°C, using a tip force of 30 ± 5 grams with 0.7 or 1.0 mil gold wire. A one mil minimum wire clearance at the passivation edge is recommended. (Ultrasonic bonding is not recommended.)

Packaging — The chip should be packaged into a clean, dry, hermetic environment.

Typical S-Parameters* V_{CE} = 18V, I_C = 30 mA

Freq. (MHz)	S_{11}		S_{21}			S_{12}			S_{22}	
	Mag.	Ang.	(dB)	Mag.	Ang.	(dB)	Mag.	Ang.	Mag.	Ang.
100	0.658	-17	18.5	8.439	170	-35.9	0.016	82	0.991	-7
200	0.656	-32	18.3	8.180	161	-30.1	0.031	75	0.965	-14
300	0.652	-47	17.8	7.792	153	-27.0	0.045	68	0.926	-20
400	0.648	-60	17.3	7.334	145	-25.0	0.056	62	0.881	-25
500	0.644	-72	16.7	6.845	138	-23.7	0.066	56	0.833	-29
600	0.641	-82	16.1	6.366	132	-22.7	0.073	52	0.787	-33
700	0.637	-91	15.4	5.911	126	-22.0	0.080	48	0.744	-36
800	0.634	-99	14.8	5.491	121	-21.5	0.085	45	0.706	-39
900	0.632	-105	14.2	5.110	117	-21.0	0.089	42	0.671	-41
1000	0.629	-111	13.6	4.764	113	-20.7	0.092	39	0.641	-43
1500	0.623	-131	10.9	3.503	98	-19.7	0.103	32	0.541	-50
2000	0.618	-143	8.8	2.739	88	-19.2	0.110	29	0.492	-54
2500	0.614	-151	7.0	2.243	79	-18.8	0.115	28	0.469	-58
3000	0.611	-156	5.6	1.899	72	-18.4	0.120	27	0.461	-62
3500	0.608	-160	4.3	1.649	65	-18.1	0.125	27	0.460	-66
4000	0.604	-163	3.3	1.458	59	-17.7	0.130	27	0.465	-70

*Values do not include any parasitic bonding inductances and were generated by use of a computer model.

For more information, call your local HP Sales Office or East (301) 948-6370 - Midwest (312) 677-0400 - South (404) 434-4000 - West (213) 877-1282. Or write: Hewlett-Packard Components, 640 Page Mill Road, Palo Alto, California 94304. In Europe, Post Office Box 85, CH-1217, Meyrin 2, Geneva, Switzerland. In Japan, YHP, 1-59-1, Yoyogi, Shibuya-Ku, Tokyo, 151.

Printed in U.S.A. Data Subject to Change 5953-4423 (4/80)

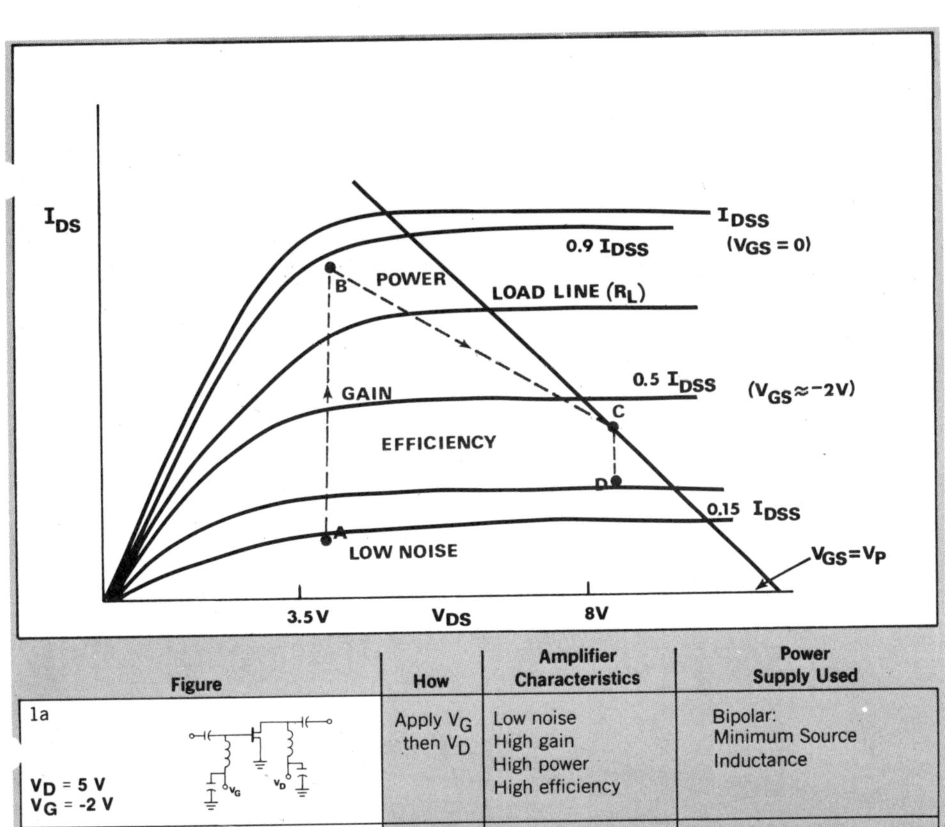

Figure	How	Amplifier Characteristics	Power Supply Used
1a V_D = 5 V V_G = -2 V	Apply V_G then V_D	Low noise High gain High power High efficiency	Bipolar: Minimum Source Inductance
1b V_D = 7 V V_S = 2 V	Apply V_S then V_D	(same as 1a)	Positive supply
1c V_G = -7 V V_S = -5 V	Apply V_S then V_D	(same as 1a)	Negative supply
1d V_D = 7 V V_S = 2 V = $I_{DS} R_S$	Apply V_D	Low noise High gain High power Lower efficiency Gain easily adjusted by varying R_S	Unipolar, incorporating R_S: Automatic transient protection
1e V_G = -7 V V_S = -5 V = $-I_{DS}R_S$	Apply V_G	(same as 1d)	Negative unipolar, incorporating R_S

FIGURE 8.17. The dc drain characteristics of a typical GaAs FET are superimposed on the graph, points *A* to *D* corresponding to its applications. Point *A* is chosen for low noise, low power; point *B* for higher gain, higher power; *C* for class A power, and *D* for class B or AB power. The five bias networks shown all yield equivalent RF performance. [*From George D. Vendelin (Dexcel, Inc., Santa Clara, (Calif.), "Five basic bias designs for GaAs FET amplifiers," Microwaves, Feb. 1978.*]

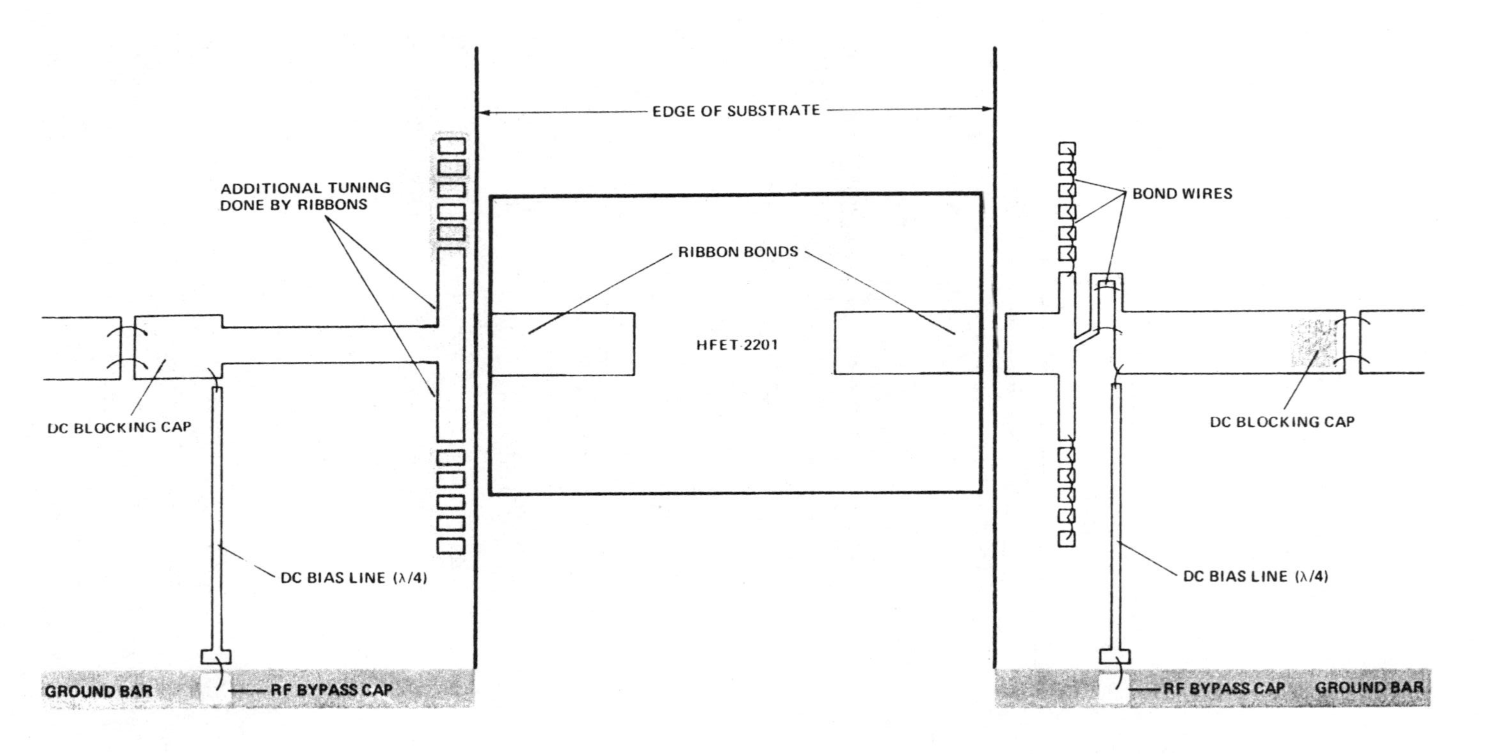

FIGURE 8.18. Completed 12-GHz amplifier with dc blocking and bias. (*Courtesy of Hewlett-Packard, AN-973.*)

above unity. As shown in Section 8.13, this can lead to instability. For example, S_{11} is quoted (*Hewlett-Packard Application Note 944-1*) as rising from 0.52 $\angle 154°$ at 4 GHz to 1.066 $\angle -8.5°$ at 0.1 GHz as a result of imperfect emitter resistor bypassing.

Variations in collector current with temperature must be guarded against, and this is usually achieved through collector dc feedback. Figure 8.19(a) shows the percentage variation in collector current expected for the three basic bias circuits shown in Fig. 8.19(b), (c), (d). The non-stabilized circuit of Fig. 8.19(b) is seldom used in practice because it has little dc stability. The other two circuits are widely used. The circuit of Fig. 8.19(c) has a potential divider added which does two things. For a given bias requirement it reduces the absolute ohmic values required for each resistor, making them easier to fabricate in thin/thick-film form. Also, the inclusion of the series base resistor makes the base current source appear as a constant-current source, and this resistor can be trimmed, on a production basis, to set collector current to its initial value.

Active current sources are also used. Figure 8.20(a) shows an active bias circuit used with a 6-GHz FET amplifier (*Hewlett-Packard Application Note 970*) and Fig. 8.20(b) shows the active bias circuit used with 2 GHz and 4-GHz bipolar transistor amplifiers (*Hewlett-Packard Application Note 972*).

8.16 MULTISTAGE AND CASCADABLE AMPLIFIERS

Single-stage amplifiers may be connected in cascade to obtain an increase in gain. Amplifiers intended for cascade connections are generally designed in modular form (Kramer et al. 1978). In the modular approach, each stage is designed to present input and output matching to 50 Ω, and the characteristics of any one stage is not unduly influenced by the others. Cascadable amplifiers can be purchased as microwave components. Figure 8.21 shows the basic circuit for a typical single-stage module (Avantek 1981) T_1 is the RF transistor, while T_2 and T_3 are part of the active biasing circuit. It will be observed that an unbypassed emitter resistor is used in the amplifier stage. The circuit is constructed as a hybrid thin-film–integrated-circuit module and is supplied in a TO-8 metal–glass hermetic package suitable for mounting on a 50-Ω microstrip PC board. Figure 8.22 shows the internal construction and package for a typical cascadable amplifier (Watkins Johnson 1978).

Integrated-circuit amplifiers have been constructed which can operate from dc up to 4.5 GHz (Hornbuckle 1980). Figure 8.23(a) shows the circuit for one of these amplifiers. Active loads are used and the source-coupled input pair feed a source follower output stage. A peaking inductor is used

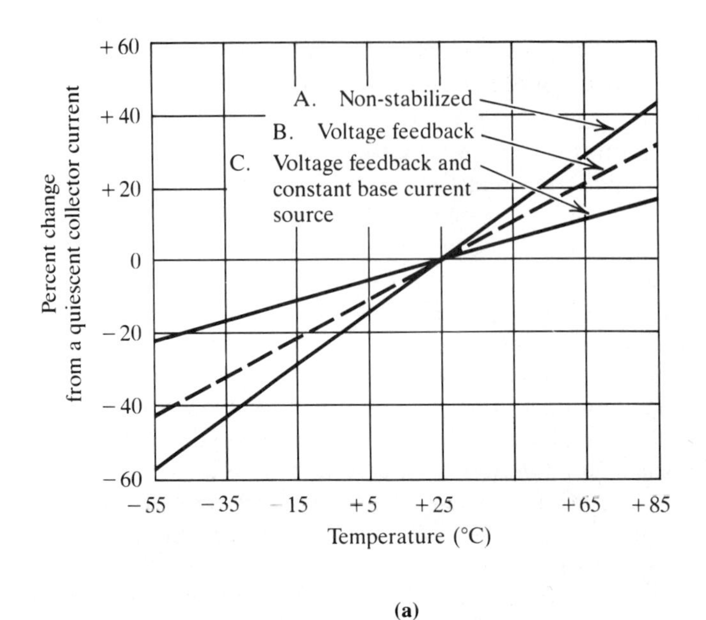

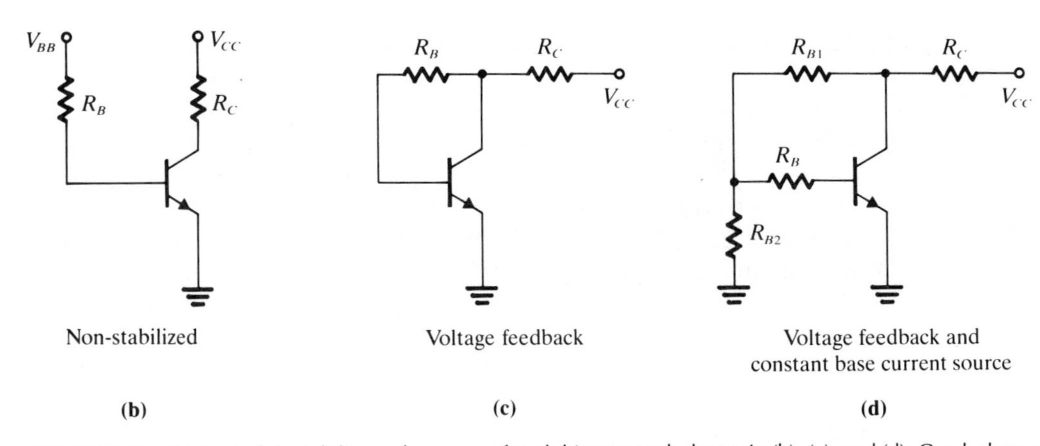

FIGURE 8.19. (a) Typical dc stability performance of each bias network shown in (b), (c), and (d). Graph shows the percent change from a nominal quiescent collector current as a function of temperature normalized to 25°C. (b) Nonstabilized. (c) Voltage feedback. (d) Voltage feedback and constant base current source. (*Courtesy of Hewlett-Packard, AN-944-1.*)

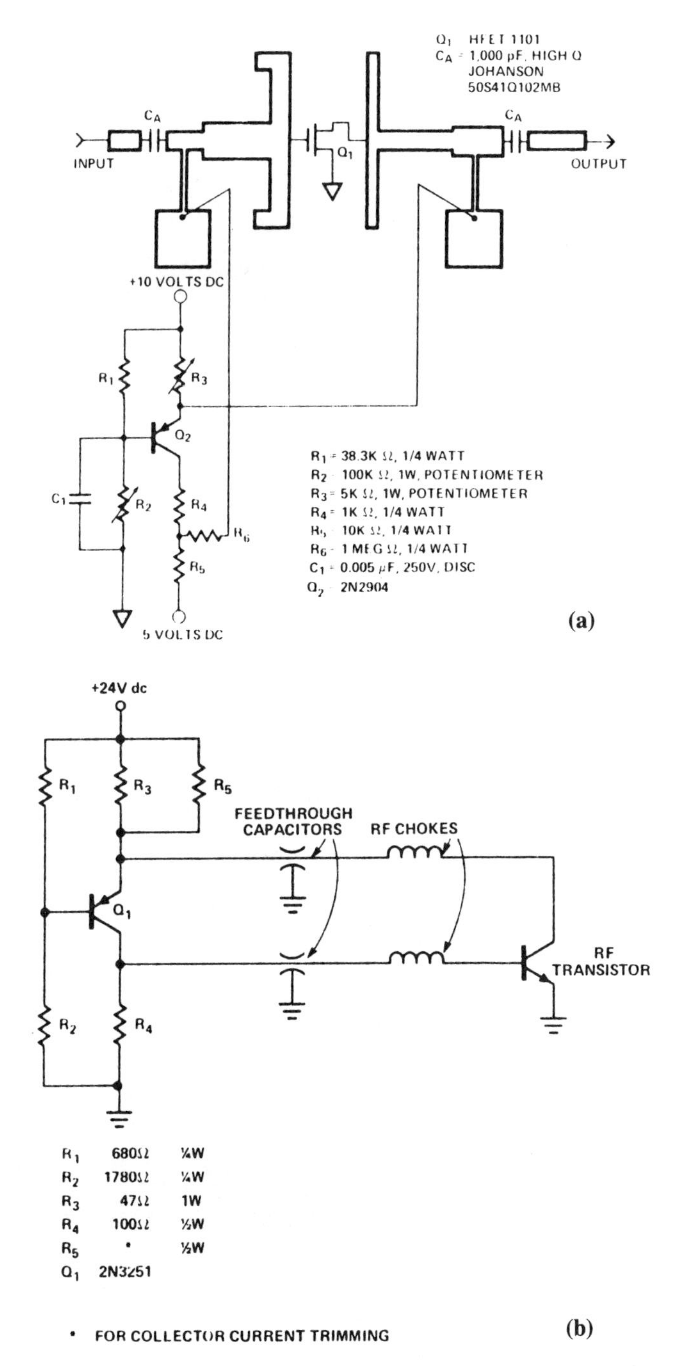

FIGURE 8.20. (a) The quiescent point is controlled by R_2 and R_3. R_2 is adjusted to provide the proper V_{ds} and R_3 is adjusted to supply the correct drain current I_{DS}. (b) Active bias circuit for both the 2-GHz and the 4-GHz amplifiers. (*Courtesy of Hewlett-Packard, An-970 and AN-972.*)

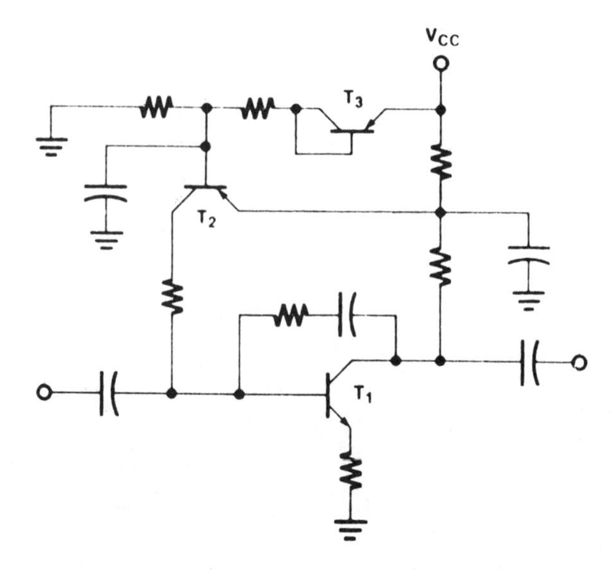

FIGURE 8.21. Basic circuit of a typical signal-stage UTO series amplifier module. T_1 is the transistor, and T_2 and T_3 are part of the active biasing circuit. (*Courtesy of Avantek Transistor Designer's 1981 Catalog, Santa Clara, Calif.*)

FIGURE 8.22. Cascadable amplifier covering the frequency range 100 to 2000 MHz. (*Courtesy of Watkins-Johnson Co., Palo Alto, Calif., RF Signal Processing Components, 1978.*)

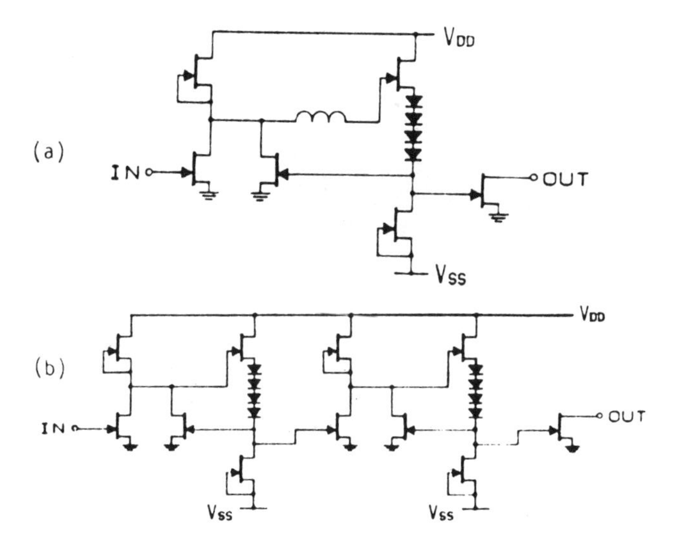

FIGURE 8.23. Circuit schematic for GaAs direct-coupled amplifiers: (a) incorporating inductor peaking (A60); (b) two-stage amplifier (A40). *(From Derry Hornbuckle, "GaAs direct coupled amplifiers," IEEE MTT-S Inter. Micro. Symp. Digest 1980; © IEEE 1980, with permission.)*

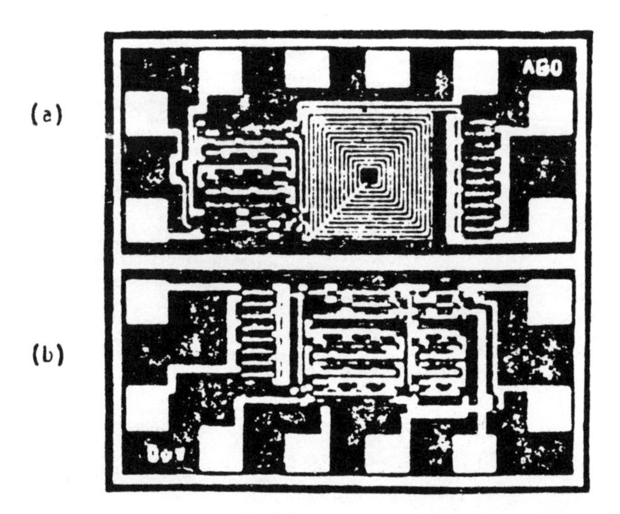

FIGURE 8.24. (a) A60 and (b) A40 amplifiers. Chip size is 300 × 650 μm. *(From Derry Hornbuckle, "GaAs direct coupled amplifiers," IEEE MTT-S Inter. Micro. Symp. Digest 1980; © IEEE 1980, with permission.)*

between stages to improve the frequency response, and the relatively large physical size of this inductor can be seen in Fig. 8.24(a). The circuit of Fig. 8.23(b) shows two such stages connected in cascade, and Fig. 8.24(b) shows the physical layout.

8.17 TRANSISTOR PACKAGING

At microwaves, parasitic reactances associated with transistor mounting cannot be ignored. To minimize these, transistors are provided in chip form, which may be mounted directly in the circuits. Some examples of transistor chips are illustrated in the data sheets for the Dexcel 2502A, the Avantek AT-8141, and the Hewlett-Packard HXTR-3002. The chip may be bonded onto the circuit substrate using an eutectic solder or a conductive

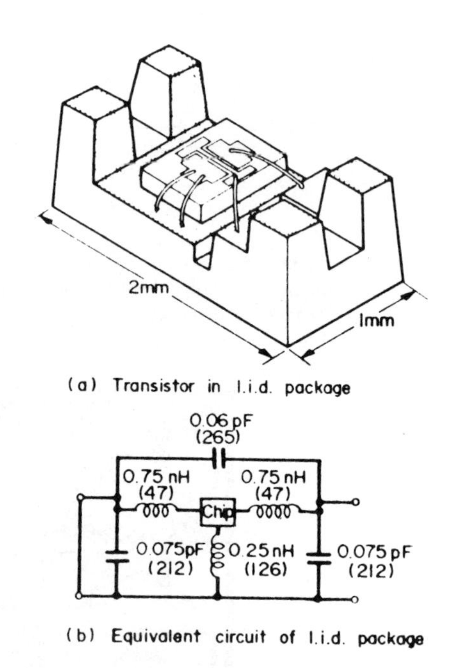

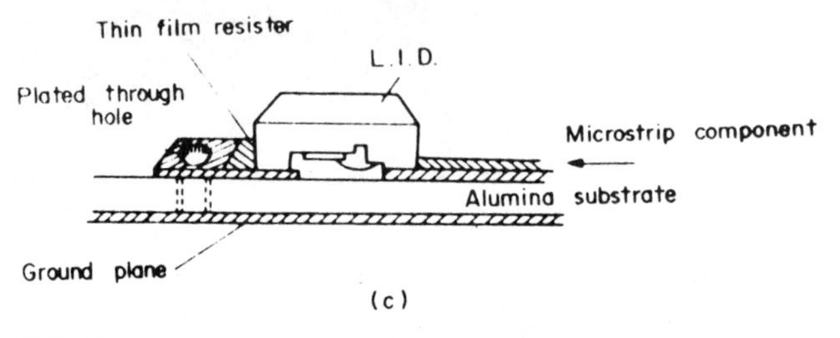

FIGURE 8.25. (a) Transistor in LID package; (b) equivalent circuit of LID package; (c) mounting of LID package and connection to ground plane by means of plated-through hole. [Parts (a) and (b) from S. J. Hewitt and R. S. Pengelly, *"Design techniques for integrated microwave amplifiers using gallium arsenide field-effect transistors," Radio Electr. Eng. Vol. 46, No. 10, 1976.*]

epoxy. Good thermal contact is required for heat removal. Bond wires are required between the chip contact pads and the circuit, and these may be made using thermal compression bonding or ultrasonic bonding. Although direct mounting of the chip in the manner described provides lowest parasitic elements, the method tends to be costly unless high-volume production is involved. A less costly method is to have the chip mounted in a carrier, as illustrated in Fig. 8.25. This is referred to as a leadless inverted device (LID). As shown, the chip is bonded into the carrier and wire bonds are made between the chip and the carrier pads. Thus these stages cannot be avoided but may be carried out in a specialized work area. The LID is inverted onto the circuit and may be attached by a simple soldering operation (see, e.g., *Hewlett-Packard Application Note 974*). Other types of carrier are also available, as shown, for example, on the DXL-2502A-CR data sheet.

Transistors are also supplied in packages specifically designed for microstrip and stripline mounting. The outline drawing for a standard 100-mil package for a bipolar junction transistor is shown in the data sheet for the AT-1845, and for a gallium arsenide field-effect transistor in the data sheet for the HFET-2202. Figure 8.26 illustrates the physical size of a 50-mil stripline package used for the AT-8060 GaAs FET. The package is filled with dry nitrogen and hermetically sealed to provide protection for the transistor.

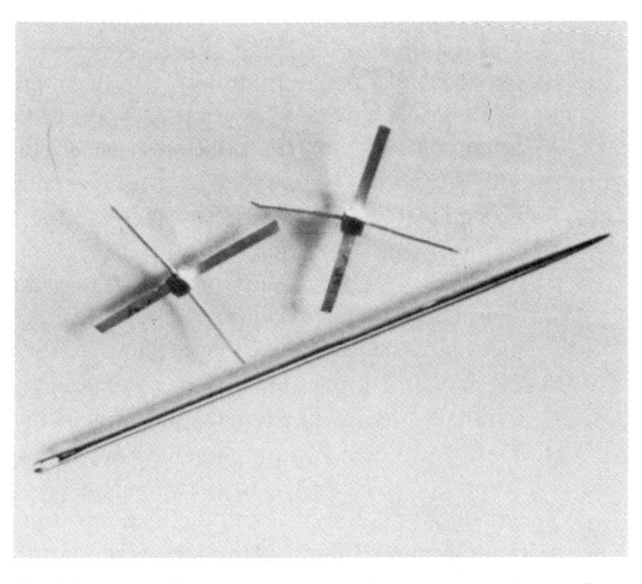

FIGURE 8.26. Fifty-mil-square metal–ceramic stripline package. *(Courtesy of Avantek, Santa Clara, Calif.)*

8.18 PROBLEMS

1. Explain what is meant by the unilateral gain of an amplifier. Which of the scattering parameters must equal zero for an amplifier to be unilateral?
2. The transistor specified in Table 8.1 is used as an amplifier at 2.0 GHz. Assuming that reverse transmission is negligible and that the source and load are reflectionless matched to the characteristic impedance of the system, calculate the power gain.
3. The transistor specified in Table 8.1 is used as an amplifier at 4 GHz in a 50-Ω system. The source impedance is 100 Ω resistive, and the load impedance is 75 Ω resistive. Calculate the power gain assuming that the reverse transmission is negligible.
4. For the amplifier in Problem 3, calculate the new power gain when the load is changed to $50 \angle 30°$ Ω, all other factors remaining unchanged.
5. The transistor specified in Table 8.1 is used as an amplifier in a 50-Ω system at 6 GHz. The source impedance is $73 + j42.5$ Ω and the load impedance consists of a 75-Ω resistor in parallel with a 0.53-pF capacitor. Calculate the power gain.
6. State the conditions that must be met by the source and the load to maximize the power gain of a unilateral amplifier. How do these conditions differ from reflectionless matching of the source and the load to the characteristic impedance of the system?
7. Determine the load and source reflection coefficients necessary to maximize the unilateral power gain of an amplifier that uses the transistor of Table 8.1 at a frequency of 10 GHz. If the characteristic impedance of the system is 50 Ω, what are the values of source and load impedances?
8. Calculate the maximum power gain for the amplifier in Problem 7.
9. The transistor of Table 8.1 is used in a unilateral amplifier at a frequency of 12 GHz. The characteristic impedance of the system is 50 Ω. Calculate the source and load impedances required for maximum power gain, and the maximum power gain.
10. Design 50 Ω input and output matching networks for the amplifier in Problem 3. Give line lengths in terms of line wavelength.
11. Design input and output matching networks for an amplifier using the transistor of Table 8.1, to operate at a frequency of 8 GHz. The source impedance is $100 \angle 0$ Ω and the load impedance $75 \angle 0$ Ω. Use fixed line lengths similar to the design in example 8.5 and determine the characteristic impedance of each line section.
12. Design 50 Ω input and output matching networks for the amplifier in Problem 4. Give line lengths in wavelengths.

13. Repeat Problem 11 for a load of $50\angle 30^\circ$ Ω.

14. Design 50 Ω input and output matching networks for the amplifier in Problem 5. Give line lengths in terms of wavelength.

15. Repeat Problem 11 for a source impedance of $74+j42.5$ Ω and a load of 60 Ω resistive in parallel with $-j10$ Ω reactive.

16. For the transistor specified in Table 8.1, draw the constant-gain circles for $G_s = 1.5$ dB and $G_L = 2.5$ dB for a frequency of 4.5 GHz. From these determine the source and load reflection coefficients which have minimum phase angles while satisfying the gain requirements.

17. For the transistor specified in Table 8.1, draw the constant-input-gain circles, $G_s = +2$, 0, and -2 dB, for a frequency of 3.0 GHz.

18. For the transistor specified in Table 8.1, draw the constant-output-gain circles, $G_L = +3$, 0, and -3 dB, for a frequency of 4.5 GHz.

19. Define the available power gain of an amplifier and explain how this differs from the transducer power gain.

20. Explain what is meant by the maximum available power gain of an amplifier, and how this is achieved.

21. Define and explain what is meant by the noise factor of an amplifier. The available power gain of an amplifier is 20 dB, and the bandwidth is 10 MHz. If the noise power output is 8 pW, calculate the noise factor.

22. An amplifier has an available power gain of 30 dB, a noise figure of 5 dB, and a bandwidth of 10 MHz. Calculate the output noise power.

23. Calculate the amplifier input noise for the amplifier in Problem 21.

24. Calculate the amplifier input noise for the amplifier in Problem 22.

25. An amplifier having an available power gain of 25 dB and noise figure of 4 dB is used to feed into a second amplifier that has an available power gain of 30 dB and a noise figure of 10 dB. What is the overall noise figure of the system?

26. Repeat Problem 25 with the positions of the amplifiers interchanged.

27. The input reflection coefficient that gives a minimum noise factor for an amplifier is $0.72\angle 80^\circ$, and the minimum noise figure is 1.5 dB. The amplifier equivalent noise resistance is 5 Ω. Draw noise circles for 2-, 3-, and 3.5-dB noise figures. $Z_0 = 50$ Ω.

28. Draw the constant-noise-figure circles $F = 2.0$, 2.5, and 3.0 dB for transistor DXL 2502A, for a frequency of 4 GHz. Use the values specified for $V_{ds} = 3.5$ V and $I_{ds} = 10$ mA.

29. Repeat Problem 28 for a frequency of 8 GHz, state why the 2-dB circle cannot be shown, and state why the 2.5-dB circle degenerates into a point.

30. Transistor DXL 2502A is operated as a unilateral amplifier (assume that $S_{12} = 0$) at 8 GHz, with $V_{ds} = 3.5$ V and $I_{ds} = 10$ mA. Draw the 3-dB noise circle and the $G_s = -1$ dB gain circle on the same Smith chart, and from these determine the source reflection coefficient required to realize both values. Given that the characteristic impedance of the system is 50 Ω, determine the source impedance.

31. Determine the maximum available power gain for the amplifier of Problem 30 when the source impedance has the value calculated.

32. Transistor HFET-2202 is operated as a unilateral amplifier (assume that $S_{12} = 0$) under minimum-noise-figure bias conditions and at a frequency of 10 GHz. Draw the 3-dB noise-figure circle and the $G_s =$ 1 dB gain circle. From these determine the source impedance required to meet both specifications, given that $Z_o = 50$ Ω.

33. Show that when the input circuit of an amplifier is adjusted for a minimum-noise figure, the gain G_s is given by

$$\frac{1 - \rho_{\min}^2}{1 + \rho_{\min}^2 |S_{11}|^2 - 2\rho_{\min} |S| \cos \theta}$$

where θ is equal to $\theta_{11} + \phi_{\min}$.

34. Transistor HFET-2202 is used in a unilateral amplifier (assume that $S_{12} = 0$) at a frequency of 8 GHz, with the bias adjusted for a minimum-noise figure. With the input matching circuit adjusted for a minimum-noise figure and the output matching circuit for maximum gain, determine the available power gain of the amplifier.

35. Show that when the input circuit of a unilateral amplifier is adjusted for maximum G_s, the noise factor F of Eq. (8.40) is given by

$$\mathrm{F} = \mathrm{F}_{\min} + 4\mathrm{r}_\mathrm{n} \frac{|S_{11}|^2 + \rho_{\min}^2 - 2\,|S_{11}|\,\rho_{\min} \cos \theta}{(1 - |S_{11}|^2)\,(1 + \rho_{\min}^2 + 2\rho_{\min} \cos \phi_{\min})}$$

where $\theta = \theta_{11} + \phi_{\min}$.

36. Transistor type DXL 2502A is operated as a unilateral amplifier (assume that $S_{12} = 0$) at $V_{ds} = 3.5$ V and $I_{ds} = 10$ mA, the frequency being 12 GHz. The input and output circuits are adjusted for maximum gain. Determine the available power gain and the noise figure.

37. Convert the following noise factors to equivalent noise temperatures: (a) 10; (b) 6; (c) 2.

38. Convert the following noise figures to equivalent noise temperatures: (a) 10 dB, (b) 7 dB; (c) 3.5 dB.

39. The minimum equivalent noise temperature for a unilateral amplifier is 438 K, and the corresponding minimum source reflection coefficient is 0.78 $\angle 45°$. The noise resistance of the transistor is 20 Ω and the characteristic impedance of the system is 50 Ω. Draw the noise circles for the following equivalent temperatures: (a) 900 K; (b) 860 K; (c) 527 K.

40. State, with reasons, which of the following constant-noise-figure circles can be drawn for the amplifier of Problem 39: (a) 4.5 dB; (b) 4 dB; (c) 3.5 dB; (d) 3 dB.

41. Show that with $S_{12} = 0$, Eq. (8.55) reduces to Eq. (8.27). (*Hint:* Take the factor k outside the brackets and expand the square-root term using the binomial expansion. Simplify the resulting equation and let S_{12} approach zero.)

42. Show that with $S_{12} = 0$, the inequalities (8.51) and (8.52) automatically imply inequality (8.53) so that the latter is redundant.

43. Using typical values from the AT-8141 data sheet determine the power output specified at the 1 dB compression point, for a frequency of 4 GHz, and the corresponding input power required. Assume V_{DS} = 9V and I_{DS} = 500 mA.

REFERENCES

AVANTEK, Transistor designer's catalog, 1981, p. 148.

COOKE, HARRY F., 1971. "Microwave transistors: theory and design," *Proc. IEE*, Vol. 59, pp. 1163–1181. [Reprinted in Fukui (1981).]

FUKUI, HATSUAKI (ed.), 1981. *Low Noise Microwave Transistors and Amplifiers*. IEEE Press.

Hewlett-Packard Application Bulletin 17, Noise Parameters and Noise Circles for the HXTR-6101, -6103, -6104, and -6106 Low Noise Transistors.

Hewlett-Packard Application Bulletin 19, Noise and Power Parameters for the HFET-1101.

Hewlett-Packard Application Note 77-1, Transistor Parameter Measurements.

Hewlett-Packard Application Note 944-1, Microwave Transistor Bias Considerations.

Hewlett-Packard Application Note 970, A 6 GHz Amplifier Using the HFET-1101 GaAs FET.

Hewlett-Packard Application Note 972, Two Telecommunications Power Amplifiers for 2 and 4 GHz Using the HXTR-5102 Silicon Bipolar Power Transistor.

Hewlett-Packard Application Note 973, 12 GHz Amplifier Designs Using the HFET-2201.

HORNBUCKLE, DERRY, 1980. "GaAs IC direct-coupled amplifiers," *IEEE MTT-S Int. Microwave Symp. Digest*, pp. 387–389. [Reprinted in Fukui (1981).]

KRAMER, B., M. PARISOT, AND A. COLLET, 1978. "Modular construction of low-noise multi-stage f.e.t. amplifiers," *Radio Electron. Eng.*, Vol. 48, No. 1/2, pp. 23–28.

KUROKAWA, K., 1965. "Power waves and the scattering matrix," *IEEE Trans. Microwave Theory Tech.*, Vol. MTT-13, pp. 194–202. [Reprinted in Fukui (1981).]

LIECHTI, CHARLES A., 1976. "Microwave field-effect transistors–1976," *IEEE Trans. Microwave Theory Tech.*, Vol. MTT-24, pp. 279–300. [Reprinted in Fukui (1981).]

VENDELIN, GEORGE D., 1978. "Five basic bias designs for GaAs FET amplifiers," *Microwaves*, February.

FURTHER READING

MEYS, R. P., 1978. "A wave approach to the noise properties of linear microwave networks," *IEEE Trans. Microwave Theory Tech.* Vol. MTT-26, pp. 34-37. [Reprinted in Fukui (1981).]

RODDY, D., AND J. COOLEN, 1984. *Electronic Communications*, 3rd ed. Reston, Va.: Reston Publishing Company, Inc.

Watkins Johnson catalog, 1978.

nine
Microwave Tubes

9.1 INTRODUCTION

Microwave tubes are generally required for the generation of moderate to high levels of microwave power, that is, power levels ranging from a few tens of watts up to pulsed powers of the order of megawatts. Tubes can be used to convert dc energy to microwave energy, and also to provide power amplification of an existing microwave signal. The basic magnetron tube was developed in the early 1920s and the basic klystron tube in the early 1930s [see Chapter 2 of Chodorow and Susskind (1964)], and the traveling-wave tube was invented by R. Kompfner in 1943 [see Chapter 1 of Gittins (1965)]. Although the basic principles of these tubes have remained unchanged, improvements in materials and refinements in design methods have led to smaller, more efficient, and more reliable tubes. Some tubes intended for satellite use have a design lifetime of 10 years minimum. Also, a better understanding of the basic principles has led to the development of new tube types.

Although tubes, as already mentioned, are generally associated with relatively high powers, they are also used for power generation and amplification in the milliwatt range.

9.2 THE CAVITY MAGNETRON

In a cavity magnetron a cylindrical cathode is surrounded by a cylindrical anode as shown in Fig. 9.1(a). The anode contains a number of resonant cavities equispaced around the circumference. The space between anode

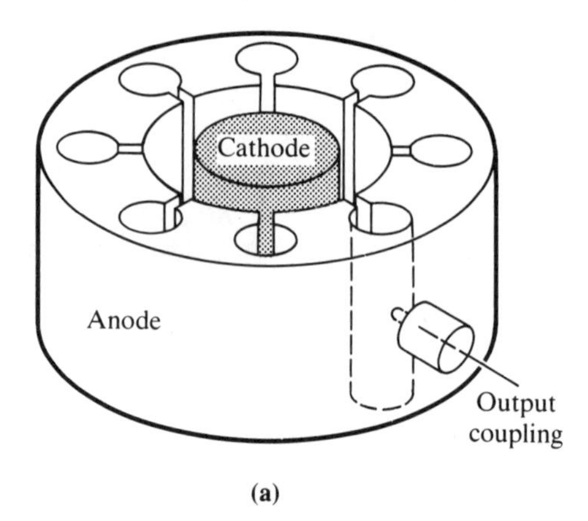

(a)

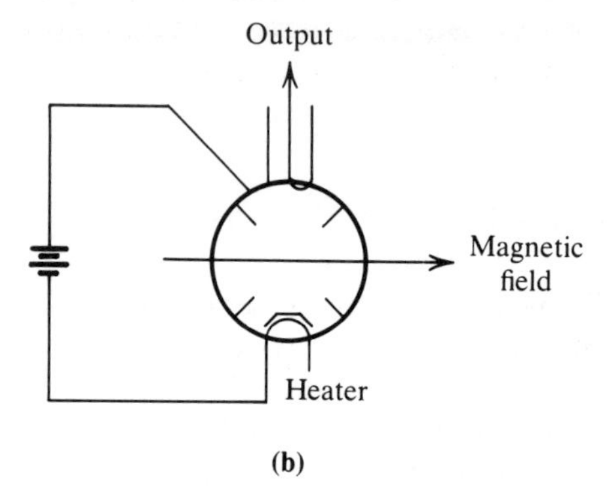

(b)

FIGURE 9.1. (a) Anode and cathode structure of a cavity magnetron; (b) magnetron circuit symbol and circuit.

and cathode, including the resonators, is totally evacuated. Electrons emitted by the cathode follow a complicated trajectory in the evacuated space, eventually being collected by the anode. During their trajectory flight the electrons excite the cavities into oscillation. The output is taken from one of the cavities, one method being to use a small coupling loop as shown. The circuit diagram for a magnetron oscillator is shown in Fig. 9.1(b). In operation the magnetron requires a steady or direct voltage to maintain the anode positive with respect to cathode, and a steady magnetic field parallel to the cathode axis to deflect the electrons into curved trajectories.

Some idea of the principles of operation can be gained by considering the motion of an electron in a planar electrode structure sketched in Fig. 9.2(a). The steady electric field by itself will tend to accelerate the electron toward the anode. The motion of the electron in the magnetic field produces a deflecting force, and the resulting path of the electron is cycloidal. Such a path is shown in Fig. 9.2(a) for conditions where the deflection is sufficient to prevent the electron from reaching the anode. For simplicity it is assumed that the initial velocity of the electron leaving the cathode is zero. An important result of such motion is that the average velocity of the electron in the z-direction is constant, and is given by [see, e.g., Chodorow and Susskind (1964)]

$$v_{z,\mathrm{av}} = \frac{E_0}{B_0} \tag{9.1}$$

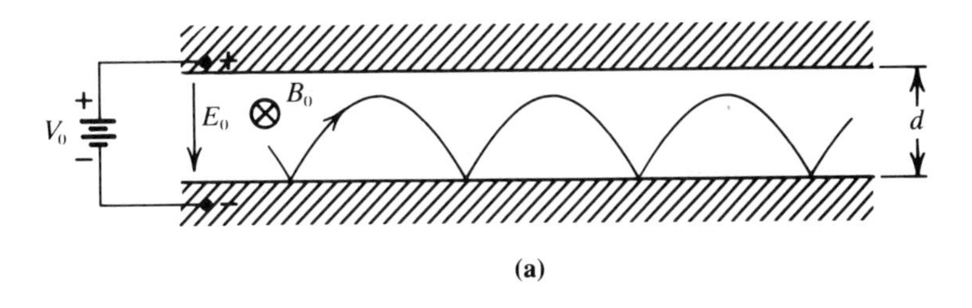

(a)

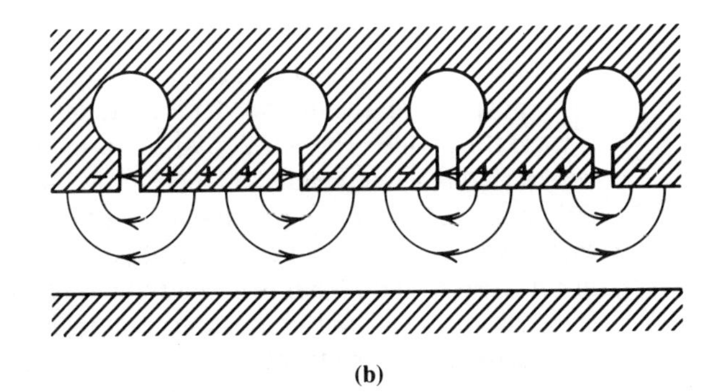

(b)

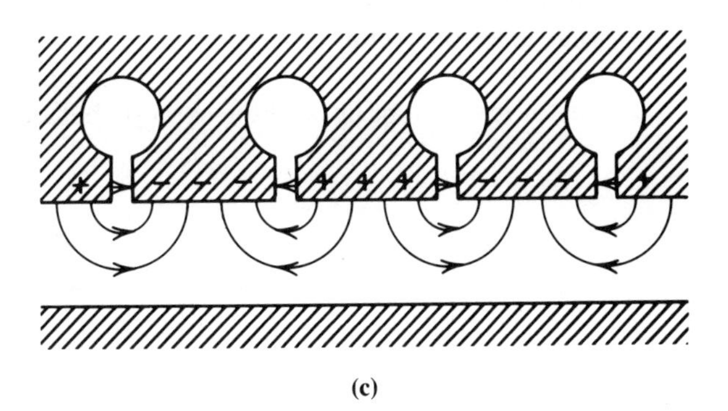

(c)

FIGURE 9.2. (a) Electron trajectory between planar electrodes for static fields; (b) electric field of resonators; (c) as part (b) but one half-cycle later.

where $E_0 = V_0/d$ is the magnitude of the steady electric field and B_0 the magnitude of the steady magnetic field. Note that up to this point, only the effects of the static fields have been considered.

An anode with resonant structures is shown in Fig. 9.2(b). Here it is assumed that oscillating fields exist in the resonators (oscillations could be started in the first instance by random electron motion). The electric field at one instant of time is shown in Fig. 9.2(b). The fields vary with time in the manner of a standing wave, for example, one half-cycle later the polarity would be reversed as shown in Fig. 9.2(c). This standing wave can be resolved into two traveling waves, one in the $+z$ or forward direction, the other in the $-z$ or reverse direction. This is similar to the standing-wave situation on transmission lines described in Section 1.12.

The speed of the forward traveling wave can be synchronized with the average electron speed, and thus the electron experiences a constant electric field resulting from the traveling wave. This is in addition to the steady applied electric and magnetic fields. Figure 9.3 shows a particular situation where the electric field component of the forward wave is along the direction of motion. This apparent steady field E_z, in conjunction with the steady magnetic field B_0, will tend to move the electron in the arc shown (just as the steady applied fields produced a cycloidal trajectory). This new electron motion has two components, a horizontal component in opposition to the forward motion, and a vertical component acting toward the anode. The horizontal component reduces the forward electron velocity, the reduction in kinetic energy appearing as an increase in the potential energy (voltage) of the forward traveling wave. At the same time, the movement toward the anode increases the kinetic energy, the energy in this case coming from the steady field. Thus the total kinetic energy of the electron stays almost constant, but an energy transfer takes place from the steady field to the traveling-wave field. Electrons that effect this transfer are referred to as *working electrons*. There will be other electrons synchronized with other regions of the traveling-wave field, but it can be shown that electrons other than working electrons are rapidly returned to the cathode. In a coaxial magnetron, working electrons may complete a number of revolutions around the cathode before being collected by the anode;

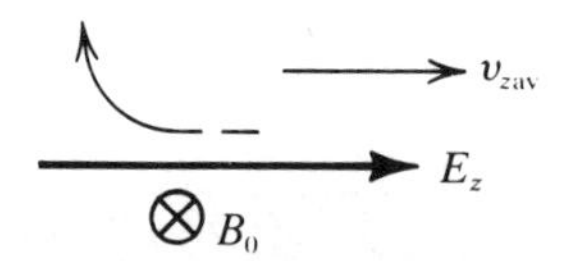

FIGURE 9.3. Electric field E_z of the forward traveling wave, moving at phase velocity $v_{z,\mathrm{Av}}$.

that is, they can reenter the anode–cathode space. This gives rise to the name *reentrant magnetron*.

The foregoing description shows in a simplified way how a single electron may transfer energy from a steady field to the resonant field of a cavity. In the magnetron a rotating space charge of electrons exists in the space between cathode and anode, and the details of energy transfer are considerably more complicated. Further complications arise because the electric field associated with the steady-state oscillations is large. However, the general principle applies, which is that the electrons provide a mechanism for transferring energy from the steady electric field to the oscillating field.

As mentioned, the standing wave within the magnetron can be resolved into forward and reverse travelling waves. The magnitude of the phase shift between pole faces for either wave is $\theta = \beta p$ radians where p is the pole pitch as shown in Fig. 9.4(a). Because of its re-entrant nature, the total phase shift around the magnetron must be some integer multiple of 2π for standing waves to exist. Let N equal the number of cavities (or pole faces) then

$$N\theta = 22\pi n \qquad n = 1, 2, \ldots N/2 \tag{9.2}$$

Both the forward and the reverse travelling wave experience this total phase shift and therefore the differential phase shift is 2θ. Values of n greater than N/2 simply increase this differential phase shift by 2π and can be ignored. With this in mind, an equivalent circuit for the magnetron can be drawn as shown in Fig. 9.4(b) (Gandhi, 1981). Each cavity is represented by an equivalent parallel LC resonant circuit and the coupling between cavities is through the coupling capacitors C_c between pole faces and cathode. The frequencies at which this circuit maintains the correct phase relationship given by Eq. (9.2) is given by (see e.g. Gandhi, 1981)

$$\omega^2 = \frac{1}{L\left[C + \dfrac{C_c}{2(1 - \cos\theta)}\right]} \tag{9.3}$$

The frequencies given by Eq. (9.3) are known as the mode frequencies, there being a different value for each value of $n = 1, 2, \ldots N/2$, contained in θ. The most important one is that for which $n = N/2$. At this value, $\theta = \pi$, and the mode is referred to as the π mode.

The mode frequencies vary with n as shown in Fig. 9.5. An eight-pole magnetron will have $n = 4$ for the π-mode, and the resonant frequency will be as indicated in Fig. 9.5. However, the adjacent mode frequency,

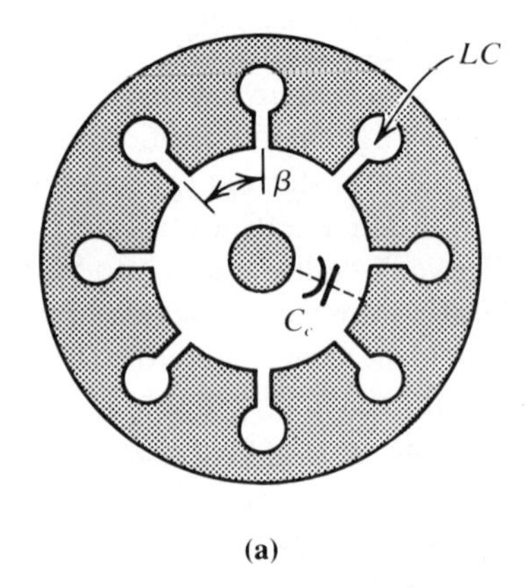

(a)

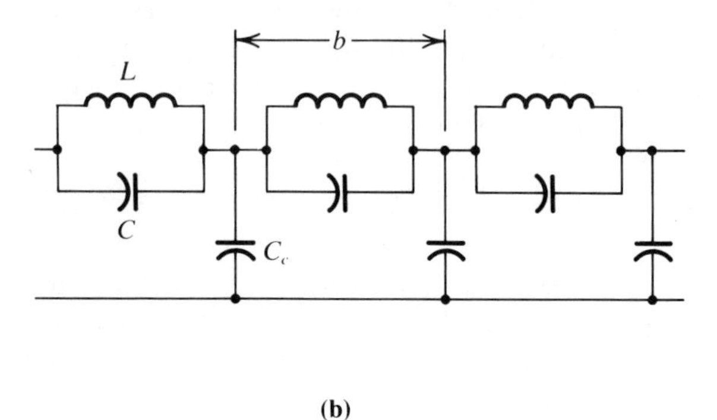

(b)

FIGURE 9.4. (a) Approximate equivalent circuit for a magnetron; (b) the circuit opened out to show its similarity to a band-pass filter.

at $n = 3$, is close to the π-mode frequency and it is possible for the frequency to jump from one mode to another, with is highly undesirable. Frequency separation between modes can be increased by strapping alternate pole faces together as shown in Fig. 9.6. Since in the π-mode alternate poles have the same polarity, no π-mode current flows in the straps and therefore strap inductance has no effect. However, the capacitance between straps does add in parallel to capacitance C at each slot because the straps are connected to opposite polarities across the slots for the π-mode. From Eq. (9.3) an increase in C is seen to decrease the π-mode frequency. For other modes, the phasing between adjacent poles in no longer π radians. The capacitive effect is reduced, and at the same time current flows in the straps, which gives rise to an inductive field. The inductance is in parallel with the slot, which effectively reduces the slot capacitance, thus raising the unwanted mode frequency. The overall effect of strapping is therefore to increase the frequency separation between the π-mode and the higher adjacent modes.

Separation of mode frequencies can also be achieved by altering the equivalent LC values of alternate slots. This gives rise to an anode structure known as a "rising sun" anode, shown in Fig. 9.7(c). A number of other commonly used resonant structures are shown in Fig. 9.7.

Although magnetrons can be operated in an amplifying mode, their main use is as a source of microwave power. They are widely used in microwave ovens, where the power output requirement has been standardized at 600 W, at one of two allocated frequencies, 915 MHz or 2450 MHz.

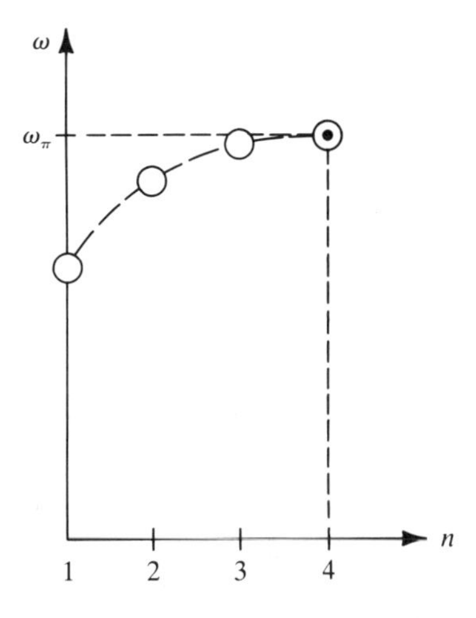

FIGURE 9.5. Mode frequencies for the circuit of Fig. 9.4. ω is shown for an eight-cavity magnetron.

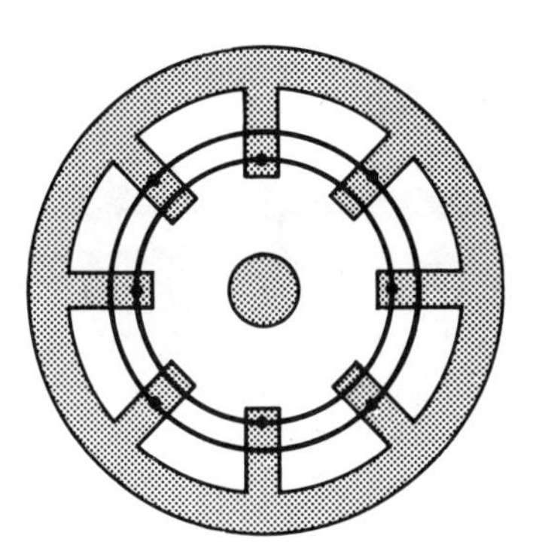

FIGURE 9.6. Strapping of anode poles.

The constructional features of a typical magnetron intended for use in microwave ovens are described in Oguro (1978).

A photograph of a 100-kW 915-MHz magnetron intended for industrial heating applications is shown in Fig. 9.8, and the recommended operating circuit in Fig. 9.9. An important part of any circuit is the filtering required to suppress unwanted microwave radiation, as shown in Fig. 9.9. The tube shown is the M1359, manufactured by the New Japan Radio Co. Ltd. Electrical characteristics for the tube are:

Heater voltage 10V

Heater current 225A

Anode voltage 27.5 kV

Anode current 4.6A

Magneto-motive-force 11400 Amp-turns

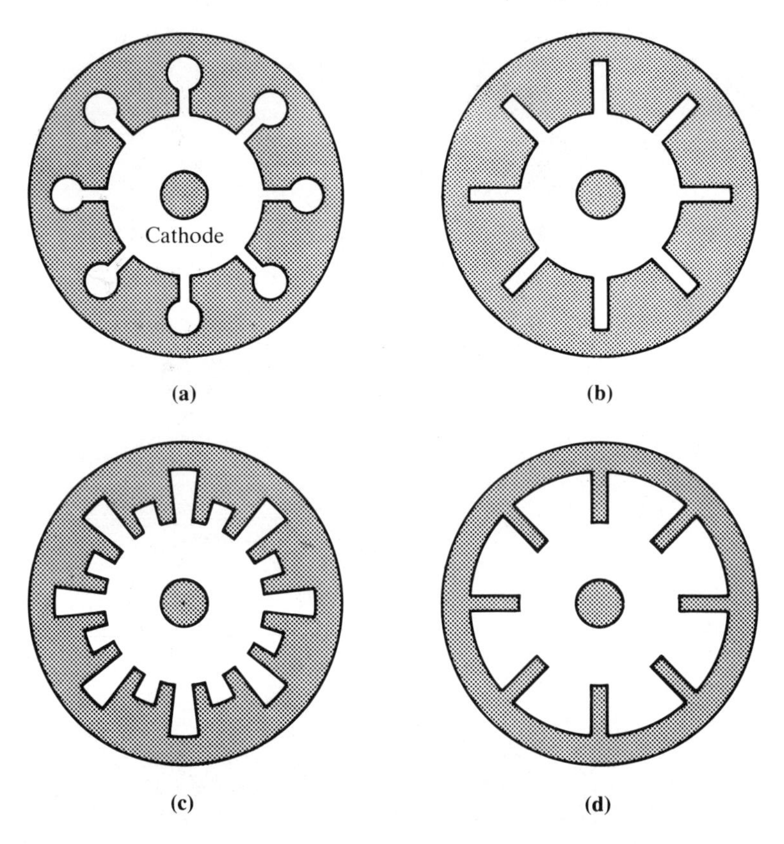

FIGURE 9.7. Cavity types used in cavity magnetrons: (a) hole and slot; (b) slot; (c) rising sun; (d) vane.

FIGURE 9.8. Photograph of a 100-kW 915-MHz magnetron. (*Courtesy New Japan Radio.*)

For a description of the developments leading to the M1359 magnetron see Shibata et al. (1978).

9.3 THE KLYSTRON

The basic klystron amplifier consists of two resonant cavities coupled by means of an electron beam. The cavities are of the reentrant type described in Section 4.17, and in the klystron the capacitive elements of the cavities are in the form of grids which allow the electron beam to pass through.

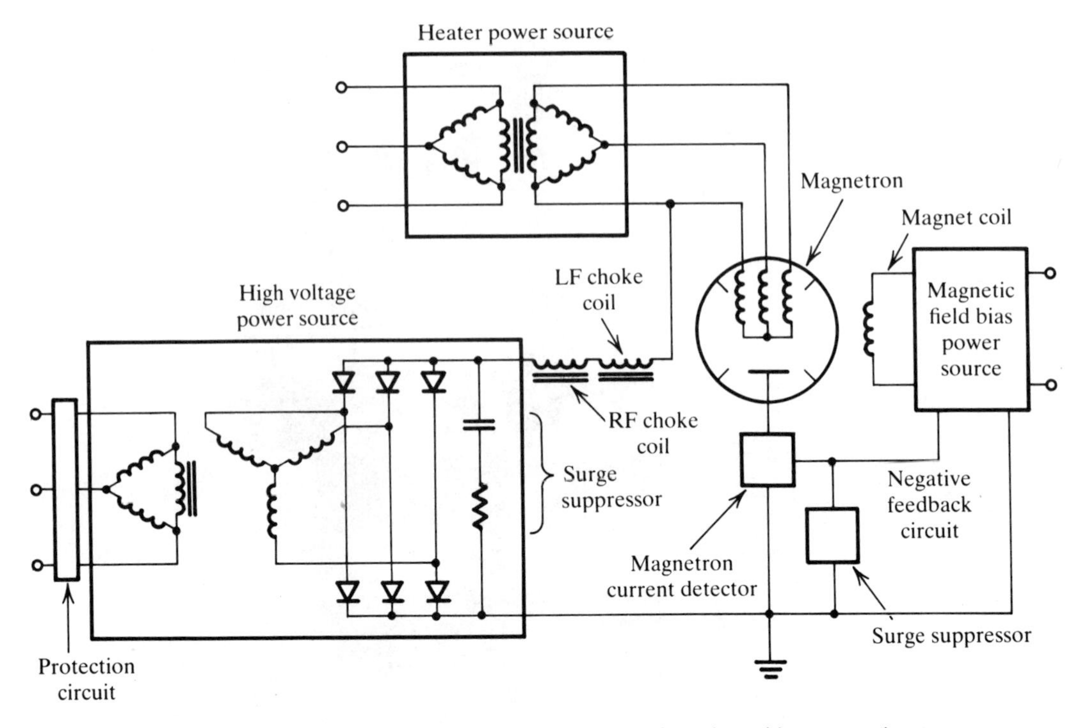

FIGURE 9.9. Recommended operating circuit for industrial heating applications.

This is illustrated in Fig. 9.10(a). The electron beam is emitted from an electron gun assembly, and in large klystrons, a longitudinal magnetic field may be used to provide additional focusing along the length of the tube. The complete assembly is sealed and evacuated.

As shown in Section 4.17, the reentrant cavity can be represented by an equivalent LC_g circuit, where C_g is the capacitance of the gap. The equivalent circuits are shown in Fig. 9.10(b), and these are resonant at the same frequency. An input signal at the resonant frequency will set up a time-varying electric field across C_{g1}. This field will alternately accelerate and decelerate electrons in the beam, with the result that the beam forms into bunches. For this reason the first cavity is sometimes termed the *buncher cavity*. As the bunches cross the capacitive gap C_{g2} of the output cavity, they induce a current flow in the output circuit. The second cavity is sometimes termed the *catcher cavity*, as it "catches" the signal energy in the bunches and transfers this to the output. Note that the catcher cavity does not catch the electrons; these carry on through to a collector electrode. The name *klystron* was introduced by the Varian brothers [see Varian and Varian (1939)] and is derived from the Greek *klyzo*, expressing the breaking of waves on a beach. The analogous situation is that of the bunches delivering up energy as they sweep through the second cavity.

The action can be seen in more detail by considering the action of the cavity grids on individual electrons, but as with the magnetron this is a coarse approximation, as it ignores the space-charge effects of the beam, which are significant.

Let V_0 volts be the voltage of the final accelerating anode of the electron gun, which is numerically equal to the potential energy of the field

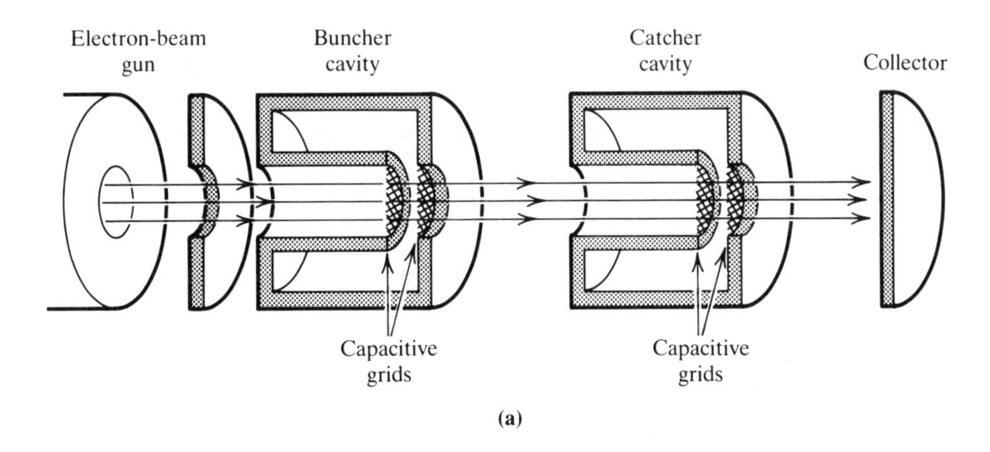

(a)

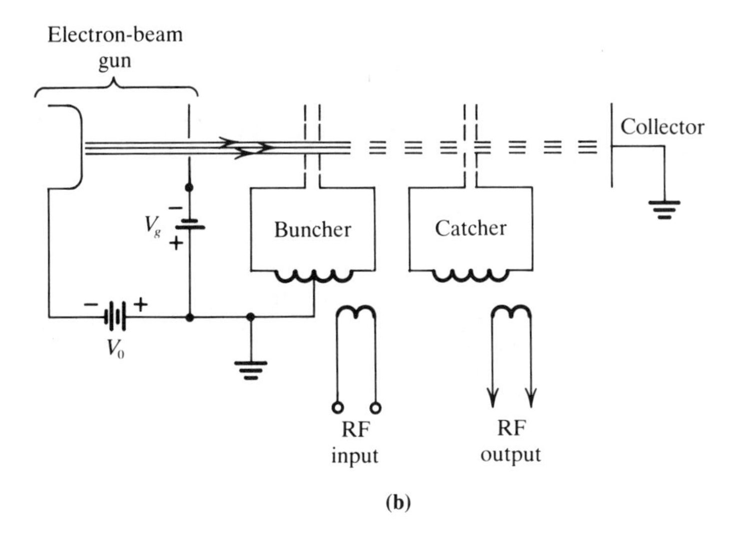

(b)

FIGURE 9.10. (a) Cutaway view of a two-cavity klystron; (b) equivalent circuit.

in electron volts. The kinetic energy gained by an electron on leaving the gun is numerically equal to this potential energy. Denoting the velocity of the electron leaving the gun by u_0, then $u_0^2 \propto V$, since kinetic energy is proportional to velocity squared. The electron enters the cavity gap with velocity u_0. Let the signal voltage across the gap be $V_1 \sin \omega t$. As the electron enters the gap at time t_1 the signal voltage acting on it is $V_1 \sin \omega t_1$, and on leaving at time t_2 is $V_1 \sin \omega t_2$. The *average* value is

$$v_{av} = \beta V_1 \sin \omega t_A \tag{9.4}$$

when $t_A = (t_1 + t_2)/2$. The *beam coupling coefficient*, denoted by β (do not confuse with phase shift coefficient), depends on the ratio of transit time $t = t_2 - t_1$ to the periodic time of the signal. At time t_A the electron is midway across the gap. Denoting by u_a the velocity of the electron at this time, then $u_a^2 \propto (V_0 + v_{av})$ and

$$\frac{u_a^2}{u_0^2} = \frac{V_0 + v_{av}}{V_0}$$

or

$$u_a = u_0 \left(1 + \frac{\beta V_i}{V_0} \sin \omega t_a\right)^{1/2} \tag{9.5}$$

The term $\beta V_i / V_0$ is termed the modulation depth, and denoting this by m, we have

$$u_a = u_0 (1 + m \sin \omega t_a)^{1/2} \tag{9.6}$$

Because m in practice is always much smaller than unity, the binomial expansion of Eq. (9.6) gives

$$u_a \simeq u_0 (1 + \frac{m}{2} \sin \omega t_a) \tag{9.7}$$

The electron leaves the buncher gap with this velocity and with a time lag equal to one-half the transit time τ:

$$u_a = u_0 \left(1 + \frac{m}{2} \sin \omega \left(t_a - \frac{\tau}{2}\right)\right) \tag{9.8}$$

Thus in the drift space between the two cavities, electrons travel with differing velocities. Faster electrons will catch up with slower ones, and bunches will form. By plotting distance as a function of time for the drift space, the bunching action can be illustrated graphically in what is known as an Applegate diagram. This is illustrated for three specific times in Fig. 9.11: time t_0 at which the velocity is u_0; one-quarter of a cycle before t_0, at which the velocity is at a minimum or u_0 $(1 - m/2)$; and one-quarter of a cycle later, at which the velocity is at a maximum or u_o $(1 + m/2)$. At distance d_1 the electron paths are seen to bunch together. The electro-

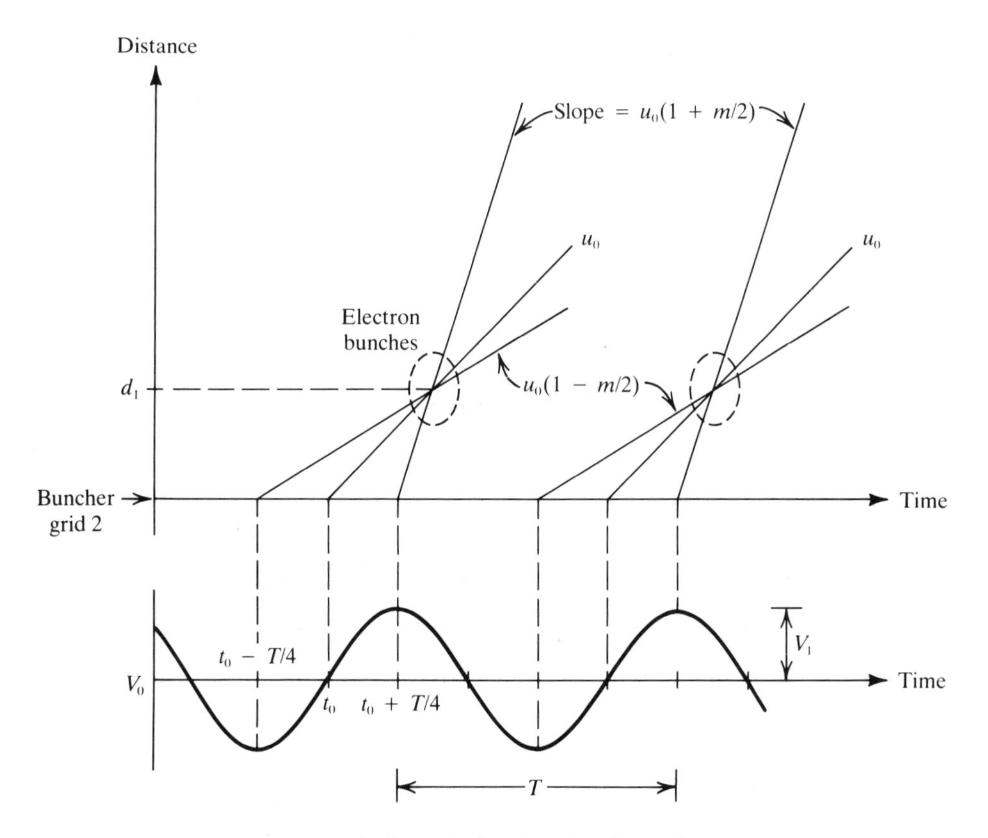

FIGURE 9.11. Distance–time graph illustrating bunching (Applegate diagram).

static force of repulsion between electrons prevents collision, and in fact causes beam spreading, which is undesirable.

The grids of the buncher cavity are maintained at approximately the same dc potential V_0 as the gun anode (there may be a small bias difference), and the C_{g1} voltage–time graph is shown below the Applegate diagram. It is seen that bunching occurs where the voltage slope is positive; that is, the field across C_{g1} is changing from a decelarating to an accelerating one.

The catcher cavity is positioned so that the electron bunching occurs within its gap. As a bunch crosses the gap it repels conduction electrons in the metallic grids. Thus an equivalent positive charge is induced, which transfers from grid 3 to grid 4 as the beam bunch traverses the gap, as shown in Fig. 9.12(a), (b), and (c). In effect, a positive current flow takes place from grid 3 to grid 4 around the external circuit. This has the form of a rectangular pulse shown in Fig. 9.12(d), the pulse width being equal to the transit time across the gap, and the repetition time of the pulses being equal to the periodic time of the input signal. Because of the high charge density in the bunches, the induced current pulses are large, and considerable power output can be achieved. The output cavity is tuned to the fundamental component of the pulse and therefore provides a sinusoidal signal at the output port.

The maximum conversion efficiency of the two-cavity klystron is about 58%, the conversion efficiency being the ratio of average output power to beam power, V_0I_0, where I_0 is the direct current in the beam. Power gains of around 30 dB are achievable, the gain being the ratio of output signal power to input signal power.

9.4 THE CASCADED KLYSTRON AMPLIFIER

The power amplification of a klystron amplifier can be increased by increasing the number of cavities. As partially bunched electrons pass through a cavity, they induce an RF voltage across the gap, and the phase of this voltage can be adjusted to increase the bunching action; that is, a feedback action occurs between the induced voltage and the electron beam. By placing a number of cavities in cascade as shown in Fig. 9.13, a very high charge density can be achieved in the bunches reaching the final, or output cavity. The electron beam along the tube has to be kept in focus by means of a longitudinal magnetic field. The Varian 936 tube shown uses permanent magnet focusing, and forced air cooling.

The tube, shown in Fig. 9.13(b), is designed to operate in the 5.925 to 6.425 GHz satellite communication uplink band. The overall length of the klystron is approximately 39 cm. Other mechanical and electrical characteristics are given in the accompanying data sheet.

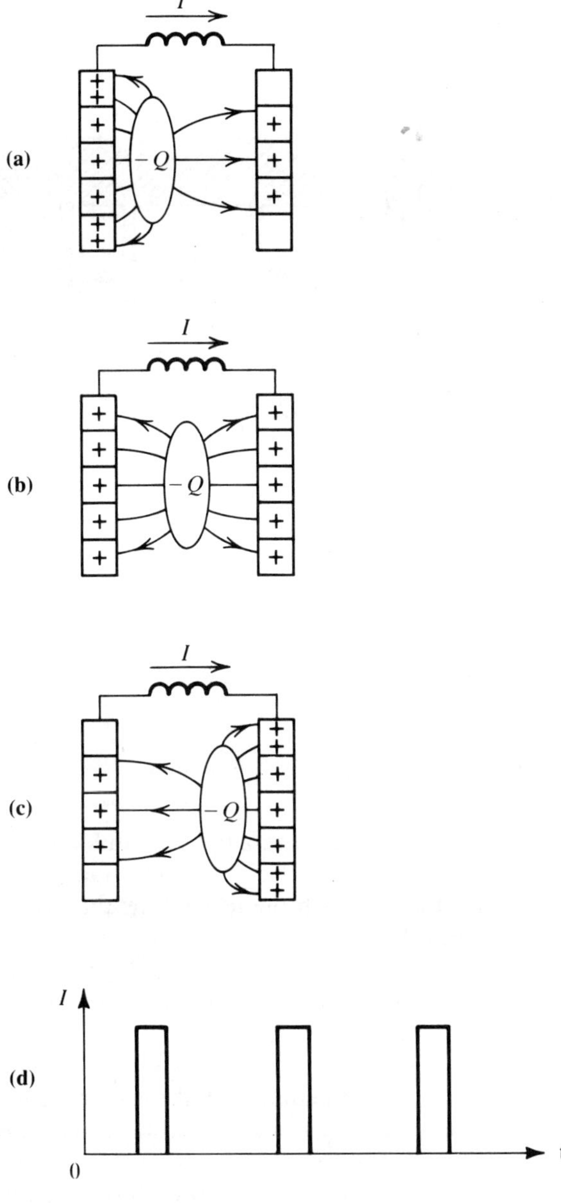

FIGURE 9.12. Electron bunch crossing the catcher gap: (a) entering; (b) midway; (c) leaving. (d) Induced current in the external circuit.

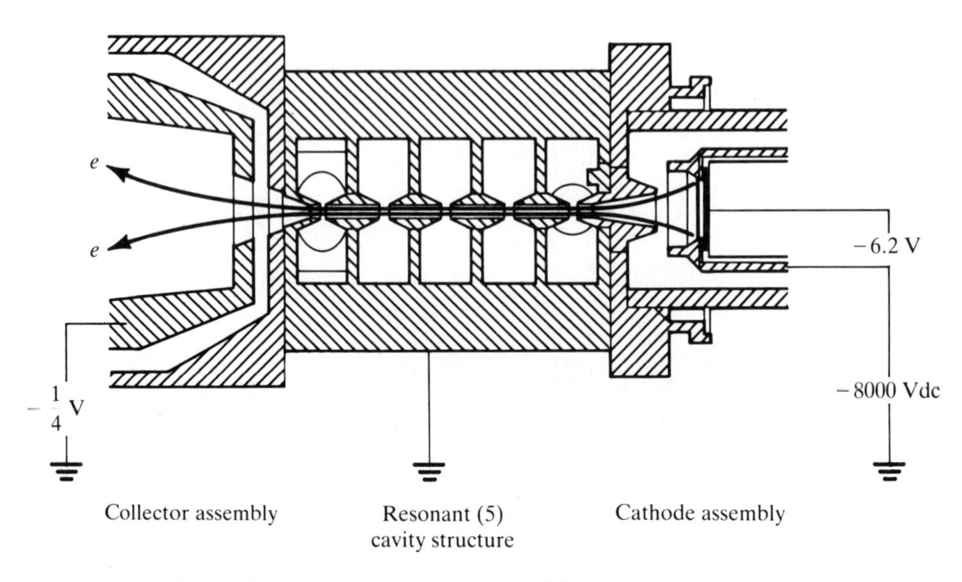

FIGURE 9.13. Sectional view of a 5-cavity klystron amplifier. (*Courtesy Varian Co.*)

9.5 THE REFLEX KLYSTRON OSCILLATOR

Although the klystron amplifier could be used as an oscillator by feeding part of the output signal back to the input in the correct phase, the practical difficulties of synchronously tuning the cavities, together with adjusting V_0 to control the transit time, makes this an unsatisfactory approach. Oscillations can be achieved in a single-cavity klystron by reflecting electrons back to the cavity by means of a reflector electrode. Such a klystron is referred to as a *reflex klystron*.

The circuit for a reflex klystron oscillator is shown in Fig. 9.14. The cavity grids are maintained at the dc potential V_0. It will be assumed that oscillating energy is present in the cavity (as with the magnetron oscillator, oscillations can start from random movement of electrons). The alternating voltage across the cavity gap will produce velocity modulation of the beam as in the klystron amplifier. However, in the reflex klystron, the electron beam, on leaving the cavity, travels toward the reflector electrode, which is at a negative potential with respect to the cathode. This is sufficiently negative to totally repel the electrons that return through the cavity, which also acts as a collector. The total retarding field in the reflector-cavity space

TECHNICAL DATA SHEETS AND
VA-936 SERIES KLYSTRON SPECIFICATIONS

TECHNICAL DATA
VA-936G6/G12
VA-936L6/L12
CW Klystron Amplifier
5.925 – 6.425 GHz
3.0, 3.35 kW

The VA-936 series of permanent-magnet focused, forced-air cooled, five-cavity klystron amplifiers with integral channel tuners are for use in commercial CW earth-to-satellite communication systems. These tubes are mechanically tunable over the 5.925 to 6.425 gigahertz band and for "G" and "L" versions deliver CW outputs of at least 3.0 and 3.35 kilowatts, respectively. The VA-936G12 and the VA-936L12 channel tuners permit tuning to any of twelve selected preset frequencies within the 500-megahertz band in seconds. The VA-936G6 and the VA-936L6 have six selected preset frequencies. The channel settings may be reset and adjusted in the field. An optional accessory, the VA-1436 Remote Servo Tuner is available for remote channel selection. Over the 500-megahertz range, the 1-dB bandwidth is at least 45 megahertz.

MAXIMUM RATINGS[1]

Beam Voltage	10 kVdc
Beam Current	1.25 Adc
Beam Power Input	10 kW
Body Current, with drive	50 mAdc
Heater Surge Current	14 A
Reflected Power[2]	120 W

GENERAL CHARACTERISTICS[3]

ELECTRICAL

Frequency Range	5.925-6.425 GHz
Tuning Range	500 MHz
Heater Voltage	6.0 V
Heater Current	6.5 ± 0.5 A
Heater Warm-up Time, minimum	5 min

PHYSICAL

Dimensions	See Outline Drawing
Weight, approximate	70 lb/31.8 kg
Mounting Position	Any
Tuning Method	Channel Selector
Cooling[4]	Forced Air
Connectors	See Outline Drawing

COOLING[4]

Air Flow, minimum	
Collector	
VA-936G6 and VA-936G12	900 lb/h(409 kg/h)
VA-936L6 and VA-936L12	1000 lb/h(454 kg/h)
Body	90 lb/h(41 kg/h)
Cathode	85 lb/h(39 kg/h)
Back Pressure, collector	
at 900 lb/h	2.7(6.9 cm) in-H_2O
at 1000 lb/h	3(7.6 cm) in-H_2O

is $V_0 + V_r$ and this results in a deceleration of magnitude

$$a = \left| \frac{q_e(V_0 + V_r)}{m_e s} \right| \tag{9.9}$$

where q_e is the charge on an electron, m_e the electron mass, and s the distance between reflector and grid 2 of the cavity.

VA-936G6, VA-936G12, VA-936L6 AND VA-936L12

TYPICAL OPERATING CONDITIONS [a]

VA-936G6 AND VA-936G12

Frequency	6.395 GHz
Power Output	3.1 kW
Bandwidth, 1-dB	48 MHz
Drive Power	250 mW
Gain	41 dB
Beam Voltage	8.0 kVdc
Beam Current	1.05 Adc
Body Current, with drive	15 mAdc
Load VSWR	1.05:1

VA-936L6 AND VA-936L12

Frequency	6.395 GHz
Power Output	3.4 kW
Bandwidth, 1-dB	48 MHz
Drive Power	250 mW
Gain	41 dB
Beam Voltage	8.2 kVdc
Beam Current	1.08 Adc
Body Current, with drive	15 mAdc
Load VSWR	1.05:1

CHARACTERISTIC CURVES

Typical performance values

VA-936G6 AND VA-936G12

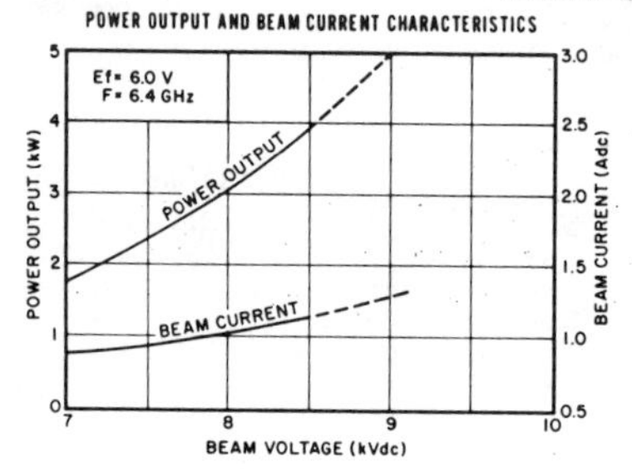

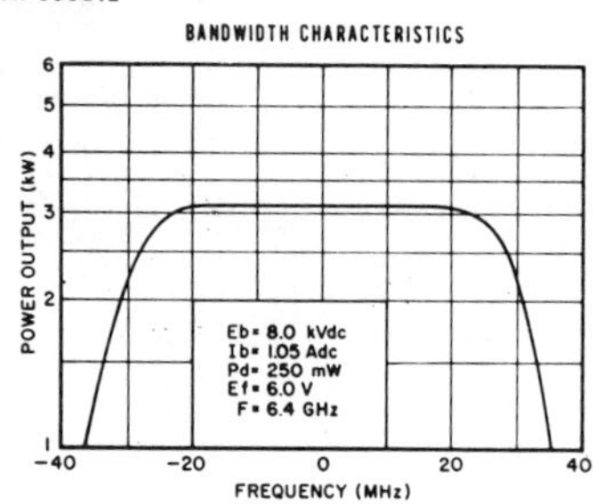

NOTES:

1. Ratings should not be exceeded under continuous or transient conditions. A single rating may be the limitation and simultaneous operation at more than one rating may not be possible. Equipment design should limit voltage and environmental variations so that ratings will never be exceeded.

2. Corresponds to a load VSWR of 1.5:1 at 3.0 kilowatts.

3. Characteristics and operating values are based on performance tests. These figures may change without notice as a result of additional information or product refinement. Varian Associates should be consulted before using this information for final equipment design.

4. Minimum air flow rate required at maximum beam power input and 20°C ambient temperature. At increased ambient temperatures, more air flow may be required to provide adequate cooling. The required increase in air flow can be determined from information published by blower manufacturers. For additional cooling information, contact the Varian Palo Alto Microwave Tube Division.

The Applegate diagram is shown in Fig. 9.15, where again, for clarity, just three electron paths are shown. For a constant retarding field, the distance–time equation describes a parabola [found by integrating Eq. (9.9) twice]:

$$d = u_a t - \frac{1}{2} a t^2 \tag{9.10}$$

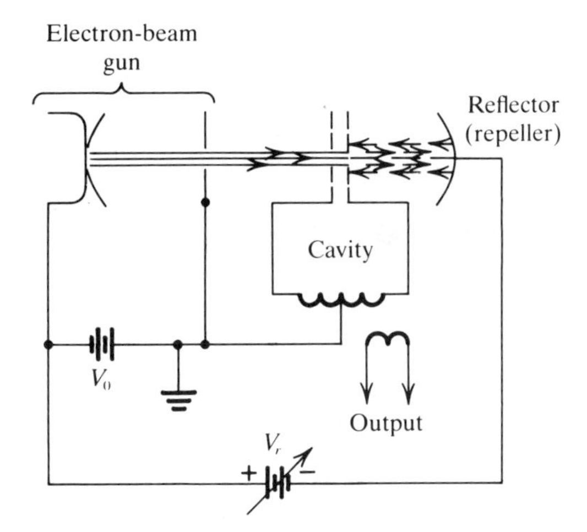

FIGURE 9.14. Circuit for a reflex klystron oscillator.

The initial velocity u_a is given by Eq. (9.8) since the bunching action is similar to that for the two-cavity klystron. Curve (a) is for maximum initial velocity u_0 $(1 + m/2)$, curve (b) for unmodulated initial velocity u_0, and curve (c) for minimum initial velocity u_0 $(1 - m/2)$. It will be seen that bunching occurs for those paths whose initial velocities are determined by the negative slope of the cavity voltage, compared to the positive slope for the two-cavity klystron. This is because the faster electrons travel farther into the retarding field and spend more time there than do the slower electrons. This makes it possible for slow and fast electrons to arrive back together at the cavity, as shown in Fig. 9.15.

The reflector voltage is adjusted so that these electron bunches enter the cavity gap around the time when the gap voltage is maximum. Since the gap field under these conditions imparts maximum initial velocity to outgoing electrons, it will act to retard the reflected electrons. The kinetic energy recovered from this process goes into increasing the potential energy in the cavity. In this way, sufficient energy feedback can be obtained to sustain oscillations.

In order for the bunches to form as they enter the cavity, the average time of flight in the reflector field is seen from Fig. 9.15 to be $1\frac{3}{4}$ cycles. In fact, the number of cycles may be $n + \frac{3}{4}$ where $n = 0, 1, 2, \ldots$ etc. and the phasing will still be correct for oscillation. However, the power output decreases as the number of cycles increases, the variation of output with reflector voltage being shown in Fig. 9.16. Although maximum power output is obtained with the 3/4 mode ($n = 0$) this requires the highest reflector voltage, with consequent danger of breakdown in the klystron.

For this reason either the 1¾ or 2¾ mode is usually chosen in practice. Frequency changes also occur with changes in reflector voltage, as shown in Fig. 9.16.

9.6 THE TRAVELING-WAVE TUBE

The basic circuit for a traveling-wave-tube amplifier is shown in Fig. 9.17. An electron beam is confined, by means of a magnetic field, to travel along the inside of a wire helix. The RF input signal enters the helix at the end nearest the electron beam gun. The electric field of the wave will have a component along the axis of the helix. At some sections this field will decelerate the electrons in the beam, and at other sections it will accelerate them. Thus electron bunching occurs as for the klystron, but in the traveling-wave tube, the interaction is continuous along the beam.

The purpose of the helix is to reduce the phase velocity of the wave to that of the beam velocity, and it is referred to as a *slow-wave structure*.

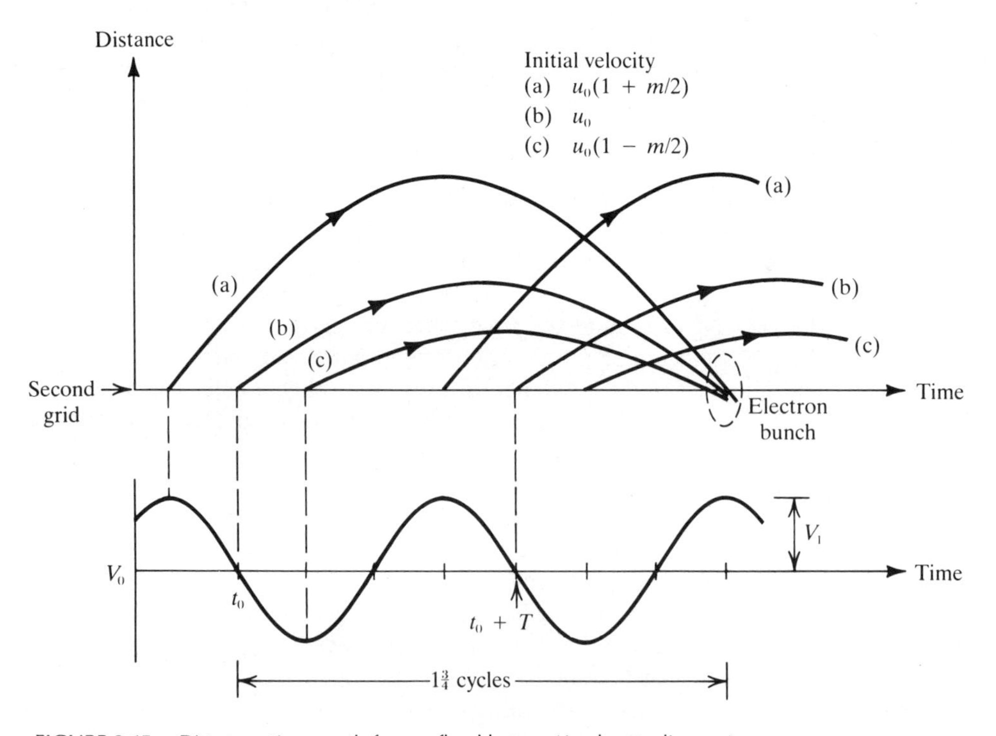

FIGURE 9.15. Distance–time graph for a reflex klystron (Applegate diagram).

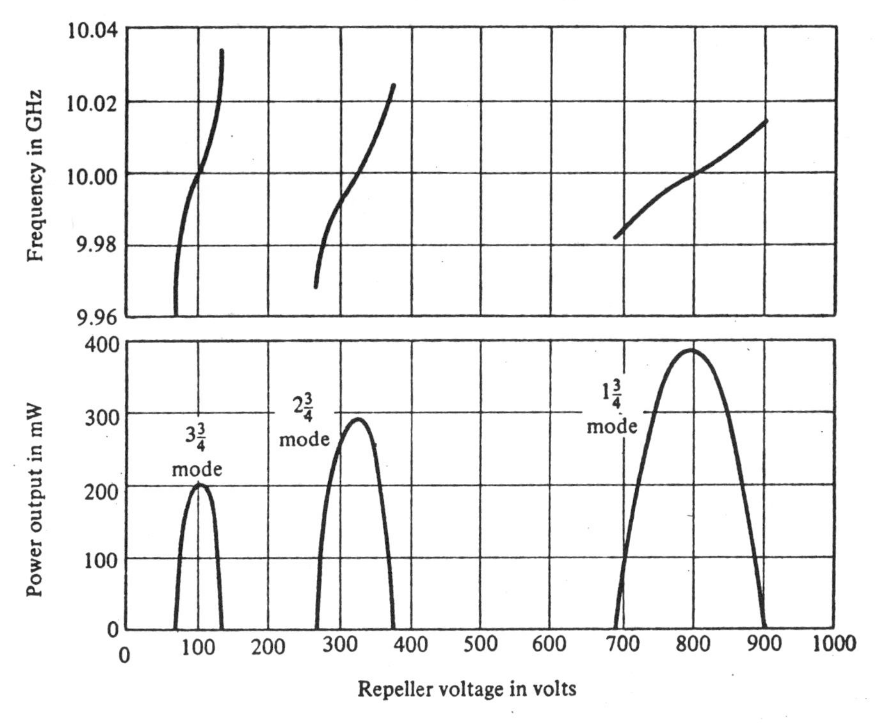

FIGURE 9.16. Power output and frequency characteristics of a reflex klystron. [*From Samuel Y. Liao, Microwave Devices and Circuits (Englewood Cliffs, N.J.: Prentice-Hall, Inc., © 1980).*]

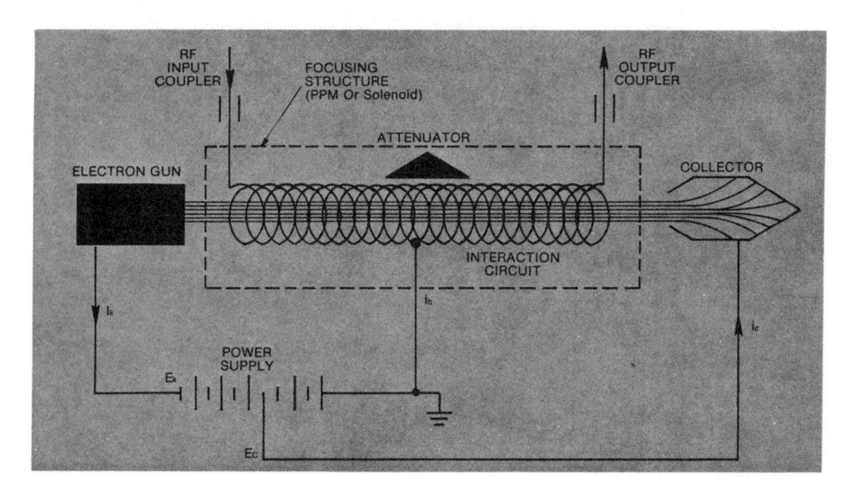

FIGURE 9.17. The major elements of a TWT are the electron gun assembly, RF interaction circuit, focusing magnets, and collector. (*Courtesy of Hughes Electron Dynamics Division, Torrance, Calif., TWT and TWTA Handbook.*)

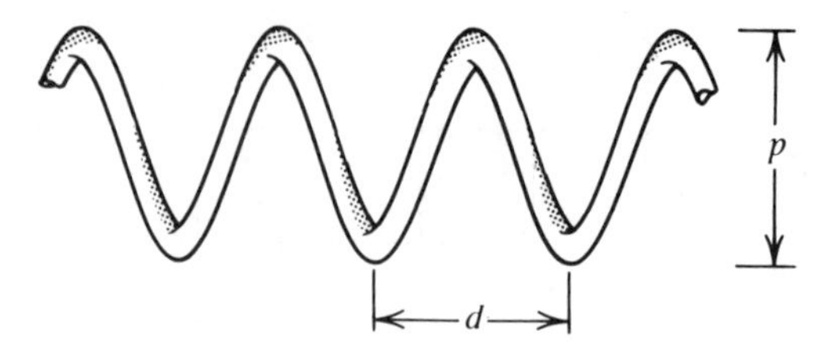

FIGURE 9.18. Helix, pitch p and diameter d.

The wave travels at approximately the velocity of light c, along the wire. Referring to Fig. 9.18, the length of one complete turn is approximately πd, the circumference of a circle of diameter d. Thc time taken for the wave to travel this length is $\pi d/c$. In the same time, the wave moves an axial distance equal to the pitch p of the helix, and therefore the axial velocity of the field v_f, is

$$v_f = \frac{p}{\pi d/c}$$

$$= c\,\frac{p}{\pi d} \tag{9.11}$$

By making the phase velocity of the wave approximately equal to the average velocity of the electron beam, continuous interaction between specific regions of wave and beam can be maintained. The electric field of the slow-wave structure will extend into the beam region as shown in Fig. 9.19. Remembering that an electric field directed against the electron flow accelerates the electrons because they are negatively charged, electrons

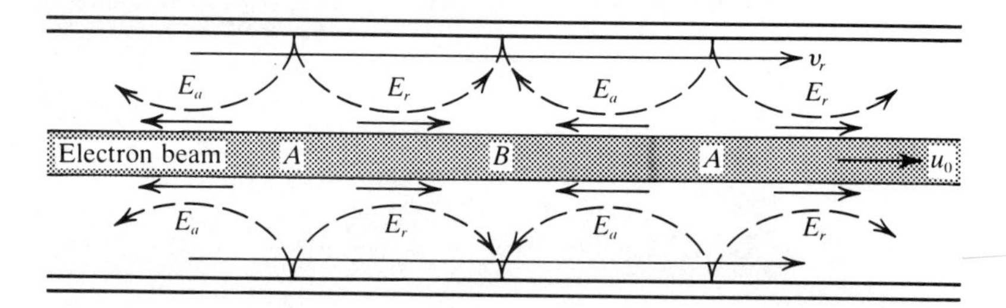

FIGURE 9.19. Electron bunching in a traveling-wave tube. Electron density is increased in regions A and decreased in regions B. The average velocity of the beam relative to the field is $(u_0 - v_f)$.

influenced by fields E_a will be accelerated, and those influenced by fields E_r will be retarded. Thus electron bunches will form in regions A and electron deficiencies in regions B, as shown in Fig. 9.19. It is assumed that $|E_a| = |E_r|$. Because the average velocity of the beam is maintained slightly greater than the phase velocity of the traveling wave, the electron bunches will tend to move into the influence of E_r, as both the beam and the wave progress along the length of the tube. The electron bunches therefore are moving against a retarding force which reduces their kinetic energy, the energy "lost" being transformed to potential energy in the traveling-wave field. The beam in effect provides a mechanism for the continuous transfer of energy, from the potential energy of the dc accelerating field in the gun via the kinetic energy of the beam to the potential energy of the RF field. It will be noted that this is similar to the transfer mechanism in the magnetron described in Section 9.2, and the cavity magnetron is also referred to as a traveling-wave magnetron.

In Fig. 9.17, an attenuator is shown midway along the helix. This attenuator affects only the waves on the helix, not the electron beam. It may simply take the form of Aquadag sprayed on the wall of the retaining tube for the helix. The purpose of the attenuator is to attenuate reflected waves which may arise from mismatch at the output port, and which, if reaching the input could cause oscillations. The attenuator will attenuate both the forward and reflected waves on the slow-wave structure. However, it has little effect on the electron beam, which, as it exits the attenuated region, reinduces the forward wave on the slow-wave structure.

The most important advantage of the traveling-wave tube compared to the klystron is that it is capable of providing wideband amplification. This is because the wave circuit is nonresonant, unlike the klystron, which uses resonant cavities. The component parts of a typical helix TWT are shown in Fig. 9.20(a), and the assembled tube is shown in Fig. 9.20(b).

Single-helix-type tubes are limited to about 3 kW peak output power. This is partly because it is difficult to remove heat from the helix, generated by the ohmic losses. In addition, to produce high powers the beam voltage, and hence beam velocity, must be increased. A corresponding increase is required in the slow-wave velocity, which, from Eq. (9.11), requires an increase in helix pitch (the diameter is fixed by the beam size). It is found that a stretched-out helix supports backward waves, which can give rise to oscillations. Backward waves, discussed in Section 9.8, must be distinguished from the reflected waves mentioned earlier. Two modified helix-type structures which help to overcome these limitations are shown in Fig. 9.21. The circuit properties of these structures do not support certain backward wave modes; also, they have higher thermal capacity and make better thermal contact through the mechanical supports, which makes for better heat removal. In some cases diamond supporting rods are used for the

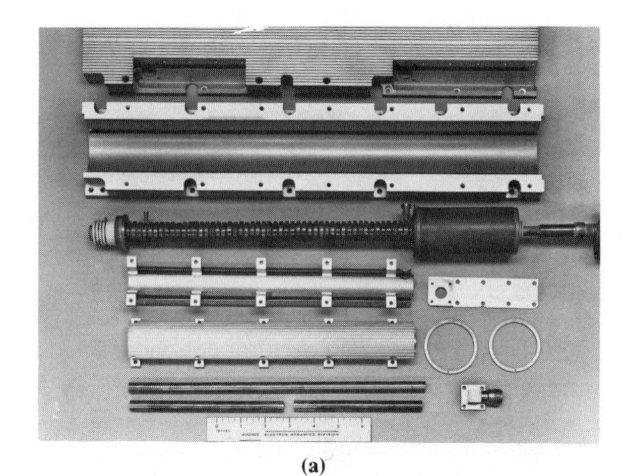

(a)

(b)

FIGURE 9.20. (a) Various components of a typical metal–ceramic helix TWT, with the metal–ceramic envelope for the PPM focusing structure shown in the center; (b) the assembled helix TWT provides broadband pulsed output from 2.5 to 8 GHz. (*Courtesy of Hughes Electron Dynamics Division, Torrance, Calif., TWT and TWTA Handbook.*)

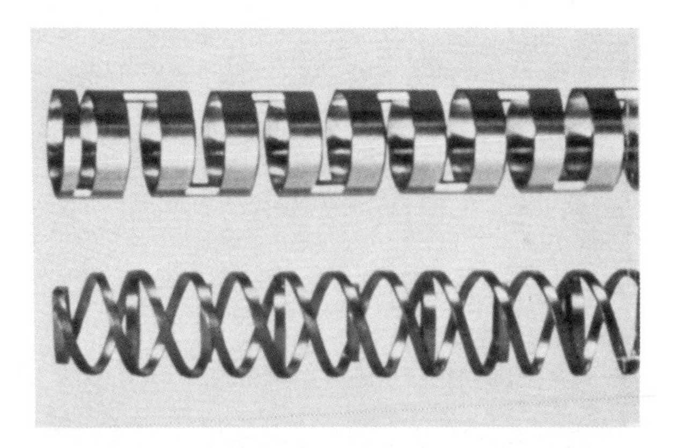

FIGURE 9.21. Two variations on the basic helix structure are (a) the ring bar structure (top) and (b) the bifilar or folded helix (bottom). (*Courtesy of Hughes Electron Dynamics Division, Torrance, Calif., TWT and TWTA Handbook.*)

helix to increase the rate of heat removal (Acker 1982). The frequency bandwidth of the single helix is wider than that of the other types shown.

Another form of slow-wave structure is the coupled cavity shown in Fig. 9.22. This is capable of handling higher powers than the helix type, but again at the expense of bandwidth.

Focusing of the electron beam, or more correctly, collimating it into a narrow beam, is achieved by means of magnetic fields. One of the simplest and most effective ways of achieving this is by means of a current-carrying solenoid, as shown in Fig. 9.23(a). The disadvantages of the solenoid are that it is relatively bulky and it consumes power, but in general, tubes rated at more than a few kilowatts RF output require solenoid focusing. For lower-power tubes such as those used in communications satellites and in other applications where weight and power consumption must be minimized, permanent-magnet focusing methods have been developed. The most important of these is what is termed *periodic permanent-magnet* (PPM) *focusing*, the principle of which is illustrated in Fig. 9.23(b). The magnetic fields in the gaps form "magnetic lenses," the focal length of which depend only on the field strength, not on the field direction [see, e.g., Gittins

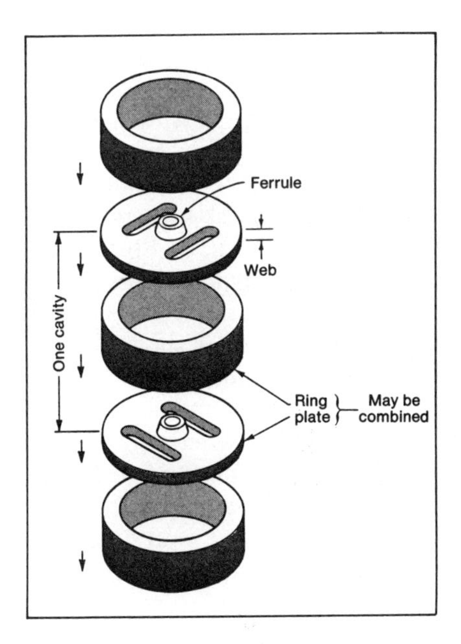

FIGURE 9.22. Axial stacking of a typical staggered-slot, round-cavity circuit. *(From Arnold E. Acker, "Interest in mm-wave spurs tube growth," Microwaves, July 1982; (with permission.)*

(1965)]. The electron beam periodically diverges and converges as it passes through the lenses, while being confined to the tube. Figure 9.23(c) shows a radial magnet PPM structure which utilizes radial segments of Alnico magnets, and Fig. 9.23(d) one that utilizes Alnico magnet rings (from *Hughes TWT & TWTA Designer's Handbook*).

In Fig. 9.24 is shown in cross section a TWT designed for high-power pulsed output. The tube is of metal–ceramic construction, weighs a total of 5 pounds, and utilizes a periodic permanent-magnet (PPM) focusing arrangement specially developed by the Hughes Company for high-power tubes. The magnetic disks are shown in cross section.

9.7 THE TRAVELING-WAVE-TUBE AMPLIFIER

The combination of power supply and TWT used as an amplifier is referred to as a *traveling-wave-tube amplifier* (TWTA). These are supplied by manufacturers as complete units to ensure correct operation of the TWT. Obtaining rated performance from a traveling-wave tube is critically dependent on having the correct dc supplies available, and in fact, use of improper dc levels or a poorly designed power supply can lead to tube damage. The power supply must be capable of supplying each electrode with its rated direct current and voltage, and some TWTs are designed with multistage collectors each stage of which requires its own dc supply. With multistage collectors, faster electrons in the beam are collected at low potential stages, while slower electrons are collected at the higher potential stages. In this way, high currents at high potentials are avoided, the collector power dissipation goes down, and tube efficiency is increased.

The voltage requirements for a 25-W CW TWTA, employing a multistage collector, are shown in Fig. 9.25. The heater supply can be ac or dc, but where dc is used as shown, the heater element must be negative with respect to the cathode. This is to prevent electrolysis of the insulation between heater and cathode, and also to prevent electron emission from cathode to heater. Referring all voltages to ground, the cathode is at −4000 V, and the gun anode is +200 V. The slow-wave structure together with the first stage of the collector are at ground potential (and therefore 4000 V positive with respect to cathode). Slower electrons will be collected by the slow-wave structure and the first collector stage, the combined current being referred to as the *helix current*, or where cavities are used, as the *body current*. Note that there is no way of separating out the current to the slow-wave structure and that to the first collector stage. The second collector stage is at a potential of +1500 V with respect to cathode, and therefore −2500 V with respect to ground. Because it is below ground

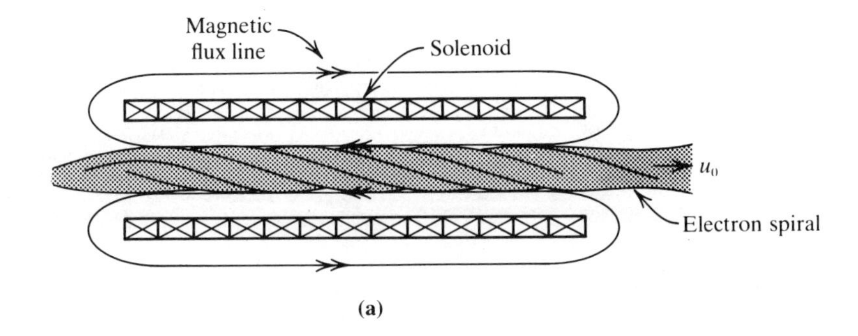

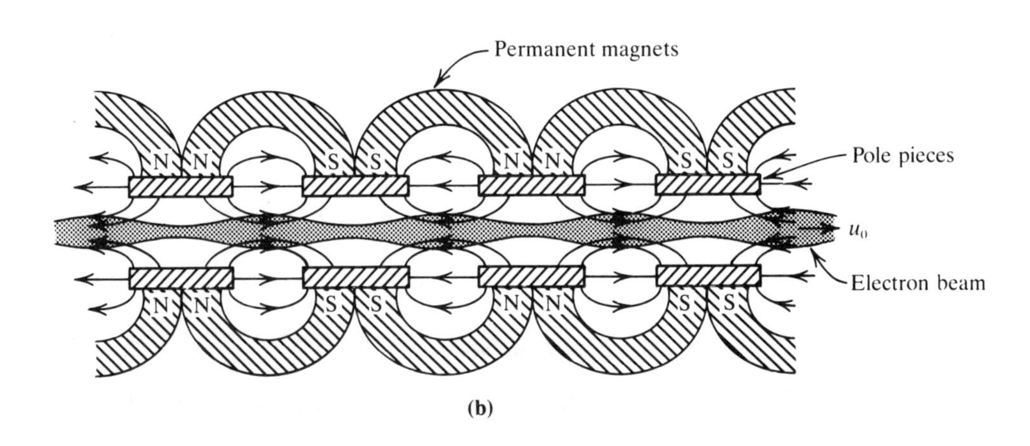

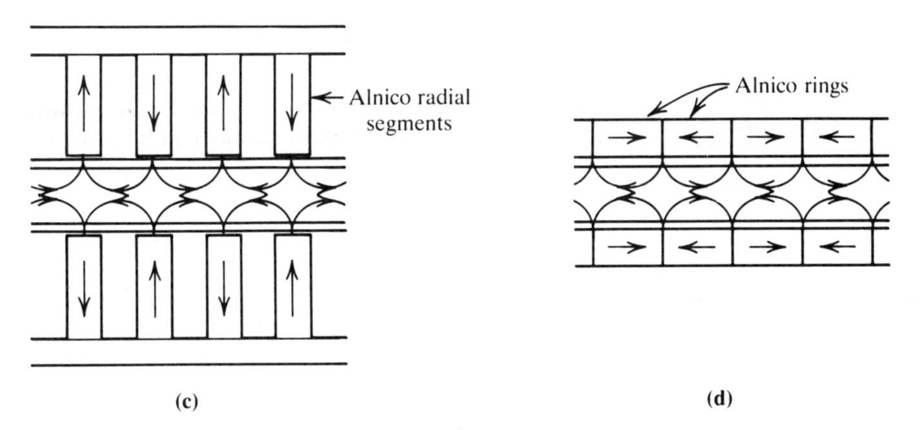

FIGURE 9.23. Magnetic focusing of electron beams: (a) solenoid field; (b) periodic permanent magnet (PPM) structure; (c) PPM using Alnico radial segments; (d) PPM using Alnico rings.

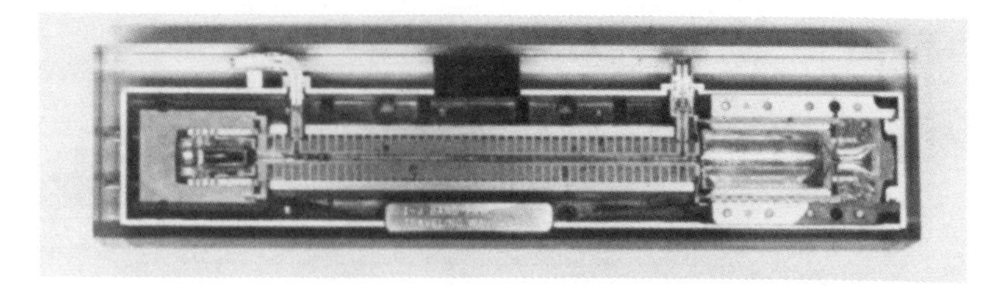

FIGURE 9.24. The Hughes 774H X-band high-power helix TWT weighs only 5lb and is rated at 1.25-kW minimum power output with a 0.04 duty cycle. (*Courtesy of Hughes Electron Dynamics Division, Torrance, Calif., TWT and TWTA Handbook.*)

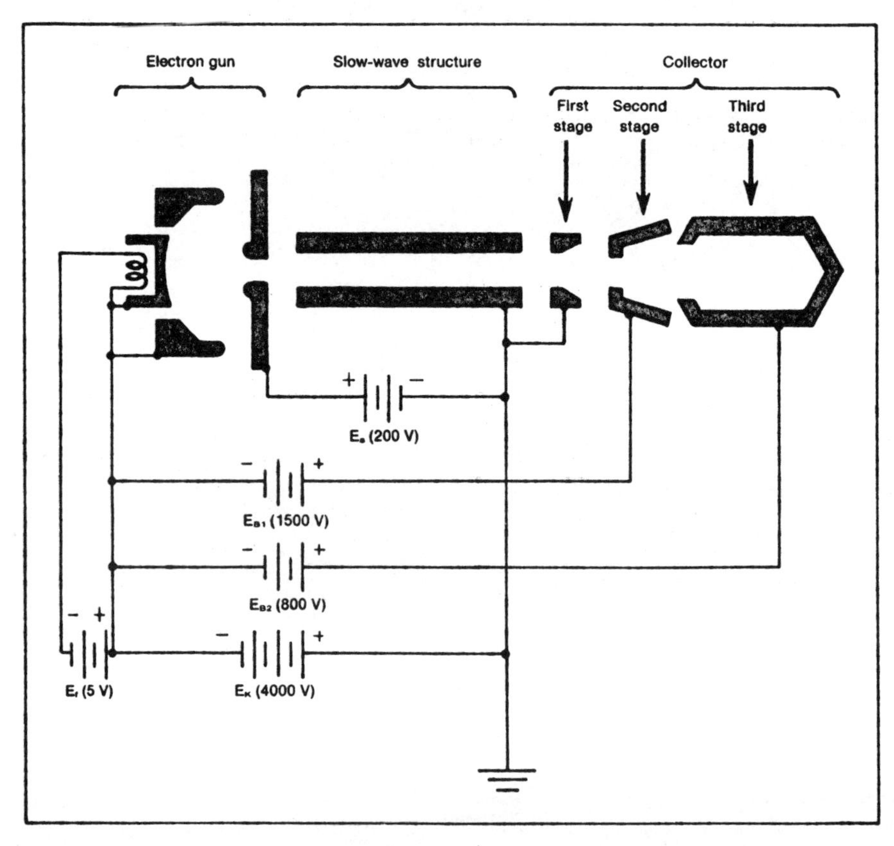

FIGURE 9.25. TWT power supply assembly consists of a mix of high- and low-voltage supplies, each of which must deliver the correct potential to the tube's cathode, anode, and collector elements. (*From James W. Hansen, "Multiple output supplies must meet TWT demands," Microwaves, July 1982; with permission.*)

potential it is referred to as a *depressed stage*, in this case the first depressed stage. The third collector stage, which is the second depressed stage, is at −3200 V with respect to ground. Further details of this power supply will be found in Hansen (1984), and on the design of power supplies in Willard and Joslin (1984). Figure 9.26(b) shows a rack-mounting TWTA designed for use in satellite earth stations, and Fig. 9.26(a) shows the European slim-line format, in which the tube and power supply are physically separate, although designed in combination.

9.8 OPERATIONAL CHARACTERISTICS OF THE TWTA

The power transfer curve for a TWTA is the graph of RF power output against RF power input, and it has the form shown in Fig. 9.27(a). This curve is sometimes referred to as the *AM/AM conversion curve*. For purposes of illustration, typical power levels are shown in Fig. 9.27. These are in dBm and it will be recalled that 0 dBm is equal to 1 mW. Thus the powers are shown in dB relative to 1 mW, and, for example, an input of −32 dBm is equivalent to 0.631 μW.

The gain is defined as the ratio of power output to power input, and in decibels this is simply

$$G(\text{dB}) = P_0\,(\text{dBm}) - P_i\,(\text{dBm}) \tag{9.12}$$

Often, such expressions are written simply as $G = P_0 - P_i$, the decibel units being understood.

For low inputs, the gain is almost constant and is termed the small-signal (or linear) gain. This is the range where a given decibel change in input results in an equal decibel change in output. As the RF power input is increased, the RF output power does not increase in proportion, but instead levels off, and finally starts to decrease. The point at which the output power levels off is termed the *saturation point*, and the gain at this point the saturation gain. The gain curve is plotted in Fig. 9.27(b).

There will be a range of input for which the output remains in saturation, and this range defines the *overdrive capability* of the TWT.

The saturation level is an important reference point for TWT operation, and often the power transfer curve is plotted with reference to the saturation level as shown in Fig. 9.28. In certain types of satellite communications, operation must be kept well below saturation, this condition being referred to as *backoff*. An input backoff of 10 dB means that the input level is −10 dB relative to the input required for saturation.

As the operating level approaches saturation, the nonlinearity of the transfer characteristic gives rise to a form of distortion known as *inter-*

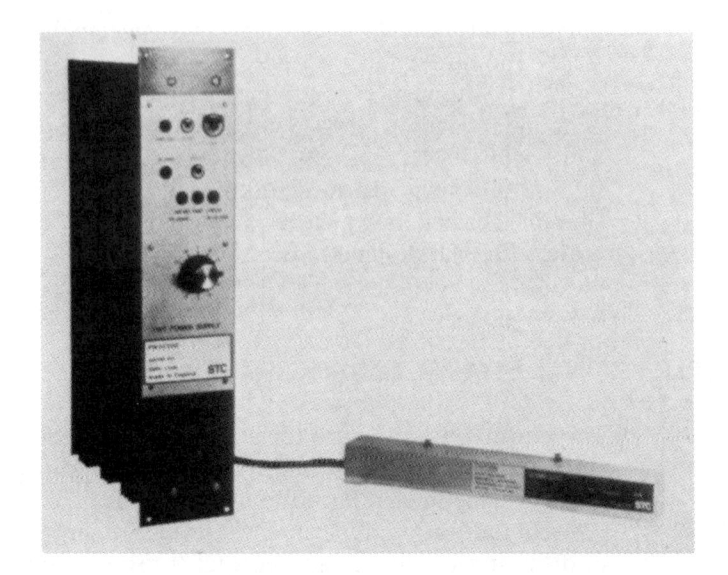

(a)

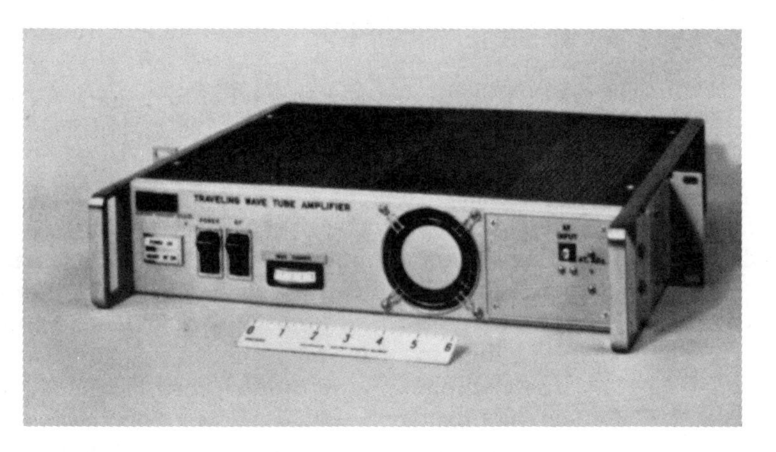

(b)

FIGURE 9.26. (a) European slimline traveling-wave-tube amplifier configuration; (b) this Hughes communications power amplifier is self-contained and includes a PPM focused TWT and solid-state power supply with integral cooling and protective circuitry. [*Part (a) from J. Willard and C. Joslin, "Modern medium-power traveling-wave-tube amplifiers for communications," Radio Electr. Eng., Vol. 54, No. 7/8, 1984; with permission; part (b) courtesy of Hughes Electron Dynamics Division, Torrance, Calif., TWT and TWTA Handbook.*]

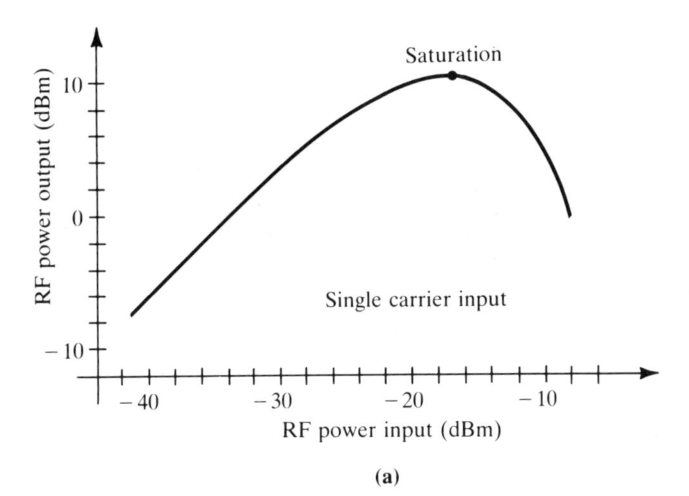

(a)

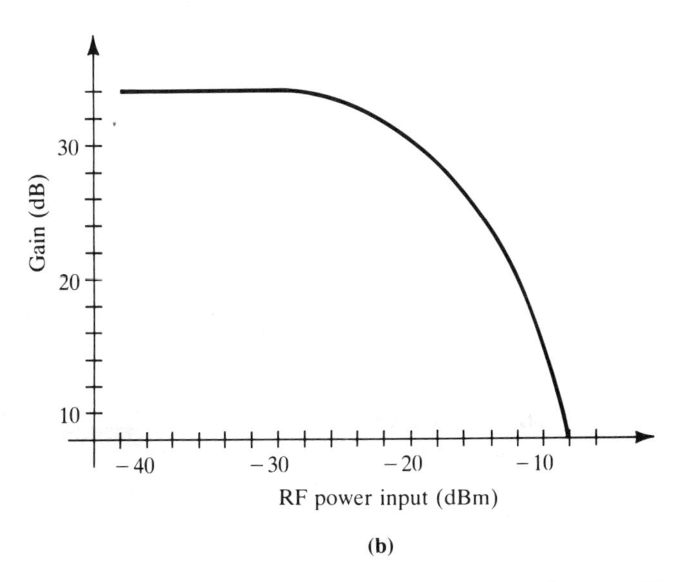

(b)

FIGURE 9.27. (a) RF power transfer curve for a TWTA; (b) gain vs. RF input.

modulation distortion. The transfer characteristic for input and output voltages can be represented by a power series, and the third-order term of this is the most troublesome. As an example, with two carriers present at the input (a common condition in satellite communications, where multiple carriers might be present), the third-order term is

$$\text{third-order term} = c(v_1 + v_2)^3 \tag{9.13}$$

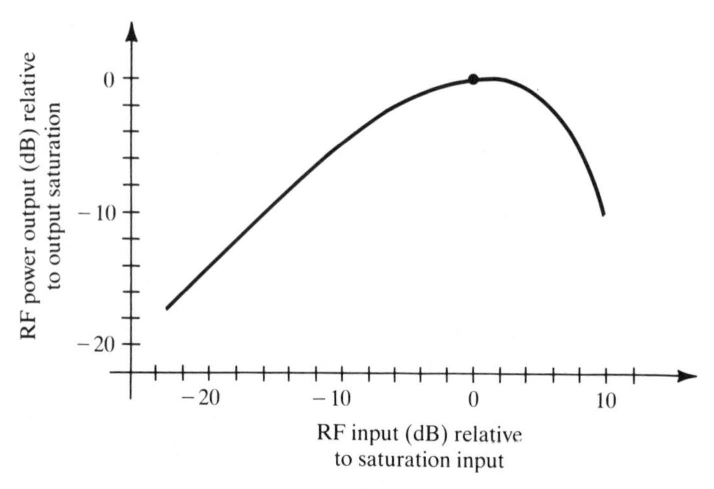

FIGURE 9.28. Power transfer curve of Fig. 9.27 plotted in dB relative to the saturation point.

where c is a third-order coefficient the value of which depends on the operating conditions of the tube. Assuming that the signals are sinusoids of unit amplitude, trignometric expansion of Eq. (9.13) yields the third-order intermodulation terms plus higher-order terms:

$$\frac{3c}{4}\left[\cos\left(2\pi\left(2f_1 - f_2\right)t\right) + \cos\left(2\pi\left(2f_2 - f_1\right)t\right)\right] + \text{higher-order terms} \quad (9.14)$$

Where multiple carriers are present, as in single carrier per channel (SCPC) operation in satellite communications, the carriers are spaced uniformly in frequency. Thus the frequency $(2f_1 - f_2)$ becomes $(f_1 - \Delta f)$, and similarly, $(2f_2 - f_1)$ becomes $(f_2 + \Delta f)$. The spectrum is shown in Fig. 9.29. The third-order intermodulation products cause interference with

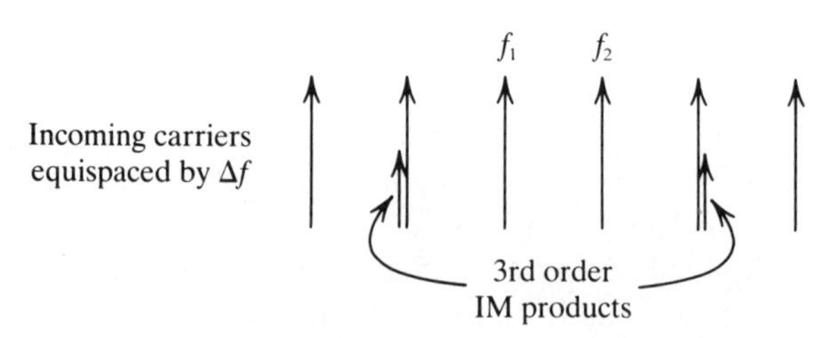

FIGURE 9.29. Third-order intermodulation interference, which can result in a TWT with multiple carriers present.

carriers neighboring on f_1 and f_2. A third-order intercept point can be determined in the same way as for mixers, shown in Fig. 7.45.

Another form of distortion occurs in a TWTA as a result of phase nonlinearities. The absolute time delay between input and output of a TWT at a fixed input level is generally not significant. However, at higher input levels, where more of the energy in the beam is converted to output power, the average beam velocity is reduced, and therefore the delay time is increased. Since phase delay is directly proportional to time delay, this results in a phase shift which varies with input level. Of importance is the phase shift at the output relative to some reference value, a convenient reference being the phase shift at saturation. Phase shift versus RF input is shown in Fig. 9.30. The slope of the phase-shift characteristic shows how amplitude modulation on the input is converted to phase modulation at the output and is termed the *AM/PM conversion*. The AM/PM conversion curve is also shown in Fig. 9.30, this being the derivative of the phase-shift curve, measured in deg/dB. This form of distortion can be troublesome where a mixture of carrier types are present at the input, such as AM, FM, and or PM, or where unwanted amplitude variations exist which could be transferred as modulation to angle-modulated carriers.

Harmonic distortion occurs in a TWT when it is operated at or near saturation. This comes about because the electron bunching is more intense, and the beam current therefore has very sharp peaks. Such peaks are rich in harmonics. In helix-type tubes operated at saturation the second harmonic output may be 9 to 11 dB below the fundamental, and the third harmonic output 16 to 18 dB below fundamental (Hansen 1984). These

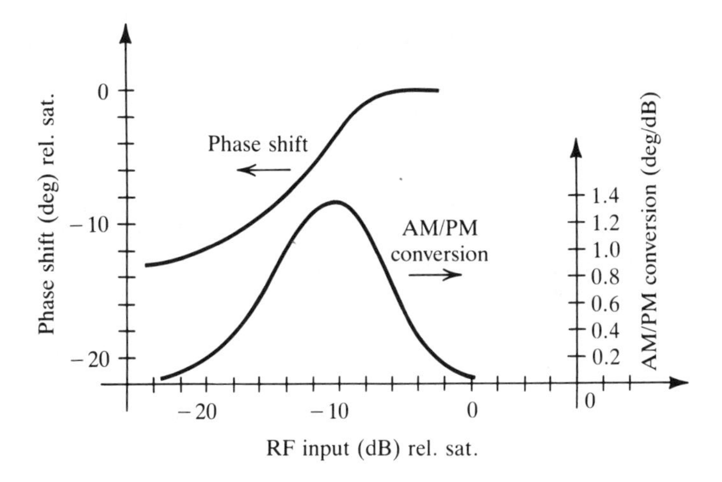

FIGURE 9.30. Phase-shift characteristic of a TWT, and the AM/PM conversion curve. The latter is the derivitive of the phase-shift curve.

harmonics must be filtered at the output. Cavity-coupled slow-wave structures do not propagate harmonics to the same extent as the helices, and harmonic levels may be as much as 30 dB below fundamental.

Where amplification of low-level signals is required, the noise generated in the TWT becomes an important limiting factor. Fluctuations in cathode emission and in electron velocities contribute to the current noise in TWTs. The noise factor F is a measure of how much the TWT-generated noise degrades the signal-to-noise ratio (see section 8.9) is defined as

$$F = (S/N)_{in}/(S/N)_{out} \tag{9.15}$$

TWTs specially designed for low-noise operation will have noise factors of less than 10 dB. Medium-power satellite TWTs have noise factors in the range 18 to 30 dB, and satellite transponder output TWTs in the range 24 to 27 dB, including the driver stage (*Hughes TWT & TWTA Designer's Handbook*).

9.9 THE CROSSED-FIELD AMPLIFIER

The crossed-field amplifier (CFA) is a development of the magnetron, the schematics for two types of CFAs being shown in Fig. 9.31. Unlike the magnetron, however, the CFA does not utilize a thermionic cathode, but uses either a cold cathode emitter as shown in Fig. 9.31(a), or an electron beam gun arrangement as shown in Fig. 9.31(b). The cathode electrode in either case is called the *sole*.

In the cold-cathode structure, emission from the cathode (or sole), made of a pure metal, depends on the electric field between cathode and anode, thus eliminating the need for thermionic emission. It is also found that cold cathode emission can be triggered on and off by the RF input signal, and therefore such tubes can lie dormant in the absence of input, which is an obvious advantage.

The tuned cavities in the anode of a magnetron form a slow-wave structure similar to that of the cavity-coupled TWT. In the CFA, the input signal is coupled into one end of the slow-wave structure. The electron beam for the continuous cathode emitting sole CFA of Fig. 9.31(a) has the shape of a "spoked wheel" and rotates in synchronism with the traveling wave in the slow-wave structure. The energy is transferred from the beam to the wave as described for the magnetron in Section 9.2.

As shown for the magnetron, forward and backward traveling waves exist on the slow-wave structure, and energy may be coupled to either of these, giving rise to either a forward wave CFA, or a backward wave CFA. More is said about backward waves in Section 9.10. The CFA finds ap-

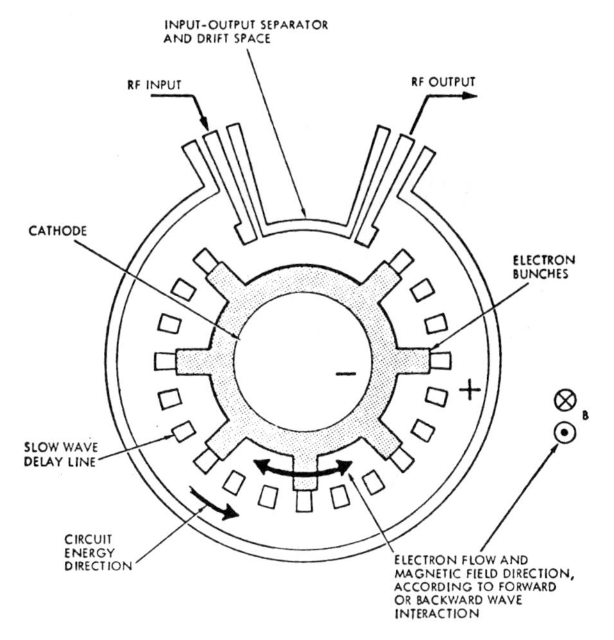

(a)

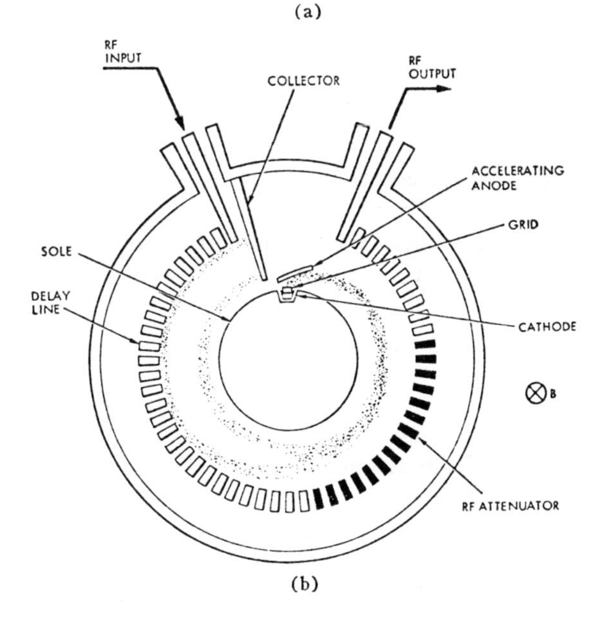

(b)

FIGURE 9.31. Schematic diagrams of two types of CFA: (a) continuous cathode emitting-sole CFA, forward or backward wave; (b) injected-beam CFA. *(From John F. Skowron. The Continuous Cathode (Emitting-Sole) Crossed-Field Amplifier. Proc IEEE vol. 61 No. 3 Mar. 1973 © 1973 IEEE. With permission.)*

plication in lightweight, transportable radar systems, where high powers at high efficiencies are required, while the tubes have to be comparatively small and light (Skowron 1973).

9.10 BACKWARD WAVES

The term *backward wave* is used to describe a wave that transmits energy in a direction opposite to the phase velocity of the wave. As shown in Section 9.2, the standing wave in a magnetron can be resolved into forward and backward wave components, and it is possible to synchronize the electron beam with the backward wave, giving rise to backward wave amplifiers and oscillators. The distinguishing characteristic of backward wave devices is that the phase velocity of the wave is approximately equal to the beam velocity, while the energy carried by the wave builds up in the opposite direction to the phase velocity. The energy is propagated at the group velocity. In terms of the angular frequency ω of the wave, and the phase shift coefficient β of the slow-wave structure, these two velocities, phase and group, are given by

$$v_p = \frac{\omega}{\beta} \tag{9.16}$$

$$v_g = \frac{d\omega}{d\beta} \tag{9.17}$$

[Phase and group velocities are also encountered in connection with waveguides (Section 4.4).]

A structure that has an ω/β relationship, as shown in Fig. 9.32(a), will support forward waves. The phase velocity is given by any two coordinates on the curve, for example, point A as shown. The beam velocity is represented by a straight line through the origin, and where this line cuts the ω/β curve for the slow-wave structure, the beam and wave will be synchronized. The slope of the ω/β curve gives the group velocity at that point.

The backward wave structure has an ω/β diagram as shown in Fig. 9.32(b). Here the phase and beam velocities are as in Fig. 9.32(a), but the slope, which gives the group velocity, is negative. This can be applied to the CC-CFA of Fig. 9.31(a) (see Problem 18). A linear version of a crossed-field backward wave oscillator, called an *M-carcinotron*, is shown in Fig. 9.33. The electron beam flows from left to right; the phase velocity of the wave on the slow-wave structure is directed from left to right also. This means that a given region of the beam will stay synchronized (or in near synchronism) with a given phase front of the wave as they both move

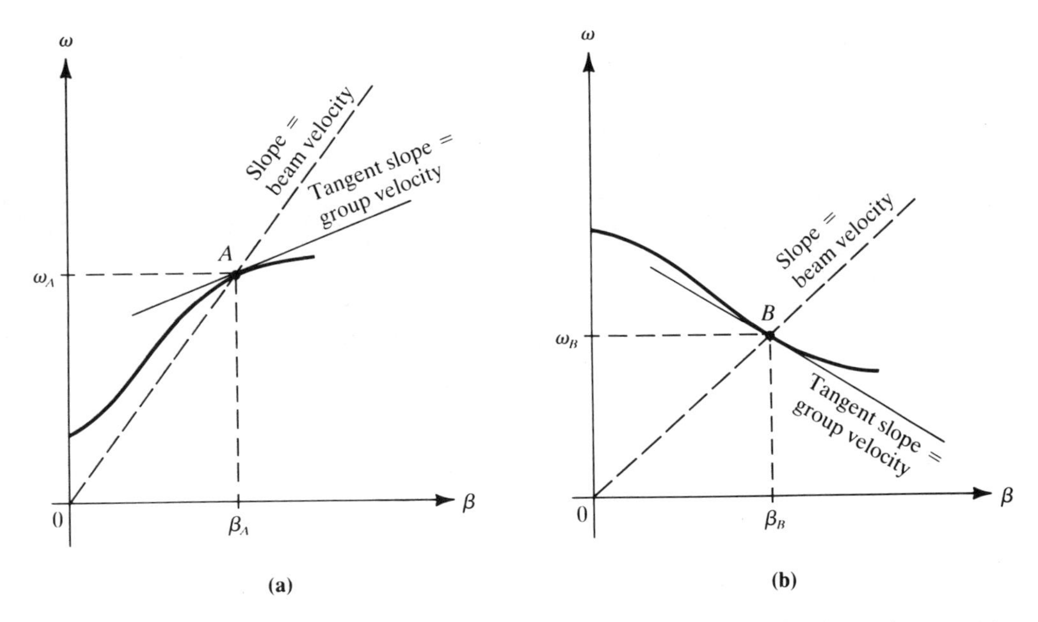

FIGURE 9.32. ω/β diagram for slow-wave structures. (a) Forward wave structure. The phase velocity at A is ω_A/β_A, and at A', ω'_A/β'_A, and is positive in both cases. The group velocity at B is positive, and at B' negative. The beam velocity in both cases in positive.

together, while the energy buildup in the wave progresses from right to left. Note that the right-hand side of the slow-wave structure is terminated in a matched load to avoid unwanted reflections, and the RF output is taken from the end nearest the cathode. The energy buildup in the wave comes from the beam, which in turn is modulated by the wave. There is

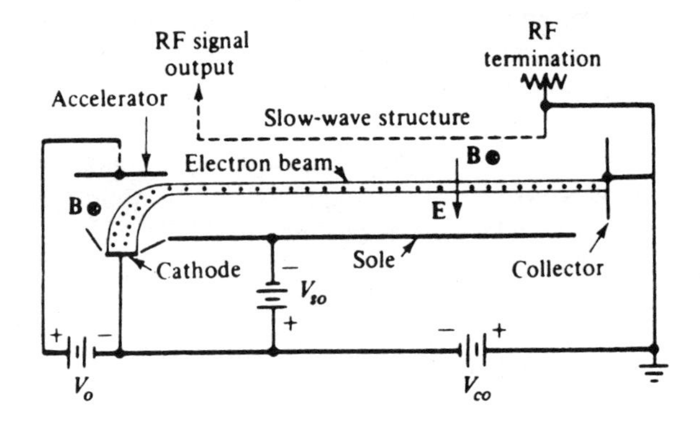

FIGURE 9.33. Linear model of an M-Carcinotron oscillator. (*From Principles of Electron Tubes by J.W. Gewartowske and H.A. Watson. Wadsworth, Inc., 1965.*)

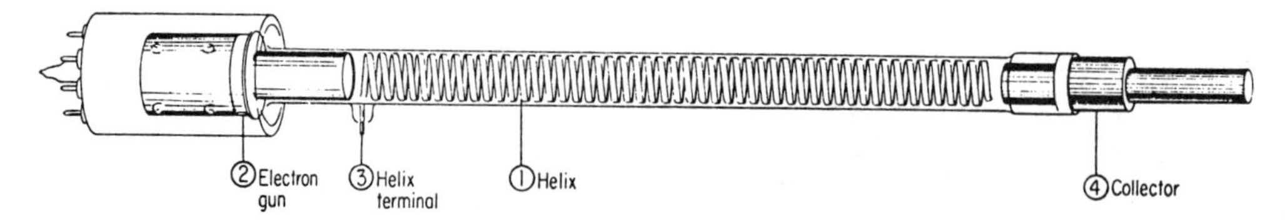

FIGURE 9.34. Backward wave oscillator tube. (*Courtesy of Hewlett-Packard, see also Lance (1964).*)

an internal feedback mechanism and the wave could be said to be "feeding on itself"—hence the name *carcinotron*, from carcinogenic or cancerous. The prefix M is explained in Section 9.11.

The slow-wave structure in a TWT is also capable of supporting backward waves under certain conditions. For example, the single-helix tube described in Section 9.6 operates well as a forward wave amplifier when there are four turns per wavelength along the beam; if conditions are changed to two turns per wavelength (e.g., by increasing the beam velocity), it can operate as a backward wave oscillator [see Gittins (1965)]. Under these conditions, the phase velocity is in the same direction as the beam velocity, but the energy flows from collector end to cathode end. A backward-wave oscillator tube is shown in Fig. 9.34. Oscillation may start from random electron motion, and as the energy is built up by interaction with the beam, the backward wave carries this toward the cathode end, where an output port may be placed. It is shown by Chodorow and Susskind (1964) that backward waves are the result of interference between two large, constant-amplitude waves that propagate in both directions along the tube. At the collector end they cancel, and at the cathode end they reinforce each other. This is in contrast to the forward-wave amplifier, where the forward wave grows exponentially with distance along the tube.

The tube may be taken below the threshold for oscillations by increasing the turns per wavelength. Under these conditions, a small signal applied as input at the collector end will determine the starting conditions for the interfering waves, which will again reinforce at the cathode or output end, and in this way a backward wave amplifier is obtained.

For both the backward wave oscillators described, the frequency is dependent on the beam voltage and therefore they can be tuned electronically, which is an advantage.

9.11 O- AND M-TYPE TUBES

The term *O-type* tube comes from the French TPO, which stands for *tubes a propagation des ondes*. The English translation is "tubes for the propagation of waves." This by itself is not very helpful since in a sense all the tubes described propagate waves in one form or another. However, a general feature of such tubes is that the beam and the magnetic field (if used) lie parallel to one another. Thus TWTs and klystrons are O-type tubes. It will be noted that small klystrons do not utilize magnetic fields in their operation. Another O-type tube is the *twystron*, which is a combination of TWT and klystron tubes [see Coleman (1982)].

The term *M-type* tube comes from the French TPOM, which stands for *tubes a propagation des ondes a champs magnétique*. The English translation is "tubes for the propagation of waves in a magnetic field." Magnetrons and tubes used in crossed-field amplifiers (CFAs) are all M-type tubes. It will be noted that these tubes need not be coaxial; they can have a linear structure as shown for the M-carcinotron in Fig. 9.33.

Another M-type tube is the *gyrotron*, a high-power device which depends for its operation on the change of mass experienced by electrons traveling at high velocities [see Smith (1981)].

9.12 PROBLEMS

1. Describe the operation of the cavity magnetron, and explain the function of the magnetic field. Explain how energy is transferred from the static field to the traveling-wave field and what is meant by working electrons.
2. Define and explain what is meant by the mode number of a cavity magnetron. Explain what is meant by the π-mode. Given that $C = C_c$ in Eq. (9.3) and that the angular frequency for the π mode is 10^{10} rad/s, calculate the angular frequency for $n = 3$ for an eight-cavity magnetron.
3. Explain the function of strapping in a cavity magnetron.
4. Describe the operation of a klystron amplifier, in particular the functions of the buncher and catcher cavities. Under what circumstances would a magnetic field be employed in a klystron?
5. Explain what is meant by modulation depth in relation to the operation of a klystron. Given that the modulation depth is 0.08, calculate the maximum and minimum electron velocities relative to the average velocity for a klystron.
6. In a klystron the final accelerating anode potential is 500 V and the modulation depth is 0.1. Assuming that the initial velocity of electrons leaving the cathode is zero, calculate (a) the steady velocity with which an electron leaves the anode, (b) the velocity at time $t_a = 0.1T$, (c) the velocity at time $t_a = 0.5T$, and (d) the velocity at time $t_a = T$, where T is the periodic time of the microwave signal being amplified.
7. A klystron amplifier operating at 5 GHz has a depth of modulation of 0.8, and the final accelerating anode voltage is 1 kV. Draw accurately to scale the Applegate diagram, using at least 20 points uniformly

distributed over two cycles of buncher voltage. Assume that the initial velocity of electrons leaving the cathode is zero.

8. Describe the operation of a cascaded klystron amplifier, showing how high power outputs are achieved.
9. Describe the operation of the reflex klystron oscillator. The reflector-to-cavity voltage for a reflex klystron is -300 V. Calculate the magnitude of the deceleration produced given that the reflector is 0.8 cm from the nearest cavity grid.
10. Plot accurately to scale the Applegate diagram for the reflex klystron of Problem 9 given that the depth of modulation is 0.1 and the frequency of operation is 1.0 GHz. Assume that $u_0 = 1.5 \times 10^7$ m/s, and use at least 20 points uniformly distributed over two cycles of oscillation.
11. Explain why bunching occurs over the positive-going slope of buncher voltage in the klystron amplifier, whereas in the reflex klystron oscillator bunching occurs over the negative-going slope.
12. Explain why the average time of flight of electrons in the reflector-cavity gap of a reflex klystron oscillator is $(n + 3/4)T$, where T is the periodic time of oscillation and n is 0, 1, 2, . . . etc. Explain also why a number of peaks in power output may be observed as reflector voltage is varied.
13. Describe the principle of operation of a traveling-wave tube. Explain why an attenuator is sometimes placed midway along the slow-wave structure. What is the main advantage of the traveling-wave-tube amplifier compared to the klystron amplifier?
14. Draw the schematic for a typical TWT power supply. Explain what is meant by a depressed collector stage in a traveling-wave tube and why depressed collector stages are used.
15. Define and explain what is meant by the saturation point of a traveling-wave-tube amplifier. Under what circumstances is it undesirable to operate close to the saturation point?
16. Describe the forms of distortion that can arise in a TWT amplifier and the effects these can have on the signals present.
17. Describe the operation of the crossed-field amplifier and explain how the cold cathode type of tube differs from the electron-gun type.
18. Referring to Fig. 9.31, indicate the electron flow and magnetic field directions required for backward wave interaction.

REFERENCES

ACKER, ARNOLD E., 1982. "Interest in mm waves spurs tube growth," *Microwaves*, July.

CHODOROW, MARVIN, AND CHARLES SUSSKIND, 1964. *Fundamentals of Microwave Electronics*. New York: McGraw-Hill Book Company.

COLEMAN, JAMES, 1982. *Microwave Devices*. Reston, Va.: Reston Publishing Company, Inc.

GANDHI, OM P., 1981. Microwave Engineering and Applications. Pergamon Press.

GITTINS, J. F., 1965. *Power Traveling—Wave Tubes*. New York: American Elsevier Publishers, Inc.

HANSEN, JAMES W., 1984. "Eliminate confusion in TWTA specifications," *Microwaves & RF*, July.

Hughes TWT & TWTA Designer's Handbook. Torrance, Calif: Hughes Electron Dynamics Division.

LANCE, ALGIE L., 1964. Introduction to Microwave Theory and Measurements. McGraw-Hill.

MALONEY, E. D., AND G. FAILLON, 1974. "A high-power klystron for industrial processing using microwaves," *J. Microwave Power*, Vol. 9, No. 3.

OGURO, T., 1978. "Trends in magnetrons for consumer microwave ovens," *Microwave Power*, Vol. 13, No. 1.

SHIBATA, C., T. ALKIOKA, Y. SATO, AND H. TAMAI, 1978. "100 kW, 915 MHz CW magnetron for industrial heating application," *Microwave Power*, Vol. 13, No. 1.

SKOWRON, JOHN F., 1973. "The continuous-cathode (emitting-sole) crossed-field amplifier," *Proc. IEEE*, Vol. 61, No. 3.

SMITH, M. J., 1981. "The gyrotron," *Electronics & Power*, May.

VARIAN, RUSSELL H., AND SIGURD F. VARIAN, 1939. "A high frequency oscillator and amplifier," *J. Appl. Phys.*, Vol. 10.

WILLARD, J., AND C. JOSLIN, 1984. "Modern medium-power traveling-wave-tube amplifiers for communications," *Radio Electr. Eng.*, Vol. 54, No. 7/8.

ten
Microwave Antennas

10.1 INTRODUCTION

The use of microwave energy generally requires, at some stage, transfer of the energy by means of radiation. In microwave heating, the energy is usually radiated from a horn antenna into the surrounding space, where it is absorbed by the body to be heated. In a communications link, the microwave signal is radiated into space by an antenna at one end, and picked out of space by an antenna at the other end. The same antenna may be used for both transmission and reception.

Antennas may be constructed from an assemblage of conducting wires or rods, a familiar example being the household TV antenna. At microwaves, however, even small discontinuities in feeder lines and mountings can give rise to reflections and phase changes which significantly alter the antenna characteristics. Because of the short wavelengths involved, microwave antennas are more efficiently designed and constructed from apertures, lenses, and mirror-type reflectors, using techniques similar to those used in optics.

10.2 ANTENNA EFFICIENCY

An antenna is used to transfer energy to and from space, and in the process of doing this, a certain amount of the energy is dissipated as heat in the antenna structure. This results both from the conduction losses in the metallic parts of the antenna, and from induced losses in dielectric supports. The efficiency can be defined generally as

$$\eta_A = \frac{\text{useful power}}{\text{useful power} + \text{losses}} \tag{10.1}$$

In the case of a transmitting antenna, the useful power is the power that gets radiated into space, assuming that the antenna is matched to the feeder. In the case of a receiving antenna, the useful power is the power that would be delivered to a matched load from a lossless antenna. With microwave antennas it may be difficult to identify the individual sources of loss, and determination of efficiency depends on being able to measure total power input and power radiated. One consequence of an important theorem known as the reciprocity theorem is that for a given antenna, the efficiency is the same for transmit and receive modes of operation.

Mismatch between the antenna and the feeder is another source of power loss, which is taken into account separately. Any mismatch between antenna and feeder can be measured in terms of the reflection coefficient, as described in earlier chapters. The matching efficiency is given by

$$\eta_\Gamma = 1 - \rho_L^2 \tag{10.2}$$

where ρ_L is the magnitude of the reflection coefficient.

Again, this applies for both transmit and receive modes of operation. In the transmit case it is assumed that the feeder is correctly matched to the transmitter, and in the receive case, that it is correctly matched to the receiver. In other words, the only mismatch is that between feeder and antenna. As shown in Chapters 1 and 4, stubs may be inserted between an antenna and feeder to reduce the mismatch, and ideally the reflection coefficient should be zero.

10.3 ANTENNA GAIN AND DIRECTIVITY

All practical antennas radiate more effectively in some directions than others. Although the directional property is defined in terms of the radiation pattern of the antenna, a result of the reciprocity theorem is that the same directional pattern applies to the antenna when receiving. A good example of this is the parabolic dish antenna used for home reception of satellite TV. The antenna has to be pointed directly at the satellite for maximum reception. Thus, although the gain of an antenna is defined in terms of radiated power, the same gain figure applies to reception.

Most antennas have a well-defined maximum in their radiation pattern, and the gain G_m is defined for this maximum. A reference antenna is required and a hypothetical radiator, known as an isotropic radiator, is used. *Isotropic* simply means "equally in all directions." Furthermore, since this is a hypothetical radiator, it may be assumed lossless. The gain of the actual antenna is then defined for the direction of maximum radiation as: the ratio of the power per unit solid angle radiated by the actual antenna to the power per unit solid angle radiated by the lossless isotropic radiator. Now, although the hypothetical nature of the reference antenna makes it impossible to realize this in practice, the theoretical gain of certain simple antenna types can be calculated. These antennas can be accurately constructed and used as standards. One such antenna is the pyramidal horn shown in Fig. 10.1. The gain of this horn is 17.2 dB at the design frequency of 9.375 GHz.

The directivity D_m of an antenna is the ratio of the maximum to the average power per unit solid angle radiated by the actual antenna. This average is equal to the isotropic value divided by the antenna efficiency and therefore

$$D_m = \frac{G_m}{\eta_A} \tag{10.3}$$

It will be seen that both gain and directivity refer to power ratios. The gain is the usual figure entered in specification sheets for antennas, and it must be understood that this refers to the gain in the direction of maximum radiation or reception. Also, it is only a gain in the sense that the antenna concentrates or focuses the power in this direction; it does not increase the total power radiated. The falloff in gain for other directions

FIGURE 10.1. Pyramidal horn antenna that can be used as a standard. The isotropic gain is 17.2 dB at a design frequency of 9.375 GHz. (*Courtesy of Marconi Instruments Ltd., Stevenage, Hertfordshire, England.*)

is taken into account through the radiation pattern $g(\theta, \phi)$, where θ and ϕ are the angular spherical coordinates of direction. The gain in any given direction, denoted by $G(\theta, \phi)$, is then

$$G(\theta, \phi) = G_m g(\theta, \phi) \tag{10.4}$$

The subscript m has been used above to denote maximum. However, following standard practice, the subscript will be dropped. The gain for a transmitting antenna will be denoted by G_T, and for a receiving antenna G_R. Where it is not necessary to distinguish between transmit and receive conditions, the symbol G will be used. For any given antenna,

$$G_T = G_R = G \tag{10.5}$$

10.4 EFFECTIVE AREA OF APERTURE

A receiving antenna may be thought of as a collector of electromagnetic energy rather as a solar collector collects energy from sunlight. As an electromagnetic wave sweeps over the antenna, the power density of the wave, measured in watts per square meter, gives rise to a power flow, measured in watts, into the receiving load connected to the antenna. The antenna may therefore be thought of as having an effective area which relates the power density to the power by an equation of the form

$$P_R = P_D A_{\text{eff}} \tag{10.6}$$

Here P_R is the power in watts delivered to a matched load, P_D the power density in W/m^2 of the wave, and A_{eff} the effective area in square meters. Equation (10.6) applies when the antenna is pointing in the direction for maximum reception. In any other direction the power will be altered in amount by the factor $g(\theta, \phi)$ as defined in Section 10.3. Thus A_{eff} is the maximum effective area analogous to maximum gain, and the effective area in any given direction is

$$A(\theta, \phi) = A_{\text{eff}} g(\theta, \phi) \tag{10.7}$$

When effective area is referred to, the maximum value A_{eff} is implied.

The effective area is proportional to the power gain, and it has been established [see e.g. Glazier and Lamont (1958)] that for any antenna,

$$\frac{A_{\text{eff}}}{G_M} = \frac{\lambda^2}{4\pi} \tag{10.8}$$

This result is of fundamental importance, as it enables the antenna as a receiver to be characterized in terms of its power gain as a transmitter. An example of its use will be seen in the derivation of Eq. (11.5).

It may be thought that the effective area of an aperture antenna should be equal to the physical area of the aperture. This is not so, because the antenna itself produces reflections that alter the way in which the wave "illuminates" the aperture. For example, the pyramidal horn (see Section 10.8) has an effective area given approximately by

$$A_{\text{eff}} \simeq 0.8ab \tag{10.9}$$

where ab is the length $\times$ breadth of the physical aperture.

10.5 RADIATION PATTERN AND SIDELOBES

The function $g(\theta, \phi)$, termed the radiation pattern, gives a three-dimensional plot of the power density in the far-field zone. In the far-field zone only one component of the total radiation retains significance. This is a plane transverse electromagnetic wave which decreases as the distance squared. Radio communication links make use of the far-field radiation. An empirical rule that gives the distance from the antenna to the far-field zone is

$$d \gtrsim \frac{2D^2}{\lambda} \tag{10.10}$$

Here d is distance, λ is wavelength, and D is the largest antenna dimension. This rule applies for $D >> \lambda$.

For most microwave antennas, $g(\theta, \phi)$ consists of a main lobe and a number of sidelobes, as sketched in Fig. 10.2. The mainlobe is usually very narrow, being referred to as a pencil beam. This is difficult to show accurately in a polar plot, and Cartesian plots are normally used to show detail. Two Cartesian planes are normally chosen, one containing the electric-field or E-vector direction, the other the magnetic-field or H-vector direction, and both planes containing the line of maximum radiation. Figure 10.3 shows both the *E*-plane and *H*-plane radiation patterns for a pyramidal horn.

One has to be careful to note whether the actual radiation pattern $g(\cdot)$, which is a power-gain ratio, is being plotted, or a field-strength ratio. Field strength is proportional to the square root of power density, and hence the field-strength radiation pattern is proportional to the square root

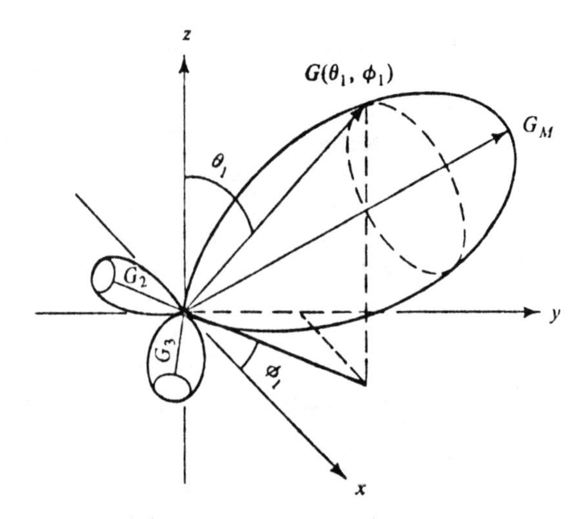

FIGURE 10.2. Gain function $G(\theta, \phi)$. The radiation pattern is $g(\theta, \phi) = G(\theta, \phi)/G_M$.

of $g(\theta, \phi)$. By plotting the radiation pattern in decibels, this difficulty is avoided:

$$g(\cdot)(\text{dB}) = 10 \log g(\cdot) \tag{10.11}$$

Figure 10.4 shows the E-plane graph of Fig. 10.3 replotted using decibels. Note that the nulls which go to zero in Fig. 10.3 cannot be shown in Fig. 10.4 since $\log 0 = -\infty$.

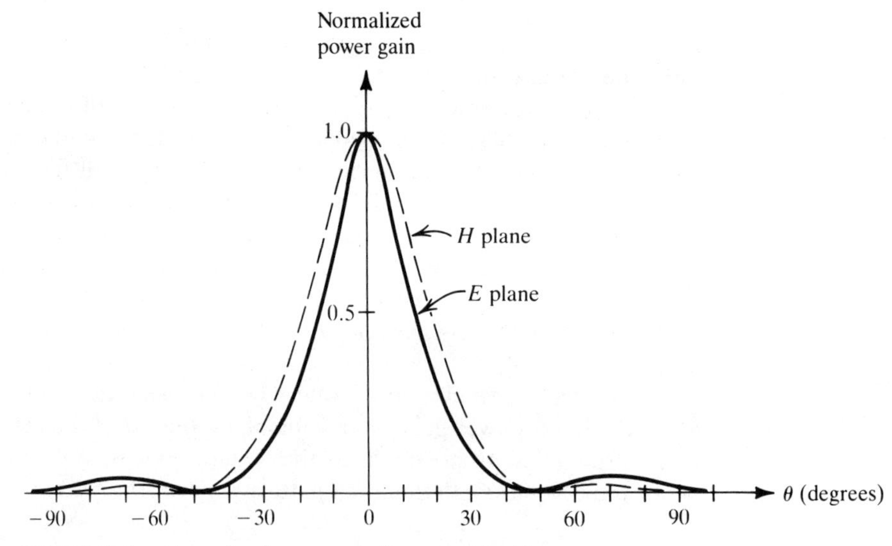

FIGURE 10.3. Normalized power gain for a pyramidal horn.

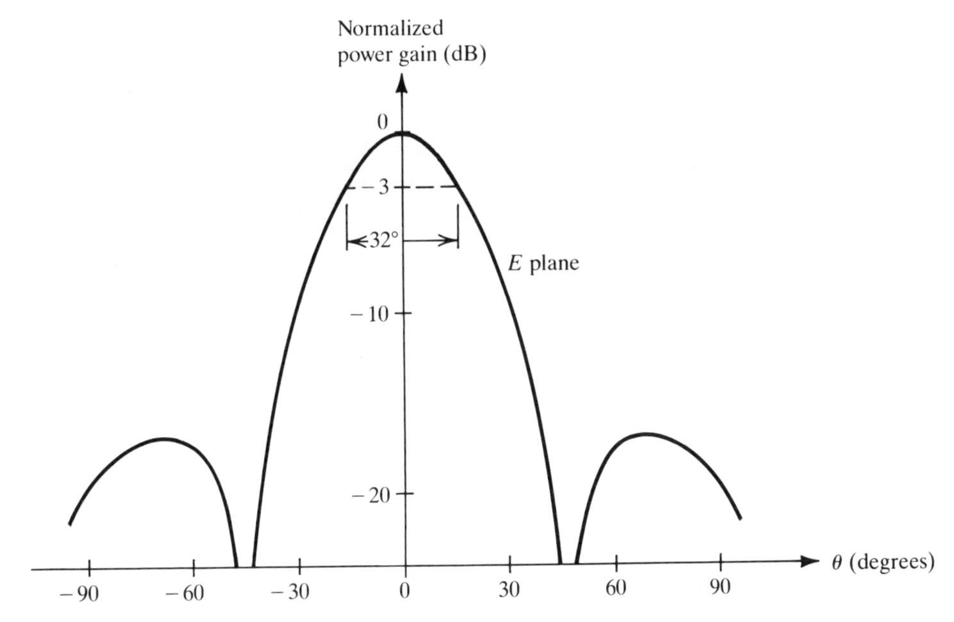

FIGURE 10.4. Normalized power gain in decibels.

The beamwidth of an antenna is usually specified as the total angle encompassed by the −3-dB levels as shown in Figure 10.4. However, other levels are frequently used, for example −10 dB and −60 dB, and sometimes the first nulls either side of the main peak are used to specify beamwidth.

Often, the detailed shape of $g(\theta, \phi)$ is not as important as the level at which the sidelobes are below the main beam. This level can be expressed by the envelope curve, which fits over the peaks as shown in Figure 10.5. For satellite systems, in particular, the sidelobe level is a major factor in determining the interference between systems. For example, with satellites spaced by 4° in the geostationary orbit, the envelope curve that is accepted by most licensing authorities is given by

$$G(\theta)(\text{dBi}) = \begin{cases} 32 - 25 \log \theta & 1^\circ < \theta < 48^\circ \\ -10 & 48^\circ < \theta < 180^\circ \end{cases} \qquad (10.12)$$

With the move to 2° orbital spacing, the envelope specifications for the sidelobes closest to the main peak have had to be tightened to

$$G(\theta)(\text{dBi}) = 29 - 25 \log \theta \qquad 1^\circ < \theta < 20^\circ \qquad (10.13)$$

It should be noted here that $G(\theta)(\text{dBi})$ is the isotropic gain in decibels, and these equations apply only to the envelope curves, not to the fine detail of the gain curve. The isolation $I(\theta)$ that the antenna provides is the

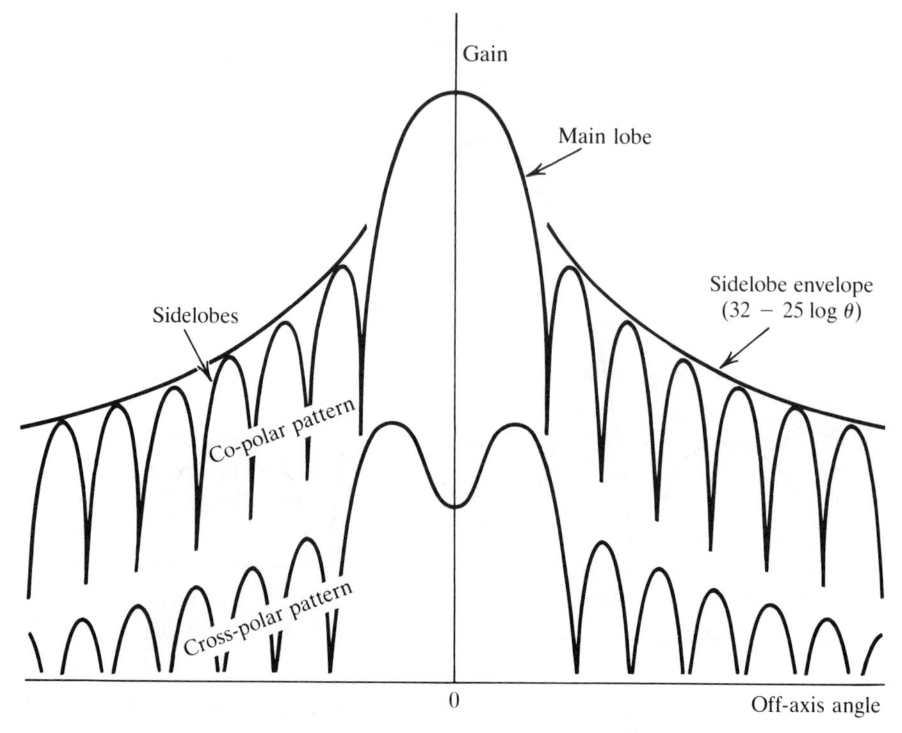

FIGURE 10.5. Antenna gain patterns and envelope curve. (*From Report FCC/OST R83-2, May 1983, prepared by George L. Sharp.*)

difference between the on-axis gain G_0 and the off-axis gain $G(\theta)$, both expressed in decibels:

$$I(\theta) = G_0 - G(\theta) \qquad \text{dB} \tag{10.14}$$

10.6 POLARIZATION

In the far-field zone, the *polarization* of the wave is defined by the direction of the electric field vector in relation to the direction of propagation. *Linear* polarization is when the electric vector remains in the same plane, as shown in Fig. 10.6(a). A linear polarized wave that is propagated across the earth's surface is said to be *vertically polarized* when the electric field vector is vertical, and *horizontally polarized* when it is parallel, to the earth's surface. For example, in North America, television transmissions are horizontally polarized, and it will be observed that receiving antennas are also horizontally mounted, whereas in the United Kingdom vertical polarization is used, and there, antennas are mounted vertically.

In certain situations the electric vector may rotate about the line of

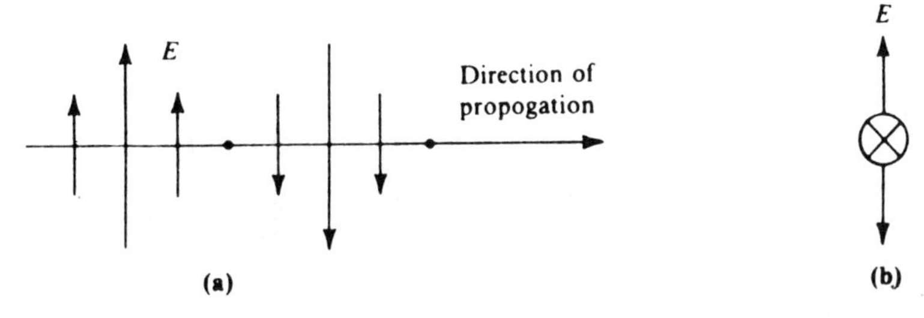

FIGURE 10.6. Linear polarization: (a) as viewed on axis of propagation; (b) as viewed along direction of propagation.

propagation. This can be caused, for example, by the interaction of the wave with the earth's magnetic field in the F_2 layer of the ionosphere. Rotation of the electric vector can also be produced by the type of antenna used, and this effect is put to good use in satellite communications. The path traced out by the tip of the electric vector may be an ellipse, as illustrated in Fig. 10.7, in which case it is referred to as *elliptical polarization*. If the rotation is in a clockwise direction when looking along the direction of propagation, the polarization is referred to as *right-handed*; if it is anticlockwise, it is called *left-handed*. In Fig. 10.7(a) the direction of propagation is into the paper, so the polarization is right-handed. A special case of elliptical polarization is *circular polarization*, as illustrated in Fig. 10.8, and both right-handed and left-handed circular polarization are used in satellite communications systems as described later. Linear polarization can also be considered to be a special case of elliptical polarization. As shown in Fig. 10.7, elliptical polarization can be resolved into two linear vectors, E_x and E_y. Linear polarization results when one of these components is zero.

In order to receive a maximum signal, the polarization of the receiving

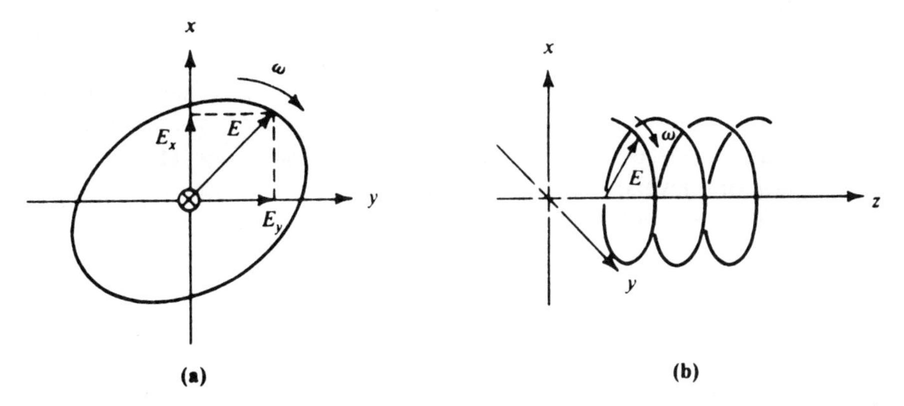

FIGURE 10.7. Elliptical polarization: (a) as viewed along direction of propagation; (b) as viewed on axis of propagation.

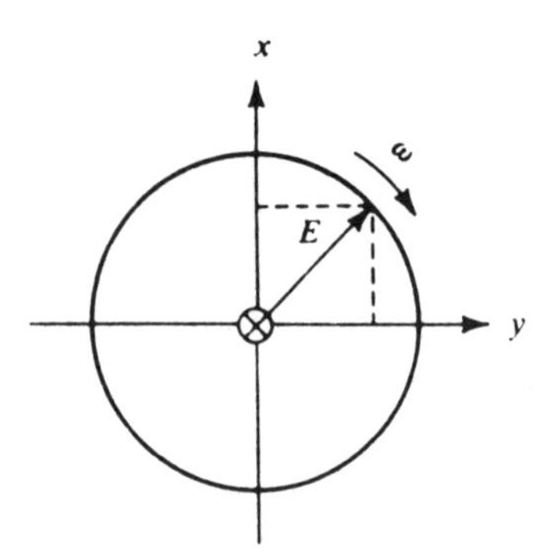

FIGURE 10.8. Circular polarization.

antenna must be the same as that of the transmitting antenna, which is defined to be the same as that of the transmitted wave. For example, a wire dipole antenna, illustrated in Fig. 10.9(a), will radiate a linear polarized wave. A similar receiving dipole must be oriented parallel to the electric vector for maximum reception. If it is at some angle ψ, as illustrated in Fig. 10.9(b), then only the component of the electric field parallel to the receiving antenna will induce a signal in it. This component is $E \cos \psi$, and therefore the polarization loss factor is

$$\text{PLF} = \cos \psi \tag{10.15}$$

A similar situation can exist with an aperture antenna, as illustrated in Fig. 10.9(c). The angle ψ is the angle between the induced field in the aperture and the incoming electric field, and the polarization loss factor is also $\cos \psi$ in this case. In both cases the direction of propagation is normal to the plane of the antenna.

Cross polarization is used in satellite communications to permit frequency reuse. If the main beam is aligned for reception of vertical polarization, it should provide isolation against horizontally polarized signals on the same frequency. A similar situation should exist for right-hand and left-hand circular polarization. Figure 10.5 shows typical cross-polarization isolation levels. It will be seen that the cross-polarization isolation at the sidelobes is not as great as that for the mainlobe.

10.7 THE HALF-WAVE DIPOLE

The half-wave dipole consists of two conductors lying in a straight line, as shown in Fig. 10.10, the overall length from tip to tip being approximately one-half wavelength. The spacing between the conductors should be negligible compared to the overall length, and this may be difficult to achieve

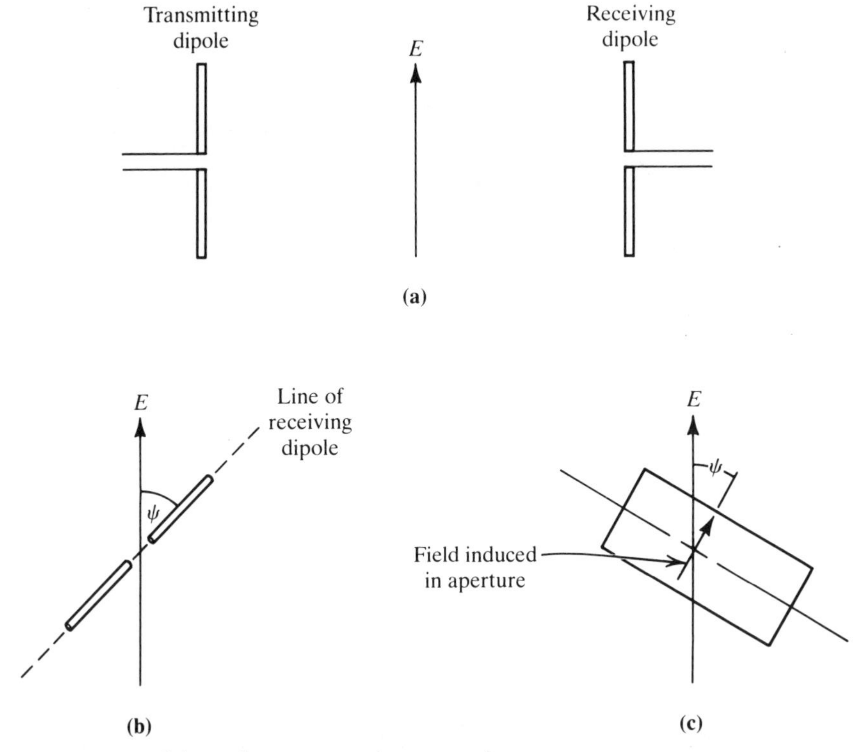

FIGURE 10.9. Two dipoles aligned with same polarization; (b) receiving dipole in same plane as E but polarization misaligned; (c) incoming wave E in same plane as aperture but aperture polarization misaligned.

at microwaves. The wave velocity along the antenna wire is slightly less than that of free space, and as a result the wavelength is slightly less than about 95% of the free-space value. The antenna must be cut to this shorter length for resonance as illustrated in Fig. 10.10. The impedance of the antenna under these conditions is 73 Ω resistive.

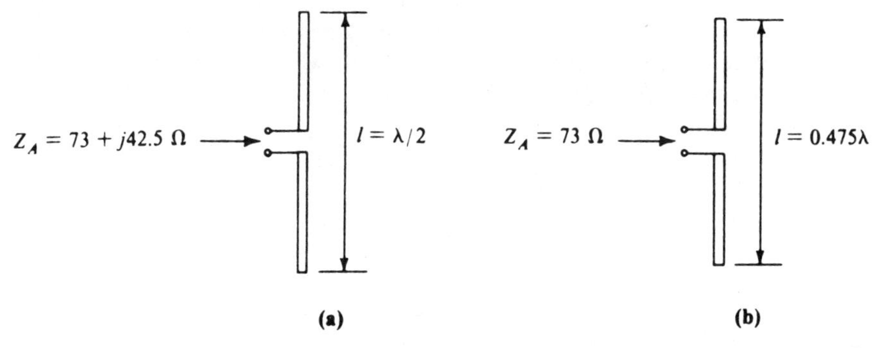

FIGURE 10.10. (a) Input impedance of half-wave dipole cut exactly to $\lambda/2$ is $73 + j42.5\ \Omega$; (b) cutting the dipole about 5% shorter reduces reactive component to zero.

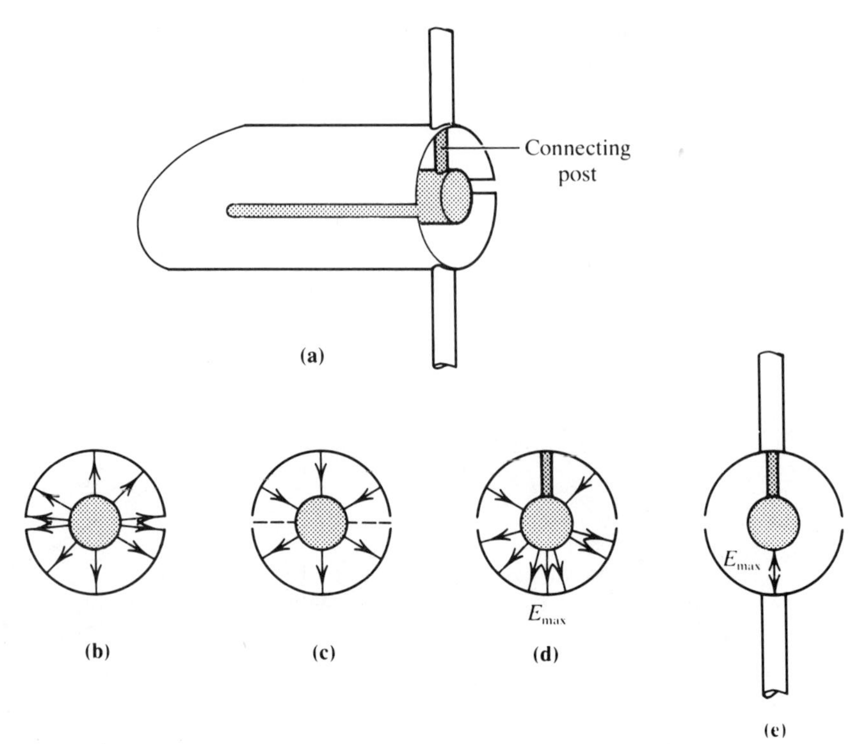

FIGURE 10.11. (a) Dipole slot fed; (b) coaxial TEM wave; (c) reflected TE_{11} wave; (d) combined wave resulting in zero field across the post and maximum field diametrically opposite; (e) excitation of the dipole through E_{max}.

The dipole is a balanced antenna and should be fed from a balanced feeder. Where coaxial feeder is used, a balanced-to-unbalanced feed is required. The slot feed provides one method of achieving this, one form of slot feed being shown in Fig. 10.11. A narrow slot is cut along the outer conductor of the coaxial feed as shown in Fig. 10.11(a), and the dipole is connected at the ends of the slotted section. The inner conductor is connected to the end of one half the slotted section, the connecting post usually being in line with the dipole. A description of how the arrangement functions is given in Silver (1949), on which the following is based. Assuming that the slot width is very much smaller than a wavelength, the TEM wave on the coaxial line will extend into the slotted section as shown in Fig. 10.11(b). Assuming also that the connecting post is a perfect conductor, the electric field along this must be forced to zero. This condition can be met by the post giving rise to a reflected wave in the TE_{11} mode. The electric field for the TE_{11} mode is shown in Fig. 10.11(c). The combined TEM field and TE_{11} field is shown in Fig. 10.11(d). It will be seen that a zero field exists along the connecting post as required by the boundary conditions, and a maximum field exists between inner and outer conductors diametrically opposite the post. Since the top element of the dipole is connected via the post to the inner conductor, and the lower element is connected to the outer conductor, E_{max} also appears across the dipole, as shown in Fig. 10.11(e).

The dipole may also be excited from a waveguide feed, as shown in Fig. 10.12(a). The dipole protrudes from a metal plate, which is fitted into the mouth of the guide as shown. The guide is tapered to improve the match between guide and dipole. The taper also decreases the coupling between the dipole and the outer wall of the guide, which improves the radiation pattern. The dipole is mounted parallel to the electric field vector of the TE_{10} mode in the guide, and with the metal plate inserted sym-

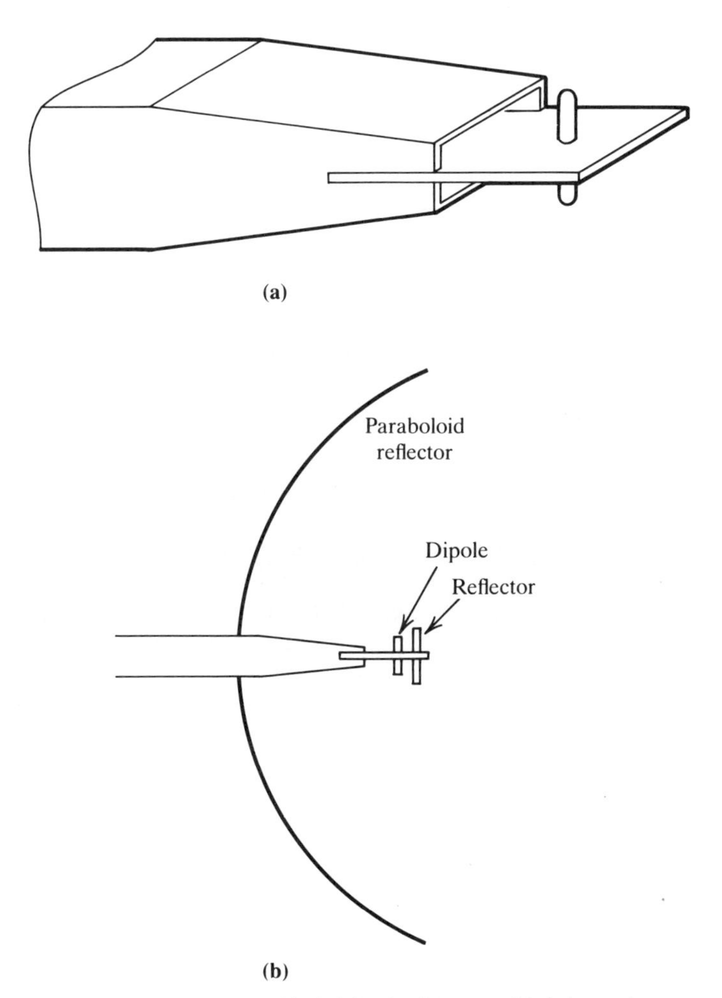

FIGURE 10.12. (a) Waveguide fed dipole; (b) waveguide fed two-element dipole array used as a primary antenna for a paraboloid reflector.

metrically as shown, radiation from the guide excites both halves of the dipole equally.

The half-wave dipole is seldom used by itself as the main antenna in microwave systems. More usually, it is used as a feed antenna for reflector systems, as shown in Fig. 10.12(b), or it may form the basic element in a microstrip array, as described later. For these practical arrangements, the feed to the dipole forms a significant part of the antenna, and the theoretical results for the half-wave dipole only provide a guide to antenna performance at microwaves. For a half-wave dipole made of thin wires or tubes, for which the overall radius is negligible compared to length, and for which the feed gap is negligibly small, the main theoretical properties are as summarized in Table 10.1.

TABLE 10.1
Properties of the Half-Wave Dipole

Resonant impedance	73 Ω
Isotropic power gain	2.16 dB
−3-dB beamwidth	78°
Effective area	$0.13\ \lambda^2$
Effective length	λ/π

The effective length of an antenna, l_{eff}, relates the open-circuit emf V_A appearing at the antenna terminals to the electric field strength E of the wave as

$$V_A = El_{\text{eff}} \tag{10.16}$$

It will be seen that the relationship of effective length to voltage and field strength is analogous to that of effective area to power and power density. At microwaves, effective area is the more useful of the two.

With the angles θ and ϕ defined as shown in Fig. 10.13, the radiation pattern in the equatorial plane is simply a circle of unity radius, or

$$g(\phi) = 1 \tag{10.17}$$

In the meridian plane, the electric field radiation pattern is given by

$$F(\theta) = \frac{\cos\left[(\pi/2 \cos\theta)\right]}{\sin\theta} \tag{10.18}$$

The power radiation pattern is this function squared:

$$g(\theta) = F^2(\theta) \tag{10.19}$$

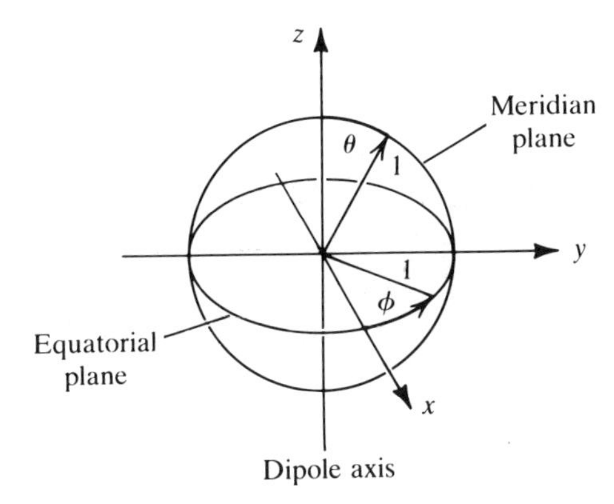

FIGURE 10.13. Planes selected for the dipole radiation patterns $g(\phi)$ and $g(\theta)$.

10.7.1 The Cylindrical Dipole

A great deal of work has been carried out both theoretically and experimentally to establish the characteristics of the cylindrical dipole, taking into account the finite radius of the conductors. The model dipole is shown in Fig. 10.14. It consists of two cylindrical tubes of negligible thickness, and radius a. The tubes are assumed to be perfectly conducting, and the feed gap is assumed negligibly small. A formula for the terminal impedance

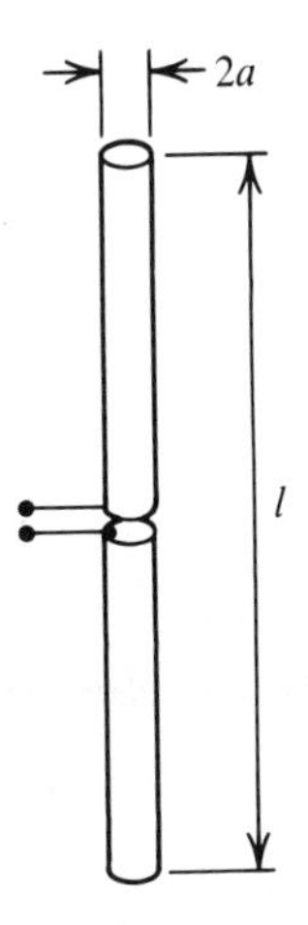

FIGURE 10.14. Cylindrical dipole. The tubes are of radius a. Tube thickness and feedgap width are assumed negligible.

of such a dipole, valid for the practical range of values $0.001588 \leq a/\lambda \leq 0.009525$ and $0.4138 \leq l/\lambda \leq 0.5411$, is given in Eliott (1981) as

$$R_a = 122.65 - 641.2\frac{l}{\lambda} + 1085.66\left(\frac{l}{\lambda}\right)^2 \tag{10.20}$$

$$X_a = 162.5 - 120\left(\ln\frac{l}{a} - 1\right)\cot\frac{\pi l}{\lambda} - 439.82\frac{l}{\lambda} + 394.78\left(\frac{l}{\lambda}\right)^2 \tag{10.21}$$

The antenna impedance is

$$Z_a = R_a + jX_a \tag{10.22}$$

EXAMPLE 10.1

Calculate the impedance of a half-wave dipole for which $a/\lambda = 0.002$.

SOLUTION From Eq. (10.20),

$$R_a = 122.65 - \frac{641.2}{2} + \frac{1085.66}{4}$$
$$= \mathbf{73.46\ \Omega}$$

From Eq. (10.21),

$$X_a = 162.5 - 0 - \frac{439.82}{2} + \frac{394.78}{4}$$
$$= \mathbf{41.28\ \Omega}$$

Therefore,

$$Z_a = \mathbf{73.46 + \mathit{j}41.28\ \Omega}$$

Note that this is close to the value shown in Fig. 10.10, for which the radius was assumed negligibly small. It will be observed from Eqs. (10.20) and (10.21) that for the range of values stated, the radius a affects only the reactance term, and then only when the length is other than half a wavelength.

10.8 HORN ANTENNAS

Electromagnetic energy can be transmitted and received through the open end of a waveguide. However, this is not an efficient way of coupling a waveguide to space, as the sudden change in cross section area results in reflections. More efficient coupling can be achieved by flaring the waveguide, the resulting structure being known as a waveguide horn. Figure 10.15(a) shows an *H*-plane sectoral horn, so named because the flare is in the same direction as the magnetic field vector at the center of the aperture. The waveguide dimension a' is expanded to a, while the waveguide dimension b remains constant. The slant length is l_H as shown in Fig. 10.15(a). Figure 10.15(b) shows an *E*-plane sectoral horn, the flare in this case being in the same direction as the electric field vector. Waveguide dimension b' is expanded to b, while dimension a remains constant. The slant length is l_E, as shown in Fig. 10.15(b). When both dimensions are flared, the horn is referred to as a pyramidal horn, as shown in Fig. 10.1.

The gain of a horn antenna depends on the wavelength, the aperture dimensions a and b, and the slant length. As shown in Jull (1970), the isotropic power gain for a rectangular horn is given by

$$G = G_0 R_E R_H \tag{10.23}$$

Here R_E and R_H are gain reduction factors which take into account the *E*- and *H*-plane flare of the horn, and also the effect of finite radiation distance. Values for R_E and R_H are tabulated in the paper by Jull and presented in graphical form in Fig. 10.16. The abscissa in Fig. 10.16 is a generalized coordinate. When using the R_E curve, it is necessary to substitute a for d, and l'_E for l, and for the R_H curve, to substitute b for d and l'_H for l. The parameters l'_E and l'_H take into account the range r and the slant heights as

$$l'_E = \frac{r l_E}{r + l_E} \tag{10.24}$$

and

$$l'_H = \frac{r l_H}{r + l_H} \tag{10.25}$$

G_0 is the far-field gain of a rectangular aperture given by

$$G_0 = \frac{32ab}{\pi\lambda^2} \tag{10.26}$$

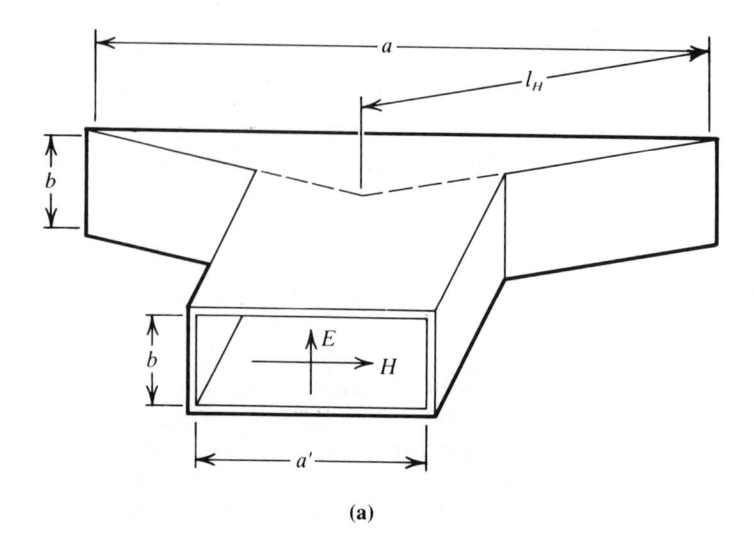

(a)

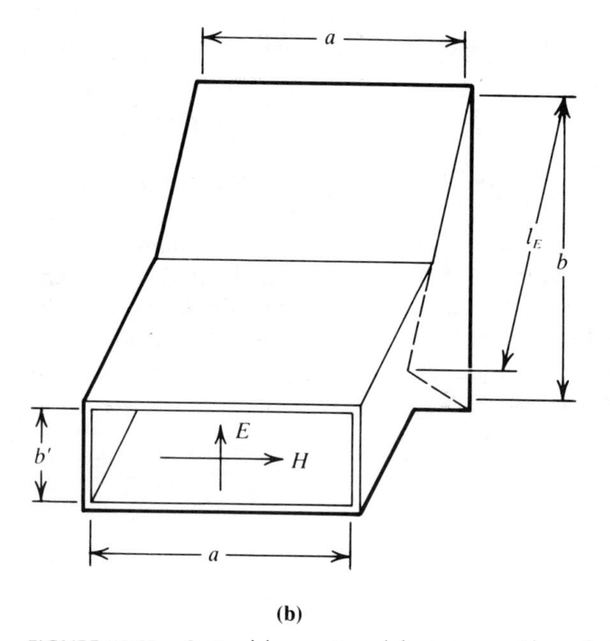

(b)

FIGURE 10.15. Sectoral horns viewed from waveguide end: (a) *H*-plane horn; (b) *E*-plane horn.

Expressed in decibels, the gain G_0 is

$$G_0(\text{dB}) = 10 \log \frac{32ab}{\pi\lambda^2} \tag{10.27}$$

Then the gain of pyramidal horn in decibels is

$$G(\text{dB}) = G_0(\text{dB}) + R_E(\text{dB}) + R_H(\text{dB}) \tag{10.28}$$

These data can also be used to find the gain of sectoral horns. For the E-plane sectoral horn,

$$G_E(\text{dB}) = G_0(\text{dB}) + R_E(\text{dB}) \tag{10.29}$$

and for the H-plane sectoral horn,

$$G_H(\text{dB}) = G_0(\text{dB}) + R_H(\text{dB}) \tag{10.30}$$

Gain calculations can be made very accurately for horn antennas, and as a result they are used extensively as standard gain antennas, to which the gain of other antennas can be compared. The pyramidal horn shown in Fig. 10.1 is such a standard antenna.

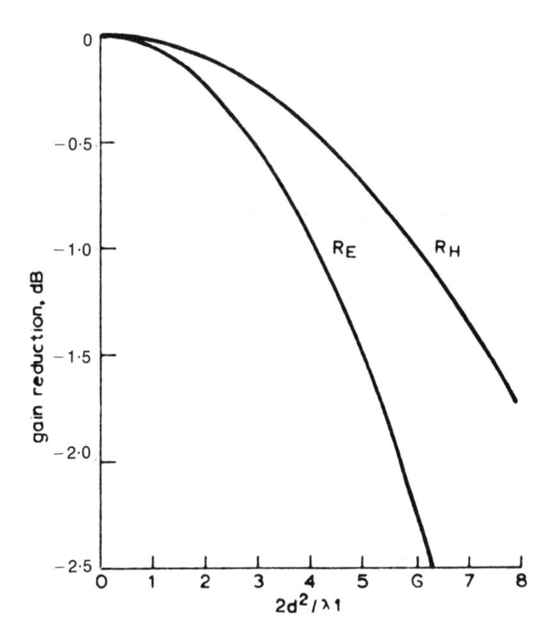

FIGURE 10.16. Gain reduction factors R_S and R_H in decibels. *(From E. V. Jull, "Finite range gain of sectional and pyramidal horns," Electr. Lett., Vol. 6, IEE, Oct. 1970; with permission.)*

EXAMPLE 10.2

A pyramidal horn antenna has the following dimensions: $a/\lambda = 3.2$, $a/l_E = 0.782$, $b/\lambda = 2.4$, and $b/l_H = 0.521$. Calculate the far-field gain.

SOLUTION For the far-field gain, $l_E = l'_E$ and $l_H = l'_H$. The abscissa values $s = 2d^2/\lambda l$ in each case are

$$s_a = 2\frac{a}{\lambda}\frac{a}{l_E} = 2 \times 3.2 \times 0.782 = 5$$

$$s_b = 2\frac{b}{\lambda}\frac{b}{l_H} = 2 \times 2.4 \times 0.521 = 2.5$$

From Fig. 10.16, $R_E = -1.52$ dB and $R_H = -0.18$ dB.

The gain G_0 is calculated from Eq. (10.26) as

$$G_0 = \frac{32}{\pi} \times 3.2 \times 2.4$$

$$= 78.23$$

In decibels this is $G_0(\text{dB}) = 18.93$ dB. Hence, using Eq. (10.28), we obtain

$$G(\text{dB}) = 18.93 - 1.52 - 0.18$$

$$= \mathbf{17.23\ dB}$$

Various other shapes of horn antennas are available, such as the conical horn, the diagonal horn, the octagonal horn, and so on. However, the sectoral horns and the pyramidal horn are the most widely used.

10.9 CORRUGATED HORN ANTENNAS

The horn antenna is frequently used as a primary feed antenna for parabolic reflector antennas. The simple horn has two disadvantages as a feed, these being that the sidelobes are relatively large and that the pattern is not symmetric in the E and H planes.

It has been found that by corrugating the insides of the horn as shown in Fig. 10.17(a), hybrid modes of propagation are set up in the horn which result in a more nearly uniform illumination of the reflector, and also a considerable reduction in the "spillover," so that both the efficiency of the antenna is increased and the sidelobe levels are reduced. Figure 10.17(b)

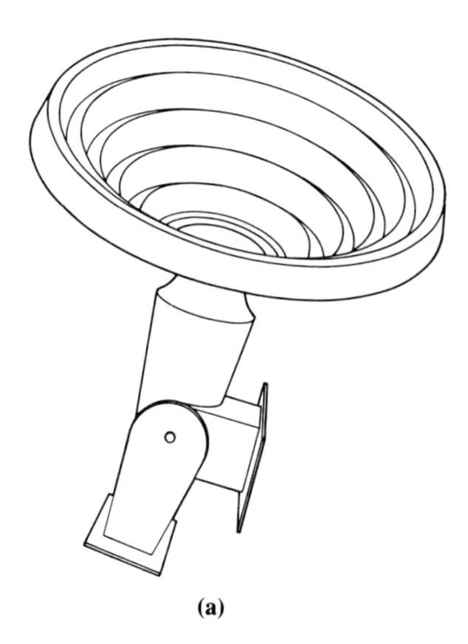

(a)

(b)

FIGURE 10.17. (a) Typical scalar feed; (b) commercial feed horn for a parabolic reflector. [*Part (a) from A. J. Simmons and A. F. Kay, "The Scalar Feed—A High Performance Feed for Large Parabolic Reflectors," in Electromagnetic Horn Antennas, ed. A. W. Low (IEEE Press, 1976); with permission; part (b) courtesy of General Instrument, RF Systems Division, Dehli, Ontario.*]

shows a commercially available version of a corrugated horn where the corrugations form a ground plane around the radiating aperture. Further details of corrugated horn design will be found in Love (1976).

10.10 THE PARABOLOIDAL REFLECTOR ANTENNA

The most widely used antenna for microwaves is the paraboloidal reflector antenna, which consists of a primary antenna such as a dipole or horn situated at the focal point of a paraboloidal reflector, as shown in Fig. 10.18(a). The mouth, or physical aperture, of the reflector is circular, and the reflector contour, when projected onto any plane containing the focal point F and the vertex V, forms a parabola as shown in Fig. 10.18(b). The path length $FAB = FA'B'$ for this curve, where the line BB' is perpendicular to the reflcctor axis. The important practical implication of this property is that the reflector can focus parallel rays onto the focal point, and conversely, it can produce a parallel beam from radiation emanating from the focal point. Figure 10.18(c) illustrates this. An isotropic point source is assumed to be situated at the focal point. In addition to the desired parallel beam being formed, it can be seen that some of the rays are not captured by the reflector, and these constitute "spillover." In the receive mode, spillover increases noise pickup, which can be particularly troublesome in satellite ground stations. Also, some radiation from the primary radiator occurs in the forward direction in addition to the desired parallel beam. This is termed *blacklobe radiation* since it is from the backlobe of the primary radiator. Backlobe radiation is undesirable because it can interfere destructively with the reflected beam, and practical radiators are designed to eliminate or minimize this. The isotropic radiator at the focal point will radiate spherical waves, and the paraboloidal reflector converts these to plane waves. Thus, over the aperture of an ideal reflector, the wavefront is of constant amplitude and constant phase.

The directivity of the paraboloidal reflector is a function of the primary antenna directivity and the ratio of focal length to reflector diameter, f/D. This ratio, known as the aperture number, determines the angular aperture of the reflector, 2Ψ [Fig. 10.19(a)], which in turn determines how much of the primary radiation is intercepted by the reflector. Assuming that radiation from the primary antenna is circularly symmetric about the reflector axis $(F - V)$ and is confined to angles ψ in the range $-\pi/2 < \psi < \pi/2$, it is found that the effective area is given by

$$A_{\text{eff}} = AI(\theta) \tag{10.31}$$

where $A = \pi D^2/4$ is the physical area of the reflector aperture and $I(\theta)$

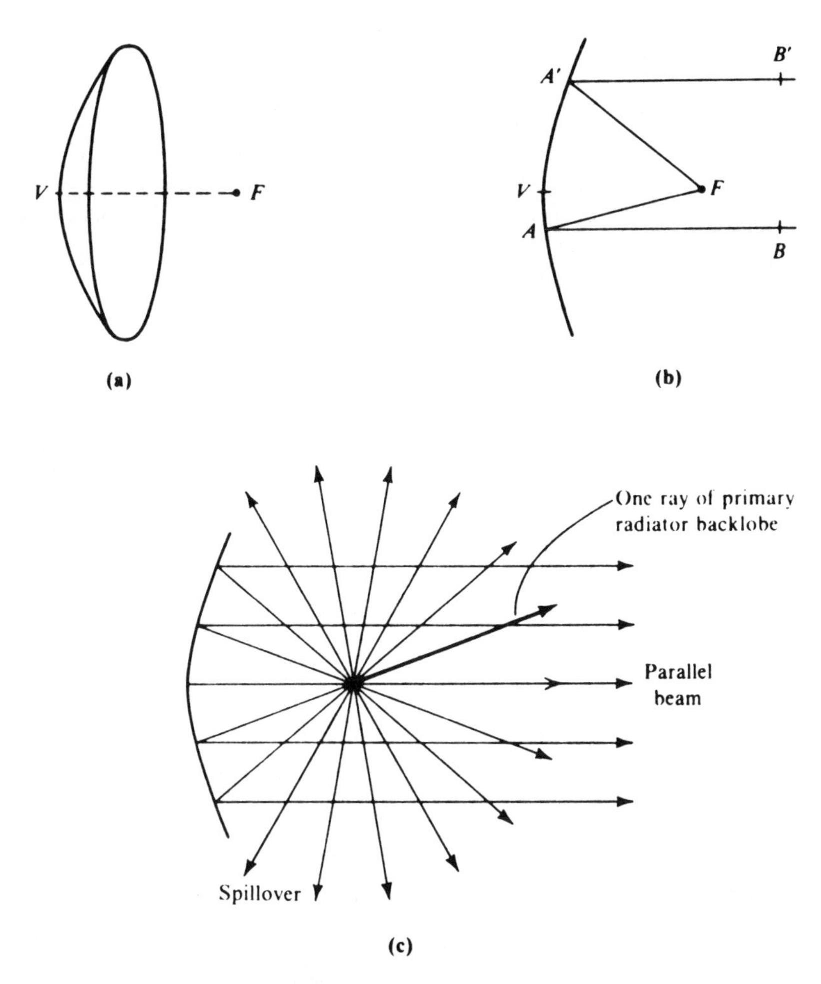

FIGURE 10.18. (a) Paraboloidal reflector; (b) the parabola; (c) radiation from the paraboloidal reflector and primary radiator.

is a function, termed the *aperture efficiency* (or *illumination efficiency*), which takes into account both the radiation pattern of the primary radiator and the effect of the angular aperture. With the focal point outside the reflector, as shown in Fig. 10.19(a) (which requires $f/D > \frac{1}{4}$), the primary radiation at the perimeter of the reflector will not be much reduced from that at the center, and the reflector illumination approaches a uniform value. This increases the aperture efficiency, but at the expense of spillover occurring. Making f/D too large increases spillover to the extent that aperture efficiency then decreases. Reducing f/D to less than $\frac{1}{4}$ places the focal point inside the reflector, as shown in Fig. 10.19(b). Here no spillover occurs, but the illumination of the reflector tapers from a maximum at the center to zero within the reflector region. This nonuniform illumination tends to reduce aperture efficiency. Also, placing the primary antenna too close to the reflector results in the reflector affecting the primary antenna impedance and radiation pattern, which is difficult to take into account. It can be shown that the aperture efficiency peaks at about 80%, with the angular aperture ranging from about 40 to 70°, depending on the primary radiation pattern. The relationship between aperture number and angular aperture is

$$\frac{f}{D} = 0.25 \cot \frac{\Psi}{2} \tag{10.32}$$

Typically, for an angular aperture of 55°, the aperture number is

$$\frac{f}{D} = 0.25 \times 1.92 = 0.48$$

This shows that the focal point should lie outside the mouth of the reflector, since f/D is then greater than $\frac{1}{4}$. Satisfactory results are obtained in practice if the main lobe of the primary antenna intercepts the perimeter of the reflector at the level -9 to -10 dB, as shown in Fig. 10.20.

On substituting $\pi D^2/4$ for A in Eq. (10.31) and using Eq. (10.8) for gain, we get

$$G = I(\theta)\left(\frac{\pi D}{\lambda}\right)^2 \tag{10.33}$$

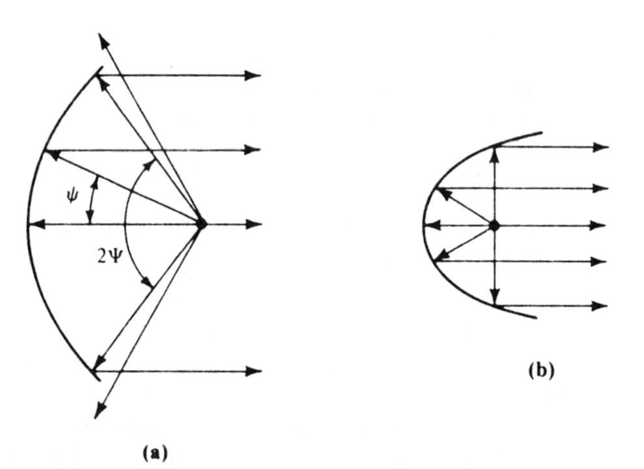

FIGURE 10.19. (a) Focal point outside reflector; (b) focal point inside reflector.

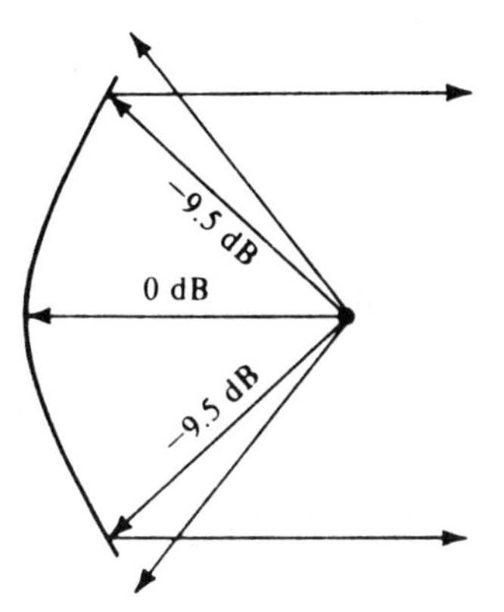

FIGURE 10.20. Edge illumination from primary antenna is 9 to 10 dB below that at vertex.

The beamwidth also depends on the primary radiator and its position. In practice, it is found that for most types of feed the −3-dB beamwidth is given approximately by

$$\mathrm{BW}_{(-3\ \mathrm{dB})} \simeq \frac{70\lambda}{D} \quad \mathrm{deg} \qquad (10.34)$$

and the beamwidth between nulls by

$$\begin{aligned} \text{nulls BW} &= 2(\mathrm{BW}_{(-3\ \mathrm{dB})}) \\ &= \frac{140\lambda}{D} \quad \mathrm{deg} \qquad (10.35) \end{aligned}$$

EXAMPLE 10.3

Find the gain, beamwidth, and effective area for a paraboloidal reflector antenna for which the reflector diameter is 6 m and the illumination efficiency is 0.65. The frequency of operation is 10 GHz.

SOLUTION

$$\lambda = \frac{c}{f} = \frac{300 \times 10^6}{10 \times 10^9} = 0.03\ \mathrm{m} = 3\ \mathrm{cm}$$

$$A = \frac{\pi D^2}{4} = \frac{3.14 \times 6^2}{4} = 28.26\ \mathrm{m}^2$$

$$A_{\text{eff}} = 0.65A = 18.4\ \text{m}^2$$

$$G_0 = \frac{4\pi}{\lambda^2}A_{\text{eff}} = 257{,}000\ (54.1\ \text{dB})$$

$$\text{BW}_{(-3\ \text{dB})} = \frac{70\lambda}{D} = \frac{70 \times 0.03}{6} = 0.35°$$

$$\text{BW}_{(\text{null})} = 2 \times 0.35 = 0.70°$$

10.10.1 Variations on the Parabolic Reflector

The *parabolic reflector* is a favorite antenna for fixed point-to-point microwave communications systems. It is relatively simple in construction, and unless large in size, it is quite inexpensive. Huge steerable parabolic dishes have been built for use with the radio telescopes, up to 200 ft in diameter, and mounted on a movable turret which allows rotation in both the horizontal and vertical directions to allow the tracking of moving targets such as satellites and radio stars.

Special feed systems are used to maintain high illumination efficiency while reducing the blockage presented by the feed antenna (blockage increases the sidelobes). Two types of feed are shown in Fig. 10.21. The first of these uses a dipole antenna, which normally radiates outside the parabola as well as onto it, but has a spherical reflector placed directly behind the dipole to prevent direct radiation. The backlobe radiation is reflected back

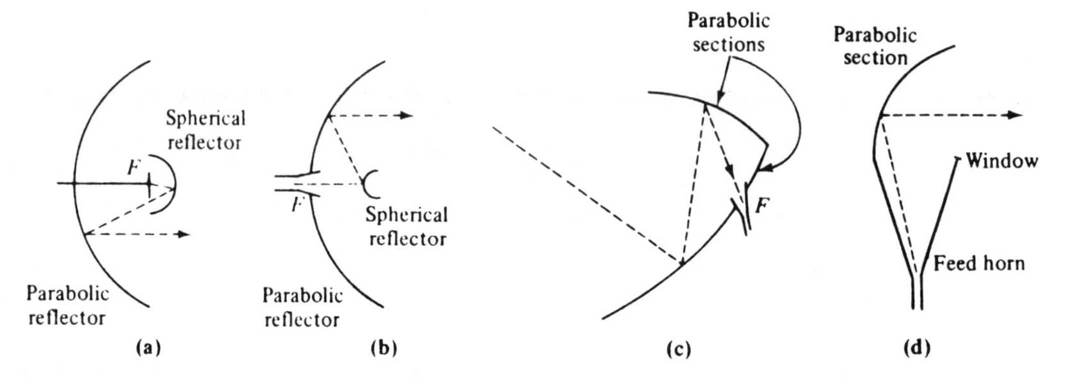

FIGURE 10.21. Methods of feeding microwave antennas: (a) parabolic reflector fed from a dipole and a small spherical reflector; (b) the Cassegrain feed: parabola fed from a horn and a small hyperboloid reflector; (c) Cass horn; (d) Hog horn.

at the parabola and is added to the main portion of the radiation. Some tuning is necessary since the reflector position is different for different frequencies.

The second method is known as the *Cassegrain feed system* [Fig. 10.21(b)]. The horn feed antenna, the paraboloid reflector, and the hyperboloid subreflector have a common axis of symmetry as shown, and the virtual focal point of the hyperboloid is coincident with the focal point of the paraboloid. Radiation reflected off the subreflector illuminates the main reflector approximately uniformly, and equally important, spillover at the edges is low. This is of particular advantage in low-noise receiving systems, where large spillover results in high noise levels.

Two types of modified horns also use parabolic surfaces. The first of these is the *Cass horn* [Fig. 10.21(c)]. Incoming radiation is reflected from a large lower parabolic section which has a rectangular sectional front and forms the bottom of the horn, up to a smaller parabolic surface which forms the top of the horn, and then to a horn antenna located at the focal point of the combined parabolas. This horn has the advantage that incoming radiation is not blocked by the feed structure, but it is much more complex to build.

The second horn is the *hog horn* [Fig. 10.21(d)]. This is basically a large horn antenna with a parabolic reflector mounted directly in front of it and oriented to radiate at right angles to the axis of the horn. It is often used for microwave communication links, such as the Bell TD-2 system, because it is compact and simple in construction while giving the high directivity of the parabola and also high efficiency. It also has the advantage that it may be rotated at the neck on a sliding joint, allowing steering of the beam without undue manipulation.

Another variant is the offset *Gregorian antenna*, which uses a paraboloidal reflector. This is a relatively small antenna designed for ground-station use in satellite communications. A 5.5-m reflector antenna is illustrated in Fig. 10.22. It will be seen that an offset feed is used to avoid aperture blockage. The main feature of this is that the sidelobe levels are reduced, and the antenna meets the new standard for sidelobe levels given by Eq. (10.13).

The antenna shown in Fig. 10.22 is designed for the 12/14-GHz band. At 14 GHz the 5-dB beamwidth is 0.17° and radiation at angles greater than 1° from the beam axis is at least 10 dB less than that from a conventional axisymmetric antenna. The offset arrangement also provides a lower angle of elevation for a given beam angle, which reduces wind loading. This antenna design is based on an optical telescope designed by James Gregory, a Scottish mathematician in the seventeenth century. Further details will be found on page 112 of *The Radio & Electronic Engineer*, Vol. 54, No. 3, 1984.

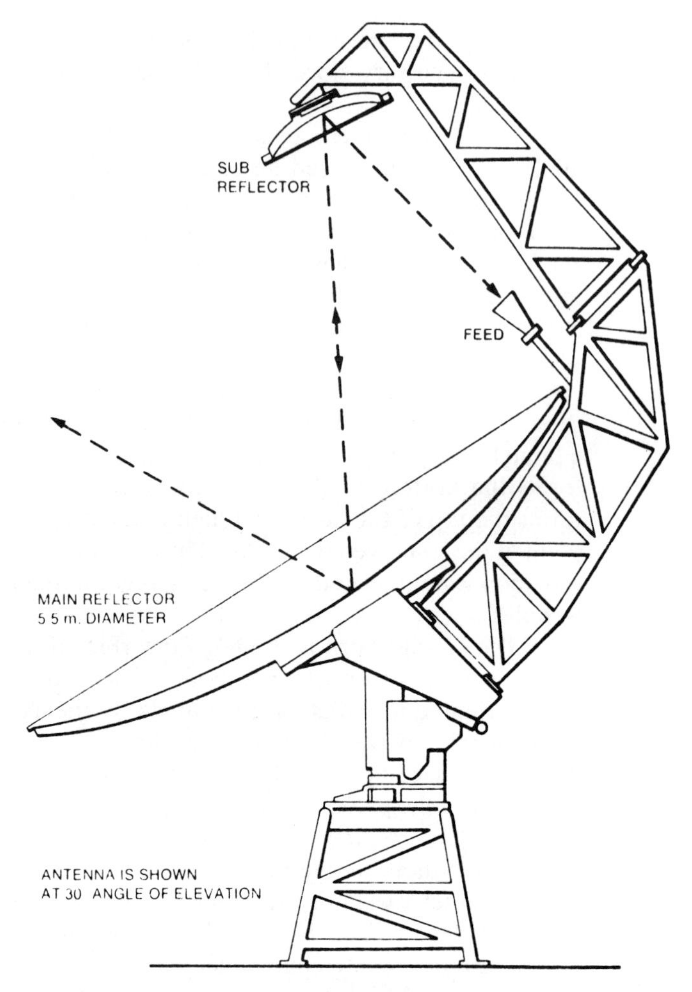

FIGURE 10.22. Offset Gregorian antenna. (*From Radio Electr. Eng., Vol. 54, No. 3, Mar. 1984, p. 112; with permission.*)

10.11 LENS ANTENNAS

10.11.1 Dielectric Lens Antennas

Electromagnetic radiation is refracted when it passes through a surface separating a zone of lower dielectric constant from one of higher dielectric constant in exactly the same manner that light is refracted. The angles of incidence and refraction are related by the modified version of Snell's law, which states [referring to Fig. 10.23(a)] that

$$\frac{\sin \phi_r}{\sin \phi_i} = \sqrt{\frac{\varepsilon_{ri}}{\varepsilon_{rr}}} = \frac{1}{n} \tag{10.36}$$

for waves entering the region of high dielectric constant, where n is the refractive index of the material. The refractive property is reciprocal, and the same relationships obtain when the radiation passes from the higher dielectric medium into the lower dielectric medium, except that the subscripts i and r are interchanged. The material used for the lens is usually one of the high-dielectric plastics, such as polystyrene or Teflon.

Figure 10.23(b) illustrates the principle of the collimating lens, which is used to make a diverging beam of radiation into one traveling in only one direction with a planar wavefront. The lens in this case is a convex one. Radiation along the axis of the lens passes through both surfaces at right angles, so no refraction takes place. Radiation at an angle from the axis is incident at the curved interface at an angle other than normal and is refracted toward the normal as it passes into the lens, at A'. The curvature of the lens is such that after refraction the rays are all parallel to the axis.

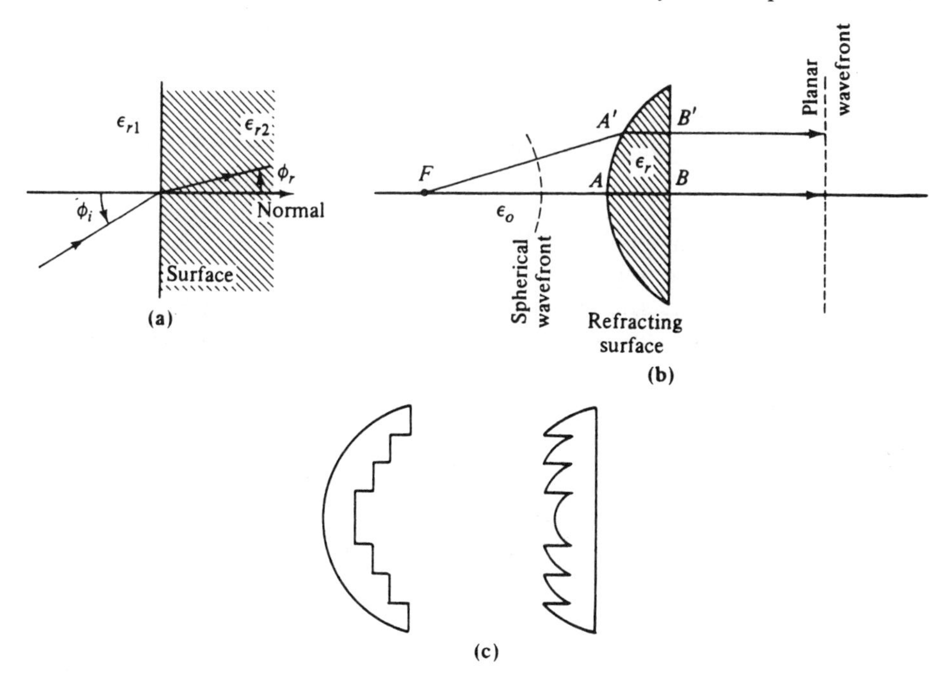

FIGURE 10.23. Dielectric lens antennas: (a) Snell's law of refraction; (b) principle of the collimating lens; (c) cross sections of two types of zoned lenses.

Radiation at the angle along FA' takes slightly longer to reach the refracting surface, arriving with a slight time lag compared to that along the axis FA. However, the velocity of propagation within the dielectric is slower than that in air, and a compensating delay occurs in the radiation along the axis AB as compared to that along $A'B'$, so that all the radiation arrives at the flat interface BB' in phase. The result is a planar wavefront leaving the lens.

Many types of lens systems can be used for the collimating or paralleling function. Two of these are shown in Fig. 10.23(c). These are *zoned lenses*, which are basically the same in function as the convex lens illustrated in Fig. 10.23(b). However, the zone structure allows a large reduction in the volume of dielectric material which must be used to make the lens, with a corresponding savings in cost and weight. This saving is made at a slight sacrifice of directivity. Also, the step size is frequency dependent [see, e.g., Glazier and Lamont (1958)], whereas the full dielectric lens is relatively frequency independent.

Figure 10.24(a) shows a lens suitable for use with a linear array, or line source, and Fig. 10.24(b) shows a circular lens illuminated from a source placed at the focal point. Uniform illumination of a lens is not easily achieved, and the actual radiation pattern will depend on the illumination as well as on the shape of the lens and the characteristics of the primary antenna feed.

Dielectric rod and horn antennas are also available, the dielectric extending out of the waveguide aperture and shaped to give the desired directional characteristics. Figure 10.25 shows a selection of rectangular dielectric antennas suitable for use with rectangular waveguide launchers. Figure 10.25(a) shows a rectangular horn antenna constructed out of a dielectric material, and Fig. 10.25(b) shows a wedge antenna. Details of these will be found in James et al. (1972).

A dielectric may also be used to guide the wave from the primary feed up to the subreflector in the Cassegrain antenna described in

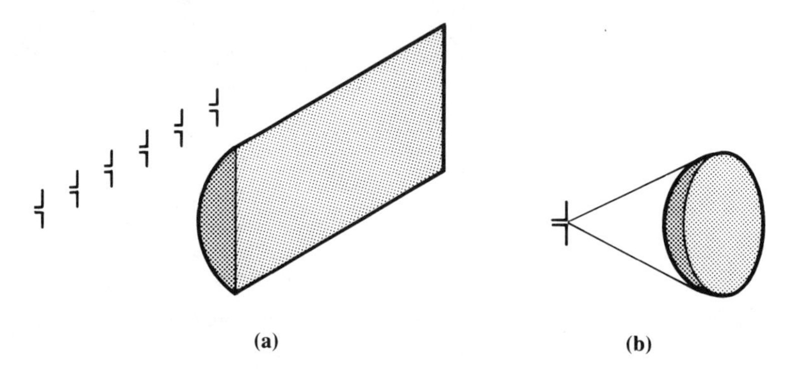

FIGURE 10.24. Examples of dielectric lens: (a) linear feed; (b) single-source feed.

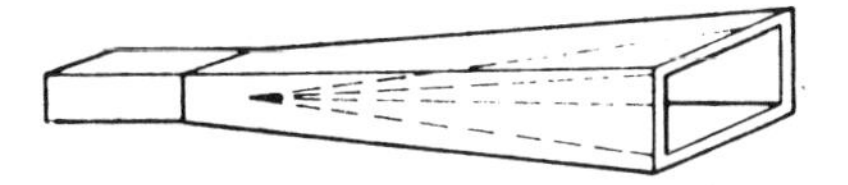

(a) rectangular horn antenna

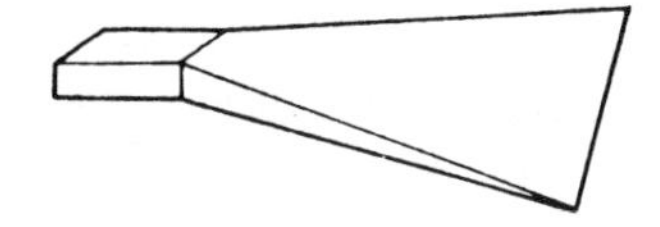

(b) wedge antenna with converging and diverging taper profiles

FIGURE 10.25. Various types of rectangular dielectric antenna suitable for rectangular waveguide launchers. (*From J. R. James, "Engineering approach to the design of tapered dielectric-rod and horn antennas," Radio Electr. Eng., Vol. 42, June 1972; with permission.*)

Section 10.10. Figure 10.26(a) shows the basic arrangement. Total internal reflection occurs inside the dielectric for ray paths leaving the primary feed, because the incident angle is greater than the critical angle for these rays. For the ray paths leaving the subreflector, the angle of incidence is less than the critical angle and therefore these rays escape from the dielectric and travel to the main reflector in the normal way. Figure 10.26(b) shows a photograph of a Cassegrain dielectric feed.

10.11.2 Metal Plate Lens Antenna

Refraction of an electromagnetic wave can also be achieved through the use of metal plates. Consider a wave passing between two parallel metal plates as shown in Fig. 10.27(a), the plate spacing being a. The arrangement is similar to the rectangular waveguide of Fig. 4.2(a), but with an infinite b dimension. Provided that the wavelength is shorter than the cutoff wavelength, a TE wave is propagated at phase velocity given by Eq. (4.3). A number of such plates may be assembled as shown in Fig. 10.27(b), and the phase velocity of the wave through the plate assembly will also be given by Eq. (4.3). The refractive index of any medium is given by the ratio of the velocity of light to the phase velocity in the medium, and hence the metal plate assembly has an equivalent refractive index, obtained from Eq. (4.3) as

$$n = \frac{\lambda}{\lambda_g} \qquad (10.37)$$

The primary source is placed at a focal point, determined by the concave shape of the plates as shown in Fig. 10.27(b).

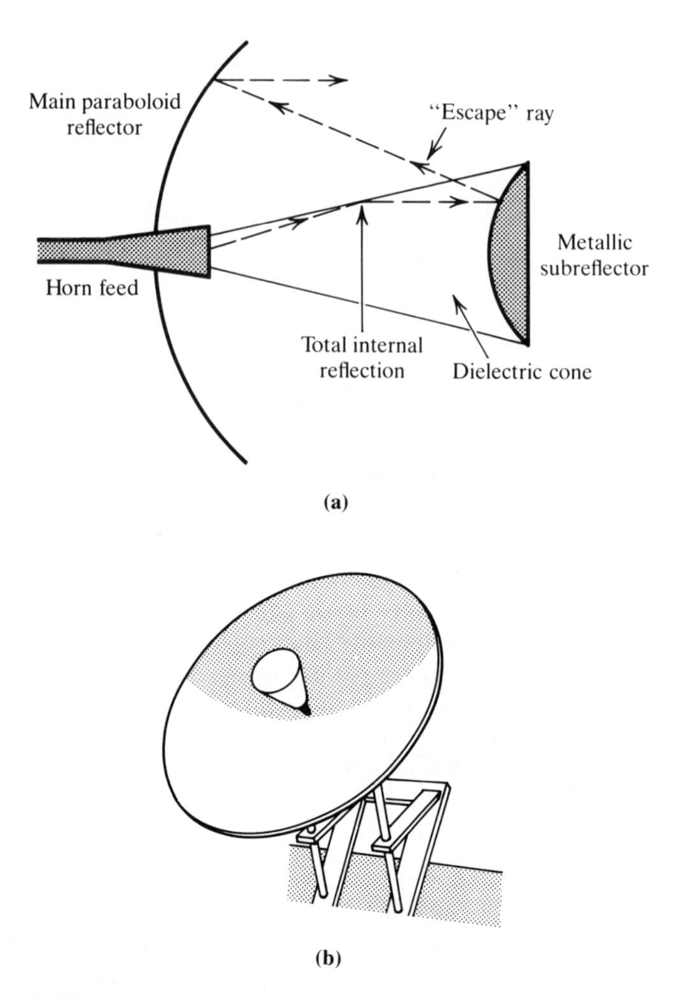

FIGURE 10.26. Dielectric cone feed for Cassegrain antenna: (a) ray path; (b) photograph of a Cassegrain feed. [*Part (b) from H. E. Barlett and E. Moseley, "Dielguides—highly efficient low noise antenna feeds," Microwave J., Vol. 9, No. 12, 1966; with permission.*]

It will be observed that the phase velocity in the metal plate lens antenna is greater than the velocity of light, whereas the phase velocity in the dielectric lens antenna is less than the velocity of light.

10.12 SLOT ANTENNAS

A slot cut in a metallic sheet allows a transmission line to be connected as shown in Fig. 10.28(a) and the resulting microwave currents flowing in the sheet give rise to a radiated electromagnetic wave. The sheet allows the currents to be distributed in space. Otherwise, if a simple rectangular loop was used, as shown in Fig. 10.21(b), the radiation from the long sides would cancel because of their proximity and the fact that they carry equal and opposite currents. Very little radiation would take place from the short ends, which are assumed to be very much smaller than a wavelength.

In optics, a principle known as *Babinet's principle* may be applied to a screen and its complement to determine the light transmitted and reflected by the screens. A screen is complementary to another when its transparent areas and opaque areas correspond exactly with the opaque areas and transparent areas of the other. The complementary screen is like a well-defined shadow of the other. Babinet's principle was first applied

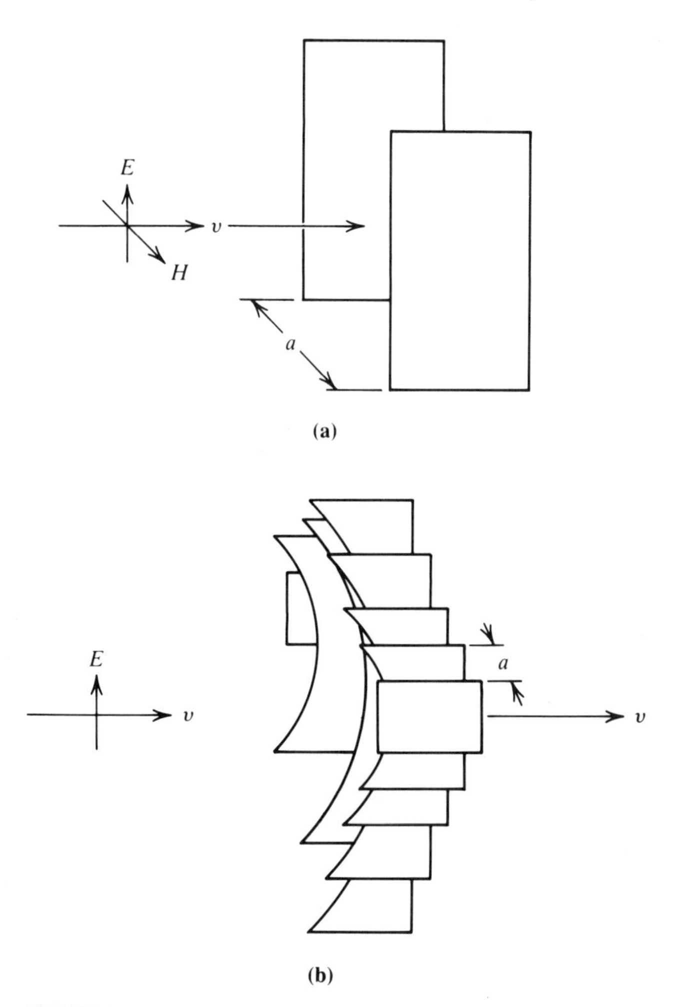

FIGURE 10.27. (a) Wave propagation between parallel metal plates; (b) concave metal plate lens.

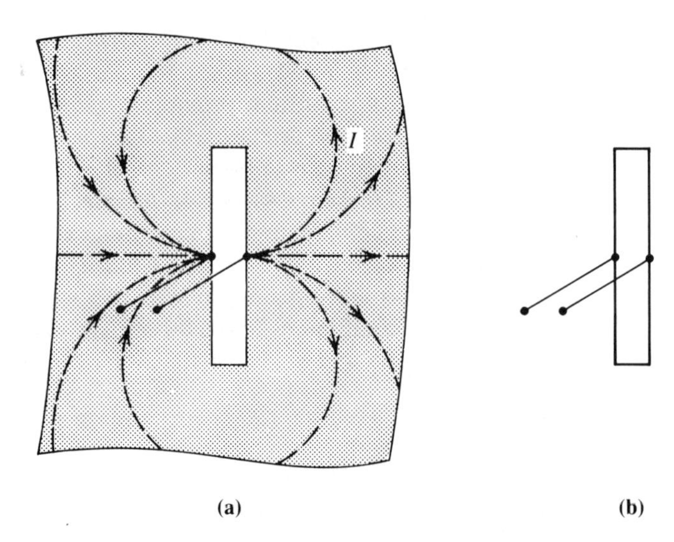

FIGURE 10.28. (a) Narrow slot antenna; (b) narrow rectangular loop antenna.

to slot antennas and their complements by H. G. Booker (*J. IEE*, Pt. IIIA, No. 4, 1946) and a detailed description will be found in Kraus (1950). For the present purpose, one of the most important results is that relating the impedance of a slot antenna to its complement. Figure 10.29(a) shows a narrow half-wavelength slot cut in a conducting screen, which is assumed

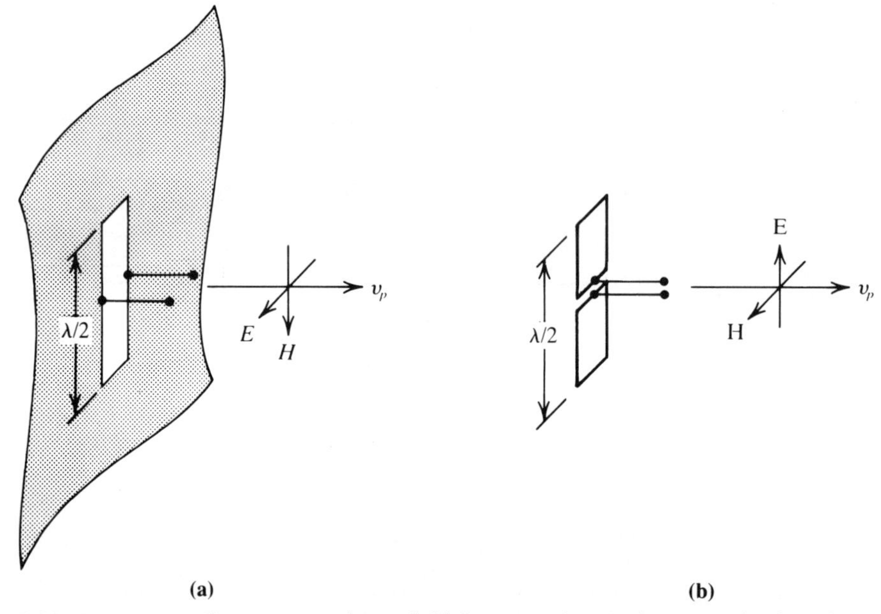

FIGURE 10.29. (a) Slot antenna and *E*, *H* field directions; (b) complementary dipole and *E*, *H* field directions.

infinite in extent. The slot is center fed. The complementary half-wave dipole is shown in Fig. 10.29(b). This is seen to be the metallic "cutout" for the slot, and when center fed it forms a strip dipole. The feed gap is assumed to be negligibly small. Letting Z_s represent the impedance of the slot and Z_d the impedance of the strip dipole, the impedance relationship derived from Babinet's principle states that

$$Z_s Z_d = \frac{Z_0^2}{4} = (60\pi)^2 \tag{10.38}$$

Here $Z_0 = 120\pi\ \Omega$ is the impedance of free space.

The radiation pattern for the strip dipole applies also to the slot antenna provided that the E and H fields are interchanged. The field directions are shown in Fig. 10.29, where it will be observed that the electric field is horizontally polarized for the slot antenna and vertically polarized for the dipole. The radiation pattern for both antennas is given by Eq. (10.19).

Impedance data for a strip dipole are usually presented in terms of an equivalent cylindrical dipole. It is shown in Elliott (1981, Sec. 7.12) that a good approximation to a strip dipole of width w and thickness t is a cylindrical dipole of radius a, where

$$a = \frac{w + t}{4} \tag{10.39}$$

EXAMPLE 10.4

A slot 1.0 mm wide is cut in a highly conducting sheet, the thickness of which is 0.27 mm. This is to form a center-fed half-wavelength slot antenna to operate at 6 GHz. Calculate the slot impedance (a) when the length is exactly a half-wavelength, and (b) when it is 95% of the half-wavelength value.

SOLUTION

(a) As pointed out in Example 10.1, the radius a affects only the reactance term, Eq. (10.21) and then only when the length is other than half a wavelength. Hence for the half-wavelength, the value calculated in Example 10.1 applies here also; that is,

$$Z_d = 73.46 + j41.28\ \Omega$$

From Eq. (10.38),

$$Z_s = \frac{(60\pi)^2}{73.46 + j41.28} = \mathbf{367.6 - j206.6\ \Omega}$$

(b) From Eq. (10.39),

$$a = \frac{1 + 0.27}{4}$$

$$= 0.3175 \text{ mm}$$

$$\lambda = \frac{3 \times 10^8}{6 \times 10^9} = 0.05 \text{ m}$$

$$= 50 \text{ mm}$$

Hence

$$\frac{a}{\lambda} = \frac{0.3175}{50}$$

$$= 0.006350$$

When the slot is cut to 95% of the half-wavelength value,

$$\frac{l}{\lambda} = 0.475$$

Hence the conditions for eqs. (10.20) and (10.21) apply. From Eq. (10.20),

$$R_d = 122.65 - 641.2 \times 0.475 + 1085.66 \times 0.475$$

$$= 63.03 \ \Omega$$

From Eq. (10.21), and noting that $\frac{l}{a} = \frac{0.475}{0.006350} = 74.80$,

$$X_d = 162.5 - 31.306 - 208.91 + 89.07$$

$$= 11.35 \ \Omega$$

From Eq. (10.38),

$$Z_s = \frac{(60\pi)^2}{63.03 + j11.35}$$

$$= \mathbf{546 - j98.3 \ \Omega}$$

It will be observed that the slot impedance is considerably greater than the dipole impedance, and where the dipole impedance is inductive, the slot impedance is capacitive.

From Eq. (10.21) it will be found that for $l/\lambda = 0.475$ and $l/a = 248.8$, the reactance is zero; that is, the slot is resonant.

EXAMPLE 10.5

For the conditions stated above, $l/\lambda = 0.475$ and $l/a = 248.8$, determine the equivalent slot antenna impedance and slot dimensions. Assume that sheet thickness may be neglected.

SOLUTION The dipole resistance will be the same as that calculated in Example 10.4(b):

$$R_d = \mathbf{63.03\ \Omega}$$

Equation (10.21) gives the dipole reactance as

$$\begin{aligned} X_d &= 162.5 - 42.656 - 208.91 + 89.07 \\ &= \mathbf{0} \end{aligned}$$

Hence

$$\begin{aligned} Z_s &= \frac{(60\pi)^2}{63.03} \\ &= \mathbf{563.7\ \Omega} \end{aligned}$$

Neglecting sheet thickness, we have

$$\begin{aligned} w &= 4a \\ &= 4 \times \frac{0.475\lambda}{248.8} \\ &= \mathbf{0.00764\lambda} \text{ (slot width).} \end{aligned}$$

The slot length is given as **0.475λ.**

In practice, a slot antenna would more likely be arranged to radiate into one-half of the surrounding space, this being achieved by enclosing one side in a cavity, as shown in Fig. 10.30(a). Both the real and reactive power radiated into the open space are approximately doubled by the presence of the cavity, assuming that the losses and stored energy in the cavity are negligible compared to the radiated levels. If the cavity dimensions are large enough not to alter the current flow in the sheet significantly, the slot impedance is effectively doubled with the cavity present. At the same time, the cavity presents a susceptance in parallel with the slot. The dimensions are such that only the TE_{10} mode propagates into the cavity, which appears as a short-circuit guide section and therefore as a pure susceptance jB_c, assuming that losses can be ignored. The total admittance seen by the feeder is

$$\begin{aligned} Y &= Y_s + jB_c \\ &= \frac{2Z_d}{Z_0^2} + jB_c \end{aligned} \tag{10.40}$$

The cavity dimensions can be designed for resonance, leaving only the

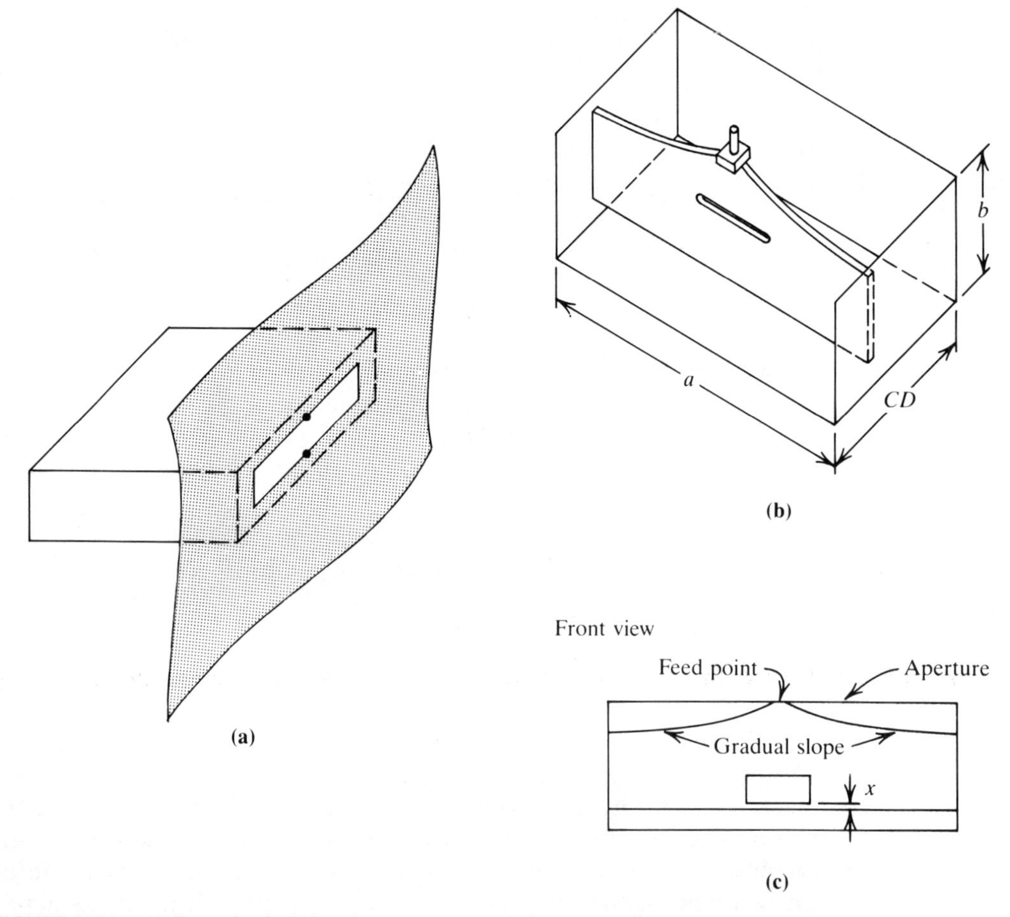

FIGURE 10.30. (a) Cavity backed slot; (b) cavity backed T-bar fed slot antenna with top removed for viewing purposes (note use of planar T-bar); (c) front view of T-bar slot antenna showing the slotted T-bar which resulted from ridge investigation. [*Parts (b) and (c) from M. R. Crews and G. A. Thiele, "On the design of shallow depth T-bar fed slot antennas," IEEE Trans., AP-25, No. 6; copyright © IEEE 1977; with permission.*]

conductance term:

$$G = \frac{2R_d}{Z_0^2} \tag{10.41}$$

If one takes R_d as typically of the order of 70 Ω, the input resistance of the cavity-backed slot is $R = 1/G = 1015\ \Omega$. This impedance level is too high for matching to coaxial lines, and matching transformer sections have to be used. One such method, known as a T-bar, is shown in Fig. 10.30(b).

Slots may also be fed from a waveguide, some examples of radiating slots in waveguides having been shown in Fig. 4.10. Waveguide slots are often used in arrays, an example being shown in Fig. 12.8. The crossed slot is excited by both the longitudinal and traverse components of the magnetic field in the guide, and as a result the radiated field is circularly polarized.

10.13 MICROSTRIP ANTENNAS

Microstrip antennas are an extension of the microstrip transmission line described in Chapter 5. The simplest antenna element is the patch antenna illustrated in Fig. 10.31. For simplicity, a rectangular patch is shown, although other shapes, including circular, may be used. The electromagnetic wave traveling along the microstrip feed line spreads out under the patch. At the perimeter it meets an open circuit and the resulting reflections set up a standing-wave pattern under the patch. The patch is resonant when the length L is made approximately equal to the half-wavelength on the line, or $L = \lambda/2$, where $\lambda = \lambda_0/\sqrt{\varepsilon_r}$ and ε_r is the effective relative permittivity of the microstrip. A detailed analysis [see, e.g., Rudge et al. (1982)] shows that the far-field radiation from ends 1 and 3 is in phase, whereas that from edges 2 and 4 tends to cancel. Because the height h of the substrate is very much smaller than the line wavelength, the patch

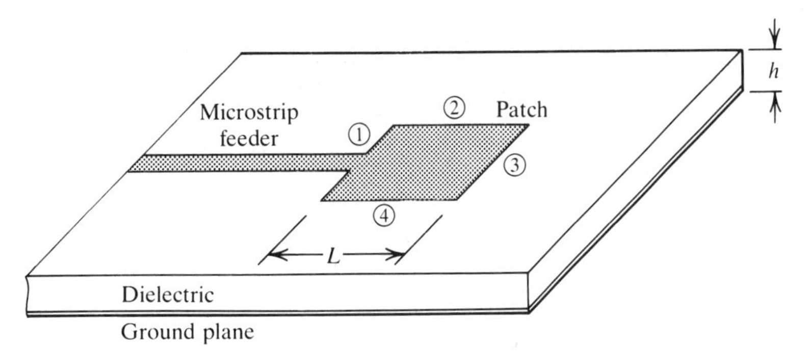

FIGURE 10.31. Patch antenna.

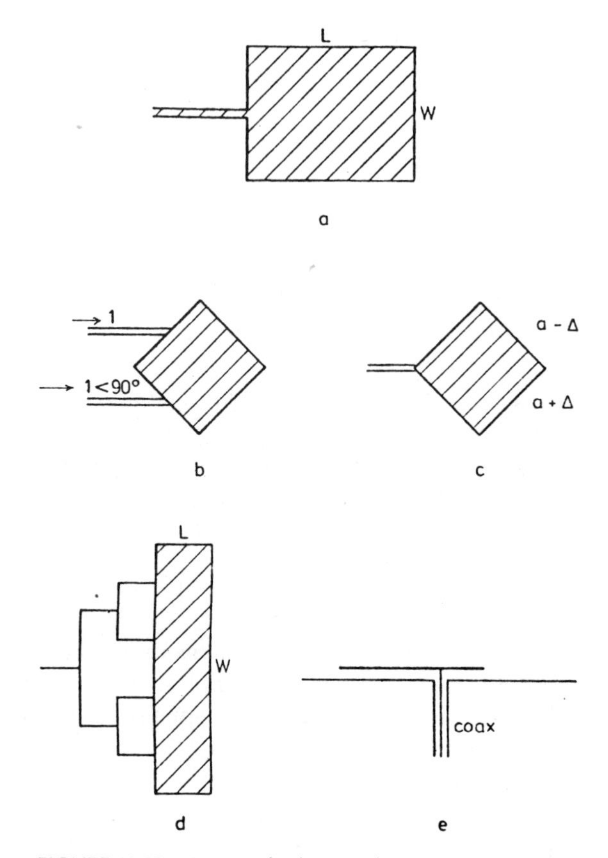

FIGURE 10.32. Various feeding mechanisms: (a) microstrip connected to edge of connector; (b) same principle, excitation of two modes to give circular polarization; (c) as part (b) but only one feed, unequal side lengths; (d) parallel feed network; (e) back fed through ground plane. [*From A. W. Rudge, K. Milne, A. D. Olver, and P. Knight, eds., The Handbook of Antenna Design, Vol. 1 (Stevenage, Hertfordshire, England: Peter Peregrinus Ltd., 1982), on behalf of the IEE; with permission.*]

antenna may be considered as a resonant slot antenna, where the resonant cavity is that formed by the volume between the patch and the ground plane, and the radiating slot is the total perimeter of the patch, with most of the radiation occurring from the far end and the feed end.

The patch antenna may be fed in a number of ways. The end feed already discussed is shown in Fig. 10.32(a). The twin feed shown in Fig. 10.32(b) is used to provide circular polarization. Circular polarization can also be achieved from the corner feed as shown in Fig. 10.32(c), but here the patch is not quite rectangular. Parallel feeds may be connected as shown in Fig. 10.32(d), and a coaxial feed as shown in Fig. 10.32(e).

Antenna arrays are easily fabricated in microstrip. These are arrays of basic elements which can be configured to give desired radiation patterns and polarization. Figure 10.33 shows an example of such an array, and an example of a planar array used in radar is shown in Fig. 12.16(b).

Microstrip antennas are particularly suited to those applications where flat profile antennas are required. Examples of such applications are found in missile guidance systems, portable radar systems, and the like. These antennas are easily made to conform to the required shape. In addition, they are lightweight, simple and inexpensive to manufacture, and they may be integrated with circuits. The disadvantages are that unwanted radiation from the feeder, and surface waves along the substrate, make it difficult to control the radiation pattern accurately; they have low efficiency, and

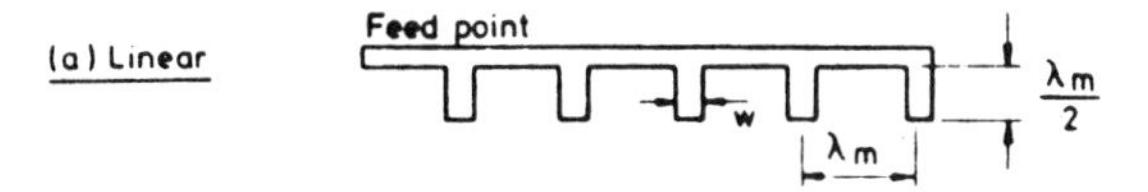

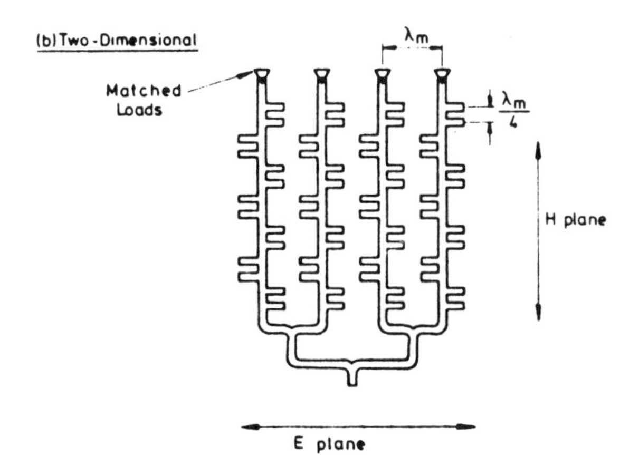

FIGURE 10.33. Comb line arrays: (a) linear resonant array; (b) two-dimensional traveling-wave array; (c) zigzag linear traveling-wave array. *(From P. S. Hall and J. R. James, "Survey of design techniques for flat profile microwave antennas and arrays,"* Radio Electr. Eng., *Vol. 48, No. 11, 1978; with permission.)*

as already mentioned, they are narrow-bandwidth devices. However, for many applications the advantages outweigh the disadvantages, and in fact often provide the only solution to an antenna requirement, this being especially true of millimeter wave systems.

A further development in millimeter wave antennas is the semiconductor dipole antenna, the basic form being shown in Fig. 10.34. Here the dipole elements are the semiconductor layers deposited on the insulating substrate. The current along the dipole is controlled by the semiconductor doping density, which can be varied along the length in such a way that the current becomes negligible at the ends. In this way, no reflected current wave occurs, and the radiation pattern is determined by the forward traveling wave along the dipole. This is quite different from the resonant dipole (e.g., the half-wave dipole), which depends on the reflected wave to set up a standing wave, which in turn determines the radiation pattern. The main advantages of traveling-wave antennas are that they are broadband, and their radiation pattern has small sidelobes. The disadvantage is that power is dissipated as heat in the resistance of the dipole structure. As shown in Figure 10.34, control electrodes similar to the control gate of the field-effect transistor can be incorporated into the antenna, which opens up the possibility of electronically controlling the radiation pattern. Further details of the semiconductor antenna will be found in Jain and Bansal (1984).

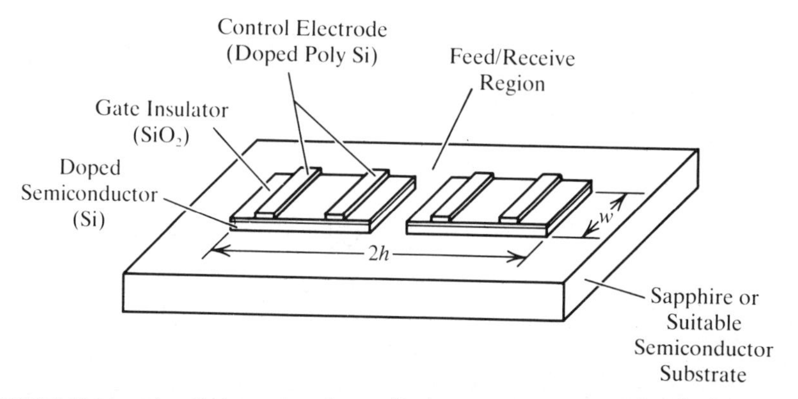

FIGURE 10.34. Monolithic semiconductor dipole antenna structure with polysilicon control gates. *(From F. C. Jain and Bansal, "Monolithic mm-wave antennas," Microwave J., Vol. 27, No. 7, 1984; with permission.)*

10.4 PROBLEMS

1. A 50-Ω feeder line is connected to a 73-Ω antenna. The power input to the line is 5 kW, and the total power radiated is 4.2 kW. Calculate (a) the antenna efficiency and (b) the power loss in the antenna.
2. Define the terms *gain* and *directivity* as applied to antennas. The isotropic gain of a certain reference antenna is 19 dB. The gain of a test antenna relative to the reference is 23 dB and the efficiency of the test antenna is 90%. Calculate for the test antenna (a) its isotropic gain and (b) its directivity.
3. Define the term *effective area* for an antenna. What is the effective area for (a) an isotropic antenna (in terms of wavelength) and (b) a pyramidal horn the aperture dimensions of which are 2 cm × 3 cm?
4. An antenna has a gain of 45 dB at a frequency of 4 GHz. Calculate its effective area.
5. An antenna used for satellite communications has a sidelobe envelope curve given by Eq. (10.12). Calculate the degradation in isolation which occurs when satellite spacing is reduced from 4° to 2°.
6. Repeat Problem 5 assuming that when the satellite spacing is reduced, the sidelobe envelope pattern is also changed to that given by Eq. (10.13).
7. Explain what is meant by the polarization of an antenna. A dipole receiving antenna is aligned so that its plane is normal to the direction of propagation of the wave. Calculate the polarization loss factor if the line of the dipole makes an angle at 25° with the electric vector of the wave.
8. Calculate the polarization loss factor for the antenna in Problem 7 when the plane of the antenna makes an angle of 10° with the electric field vector, all other factors remaining unchanged.
9. Using Eqs. (10.18) and (10.19) plot the power radiation pattern for a half-wave dipole in the first quadrant, $0 \leq \theta \leq 90°$. Verify that the −3-dB beamwidth is 78°.
10. For the dipole antenna of Problem 9, calculate the gain for $\theta = 30°$.
11. A half-wave dipole is operated at a frequency of 500 MHz. Calculate (a) the effective area and (b) the effective length. State any assumptions made.
12. The antenna in Problem 11 lies parallel to an electric field of 100 μV/m, the frequency of the wave being 500 MHz. Calculate (a) the induced EMF in the antenna, (b) the power delivered to a matched load, and (c) the power density in the wave.
13. A cylindrical dipole of radius 0.3 cm and length 30 cm is operated at a nominal resonant frequency of 500 MHz. Calculate the impedance at 500 MHz and at frequencies ±5% from 500 MHz.

14. Repeat Problem 13 for an antenna of radius 0.5 cm.
15. Calculate the effective area of a rectangular aperture, 2 cm × 3 cm.
16. A pyramidial horn has dimensions $a = 2$ cm, $b = 3$ cm, $l_E = 2.8$ cm, and $l_H = 6$ cm. Calculate the far-field gain in decibels for a frequency of 10 GHz.
17. An H-plane sectoral horn has dimensions $a = 10$ cm, $b = 3$ cm, and a slant length 7.07 cm. Calculate the far-field gain in decibels for a frequency of 3 GHz.
18. An E-plane sectoral horn has dimensions $a = 3$ cm, $b = 10$ cm, and a slant length of 7.07 cm. Calculate the far-field gain in decibels for a frequency of 3 GHz.
19. Calculate the gain in decibels for a 3-m-diameter parabolic dish which has an aperture efficiency of 0.55 and is operating at a frequency of 4 GHz.
20. Calculate (a) the −3-dB beamwidth and (b) the beamwidth between nulls for the parabolic antenna of Problem 19.
21. The −3-dB beamwidth of a parabolic antenna is 0.5° and the illumination efficiency is 0.55. Calculate the antenna gain in decibels.
22. Calculate (a) the −3-dB beamwidth and (b) the gain for the antenna in Problem 21 when the frequency is increased by 5%.
23. Explain the principles of operation of (a) dielectric lens and (b) metal plate antennas.
24. A center-fed, half-wavelength slot antenna is constructed in a highly conducting metal sheet 2 mm thick. The width of the slot is 1 cm and the nominal center frequency is 1000 MHz. Calculate the impedance (a) at resonance and (b) at frequencies 5% above and below resonance.
25. A center-fed half-wave slot antenna is designed to operate at a wavelength of 3 cm, the length of the slot being 1.425 cm and the width 229.1 μm. Assuming that the sheet thickness can be neglected, calculate the impedance (a) at the design wavelength and (b) at 5% above and below the design wavelength.
26. Discuss the principles of operation of microstrip antennas and their feed arrangements. State the principal advantages and disadvantages of microstrip antennas.

REFERENCES

Barlett, H. E., and R. E. Moseley, 1966. "Dielguides—highly efficient low noise antenna feeds," *Microwave J.*, Vol. 9, pp. 53–58.

ELLIOTT, ROBERT S., 1981. *Antenna Theory and Design.* Englewood Cliffs, N.J.: Prentice-Hall, Inc.

GLAZIER, E. V. D., and H. R. L. LAMONT, 1958. *Transmission and Propagation.* (London: Her Majesty's Stationery Office.) (*The Services' Textbook of Radio*, Vol. 5.)

JAIN, F. C., and R. BANSAL, 1984. "Monolithic mm-wave antennas," *Microwave J.*, July.

JAMES, J. R., 1972. "Engineering approach to the design of tapered dielectric-rod and horn antennas," *Radio & Electron. Eng.*, Vol. 42, pp. 251–259.

JAMES, J. R., P. S. HALL, and C. WOOD, 1981. *Microstrip Antennas.* Stevenage, Hertfordshire, England: Peter Peregrinus on behalf of the IEE.

JULL, E. V., 1970. "Finite-range gain of sectoral and pyramidal horns," *Electron. Lett.*, Vol. 6, pp. 680–681.

KRAUS, JOHN D., 1950. *Antennas.* New York: McGraw-Hill Book Company.

LOVE, A. W. (ed.), 1976. *Electromagnetic Horn Antennas.* IEEE Press.

RUDGE, A. W., K. MILNE, A. D. OLVER, and P. KNIGHT (eds.), 1982. *The Handbook of Antenna Design.* Stevenage, Hertfordshire, England: Peter Peregrinus on behalf of the IEE.

SILVER, SAMUEL. editor 1949. Microwave Antenna Theory and Design. McGraw-Hill. (Republished by Dover Publ. Inc., 1965).

eleven
Microwave Radio Systems

11.1 INTRODUCTION

Microwave radio systems can be broadly classified as terrestrial systems and satellite systems. Many aspects of radio transmission are common to both kinds of systems, but parameters can differ greatly, and planning methods have to take into account the specifics of the particular system.

Figure 11.1 shows the essential features of a basic microwave radio system. A transmit–receive terminal unit is required at each end for the purpose of transferring the baseband signal to and from the microwave carrier. The baseband signal itself will usually be a multiplexed signal, carrying a number of individual telephony signals and possibly video signals. Transmission from one terminal station to the other is by means of radio waves, but except for very short distances, it is usually necessary to provide one or more repeater stations in between. These repeater stations amplify the signal to make up for transmission losses.

Both analog and digital methods are employed in microwave radio systems. Analog systems utilize frequency-division multiplexing to form the multiplexed baseband signal, and this is then frequency modulated onto the microwave carrier. This is referred to as a FDM/FM system. In theory it should be possible to omit the frequency-modulation stage and implement the frequency-division multiplexing directly with microwave carriers. It is found that this requires very low distortion in the radio equipment, and although some experimental systems are under investigation, the method is not used in practice. Digital systems utilize time division multiplexing to form the multiplexed baseband signal, and this is

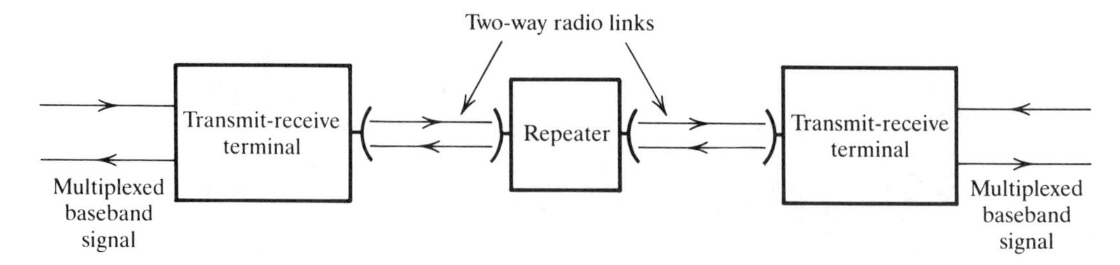

FIGURE 11.1. Basic microwave radio relay system.

then used to phase modulate the microwave carrier, phase-shift keying being used. Interestingly enough, early microwave systems (in use before 1950) utilized digital methods because narrow bandwidths and comparatively high distortion levels in equipment made analog methods impractical. These early digital systems were not very efficient, having very low channel capacity. During the early 1950s, FDM/FM systems were developed and improved steadily, with the result that the early digital systems were displaced. Now with the very rapid developments in digital integrated circuits, highly efficient digital systems are possible, and the trend is toward microwave digital radio systems.

11.2 TERMINAL STATIONS

Figure 11.2 shows in more detail some of the equipment required at a terminal station. For medium- to large-capacity systems (capacity referring to the number of baseband channels), more than one microwave or RF carrier is usually employed. In Fig. 11.2 are shown three separate baseband inputs, and these are used to modulate carriers f_1, f_2, and f_3. The three stages are similar to that shown for f_1. The baseband input is used to modulate a 70-MHz IF carrier. This modulated signal in turn is passed into an up-converter (U.C. in Fig. 11.2) or frequency changer, which converts the carrier frequency to the desired value f_1. The microwave carrier f_1 is filtered and amplified and passed to a branching filter, where it connects with the other microwave carriers, f_2 and f_3. Thus the carriers f_1, f_2, and f_3 form a frequency-division-multiplexed signal at the output of the branching filter, and it should be kept in mind that each carrier itself may carry a frequency-division-multiplexed or time-division-multiplexed baseband signal. The output from the branching filter will be polarized in one particular direction, either horizontal or vertical, horizontal polarization being shown in Fig. 11.2. This horizontally polarized, multicarrier signal is then passed through a further polarizing filter, which ensures that only a horizontally polarized wave (in this example) is passed to the antenna.

The antenna is common to both transmit and receive signals, and the multicarrier received wave consists of horizontally polarized signals at f_1', f_3', and f_5', each of these carrying a multiplexed baseband signal. The polarization filter separates the receive carriers from the transmit carriers and passes these onto the receive branching filter. The paths for all received carriers are similar to the one shown for f_1'. The carrier is selected by the bandpass filter and down-converted (shown as D.C. in Fig. 11.2) to the 70-MHz intermediate carrier frequency. This is amplified and further filtered, and the baseband signal is recovered in the demodulation stage.

Figure 11.3 shows the preferred RF frequency scheme for the 4000-MHz band, as set out by the International Radio Consultative Committee (CCIR) and the International Telegraph and Telephone Consultative Committee (CCITT). This arrangement is intended for 600- to 1800-channel FDM systems or equivalent. To simplify filtering, all transmit channels are kept in one 200-MHz half of the spectrum, and all receive channels in the other 200-MHz half, as shown in Fig. 11.3, as this keeps the transmit and

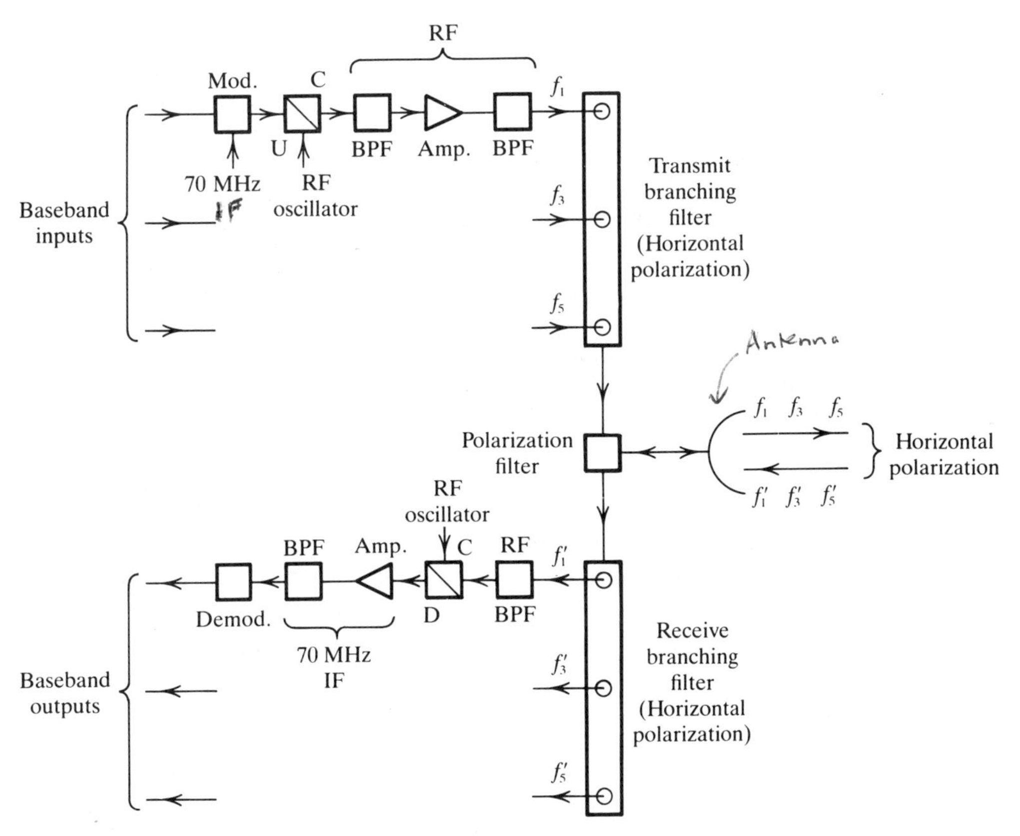

FIGURE 11.2. Block schematic of terminal station equipment.

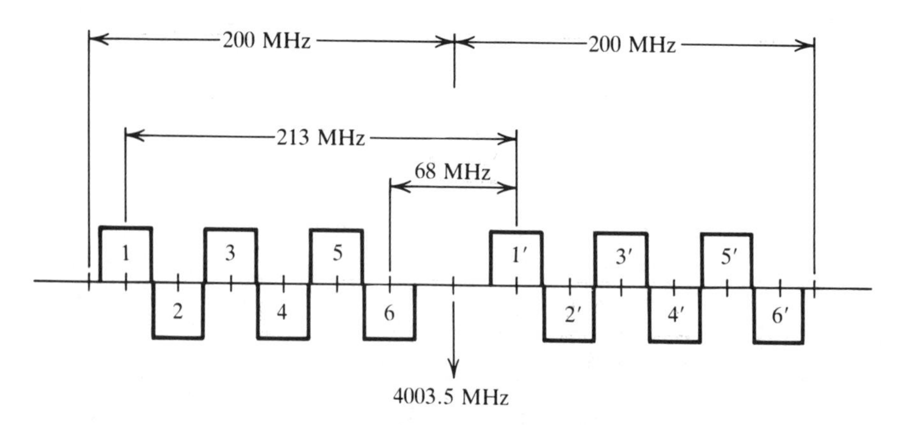

FIGURE 11.3. Preferred microwave frequency scheme for the 4000-MHz band.

receive channels widely separated. Two frequencies are allocated per channel, for example f_1 and f_1'. On any given section of the route, f_1 might be used as the carrier for left to right, and f_1' would then be used as the carrier from right to left. On the following section of the route, f_1' would be used as the left-to-right carrier, and f_1 as the right-to-left carrier. Carrier assignments would alternate in this way for any particular channel along the route. It is also usual to alternate the polarization of channels as shown in Fig. 11.3. Thus if channel 1 is horizontally polarized, channel 2 will be vertically polarized, and so on. The center frequency is shown as 4003.5 MHz, and this is the preferred value because this choice is found to minimize interference from harmonics of the shift frequency $f_n - f_n'$ (where n is the channel number).

The antenna branching arrangement for this frequency scheme is shown in Fig. 11.4. Each unit is connected into the system through a microwave shutter switch. This type of switch allows the unit to be disconnected for maintenance or testing without disrupting the rest of the system. The signal from transmitter Tx_1 passes through a channel filter and a circulator into the upper branch. It must also pass through the other circulators in its particular path and at point A, the transmissions for Tx_1, Tx_3, and Tx_5 appear as a frequency-division-multiplexed RF signal. This multiplexed signal passes through a duplexing circulator. The duplexing circulator is connected to the antenna through a feeder which is common for both transmit and receive signals in this branch. The function of the duplexing circulator is to separate the transmit and receive signals into their respective branches. Another important point is that this duplexing circulator passes only horizontally polarized waves. It will be seen that there is another duplexing circulator for channels Tx_2, Tx_4, Tx_6, and Rx_2, Rx_4, Rx_6, these being vertically polarized. The polar tee combines both

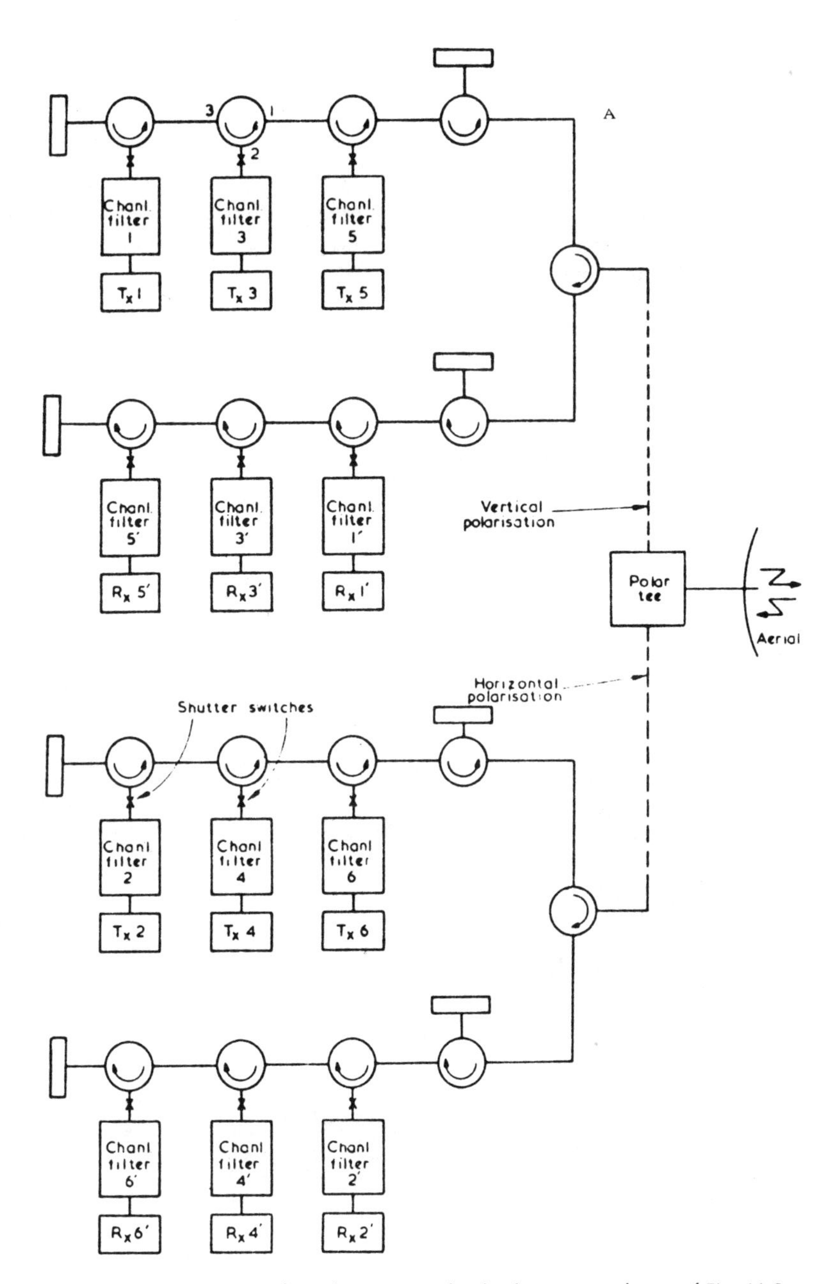

FIGURE 11.4. Equipment branching system for the frequency scheme of Fig. 11.3. [*From H. B. Wood, "Microwave Radio Systems," in Advanced Communication Systems, ed. B. J. Halliwell (Newnes: Butterworth, 1974).*]

horizontally and vertically polarized multiplexed RF signals and allows them to be fed to and from a common antenna.

11.3 REPEATERS

Repeaters may generally be classified as active and passive. Active repeaters involve transmitting and receiving equipment in addition to antennas, and require a power supply for operation. Passive repeaters are an arrangement of reflecting surfaces or "radio mirrors" which do not require any power supply for operation.

11.3.1 Active Repeaters

In Fig. 11.5 is shown one type of repeater, known as a heterodyne repeater, because most of the amplification takes place at the heterodyne beat frequency or intermediate carrier frequency, which is 70 MHz.

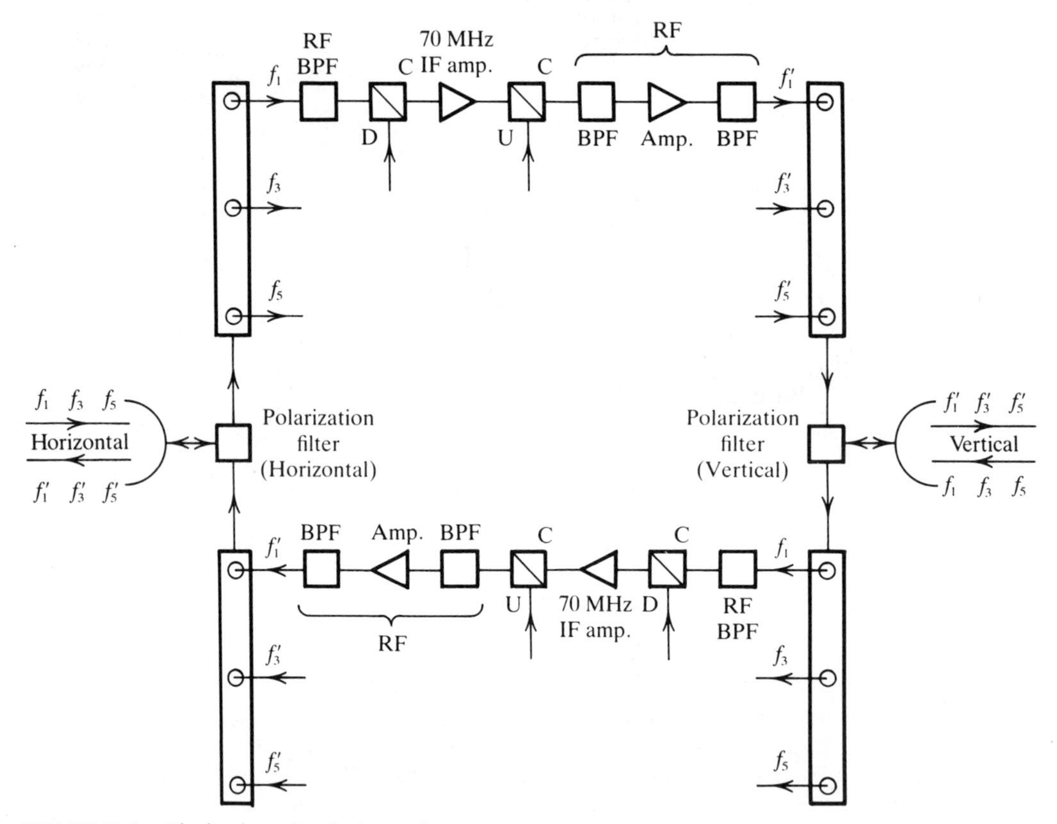

FIGURE 11.5. Block schematic of a heterodyne repeater.

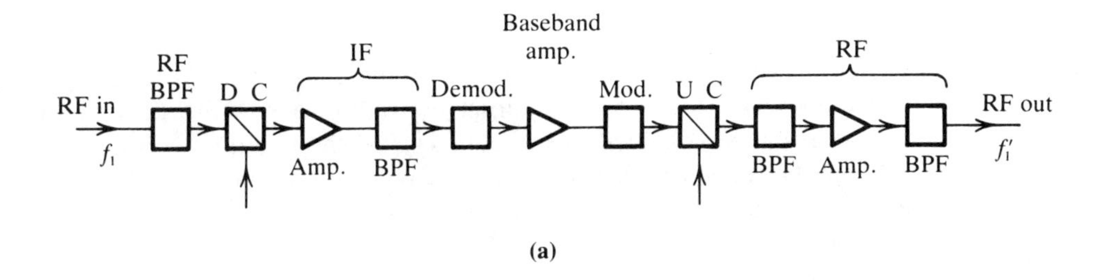

(a)

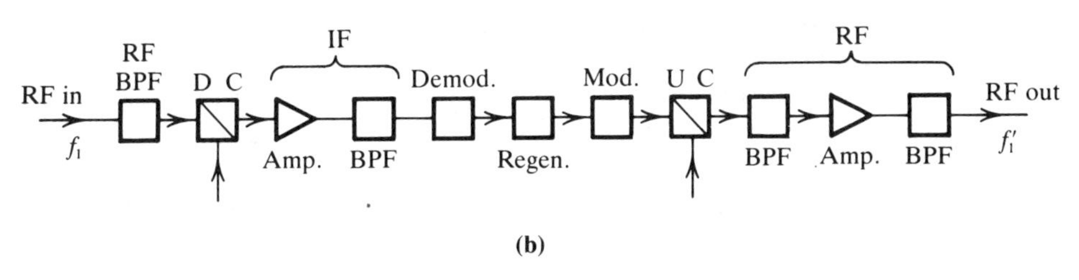

(b)

FIGURE 11.6. Block schematics: (a) baseband repeater; (b) regenerative repeater.

As shown in Fig. 11.5, the incoming carriers on the left-hand side are f_1, f_3, and f_5, horizontally polarized. At the repeater, each carrier is converted to an intermediate frequency of 70 MHz, which is then upconverted to frequency f'_1, f'_3, f'_5. These are transmitted on the next hop as vertically polarized signals. On the right-to-left transmission, the incoming signals at the repeater are vertically polarized at frequencies f_1, f_3, f_5, and these are converted to f'_1, f'_3, f'_5, horizontally polarized, for onward transmission.

Baseband repeaters are also used, in which the incoming RF carrier signal is demodulated, and the recovered baseband signal is amplified and used to modulate the new carrier. This arrangement is shown in Fig. 11.6(a). The baseband repeater is less costly than the heterodyne repeater, but in an analog system it introduces more noise and distortion on the baseband signals. Heterodyne repeaters are used on long-distance trunk routes, and baseband repeaters on short-distance spur routes connected to the main trunk route. The heterodyne repeater can also be used with digital systems, but a much better arrangement, and the one most often used, is the regenerative repeater shown in Fig. 11.6(b). Here the incoming carrier is demodulated, and the time-division-multiplexed (TDM) baseband signal, which is digital, is used to trigger a regenerator circuit. The regenerator circuit produces a digital output stream which is a clean, amplified version of the input digital stream. The output from the regenerator is then used to modulate the new carrier as shown.

11.3.2 Passive Repeaters

Passive repeaters may also be used in certain situations. These are simply reflectors which are used to redirect the radio wave in the same manner as a mirror reflects light waves. Figure 11.7(a) shows one situation in which a parabolic reflector antenna is mounted at the bottom of a tower and an elliptically shaped plane reflector at the top. The plane reflector is mounted at an angle of 45° to the beam, and thus bends it through 90°. The elliptical perimeter is chosen such that the effective area as seen by the primary antenna is circular. This type of reflector system is known as a *periscope system* and is used with high towers, where losses in a long waveguide feed would be excessive. Large, flat "billboard"-type reflectors are also used to overcome obstacles, as shown in Fig. 11.7(b). Many other radio mirror arrangements are possible. The obvious advantages of such systems are that they do not require power and are easier to install in difficult terrain. The disadvantage is they do not provide gain in the way that active repeaters do.

11.4 FREE-SPACE PROPAGATION

Free-space propagation refers to the transmission of radio waves through empty space. Obviously, such a condition cannot be realized where one terminal of the radio link is earth-based. However, it is possible to establish a model for free-space transmission which can then be modified to account for such effects as transmission through the earth's atmosphere, and reflections from surfaces.

Consider first power P_T, assumed to be radiated from a lossless isotropic radiation (see Section 10.3). This will spread out spherically as it travels away from the source, so that at distance d, the power density in the wave, which is the power per unit area of wavefront, will be

$$P_{Di} = \frac{P_T}{4\pi d^2} \text{ W/m}^2 \tag{11.1}$$

This is so because $4\pi d^2$ is the surface area of the sphere of radius d, centered on the source. P_{Di} stands for isotropic power density.

It is known that all practical antennas have directional characteristics; that is, they radiate more power in some directions at the expense of less in others. The gain is the ratio of actual power density along the main axis of radiation of the antenna to that which would be produced by a lossless isotropic antenna at the same distance fed with the same input power (see Section 10.3). Let G_T be the *maximum* gain of the transmitting antenna:

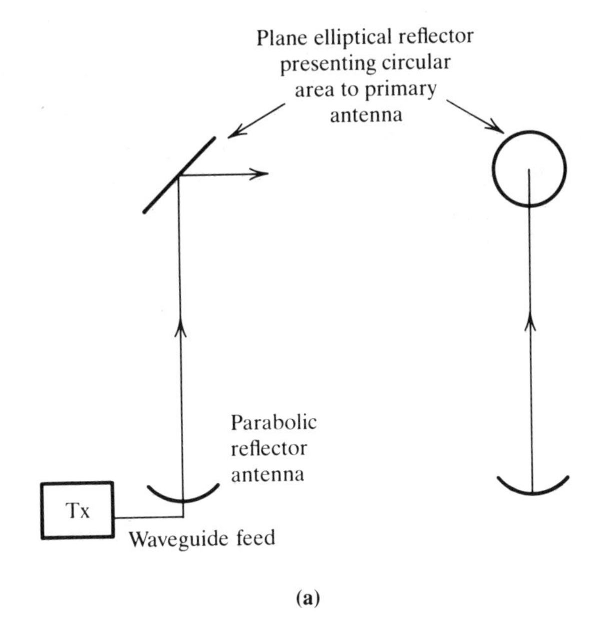

(a)

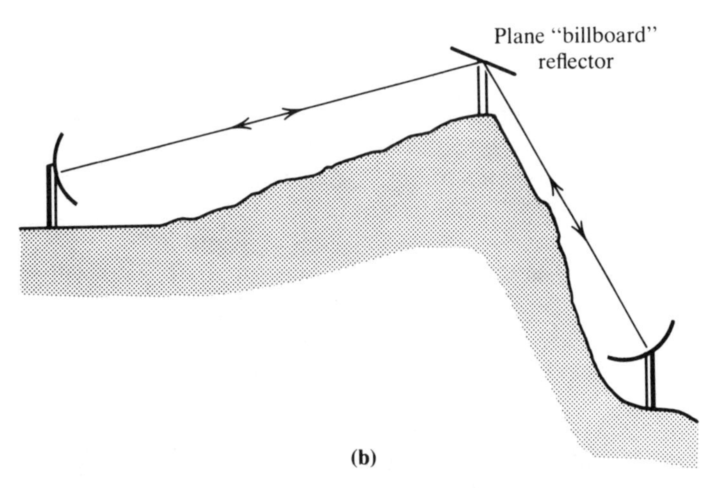

(b)

FIGURE 11.7. Passive repeater arrangements: (a) as a replacement for a waveguide feeder on tower; (b) as an intermediate repeater station.

then the power density along the direction of maximum radiation will be

$$P_D = P_{Di}G_T$$

$$= \frac{P_T G_T}{4\pi d^2} \tag{11.2}$$

A receiving antenna can be positioned so that it collects maximum power from the wave. When so positioned, let P_R be the power delivered by the antenna to the load (receiver) under matched conditions; then the antenna can be considered as having an effective area (or aperture) A_{eff}, where

$$\begin{aligned} P_R &= P_D A_{\text{eff}} \\ &= \frac{P_T G_T}{4\pi d^2} A_{\text{eff}} \end{aligned} \tag{11.3}$$

The ratio of maximum directivity gain to effective area (see Section 10.4) is

$$\frac{A_{\text{eff}}}{G} = \frac{\lambda^2}{4\pi} \tag{11.4}$$

Here, λ is the wavelength of the wave being radiated. Letting G_R be the maximum gain of the receiving antenna, we have from Eqs. (11.3) and (11.4),

$$\frac{P_R}{P_T} = G_T G_R \left(\frac{\lambda}{4\pi d^2}\right)^2 \tag{11.5}$$

This is the fundamental equation for free-space transmission. Usually, it is expressed in terms of frequency f, in MHz, and distance d, in kilometers. Substituting $\lambda = c/f$ in Eq. (11.5) and doing the arithmetic, which is left as an exercise for the reader, the result obtained is

$$\frac{P_R}{P_T} = G_T G_R \frac{0.57 \times 10^{-3}}{(df)^2} \tag{11.6}$$

By expressing the power ratios in decibels, Eq. (11.6) can be written as

$$\left(\frac{P_R}{P_T}\right)_{\text{dB}} = (G_T)_{\text{dB}} + (G_R)_{\text{dB}} - (32.5 + 20 \log_{10} d + 20 \log_{10} f) \tag{11.7}$$

The third term in parentheses on the right-hand side of Eq. (11.7) is the loss, in decibels, resulting from the spreading of the wave as it propagates outward from the source. It is known as the transmission path loss, L. Thus

$$L = (32.5 + 20 \log_{10} d + 20 \log_{10} f)_{\text{dB}} \tag{11.8}$$

where d is in kilometers and f in megahertz.

Equation (11.7) then becomes

$$\left(\frac{P_R}{P_T}\right)_{\mathrm{dB}} = (G_T)_{\mathrm{dB}} + (G_R)_{\mathrm{dB}} - (L)_{\mathrm{dB}} \tag{11.9}$$

EXAMPLE 11.1

Calculate the free-space loss for a terrestrial microwave link 50 km long, operating at a frequency of 4000 MHz.

SOLUTION From Eq. (11.8),

$$\begin{aligned} L &= 32.5 + 20 \log 50 + 20 \log 4000 \\ &= 138.5 \text{ dB} \end{aligned}$$

11.5 ATMOSPHERIC REFRACTION

Atmospheric refraction refers to the bending of the ray path of a radio wave as it passes through the earth's atmosphere. The atmosphere is essentially a gaseous dielectric and as such it has a dielectric constant (or relative permittivity) which affects the phase velocity of the wave in exactly the same manner as the solid dielectric in a transmission line does. The free-space velocity is that of light, or c = 300 Mm/s. In the atmosphere this will be reduced to

$$v_p = \frac{c}{\sqrt{\varepsilon_r}} \tag{11.10}$$

The square root of the relative permittivity is the same thing as the refractive index of the medium. Now if the refractive index were constant, it would have negligible effect on the propagation. However, in temperate regions, the refractive index decreases with increasing height above the earth's surface and this causes higher points on the wavefront to travel at a correspondingly greater phase velocity. Since the wavefront is always normal to the ray path, the ray path must bend in toward the earth. This is illustrated in Fig. 11.8(a). Although the change in refractive index is very small, and the refractive index itself is only slightly greater than unity, the curvature must be taken into account. Rather than deal with curved ray paths over a spherical earth, it is more convenient to assume that the earth's radius is greater than it is, sufficiently so to "straighten" the ray

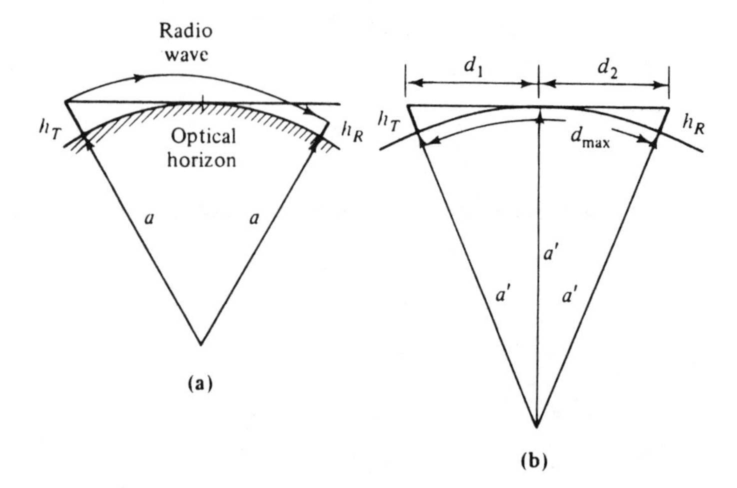

FIGURE 11.8. (a) Curvature of ray path resulting from change of the refractive index of air; (b) equivalent straight-line ray path for fictitious earth's radius a' = 4a/3.

path. This is possible where a linear negative gradient exists for the refractive index and this assumption is usually made in modeling the propagation path for a so-called standard atmosphere. Figure 11.8(b) shows this, in which the earth's radius is increased to $\frac{4}{3}$ times the actual radius of 6376 km; that is, the modified radius is 8501 km.

11.6 THE RADIO HORIZON

Based on straight-line propagation and a smoothly curved earth of radius a', the radio horizon is determined by the point where the ray path just grazes the earth's surface, as shown in Fig. 11.8(b). For this:

$$a' = \tfrac{4}{3}a \tag{11.11}$$

where a is the earth's actual radius and a' is the fictitious radius which accounts for refractions. From Fig. 11.8(b),

$$(a')^2 + d_1^2 = (a' + h_T)^2$$

Therefore,

$$d_1^2 = 2a'h_T + h_T^2 \tag{11.12}$$

But since $a' >> h_T$,

$$d_1^2 \approx 2a'h_T \tag{11.13}$$

Similarly,

$$d_2^2 \approx 2a'h_R \tag{11.14}$$

The maximum radio range d_{max} is

$$\begin{aligned} d_{max} &\approx d_1 + d_2 \\ &= \sqrt{2a'h_T} + \sqrt{2a'h_R} \end{aligned} \tag{11.15}$$

Substituting in the numerical values, $a' = \frac{4}{3} \times 3960$ miles, and expressing h_T and h_R in feet, results in the useful expression

$$d_{max}\text{ (miles)} = \sqrt{2h_T(\text{ft})} + \sqrt{2h_R(\text{ft})} \tag{11.16}$$

Alternatively, in metric units,

$$d_{max}\text{ (km)} = \sqrt{17h_T(\text{m})} + \sqrt{17h_R(\text{m})} \tag{11.17}$$

The phenomenon of *diffraction* will extend the range in practice somewhat beyond the radio horizon.

Example 11.2

Calculate the maximum range for a tropospheric transmission for which the antenna heights are 100 ft and 60 ft.

SOLUTION

$$\begin{aligned} d_{max}\text{ (miles)} &= \sqrt{200} + \sqrt{120} \\ &= 25.1 \text{ miles} \end{aligned}$$

11.7 CONTOUR MAPS

The relationship $d = \sqrt{17h}$, which is one part of Eq. (11.17), is shown plotted in Fig. 11.9(a). Now an interesting geometric property of this curve is that the distance h' between the tangent at any point P_1 and the curve is related to the distance d' [see Fig. 11.9(a)] by the same equation, that is, $d'(\text{km}) = \sqrt{17h'(\text{m})}$. By inverting the curve it can be used as a baseline. The height measured vertically upward from this baseline, when projected to the tangent of the curve, gives the straight line-of-sight distance to the horizon. Graph paper constructed using this principle is shown in Fig. 11.9(b). It will be recalled that the factor 17 in Eq. (11.17) takes into account the modified radius of the earth, and therefore the graph paper also takes this into account. The profile of the terrain between transmit and receive stations may then be plotted on this graph paper as shown in Fig. 11.10, which is based on Eq. (11.17) giving height in meters and distance in kilometers. Similar graph paper can be constructed for height in feet and distance in miles (see Problem 6).

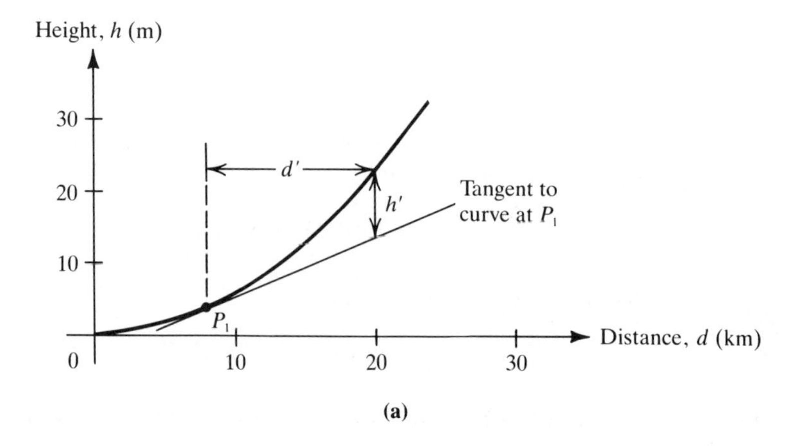

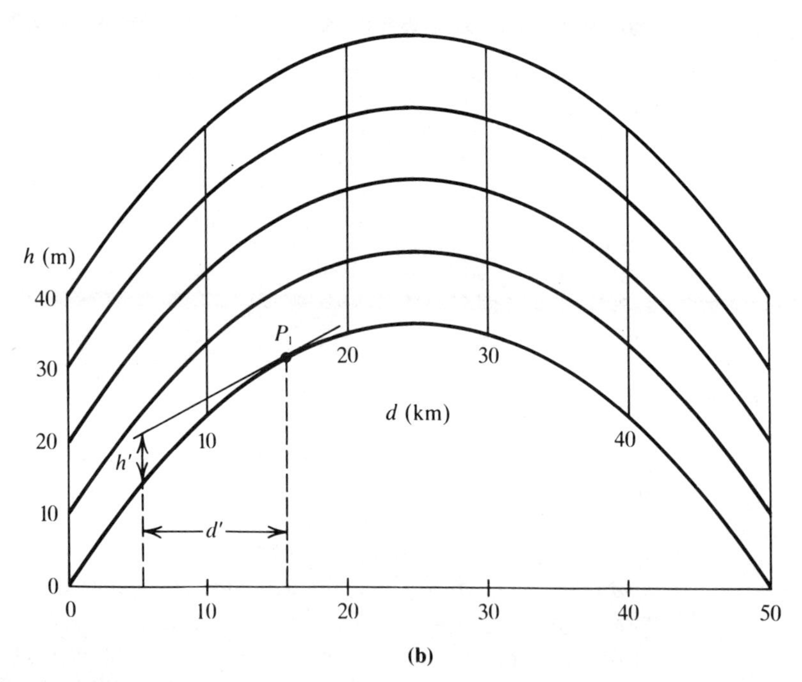

FIGURE 11.9. (a) Construction used for contour graph paper; (b) example of contour graph paper.

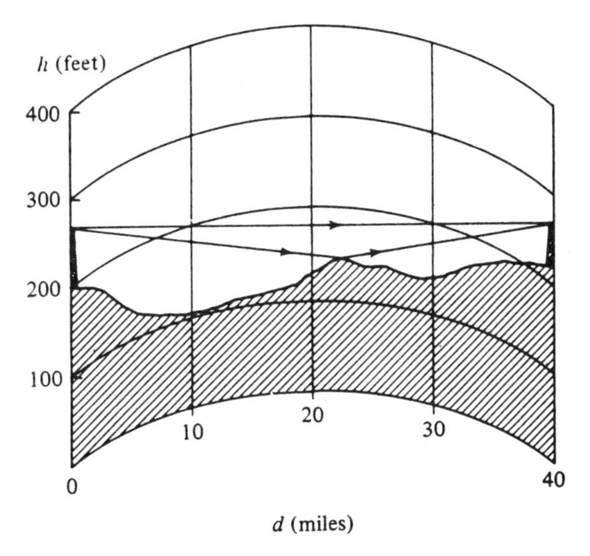

FIGURE 11.10. Example of contour map for radio path planning.

As shown in Fig. 11.10, the radio path may be chosen to avoid obstacles. Although it would appear that it is necessary only for the direct ray path to clear obstacles for line-of-sight transmission to be achieved, practical requirements are somewhat more stringent than this. The wavefront is spherical, as illustrated in Fig. 11.11(a) and energy from a point such as P will also reach the receiver (with a phase lag). An obstacle may interfere with this secondary wavelet generation and therefore reduce the energy flow, even though it does not block the direct ray path. In practice it is found that the zone defined by the indirect path $TP + PR = TR + \lambda/2$ must be cleared for unobstructed transmission to take place. The path traced out by the point P for $TP + PR = TR + \lambda/2$, is an ellipse, and extended to the space around the direct ray path forms an ellipsoid of revolution known as the first Fresnel zone. This is the zone that must clear any opaque obstacles for line of sight, or equivalently, free-space propagation conditions to exist. From the geometry of Fig. 11.11(b) it is readily shown that the radius r that must be cleared is given by

$$r = \sqrt{TP' \times P'R \frac{\lambda}{TR}} \tag{11.18}$$

EXAMPLE 11.3

Calculate the clearance required at the midpoint of a 50-km path for a microwave link operating at 6 GHz.

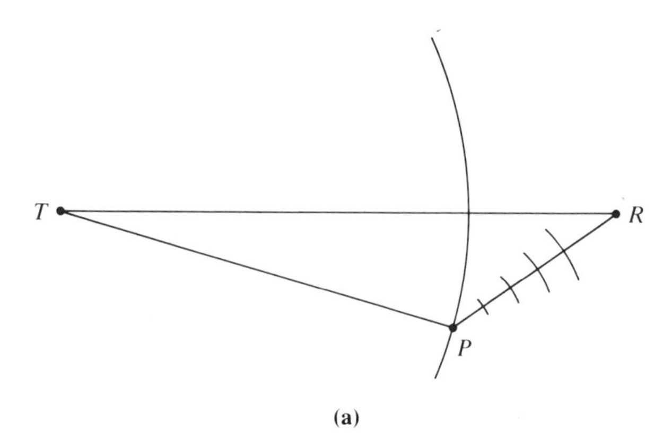

(a)

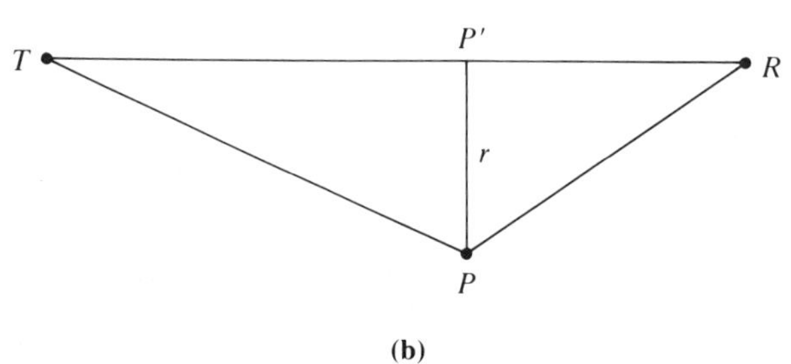

(b)

FIGURE 11.11. (a) Secondary wave generation at point P on spherical wavefront; (b) geometry used in determining Fresnel zone.

SOLUTION

$$\lambda = \frac{3 \times 10^8}{6 \times 10^9} = 5 \text{ cm}$$

From Eq. (11 18), for $TP' = P'R = TR/2$,

$$r = \sqrt{\frac{TR \cdot \lambda}{4}}$$

$$= \sqrt{\frac{50 \times 10^3 \times 0.05}{4}}$$

$$= \mathbf{25\ m}$$

11.8 SURFACE REFLECTIONS

A radio wave reflected from the earth's surface undergoes an amplitude and a phase change in general. This is similar to the reflections that occur for transmission lines as described in Section 1.10, but in this case the reflection coefficient depends on the nature of the reflecting surface (e.g., soil, water, etc.), on the frequency, on the polarization, and on the angle between the ray path and the reflecting surface. In most practical situations encountered in microwave terrestrial links, the phase angle of the reflection coefficient is 180° for both horizontally and vertically polarized waves. The magnitude of the reflection coefficient will depend on the particular conditions at the reflection surface, but of course can never exceed unity. The nature of the reflection also depends on the degree of surface roughness. The Rayleigh criterion for surface roughness is that the height of surface irregularities should not exceed 3.6 λ/ψ, where ψ is the grazing angle in radians, that the ray path makes with the surface:

$$h \leq 3.6 \frac{\lambda}{\psi} \tag{11.19}$$

For example, calm water would act as a smooth or specular reflector (mirrorlike).

EXAMPLE 11.4

A ray path makes an angle of 2° with the surface of the sea. The frequency of transmission is 6 GHz. Determine the upper limit for average wave height for which the surface may be considered smooth.

SOLUTION A frequency of 6 GHz has a wavelength of 5 cm. Using Rayleigh's criterion, the height is given by

$$\text{height} = 3.6 \times 5 \text{ cm} \times \frac{57.3}{2}$$

$$= \mathbf{516\ cm}$$

Vegetation tends to absorb microwave radiation (this being the principle behind microwave cooking), and this is particularly true for trees. Thus tree-covered terrain will attenuate microwave signals unless the path is well clear, by at least the first Fresnel zone.

Reflections from buildings will vary widely depending on the shape and the proximity of the building to one or other of the antennas in the link. Large flat surfaces such as roofs and walls close to an antenna produce specular reflections, but otherwise buildings tend to scatter the energy rather as rough surfaces do.

11.9 SUPER- AND SUBREFRACTIONS

Irregularities in the earth's atmosphere also affect tropospheric transmissions. A condition known as *superrefraction* occurs when the refractive index of the air decreases with height much more rapidly than normal, so

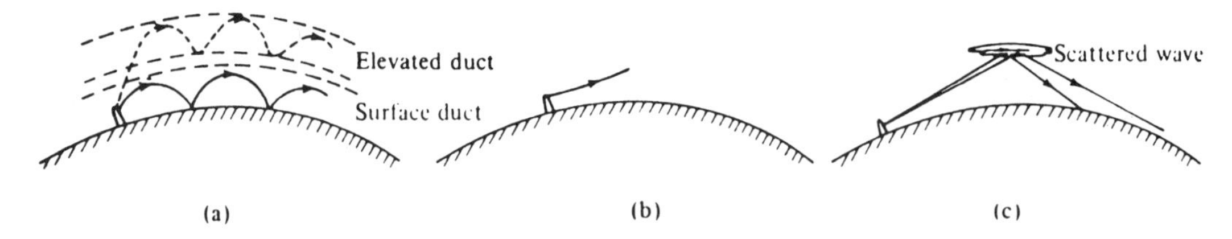

FIGURE 11.12. (a) Super-refraction; (b) tropospheric scatter propagation.

that the bending of the radio wave is much more pronounced than that shown in Fig. 11.8(a). The radio wave may then be reflected back from the earth to follow a path as sketched in Fig. 11.12(a). In this way, the radio range is considerably increased. Unfortunately, the effect is not sufficiently reliable for it to be utilized for commercial communications systems, but it does account for some of the abnormally long distance interference which has been observed at VHF.

An increase of temperature with height (known as temperature inversion) gives rise to superrefraction, as does an increase of humidity with height. It is most noticeable when both of these effects occur together. The region in which superrefraction occurs is termed a *duct*, which can be formed both at the earth's surface and in elevated strata, as sketched in Fig. 11.12(a).

It is also possible for the opposite effects to occur, giving rise to *subrefraction*, which reduces signal strength by bending the ray away from the receiving point [Fig. 11.12(b)].

Inhomogeneities in the atmosphere can give rise to a scattering of radio signals, and by using highly directional high-gain antennas, and large transmitted power, reliable communication links well beyond the radio horizon can be established [Fig. 11.12(c)]. The method is referred to as *tropospheric scatter propagation*; ranges up to 400 miles, in the frequency band 40 to 4000 MHz, have been achieved.

11.10 ATTENUATION IN THE ATMOSPHERE

The space wave is affected by atmospheric conditions, but only seriously at frequencies in the microwave region, above about 10 GHz. Heavy rain such as occurs in the tropics results in serious attenuation of electromagnetic waves at frequencies above about 10 GHz, while moderate rain, cloud, and fog will seriously attenuate electromagnetic waves at frequencies above about 30 GHz. Hail has little effect except at extra-high frequencies (e.g., above about 100 GHz), and the effect of snow is negligible at all frequencies.

The gas molecules in the air can result in attenuation of electromagnetic waves. Vibrational resonances in the water vapor molecule (H_2O) result in absorption peaks at wavelengths of 1.35 cm and 1.7 mm; the oxygen molecule (O_2) exhibits similar absorption peaks at 5 mm and 2.5 mm. Clearly, frequencies at which these resonances occur must be avoided in space-wave transmissions. Some of these results are summarized in Fig. 11.13.

11.11 FADING

Fading refers to the reduction, or loss, of signal which can occur through changes in the propagation path. For example, subrefraction, described in Section 11.9, may cause the signal to bend away from the receive antenna, as sketched in Fig. 11.14(a), which results in a reduction in signal strength. Multipath propagation can exist between two antennas, as illustrated in Fig. 11.14(b). Normally, the path will be chosen to avoid ground reflections, but this is not always possible. For example, over large stretches of water it is not possible to use intermediate repeaters to keep path lengths short, and both the direct wave and the ground reflected wave will be received. Under normal conditions the combined signal strength may be satisfactory, but changes may occur in the refractive index of the air which change the relative path lengths. The corresponding phase change can result in signal cancellation. This type of fading is referred to as *reflection multipath fading*. Superrefraction can also produce a second atomospheric ray path, as sketched in Fig. 11.14(c). Again, the phase difference caused by the path difference between the direct ray and the atmospheric refracted ray can be such as to result in signal cancellation, or fading. This is known as *atmospheric multipath fading* and is the most serious cause of fading at frequencies below about 8 GHz. Above about 8 GHz, attenuation resulting from rain is the most serious cause. To combat fading, special schemes referred to as *diversity schemes* are used, and these are described in Section 11.13. In any event, fading cannot be eliminated, and a fading margin must be allowed for in the system design. There are equations available, known as the *Barnett–Vignant equations* [see, e.g., Feher (1981)] which allow the fading margin to be calculated for a given type of terrain and path length, frequency, and reliability objective. The latter is a figure that specifies the maximum amount of time the system may be out of service; for example, utility companies specify a reliability objective of 0.0001 for worst-month operation, meaning that the system must be available 99.99%– of the time during the worst fading month. A fading margin of 30 dB is typical for a 99.99% reliability objective.

The received carrier power must be sufficient to keep down the noise in the receiver, since a high noise level would degrade the signal-to-noise performance in analog systems and introduce errors in a digital system. Equation (11.9) gives the basic relationship between transmit and receive

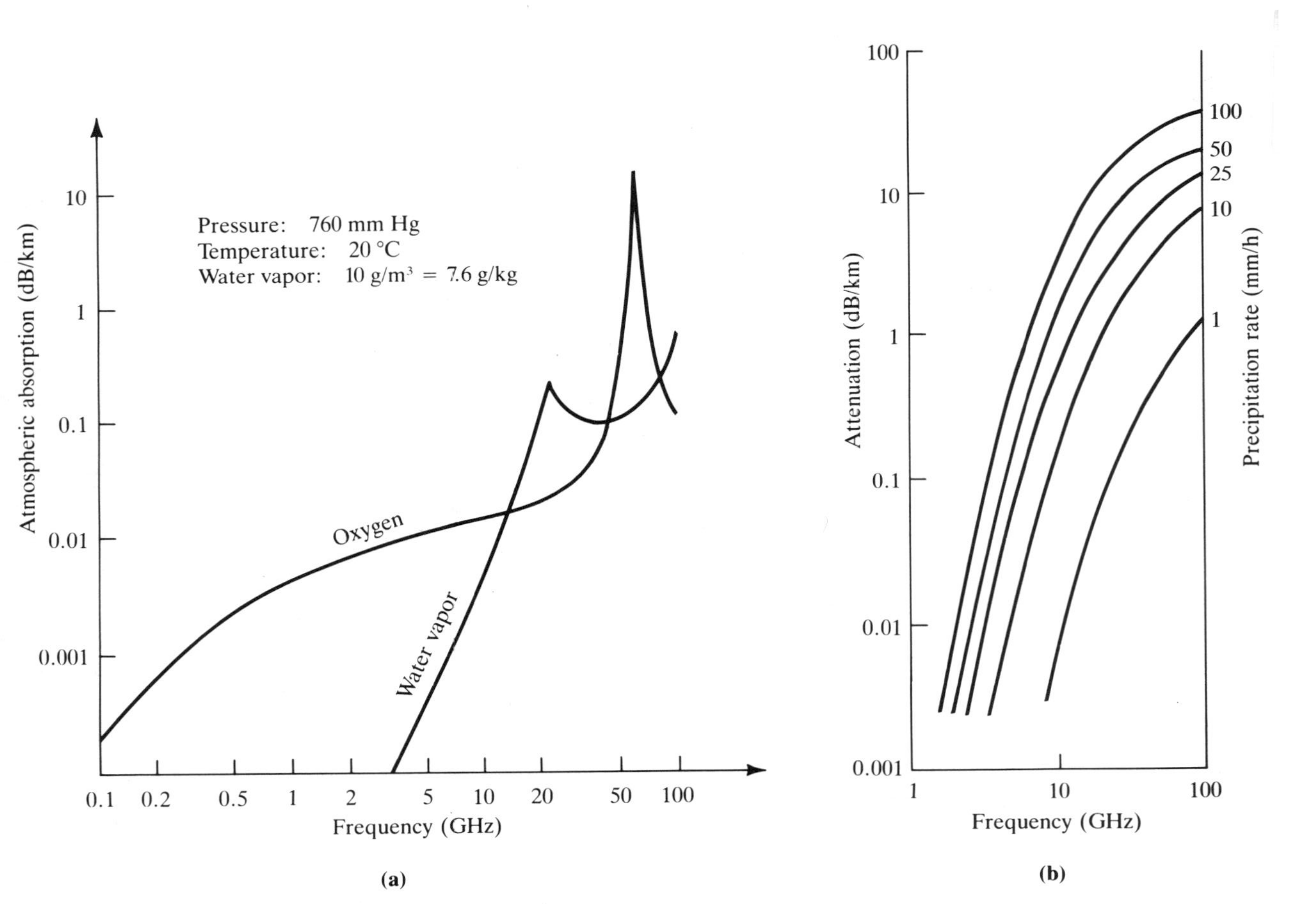

FIGURE 11.13. (a) Path loss caused by absorption of oxygen and water vapor molecules; (b) variation of path loss due to rain attenuation. [*From H. B. Wood, "Microwave Radio Systems," in Advanced Communication Systems, ed. B. J. Halliwell* (*Newnes-Butterworth, 1974*).]

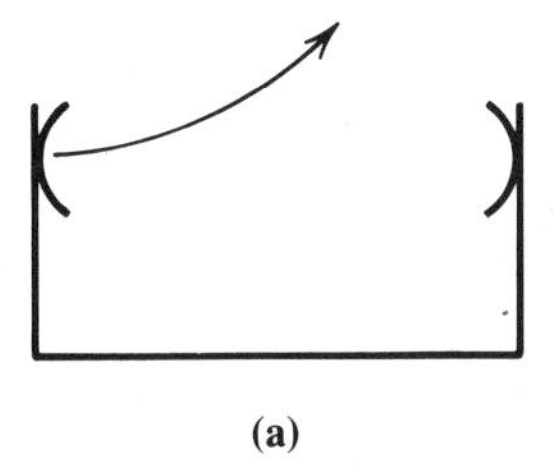
(a)

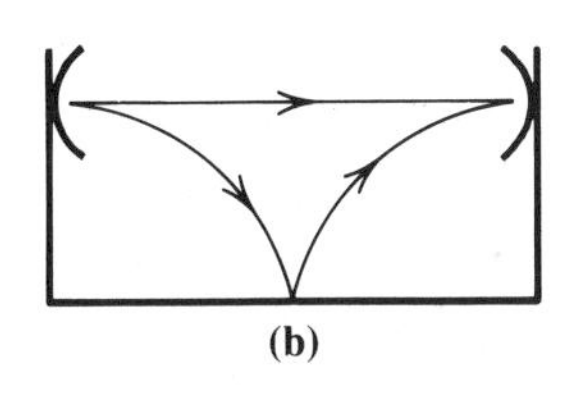
(b)

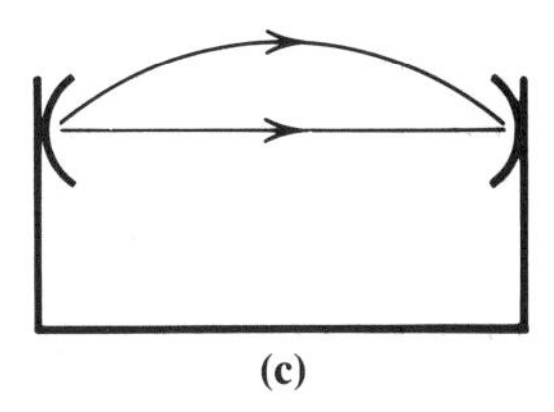
(c)

FIGURE 11.14. Possible fading mechanisms: (a) subrefraction; (b) multipath; (c) super-refraction multipath.

powers. However, it is usually the output power of the final transmitter amplifier that is a design parameter, and this will be denoted by P_{TO}. Also, the minimum acceptable receiver carrier power will be denoted by $C_{\min}$. In addition to the free-space path loss, allowance must be made for a fading margin as discussed in the preceding section, and this will be denoted by FM(dB). Also, as shown in Fig. 11.4, transmitters and receivers may be connected to a single line through filters and circulators, and the losses in these are termed *branching losses*. Branching losses will be denoted by L_b(dB). Finally, there will be losses in the feeders connecting the transmitter and the receiver to the antennas, and these will be denoted by L_f(dB). Therefore, allowing for the fading margin and the various losses, Eq. (11.9) can be rearranged as

$$P_{\mathrm{TO}} \geq C_{\min} + \mathrm{FM} + L + L_f + L_b - G_T - G_R \qquad (11.20)$$

where all quantities are in decibels.

EXAMPLE 11.5

The required receiver power to meet a given reliability objective is -80 dBm. The system operates at a frequency of 6 GHz, and the antenna gains are 44.8 dB for both transmitting and receiving antennas. The path distance is 40 km and the fading margin is calculated as 39 dB, the branching losses as 2 dB, and the feeder losses as 2 dB. Calculate the transmitter power output required to meet the minimum received carrier objective.

SOLUTION From Eq. (11.8),

$$L = 32.5 + 20 \log 40 + 20 \log 6000$$
$$= 140.1 \text{ dB}$$

From Eq. (11.20),

$$P_{\mathrm{TO}} = (-80) + 39 + 140.1 + 2 + 2 - 44.8 - 44.8$$
$$= \mathbf{13.5\ dBm}$$

11.12 SYSTEM GAIN

System gain is a measure of overall system performance and in terms of the quantities derived in the preceding section is defined as

$$G_s = P_{\mathrm{TO}} - C_{\min} \quad \mathrm{dB} \tag{11.21}$$

The system gain may be of the order of 100 dB for a well-designed system. The higher the system gain, the better, and as reported (Feher 1981, p. 137) for digital systems operating in the 11-GHz band at a transmission rate of 90 Mb/s, system gain is expected to reach 115 dB.

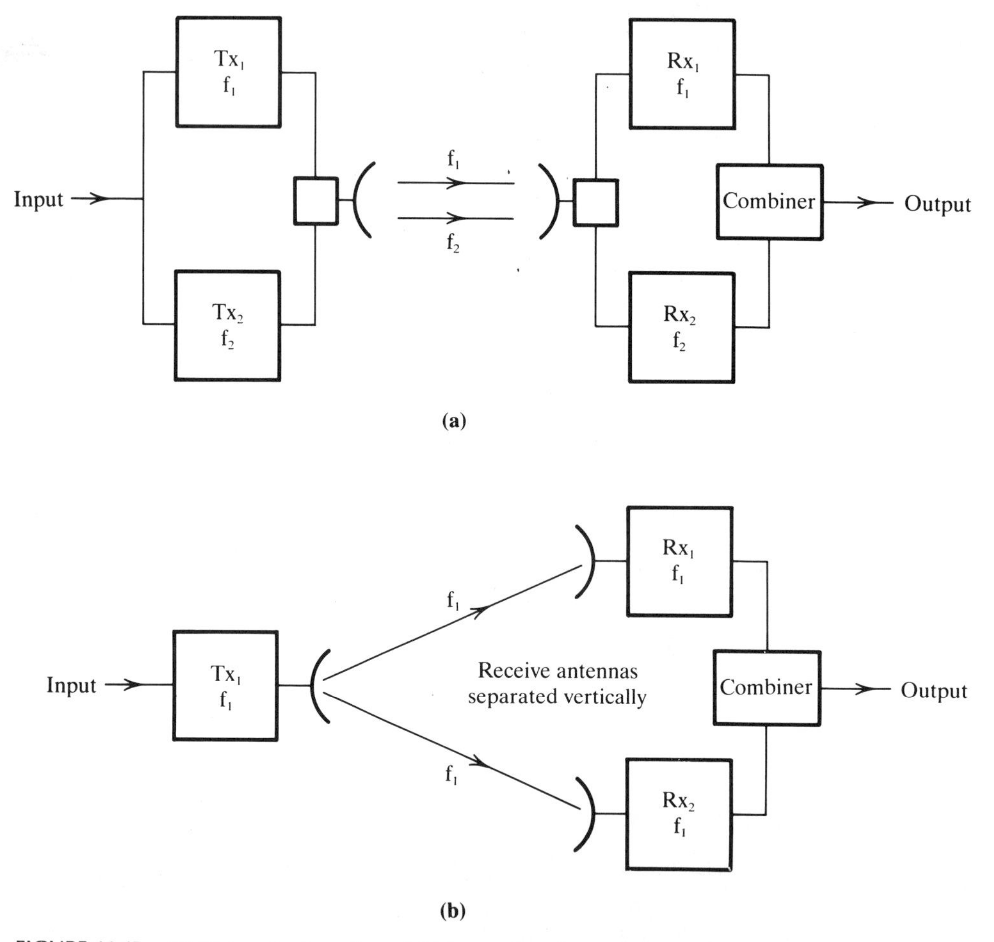

FIGURE 11.15. (a) Frequency diversity and (b) space diversity arrangements.

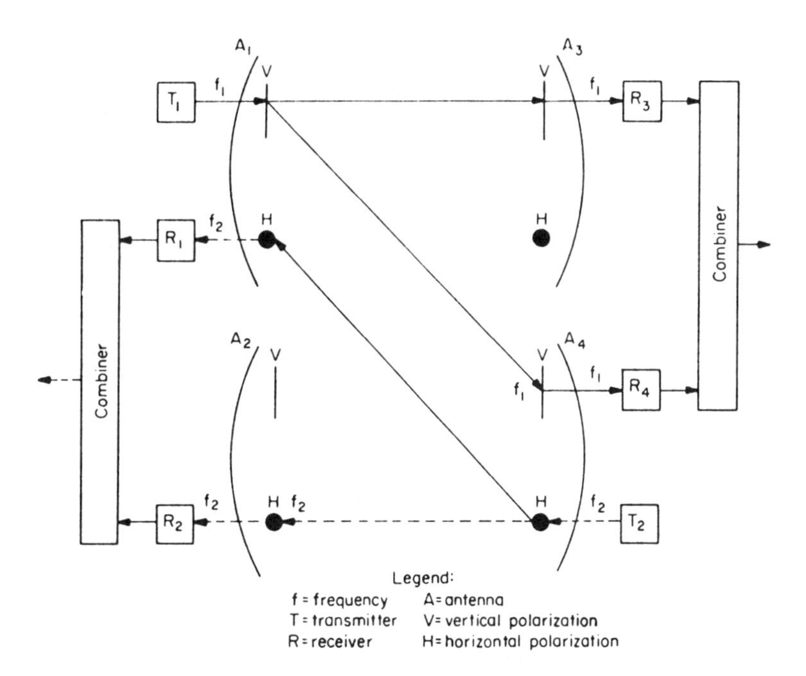

FIGURE 11.16. Twofold space diversity configuration. [*From Philip F. Panter, Communication Systems Design (New York: McGraw-Hill, © 1972); with permission.*]

11.13 DIVERSITY SYSTEMS

Diversity methods are used to combat short-term or rapid fading. As the name suggests, the signal may be sent over different, or diverse, paths. More specifically, diversity may be achieved by setting up different paths in space, or it may be achieved by using different frequencies over the same space path. These are known as *space diversity* and *frequency diversity*, respectively, and are illustrated in Fig. 11.15.

Diversity systems operate on the principle that statistically, separate radio paths as decribed are highly unlikely to fade simultaneously for signals that are well separated in frequency or space. For frequency diversity, the frequency separation required can be as great as 100 MHz or more. Because of the very high demand for channels in the microwave bands, licensing restrictions prevent widespread use of frequency diversity. Also, of course, double the number of transmitters and receivers are required, which increases system costs.

With space diversity, at least two antennas are required at each receiver site. These antennas must be spaced vertically a distance of at least 50 wavelengths. Figure 11.16 shows the arrangement for what is termed *twofold space diversity*. Here the left-to-right transmission is at frequency

f_1 vertically polarized. This transmission is received on two antennas spaced vertically, and the outputs of receivers R_3 and R_4 are combined to provide a single output. Similarly, the right-to-left transmission is on frequency f_2 horizontally polarized, and the outputs of receivers 1 and 2 are combined. Many methods of combining the signals are in use, and in fact it is possible to combine the signals at the antennas, thus avoiding the need for separate receivers. Such a method is described in Halliwell (1974). Although some improvement in signal-in-noise ratio may be achieved by using diversity, it should be kept in mind that the reason for its use is to reduce short-term fading, which is particularly troublesome for paths that extend over long stretches of water.

11.14 SATELLITE LINKS

A communications satellite is a spacecraft placed in orbit around the earth which carries on board microwave receiving and transmitting equipment capable of relaying signals from one point on earth to other points. Microwave frequencies must be used to penetrate the ionosphere, since all practical satellite orbits are at heights well above the ionosphere. Also, microwave frequencies are required to handle the wideband signals encountered in present-day communications networks and to make practical the use of high-gain antennas required aboard the spacecraft.

The first commercially operated satellite was launched in August 1965. Since that time numerous satellites have been launched for communications purposes. Such communications services include point-to-point telecommunications circuits; wide-area TV coverage, often referred to as direct broadcasting by satellite (or DBS); and navigational and communications services to ships and aircraft. Table 11.1 lists present and proposed frequencies to be used for satellite communications.

Satellites presently in use are active satellites, that is, the signal received by the satellite is retransmitted rather than being simply reflected back to earth. This means that the satellite has on-board, highly directional transmitting and receiving antennas, and complex interconnecting circuits. Accurate positioning and control mechanisms are required for the satellite. The power requirements for the on-board equipment are usually obtained from arrays of solar cells, with backup nickel–cadmium batteries for periods of solar eclipse.

Most communications satellites are placed in the geostationary orbit. The rotational period of the earth about its own axis is 23 hours and 56 minutes, and a satellite in geostationary orbit, traveling in the same direction as the earth's rotation, completes one revolution about the earth's axis in this time. The satellite therefore appears stationary to an observer on earth—hence the name *geostationary*. Keeping track of a geostationary

TABLE 11.1
Satellite Frequencies

Frequency (GHz)	*Direction*
1.530–1.559	Down
1.6265–1.6605	Up
3.400–4.200	Down
5.850–7.075	Up
7.250–7.750	Down
7.900–8.400	Up
10.70–12.70	Down
12.70–13.25	Up
14.00–14.80	Up
17.30–17.70	Up
17.70–18.10	Both
18.10–20.20	Down
27.00–30.00	Down

satellite is relatively easy, and the satellite is continually visible from within its service area on earth. Another advantage of the geostationary orbit is that the Doppler shift of frequency is negligible. Doppler shift in frequency results when there is relative movement between source and receiver. The height required for geostationary orbit may be deduced from the dynamics of motion and is approximately 36,000 km.

In addition to the path loss given by Eq. (11.8), absorption and scattering of the signal will occur as it passes through the troposphere and ionosphere, as shown in Fig. 11.17. In this case, the loss will be proportional to the path length in the attenuating medium, and this in turn depends on the angle of elevation of the ground antenna. The attenuation in the atmosphere varies with frequency as shown in Fig. 11.18. These results are for transmission through a moderately humid atmosphere, measured at sea level. Two absorption peaks are observed. The first, at a frequency of 22.2 GHz, results because water vapor molecules go into vibrational resonance at this frequency, and in so doing, absorb energy from the wave. The second peak at 60 GHz is similar, and is caused by resonant absorption of oxygen molecules. The curves show the effect of the angle of elevation on attenuation resulting from the greater path length. At 4 GHz, for example, the total atmospheric attenuation for vertical incidence is just over 0.04 dB, whereas for a 5° angle of elevation, it is about 0.1 dB. Free-electron absorption which occurs in the ionosphere is negligible at microwave frequencies.

Attenuation will also occur with precipitation, being worse for heavy rain. In system design, a fading margin would be allowed for rainfall, the value depending on the geographical location of the ground station.

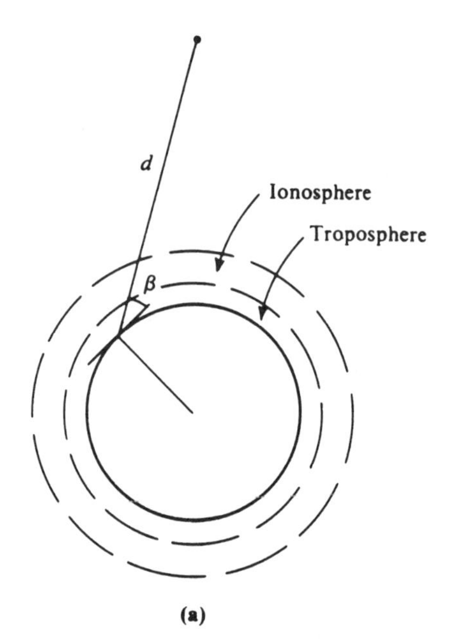

FIGURE 11.17. Transmission path, including atmosphere and ionosphere.

A satellite acts as a central or "star" point in a network system. This is illustrated in Fig. 11.19. The satellite network differs from the terrestrial network, which is an "in-line" system, although much of the equipment may be similar. In particular, access to the satellite network requires special schemes, some of which are described in Roddy and Coolen (1984). A central control station is shown by way of illustration, although in practice, various control schemes are in use. Satellite attitude-control (station-keeping) signals may be generated by means of satellite "on-board" equipment. Also, frequency allocation schemes such as are used for multiple access [see Roddy and Coolen (1984)] may be assigned through a central control station or by control functions at each ground station, depending on the particular network.

The frequency ranges most commonly used are 4 to 6 GHz and 12 to 14 GHz. The first number specified in each range gives the downlink frequency (to the nearest GHz), the second figure the uplink frequency. The Canadian Anik-B satellite system provides a good illustration of satellite technology, and a description of it is given here to bring out some of the techniques used.

The Canadian Anik-B satellite is dual-band, capable of operating in both ranges simultaneously. The 12- to 14-GHz channeling scheme for the Anik-B satellite is shown in Fig. 11.20. Each RF channel is 72 MHz wide, and there is a guard band of 8 MHz between channels. The satellite transmits in the range 11.7 to 12.2 GHz (downlink), and receives the range 14 to 14.5 GHz (uplink).

As shown in Fig. 11.20, the downlink carriers are horizontally polarized, the uplink carriers vertically polarized. Polarization in this case is with reference to the earth's N–S axis, vertical being parallel to this axis, and horizontal orthogonal to it.

The various blocks of equipment aboard the satellite are shown in Fig. 11.21(a) for the range 12 to 14 GHz. Four antennas are employed which, when transmitting, produce spot beams over Canada in the west

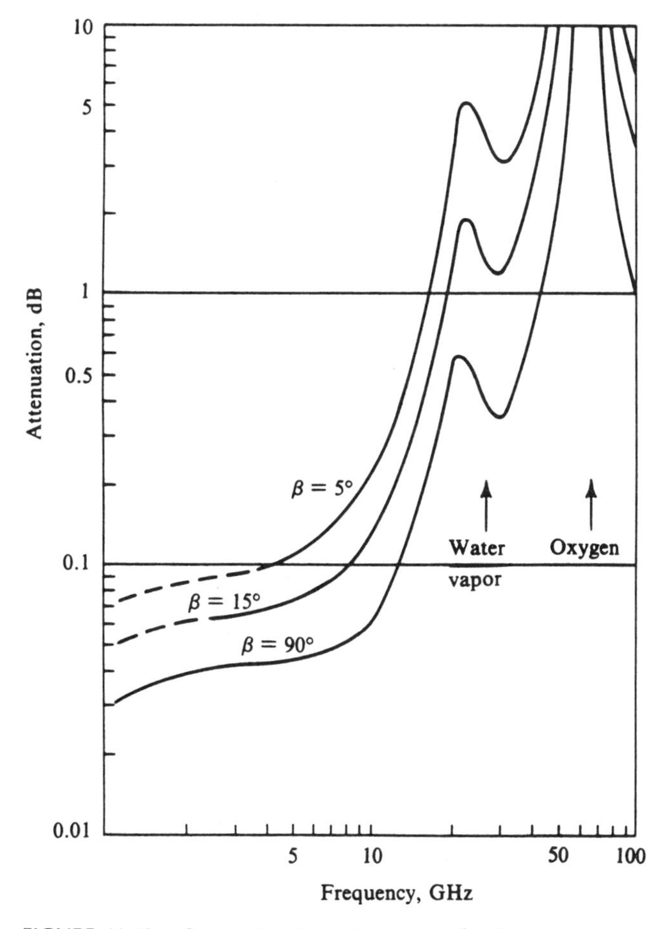

FIGURE 11.18. Composite attenuation curves for the atmosphere, showing both oxygen and water vapor absorption effects. The antenna angle of elevation is β.

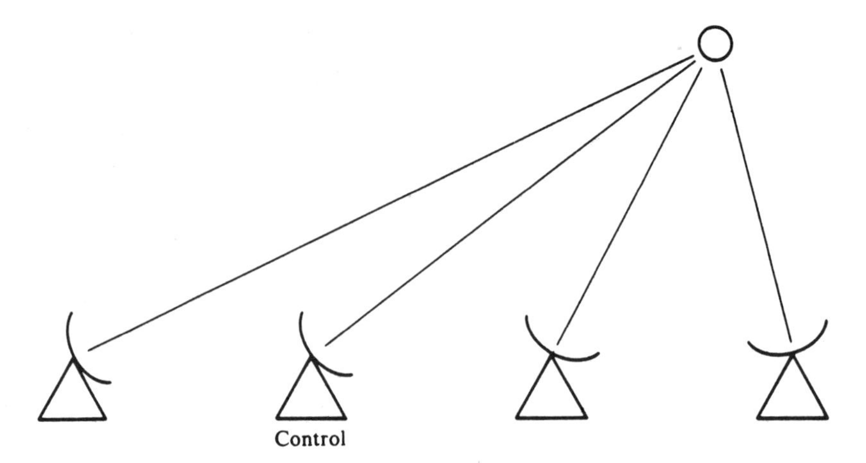

FIGURE 11.19. Satellite network showing the satellite as the center of the "star."

(W), central west (CW), central east (CE), and east (E) regions. When receiving, the antennas are coupled together by means of the orthomode couplers, which respond to the received vertical polarization. The combined received signal consists of all six channels (i.e., a wideband signal ranging from 14 to 14.5 GHz). The received signal passes through a number

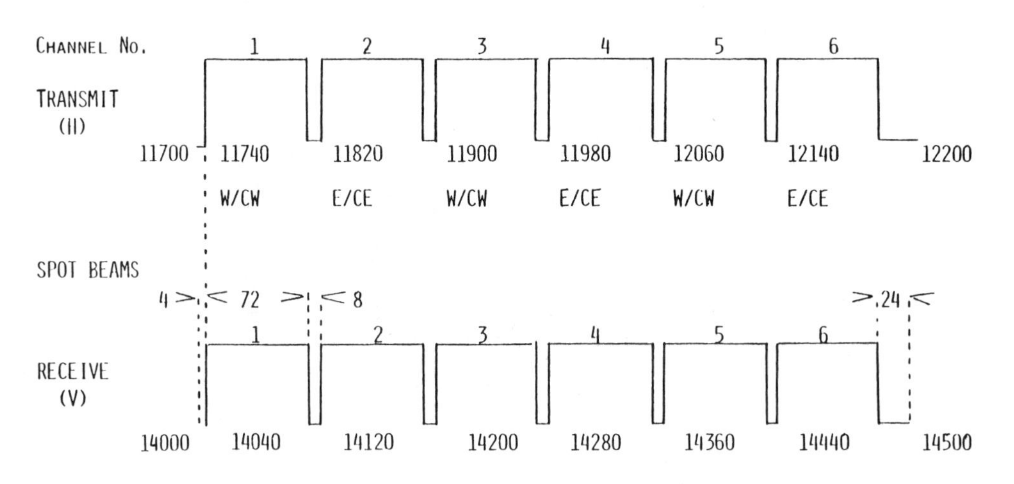

FIGURE 11.20. Anik-B channel plan. *(From J. W. B. Day, Anik-B DOC Communication System and Ground Terminals, Communications Research Centre, Department of Communications, Ottawa, Canada, 1980; with permission.)*

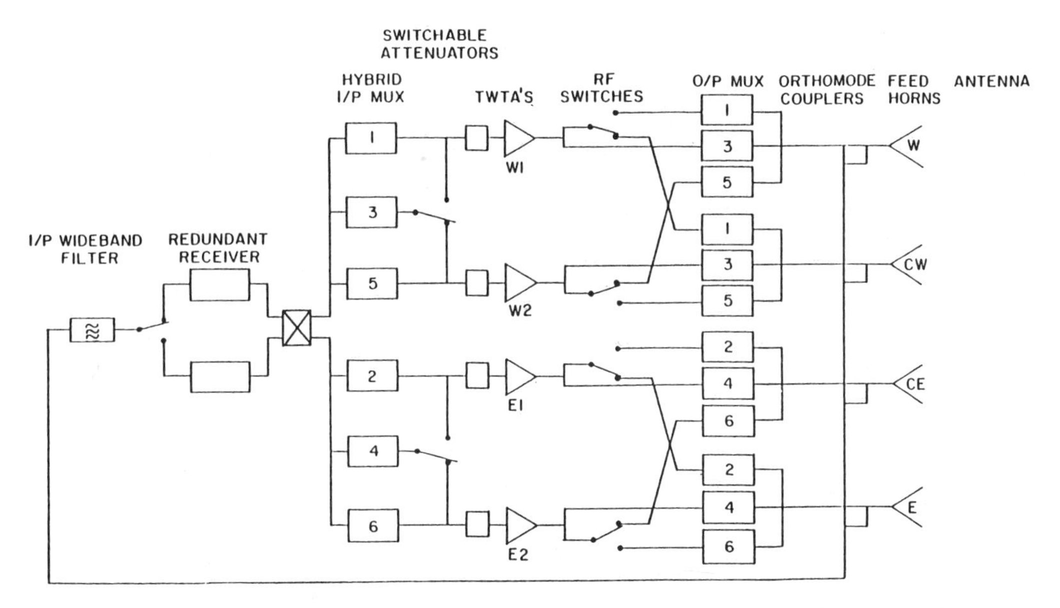

FIGURE 11.21. (a) ANIK block diagram; (b) satellite single-conversion receiver. [*Part* (a) *from J. W. B. Day, Anik-B DOC Communications System and Ground Terminals, Communications Research Centre, Department of Communications, Ottawa, Canada, 1980; with permission.*]

of units which collectively form what is termed the satellite transponder. In the case of the Anik-B satellite, it can be seen from Fig. 11.21(a) that there are six transponder channels. These share a common front-end receiver section, and the four traveling-wave-tube amplifiers (TWTAs) are shared between all six channels on a switched basis. For example, a telephony signal in the channel 4 frequency band received on the W antenna is routed through the common front end, through the input hybrid multiplexer, and, as shown in the figure, is then switched to E2 TWTA, which then feeds it to the E antenna for retransmission. This channel could be switched to the El TWTA and onto the CE antenna.

The received wideband signal is amplified and down-converted to the frequency range 11.7 to 12.2 GHz in the common front end of the receiver, comprising a low-noise amplifier (LNA), a downconverter, and a local oscillator at 2.3 GHz, as shown in Fig. 11.21(b).

RF switching, controlled from the ground station, enables various combinations of channels to be directed to various antennas. On transmission, RF channels 1, 3, and 5 can be switched between the W and the CW antennas. RF channels 2, 4, and 6 between the E and the CE antennas. Figure 11.21(a) illustrates the situation where channels 1 and 3 are switched to the CW antenna and channel 5 is switched to the W antenna. Channels 2 and 4 are switched to the E antenna, channel 6 to the CE antenna.

11.14.1 Link Budget Calculations

The free-space propagation equations developed in Section 11.4, together with a knowledge of receiver noise, may be used to calculate the power requirements for a satellite link. The concept of noise temperature of an amplifier is discussed in Section 8.11. In addition to amplifier noise, the receiving antenna, and the various waveguide and branching losses also contribute to the input noise of a receiver. The system noise temperature T_s takes all these sources into account. The noise power density at the receiver input is then kT_s (watts/hertz or joules, where k is Boltzmann's constant, equal to 1.38×10^{-23} J/K. The received signal power is P_R in Eq. (11.5). Thus the received signal-to-noise density ratio, denoted by C/N_0, is equal to P_R/kT_s. The parameter of interest in a satellite receiving system is the ratio of receiving antenna gain to system noise temperature, G_R/T_s, usually expressed in decilogs as 10 log (G_R/T_s) dB-K, and as will be shown, this cnters into the C/N_0 ratio.

The parameter of interest at the transmitter is the equivalent isotropic radiated power, EIRP. In terms of the factors in Eq. (11.5), the EIRP is given by $P_T \cdot G_T$. It is left as an exercise (Problem 13) for the student to show that Eq. (11.9) can be rewritten as

$$\frac{C}{N_0} = \text{EIRP} - M - L + \frac{G_R}{T_s} + 228.6 \qquad (11.22)$$

All quantities are in decibels or decilogs, and the number 228.6 is equal to $-10 \log k$. The propagation loss L in decibels is given by Eq. (11.8), and M is the total margin, in decibels, to allow for antenna pointing errors, fading, and waveguide and branching losses.

EXAMPLE 11.6

In a satellite communications link the satellite EIRP is 54 dBW and the earth station G_R/T_s is 17.7 dB-K. The propagation loss is 210 dB. The margins are: fading margin, 1 dB; antenna pointing margin, 1 dB; and equipment implementation margin, 0.95 dB. Calculate the received C/N_0 ratio.

SOLUTION From Eq. (11.22),

$$\frac{C}{N_0} = 54 - (1 + 1 + 0.95) - 210 + 17.7 + 228.6$$

$$= \mathbf{87.35\ dB\text{-}Hz}$$

In digital systems the ratio of energy per bit to noise density, E_b/N_0, is usually a specified quantity. It may be shown (see Problem 15) that knowing the transmission rate R, the C/N_0 ratio can be found from

$$\frac{C}{N_0} = \frac{E_b}{N_0} + R \tag{11.23}$$

All quantities are in decibels or decilogs.

EXAMPLE 11.7

Determine the E_b/N_0 ratio for the link calculation in Example 11.6 given that the satellite is transmitting a digital signal at a rate of 61 Mbps.

SOLUTION The transmission rate in decilogs is 10 log (61 × 10^6) = 77.85 dB. Hence, from Eq. (11.23),

$$\frac{E_b}{N_0} = 87.35 - 77.85 = \mathbf{9.5\ dB}$$

11.15 PROBLEMS

1. Describe the general features of terminal stations used in terrestrial microwave radio links. For the frequency scheme shown in Figure 11.3, determine (a) each carrier frequency, (b) the channel width for each carrier, and (c) the guard bandwidth on either side of the 4003.5-MHz frequency.
2. Describe the general features of (a) heterodyne, (b) baseband, and (c) regenerative repeaters. Indicate where each type is likely to be used and give the main advantages and disadvantages of each.
3. Explain what is meant by a passive repeater and how this differs from an active repeater. State where passive repeaters are likely to be used.
4. A 6000-MHz 1-W microwave carrier is fed into the transmitting antenna of a terrestrial microwave link. The antenna has a gain of 48 dB, and the link distance is 50 km. Calculate the power density at the receiving antenna, assuming that spreading losses only are present. (b) Given that the gain of the receiving antenna is also 48 dB, calculate the power delivered to a matched load.
5. A microwave transmission path is to be established over smooth terrain between transmit and receive antennas mounted on 20-m towers. Cal-

culate the maximum range possible and state the conditions for which the calculation is valid.

6. Using Eq. (11.16) and following the construction outlined in Section 11.7, construct contour graph paper showing height in feet and distance in miles. Limit the distance scale to 20 miles.
7. Explain what is meant by (a) the Fresnel zone and (b) the Rayleigh criterion, and the effect these have on terrestrial microwave communications.
8. Calculate the clearance required at the midpoint of a 40-km microwave transmission path, the frequency of the signal being 4 GHz.
9. Discuss the factors that can give rise to fading of microwave signals. How is fading allowed for in the design of microwave links?
10. A microwave radio link requires a receiver power of -85 dBm to meet the specified reliability objectives. Branching losses are 1.5 dB, feeder losses are 2 dB, and the transmit and receive antenna gains are each 48 dB. The fade margin is calculated as 37 dB. The path distance is 45 km and the frequency is 4 GHz. Calculate (a) the minimum transmitter power output that will meet the requirements and (b) the system gain.
11. Explain what is meant by (a) space diversity and (b) frequency diversity, and why diversity systems are used.
12. Discuss the main operational features of a satellite communications circuit. Describe the major differences between the satellite as a repeater and a terrestrial repeater station.
13. Using the expressions for EIRP and G_R/T_s developed in the text and, allowing for margins, show that Eq. (11.9) leads to Eq. (11.22).
14. The uplink requirements for a satellite circuit are $C/N_0 = 71.97$ dB-Hz, fade margin 0.5 dB, antenna pointing margin 0.5 dB, waveguide losses 0.2 dB, equipment implementation margin 1.0 dB, and propagation loss 207 dB. Satellite receiver $G_R/T_s = 23.5$ dB-K. Calculate the earth station EIRP in dBW.
15. Given that the average power in a digitally modulated (PSK) carrier is equal to the energy per bit divided by the bit period, and that the bit period is equal to the reciprocal of the transmission rate, derive Eq. (11.23).
16. The E_b/N_0 ratio required to meet a specified reliability figure is 9.5 dB, at a transmission rate of 1.764 Mbps. The transmission loss is 214 dB, fade margin 2.1 dB, antenna pointing margin 1.0 dB, equipment implementation margin 1.0 dB, and waveguide losses 2 dB. The earth station EIRP is 50 dBW. Calculate the satellite G_R/T_s ratio in dB-K.

REFERENCES

FEHER, KAMILO, 1981. *Digital Communications*. Englewood Cliffs, N.J.: Prentice-Hall, Inc.

HALLIWELL, B. J. (ed.), 1974. *Advanced Communication Systems*. Sevenoaks, Kent, England: Newnes-Butterworths.

PANTER, PHILIP F., 1972. *Communication System Design*. New York: McGraw-Hill Book Company.

RODDY, D., and J. COOLER, 1984. *Electronic Communications*, 3rd ed. Reston, Va.: Reston Publishing Co., Inc.

FURTHER READING

"ARGA Single-Sideband Microwave Radio System," *Bell Syst. Tech. J.*, Vol. 62, No. 10, Pt. 3, 1983.

BRAY, W. J. 1958. "A survey of microwave radio communication," *Electron. Eng.*, Vol. 30.

BRAY, W. J., 1961. *The Standardization of International Microwave Radio Relay Systems*.

GLAZER, E. V. D., and H. R. L. LAMONT, 1958. *The Services' Textbook of Radio,* Vol. 5, *Transmission and Propagation*. London: Her Majesty's Stationery Office.

YEH, LEANG P., 1961. "Recent progress in microwave radio communications," *Microwave J.*, July.

twelve
Microwave Radar Systems

12.1 INTRODUCTION

The word *radar* originated as an acronym for *r*adio *d*etection *a*nd *r*anging. However, with the development of new radar systems at frequencies well above the microwave band, and approaching optical frequencies in some instances, the IEEE definition of radar has been extended to: "An electromagnetic device for detecting the presence and location of objects. The presence of objects and their distance (range) are determined by the transmission and return of electromagnetic energy; direction is usually obtained also, through use of a movable or rotating directive antenna pattern."

With primary radar, a radio wave is transmitted from the radar station and is reflected by the object to be detected. This reflected wave is picked up at the radar station, and from a knowledge of the echo time, the range can be estimated. It is usual to have the radar transmitter and receiver in the same station, often using the same antenna, although separate transmitting and receiving stations may also be used. Military applications of primary radar include detection of the position and flight path of hostile aircraft, ships, and missiles. Using primary radar, the flight path of shells may be projected back on the battlefield, enabling positions of enemy guns to be pinpointed. Weather radar uses primary radar to detect the position and movement of storm centers.

Secondary radar is where the object to be detected receives the radar transmission and transmits a coded signal in reply. Identification of commercial airliners in flight and detection of their position and speed is achieved in this manner.

TABLE 12.1
Radar Frequency Spectrum

Band Designation	*Nominal Frequency Range*	*Specific Radio Location (radar) Bands Based on ITU Assignment for Region 2*
HF	3–300 MHz	
VHF	30–300 MHz	138–144 MHz
		216–225
UHF	300–1000 MHz	420–450 MHz
		890–942
L	1000–2000 MHz	1215–1400 MHz
S	2000–4000 MHz	2300–2500 MHz
		2700–3700
C	4000–8000 MHz	5250–5925 MHz
X	8000–12000 MHz	8500–10680 MHz
Ku	12.0–18 GHz	13.4–14.0 GHz
		15.7–17.7
K	18–27 GHz	24.05–24.25 GHz
Ka	27–40 GHz	33.4–36.0 GHz
mm	40–300 GHz	

Source: Bradsell (1981).

Microwaves are used for radar because very compact, high gain, and therefore highly directional antennas can be used. The radar frequency bands as recommended by the International Telecommunication Union for region 2 are given in Table 12.1.

12.2 THE RADAR EQUATION

For the moment it will be assumed that the radar setup is as shown in Fig. 12.1, in which a transmitter sends out a sinusoidal wave, which hits the target and is reflected back to the common antenna, where it is picked up by the receiver. The range of the target is denoted by R. The system is therefore rather like the communications link discussed in Section 11.4, and the power density at the target is given by Eq. (11.2), which is

$$P_D = \frac{P_T G_T}{4\ \pi\ R^2} \qquad \text{W/m}^2 \tag{12.1}$$

In this case distance d becomes the range R. P_T is the transmitted power and G_T is the antenna power gain relative to an isotropic radiator, as described in Section 10.3.

As the wave passes over the target, a certain amount of power will be reflected back in the direction of the incident wave. The target therefore appears to have an effective echoing area which converts the incident power density (W/m^2) into reflected power (W). Let P_{ref} represent the reflected power and σ the effective echoing area; then

$$P_{\text{ref}} = P_D \sigma \tag{12.2}$$

The echoing area is obviously an important parameter for targets. A target that is designed to evade radar may have an effective area of the order of 0.01 m^2, while commercial aircraft may have effective radar areas up to 10 m^2 and ships up to 10000 m^2. In most cases of practical interest, the echoing area has to be determined experimentally.

The power, P_{ref}, is reradiated back toward the receiver and therefore undergoes the same spreading with range as the incident wave. Letting A_{eff} represent the effective area of the receiving antenna, as was done in Eq. (11.3), the received power becomes

$$P_R = \frac{P_D \sigma}{4\pi R^2} A_{\text{eff}} \tag{12.3}$$

Substituting for P_D from Eq. (12.2), using the relationship ($\lambda^2/4\pi$) between antenna gain and effective area as given by Eq. (11.4), and noting that in this case the same antenna is used for transmitting and receiving, the equation for received power becomes

$$P_R = \frac{P_T G^2 \lambda^2 \sigma}{(4\pi)^2 R^4} \tag{12.4}$$

This equation shows that the received power is inversely proportional to the fourth power of the range, which means that it drops off very rapidly with increase in range.

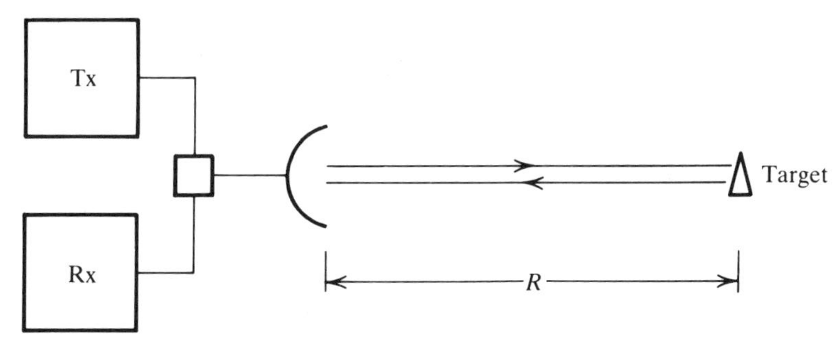

FIGURE 12.1. Basic radar system.

The received power has to compete with the noise at the receiver input. The receiver noise is given by

$$N = FkT_0B_n \tag{12.5}$$

Here F is the noise factor of the receiver, k is Boltzmann's constant, and T_0 is room temperature in kelvin. B_n is the noise bandwidth of the receiver. The factor kT_0 is a constant equal to 4×10^{-21} J/K. The signal-to-noise ratio at the input to the radar receiver is therefore

$$S/N = \frac{P_T G^2 \lambda^2 \sigma}{(4\pi)^3 R^4 FkT_0B_n} \tag{12.6}$$

In this equation, no account is taken of the various losses that are known to occur, such as atmospheric attenuation and scattering of the radio wave by rain or precipitation, as described in Sections 11.10 and 11.11. All the losses may be grouped into one attenuation factor L_A, allowing Eq. (12.6) to be written as

$$S/N = \frac{P_T G^2 \lambda^2 \sigma}{(4\pi)^3 R^4 FkT_0B_nL_A} \tag{12.7}$$

This equation is the basic radar equation, as it shows all the parameters of the radar system which affect performance. These are discussed in more detail in the following sections, but it can be noted generally that the S/N is directly proportional to transmitted power and inversely proportional to the receiver noise factor, which is to be expected and, as previously noted, it is inversely proportional to the fourth power of the range. The effect of wavelength is not quite straightforward. Equation (12.7) suggests that the S/N is proportional to λ^2, but antenna gain is also a function of wavelength, and these interrelated effects will be considered later.

12.2.1 Decibel Notation

As pointed in Section 11.11, decibel units are often used with signal-to-noise equations, and since this may not be stated explicitly, care must be taken to observe the correct use of such units. A quantity X expressed in decilogs is by definition:

$$X(\text{dB}) = 10 \log X \tag{12.8}$$

Thus noise bandwidth B_n hertz becomes $10 \log B_n$ dB, and wavelength λ meters becomes $10 \log \lambda$ dB. The abbreviation dB for decibel will also be used here for decilogs since dB is the more familiar unit, and decibels and decilogs add together directly. The power P_T must be expressed in decibels relative to some reference; for example, dBW means decibels relative to 1 W. The constant factor $(4\pi)^3 k T_0$ expressed in decilogs is equal to -171 dB. The radar equation, Eq. (12.7), may therefore be written as

$$S - N = P_T + 2G + 2\lambda + \sigma - (-171) - 4R - F - B_n - L$$

$$= P_T + 2G + 2\lambda + \sigma + 171 - 4R - F - B_n - L \qquad (12.9)$$

EXAMPLE 12.1

A radar system operates at a frequency of 3000 MHz. The common antenna gain is 32 dB, the receiver noise bandwidth is 1 kHz, and the noise factor 4.4 dB. The transmitted power is 1 kW, and the path losses amount to 10 dB. The target echoing area is 100 m^2. Calculate the range for unity signal-to-noise ratio.

SOLUTION

$$\text{Wavelength } \lambda = \frac{3 \times 10^8}{3 \times 10^9} = 0.1 \text{ m}$$

Expressing this in decilogs yields

$$10 \log 0.1 = -10 \text{ dB}$$

The bandwidth B_n expressed in decilogs is

$$10 \log 1000 = 30 \text{ dB}$$

The power P expressed in decibels relative to 1 W is

$$10 \log P = 30 \text{ dBW}$$

The echoing area expressed in decilogs is $10 \log 100 = 20$ dB.

A unity signal-to-noise ratio is equivalent to 0 dB S/N. Denoting by R_0 the range at which 0 dB is signal-to-noise ratio occurs, Eq. (12.9) gives

$$0 = 30 + 2(32) + 2(-10) + 20 + 171 - 4R_0 - 4.4 - 30 - 10$$

Therefore,

$$R_0 = \frac{30 + 64 - 20 + 20 + 171 - 4.4 - 30 - 10}{4}$$

$$= \mathbf{55.15\ dB\ (=327.3\ km)}$$

The range for which the signal-to-noise drops to 0 dB is a useful reference value, as for a given target echoing area, it lumps together all the factors of the radar equation exclusive of range. Thus from Eq. (12.9), with $S - N = 0$, the range R_0 in decilogs is seen to be

$$R_o = \frac{P_T + 2G + 2\lambda + \sigma + 171 - F - B_n - L}{4} \tag{12.10}$$

Once R_0 is determined, the radar equation may be written as

$$S - N = 4(R_0 - R) \tag{12.11}$$

EXAMPLE 12.2

For the radar system in Example 12.1, determine the signal-to-noise ratio for a range of 100 km.

Solution

100 km expressed in decilogs is $10 \log 10^5 = 50$ dB. Hence from Eq. (12.11) and the results of Example 12.1, we have

$$S - N = 4(55.15 - 50) = 20.6 \text{ dB}$$

12.2.2 The Range Equation

The radar range R_0 at which a 0-dB signal-to-noise ratio occurs is given in decilogs by Eq. (12.10). In terms of the more familiar distance units, the range R_0 may be written as

$$R_0 = \sqrt[4]{\frac{P_T G^2 \lambda^2 \sigma}{(4\pi)^3 F k T_0 B_n L_A}} \tag{12.12}$$

This can also be derived from Eq. (12.7) on setting $S/N = 1$. Inspection of Eq. (12.12) shows that the units for R_0 are determined by λ and σ, and the units for these must be consistent. Thus, if λ is in meters, σ should be in square meters, and R_0 will be in meters also. Navigational distances are often measured in nautical miles and the conversion from meters to nautical miles is

$$1 \text{ nautical mile} = 1852 \text{ m} \tag{12.13}$$

This is the international nautical mile. The British nautical mile is 6080 ft.

EXAMPLE 12.3

Calculate the range, as found in Example 12.1 in the nautical miles.

SOLUTION

$$R_0 = 327.3 \text{ km}$$

$$= \frac{327.3 \times 1000}{1852}$$

$$= \mathbf{176.73 \text{ nautical miles}}$$

The range R_0 is seen to be proportional to the fourth root of the transmitted power. Thus a 16:1 increase in power would be required to increase the range by 2:1.

Once R_0 is determined for a radar system, the range for any other S/N value may be obtained from

$$R = \frac{R_0}{\sqrt[4]{\frac{S}{N}}} \tag{12.14}$$

12.3 TRANSMIT–RECEIVE SWITCHING

When a common antenna is used with a radar system, some means must be used to isolate the transmitter from the receiver; otherwise, the transmitter power could burn out the front end of the receiver. Even if this were not to happen, the direct coupling of the two units would result in inefficient power transfer to and from the antenna. A unit known as a *duplexer* is used to couple the transmitter and the receiver to the antenna while reducing to a negligible level the transmitter-to-receiver coupling. The hybrid-T can be used as duplexer as shown in Fig. 4.18, but as pointed out in Section 4.14, a 3-dB loss occurs. A common type of duplexer used with narrowband pulsed radar is illustrated in Fig. 12.2(a). The system is narrowband because it relies on the quarter-wavelength relationships of waveguide sections connecting the units together. For ease of description transmission-line sections are shown in Fig. 12.2, but of course waveguide sections could also be used. When the transmit power is present on the line the TR (transmit–receive) switch and the ATR (anti-TR) switch both switch to the closed, or short-circuit, condition. These switches may be PIN diodes, as described in Section 7.6. A *passive* PIN switch is one that switches to its high-conducting state when the incident power is high. Al-

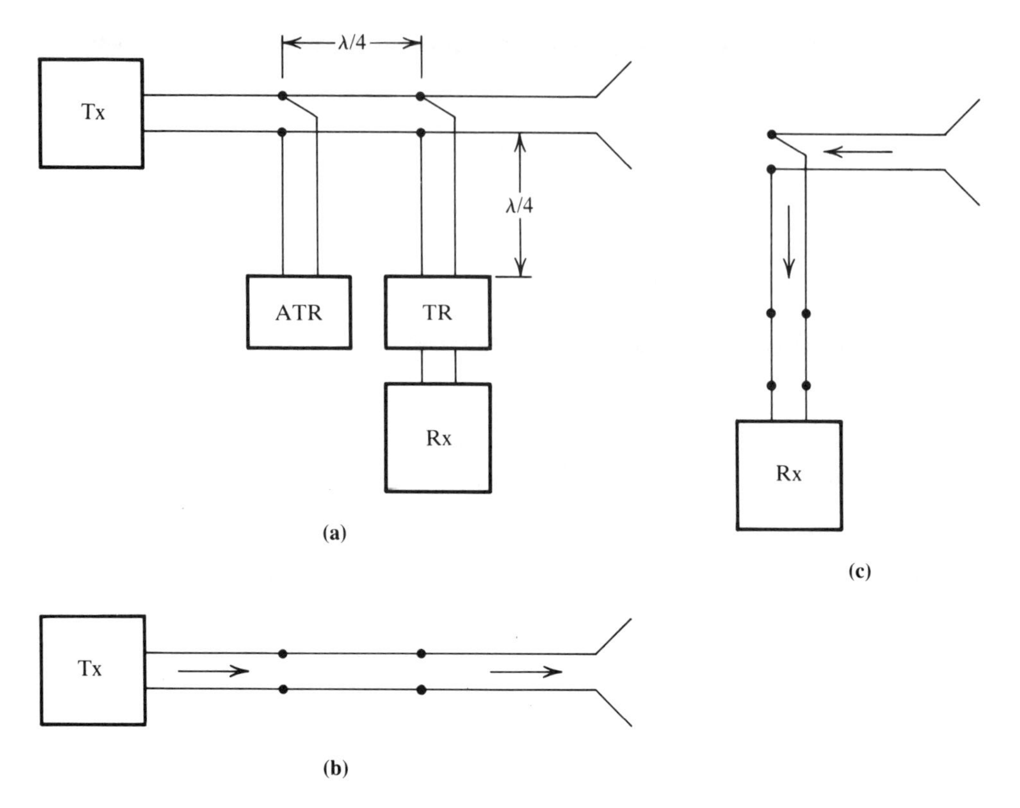

FIGURE 12.2. (a) Narrowband duplexer; (b) the duplexer under transmit conditions; (c) the duplexer under receive conditions.

ternatively, an *active* PIN switch may be used, in which the diode is normally reverse biased and therefore is in a low-conductance state. Simultaneously with transmission, the bias is changed to forward bias and the diode becomes a high conductance. For high powers a number of diodes may be connected in parallel. Gas discharge tubes are also used for switching. These consist of a section of waveguide sealed and filled with a gas such as hydrogen or argon at low pressure. Tapered electrodes are positioned top and bottom at the center of the broad wall of the guide, in the high-field region. When the high transmit power is present, the gas ionizes and acts as a short circuit. Otherwise, the gas is transparent to the microwave.

Under transmit conditions the TR and ATR switches in Fig. 12.2(a) are closed. The quarter-wavelength sections transform these short circuits to open circuits across the main guide, and therefore the power flow is directly from transmitter to antenna, as shown in Fig. 12.2(b). The receiver is protected in this way. The ATR switch plays no particular role in protecting the receiver, but is required to prevent the transmitter from absorbing any of the received signal, as this would seriously degrade the signal-to-noise ratio. In the receive condition both switches are "open." In the case of the ATR switch this means that the quarter-wave guide is terminated in an open circuit, which is transformed to a short circuit at the connection to the main guide. The interconnecting quarter-wavelength section transforms this short circuit back to an open circuit at the connection to the receiver branch and therefore the transmit line appears as an open circuit to the antenna. The TR switch in the receive line appears as a through connection to the receiver, which provides a matched load to the line, with all the received power being transferred to the receiver as shown in Fig. 12.2(c)

Ferrite circulators of the type described in Section 6.14 may also be used as TR switches. Figure 12.3 shows how TR switching can be achieved using two ferrite circulators. It will be recalled that the direction of signal travel around the circulator depends on the direction of the steady magnetic field activating the ferrite. For the transmit condition shown in Fig. 12.3(a), the transmit wave travels from port 1 to port 2 of the first circulator, and then on to the antenna. Most of the energy leaves by port 2, but any that continues around enters the second circulator through port 3 to the port 1′ branch. The magnetic field to the second circulator is in such a direction as to direct this energy to port 3′, where most of it is dissipated in the matched load. Thus a negligible amount of the transmit energy circulates around to the receiver port. Under receive conditions, Fig. 12.3(b), the magnetic field to the second circulator is reversed. The received wave travels around the first circulator from port 2 to port 3 and enters the second circulator at port 1′. The energy travels around the second circulator to port 2′, where it goes to the receiver. Any leakage signal that continues around the second circulator is mostly dissipated in the matched load at port 3′. Thus the received signal is directed to the receiver only.

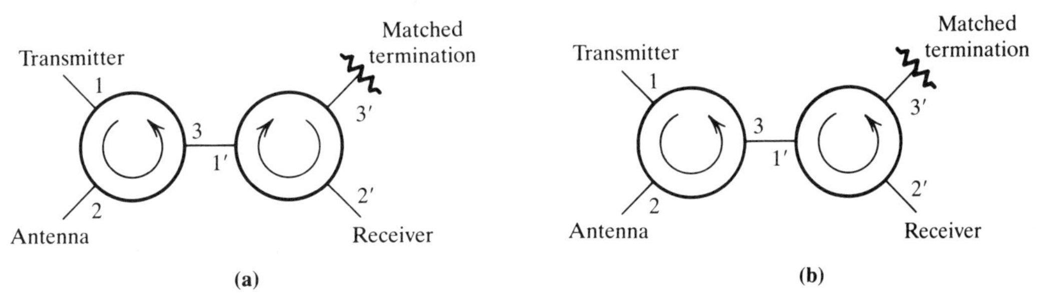

FIGURE 12.3. Ferrite circulators used as duplexers: (a) signal travels from transmitter to antenna; (b) signal travels from antenna to receiver.

12.4 PULSED RADAR

In deriving the radar equation, the radio wave was assumed to be sinusoidal. A sine wave is a mathematical abstraction; by definition it is assumed to exist from time equal minus infinity to time equal plus infinity. No radar information can be extracted from such a wave, as some change has to occur to signify detection of the target, and any change immediately implies a departure from sinusoidal. In practice, of course, even a sine wave would have to be "switched on" and thus really becomes a sine wave modulated by a step function. Now even if the switching-on part of the radio wave were used in a radar system, it would be difficult to abstract information from this, particularily when two targets are present and it is desired to differentiate between them. This is because when two sinusoids of the same frequency are added together, the resultant is a third sinusoid of the same frequency. Suppose that two targets are separated by a range distance ΔR; then the time delay of the furthest target will be $(\Delta R/c)$ relative to the nearest one, and the returned signals will be $A \sin \omega t$ and $B \sin \omega(t - \Delta R/c)$. These are shown in Fig. 12.4 together with the resultant wave, which is the sum of the two. Apart from the discontinuity at the point where the second signal "comes in," there is no way of separating the two echoes from the resultant.

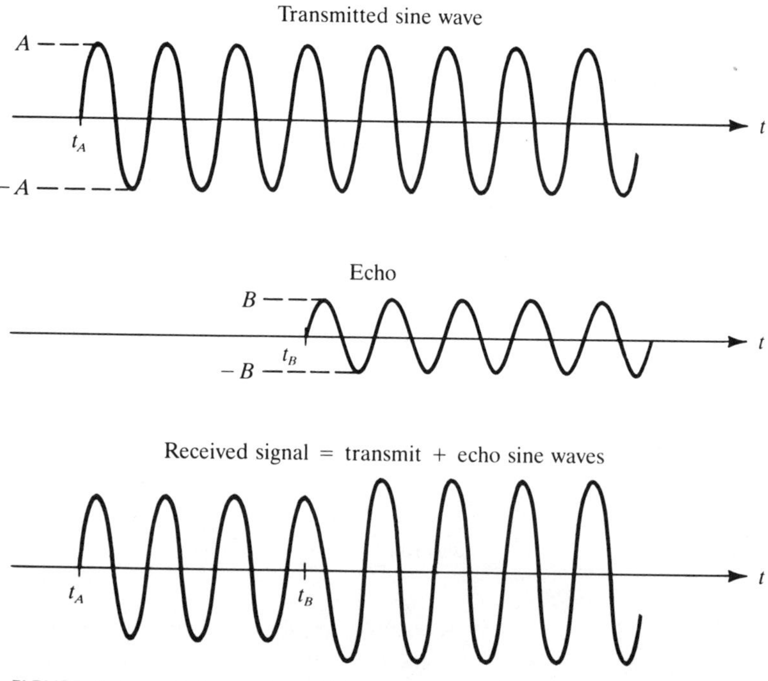

FIGURE 12.4. Addition of sinusoidal echo to transmit wave.

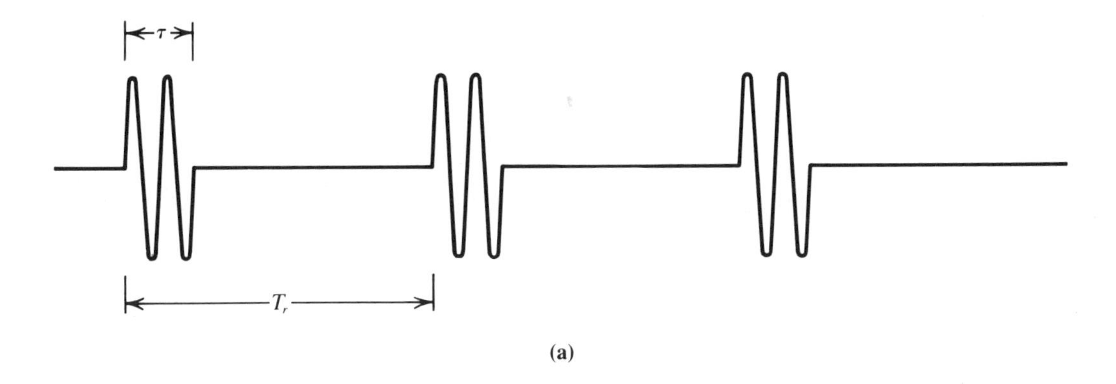

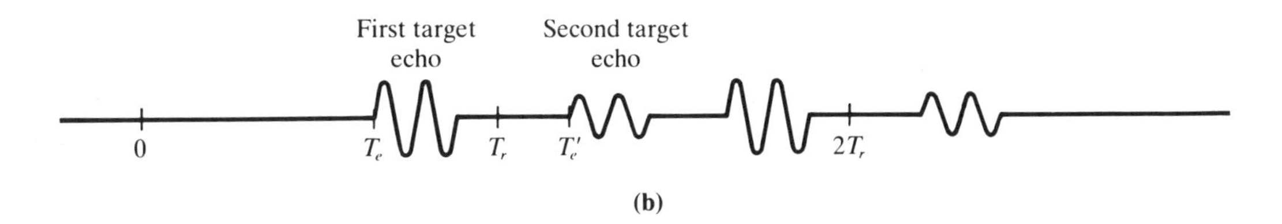

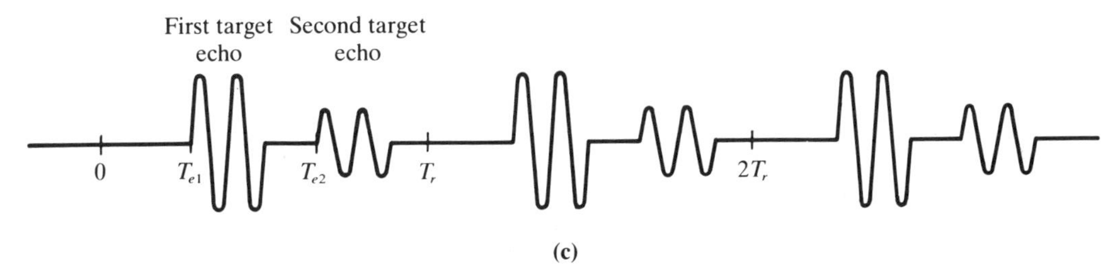

FIGURE 12.5. (a) Pulsed radar wave; (b) received echo; (c) combined transmitted and echo signals as received by the receiver.

One way of overcoming this problem is to amplitude modulate the sinusoidal carrier with a repetitive rectangular pulse wave as shown in Fig. 12.5(a), where the pulse repetition period is T_r and the pulse width is τ. Figure 12.5(b) shows the situation where an echo pulse arrives back at the receiver. It is clear that the echo time T_e must be less than the pulse repetition period T_r; otherwise, there would be ambiguity. For example, an echo at T_e' would be interpreted as $(T_e' - T_r)$. Let c represent the velocity of propagation of the radio wave; then the maximum range, as determined by T_r, is

$$R_{\max} = \frac{cT_r}{2} \tag{12.15}$$

For an echo time T_e less than T_r, the corresponding range R is

$$R = \frac{cT_e}{2} \tag{12.16}$$

A second constraint set by the system is the target resolution. As shown in Fig. 12.5(c), the separation between pulses should be such that

$$T_{e2} - T_{e1} \geqslant \tau \tag{12.17}$$

The power transmitted by a radar system also depends on the ratio of τ/T_r, this ratio being known as the *duty factor*. Figure 12.6 shows the power wave associated with the pulsed current wave of maximum value I_m. It will be seen that the peak or maximum power, denoted by P_m, is given by $I_m^2 R_{\text{rad}}$, where R_{rad} is the radiation resistance of the antenna. The average power over the pulse width is one-half of this. The average power over a complete period T_r is just P spread over a base of T_r, as shown in Figure 12.6(c). This is the average power that the radar transmitter must deliver to the antenna. Referring to Fig. 12.6(c), the power P, which is the average value over the pulse width τ, is referred to as the peak power in radar engineering. This must not be confused with the true peak power, which occurs at the peak of the RF cycle, shown as P_m in Fig. 12.6(b).

The question of power is intimately tied in with the frequency bandwidth occupied by the echo signal. From modulation theory it is known that a sinusoidal carrier, of frequency f_0, modulated with a rectangular pulse wave of repetition frequency f_r (where $f_r = 1/T_r$) has an infinite number of side frequencies extending over $f_0 \pm nf_r$, where n is integer. If the pulse repetition frequency f_r is low, such that most of the side frequencies are close to the carrier, the complete spectrum will be received. In this case, the power to be used in the radar equation (12.7) is the peak power P (i.e., the peak average power).

At the other extreme, where the pulse repetition frequency is high, the receiver may have only sufficient bandwidth to pass the center or carrier component of the spectrum f_0. A Fourier analysis of the waveform of Fig. 12.6(a) shows that the carrier amplitude is

$$I_{0m} = I_m \frac{\tau}{T_r} \tag{12.18}$$

The rms current corresponding to this is $I_0 = I_{0m}/\sqrt{2}$, and hence the average power transmitted in the carrier component only of the modulated wave, is

$$P_0 = I_0^2 R_{\text{rad}} \tag{12.19}$$

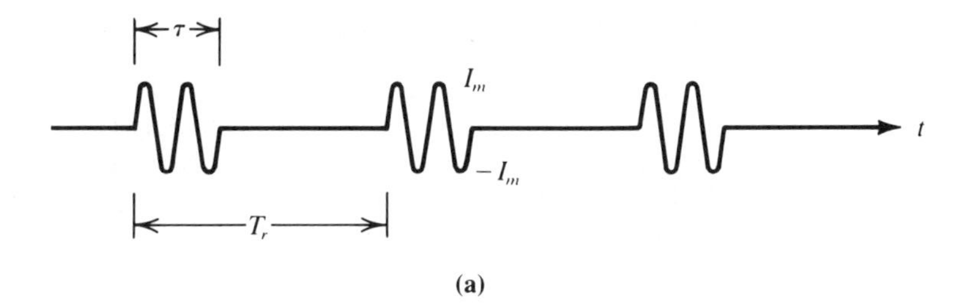

(a)

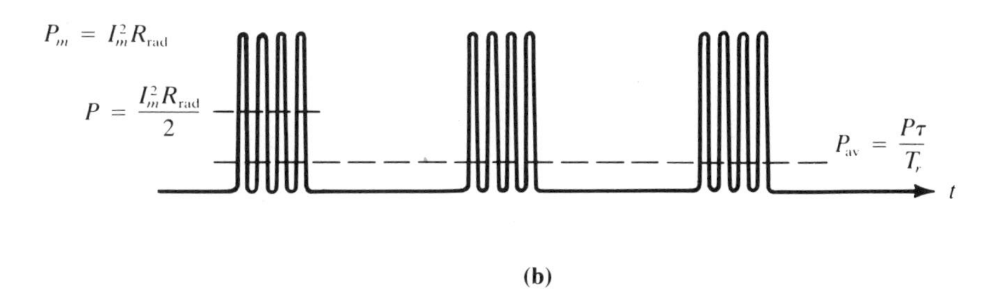

(b)

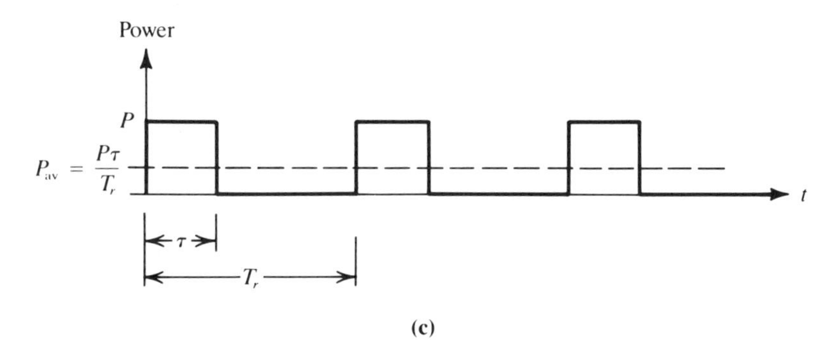

(c)

FIGURE 12.6. (a) Pulsed radar wave; (b) corresponding instantaneous power wave; (c) peak average power P and true average P_{av}.

Substituting from Eq. (12.18) gives

$$P_0 = \left(\frac{I_m}{\sqrt{2}}\frac{\tau}{T_r}\right)^2 R_{\text{rad}}$$

$$= P\left(\frac{\tau}{T_r}\right)^2$$

$$= P_{\text{av}}\frac{\tau}{T_r} \tag{12.20}$$

This is the power that must be used in the radar equation when the pulse repetition frequency is high.

EXAMPLE 12.4

The pulse repetition frequency for a radar is 1 kHz, and the pulse width is 10 μs. The peak power is 1 kW. Calculate (a) the maximum unambiguous range, (b) the minimum separation between targets which might be resolved, (c) the average power transmitted, and (d) the power in the carrier component only.

SOLUTION

(a) From Eq. (12.15),

$$R_{max} = \frac{3 \times 10^8 \times 10^{-3}}{2} = \mathbf{150\ km}$$

(b) Assuming that the minimum separation is set by the pulse width, then

$$\Delta d_{min} = \frac{c\tau}{2} = \frac{3 \times 10^8 \times 10 \times 10^{-6}}{2}$$

$$= \mathbf{1.5\ km}$$

(c) The duty factor is

$$\frac{\tau}{T_r} = f_r\tau = 10^3 \times 10 \times 10^{-6}$$

$$= 0.01$$

$$P_{av} = P\frac{\tau}{T_r} = 10^3 \times 0.01 = \mathbf{10\ W}$$

(d) From Eq. (12.20),

$$P_0 = P_{av} \times \frac{\tau}{T_r} = 10 \times 0.01 = \mathbf{0.1\ W}$$

As mentioned previously, the transmitted power, and hence range, increases with pulse length, but target resolution decreases with pulse length. Pulse compression is a method used to overcome the problem of obtaining required resolution together with range. In pulse compression, the carrier frequency is swept over a given range during transmission of the pulse. The receiver contains a filter that has an inverse frequency characteristic to the transmitted pulse. With such an arrangement it can be shown that

TABLE 12.2
Plessey AR3D Parameters

Radio frequency	In 3-GHz band
Transmitter	
Mean power	10 kW
Peak power	> 1 MW
Pulse length	36 μs
Pulse repetition frequency	250 Hz
Sweep bandwidth	140 MHz
Antenna rotation rate	6 rev/min
Horizontal beamwidth	1°
Vertical beamwidth	2°
Vertical coverage	to 30°
Compressed pulse length	0.1 μs

an echo from a point target produces a short pulse at the output of the filter even though a relatively long pulse is transmitted.

An example of a mobile high-power radar is the Plessey AR3D tactical radar, which will be described briefly. Table 12.2 lists the main parameters of this radar.

Pulse compression is used, and this is combined with frequency scanning, which causes a narrow beam from the antenna to scan an elevation sector during one pulse. Thus electronic scanning is used for elevation, while mechanical rotation of the antenna takes place in the horizontal plane. More details will be found in Bradsell (1981).

Figure 12.7 shows the antenna for this radar. The reflector is a parabolic cylinder which can rotate around the vertical axis. The antenna feed, which can be seen in Fig. 12.7, is shown in more detail in Fig. 12.8. This consists of an end-fed linear array of crossed slots which result in circular polarization. The overall coverage of the radar for echoing areas of 2 m^2 and 15 m^2 is shown in Fig. 12.9.

12.5 FM RADAR

Previously, it was shown that range is determined by measurement of the echo time, as given by Eq. (12.16), it being necessary to pulse amplitude modulate the carrier to allow separation of echo signals from the outgoing signal and from each other. An alternative to pulse amplitude modulation which is widely used in certain applications is to frequency modulate the carrier. Let the instantaneous carrier frequency as a function of time be represented by f_i, the unmodulated carrier frequency by f_0 and the frequency modulation by $f_m(t)$, then the frequency equation is

$$f_i = f_0 + f_m(t) \tag{12.21}$$

FIGURE 12.7. AR3D antenna. (*From P. Bradsell, "Microwave radar," Electr. Power, Vol. 27, No. 5, IEE, 1981.*)

While this carrier is being transmitted from the antenna, the received carrier frequency is

$$f_{i,refl} = f_0 + f_m(t - T_e) \tag{12.22}$$

This is because the received signal is really that which was transmitted T_e seconds previously. Hence the frequency difference between two signals present at the radar is

$$\begin{aligned} f_d &= f_i - f_{i,\,refl} \\ &= f_m(t) - f_m(t - T_e) \end{aligned} \tag{12.23}$$

Figure 12.10 shows one commonly used form of frequency modulation in which the modulation wave shape is triangular. Over any straight section of the modulation waveform, the frequency as a function of time can be expressed by a straight-line law of the form

$$f_m(t) = at + b \tag{12.24}$$

where a is the slope of the line in any given time interval, and b the corresponding intercept. Only the slope need be known, since substituting Eq. (12.24) in Eq. (12.23) gives

$$f_d = aT_e \tag{12.25}$$

The basic range equation is Eq. (12.16), which is $R = cT_e/2$ and hence

$$R = \frac{cf_d}{2a} \tag{12.26}$$

Thus the range can be determined from the measurable quantities f_d and a.

FIGURE 12.8. AR3D feed array. (*From P. Bradsell, "Microwave radar," Electr. Power, Vol. 27, No. 5, IEE, 1981.*)

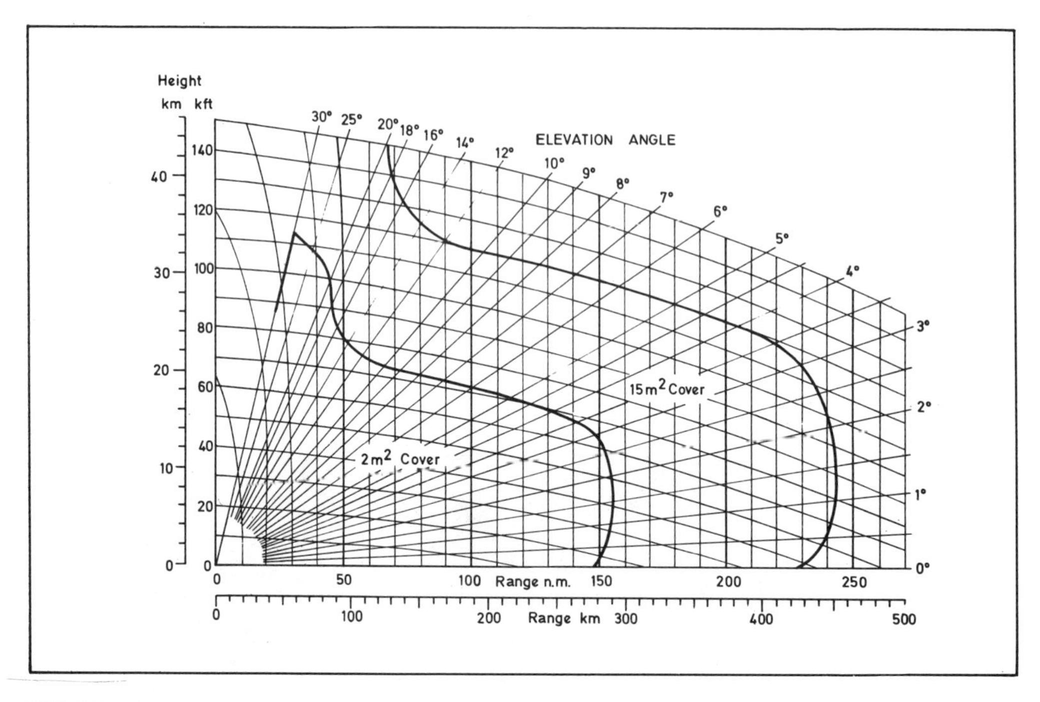

FIGURE 12.9. AR3D coverage. (*From P. Bradsell, "Microwave radar," Electr. Power, Vol. 27, No. 5, IEE, 1981.*)

EXAMPLE 12.5

An FM-CW radar employs triangular-wave modulation, as shown in Fig. 12.10, for which the rate of change of frequency is 1 GHz/s (this is 1000 Hz every microsecond). The echo from a certain target produces a beat frequency of 10 kHz. Calculate the target range.

SOLUTION

$$R = \frac{3 \times 10^8 \times 10^4}{2 \times 10^9}$$

$$= \mathbf{1.5\ km}$$

Because it operates at continuous power level, FM-CW radar tends to be relatively low power. Also, its minimum range is not restricted (as it is by the pulse width in pulsed radar), and therefore FM-CW radar is used mainly for short-range measurements, where fairly compact equipment is required. Figure 12.11(a) shows in block-diagram form a FMCW radar used for subsurface profiling. This type of radar can be used to detect buried objects and to map out interfaces between various layers, for example overburden and bedrock. Figure 12.11(b) shows the frequency modulation used, the carrier frequency being swept linearly from 2 to 4 GHz over a period of 10 ms, at a repetition rate of 10 Hz. Figure 12.11(c) shows the beat frequency as a function of time when a single target is present. It will be observed that the beat frequency pulse width is a function of the echo time T_e. This in turn is related to the target range by an equation similar to Eq. (12.16) except that the phase velocity of the wave in the medium must be used rather than c, the free-space velocity. Because the pulse width is a function of the echo time, the echo time can be obtained from a Fourier spectrum analysis of the waveform. With multiple targets the situation is much more complicated, and digital signal processing is used to separate out the target returns as described in the paper by Carr et al. (1981).

FM radar is widely used in conjunction with the Doppler effect to provide information regarding range and velocity of a moving target as described in the Section 12.8.

12.6 THE DOPPLER EFFECT

Where moving targets are involved, the rate at which the range is changing, known as the *range rate*, is required as well as the range. The range will be denoted by $\dot{R}$, the dot above the R signifying the time rate of change of R. Range-rate information is obtained from the Doppler shift in carrier frequency. When the target is moving toward the radar receiver, the apparent carrier frequency is increased, the amount of increase being given to a close approximation by

$$
\begin{aligned}
f_{dop} &= \frac{2\dot{R}f_0}{c} \\
&= \frac{2\dot{R}}{\lambda}
\end{aligned}
\tag{12.27}
$$

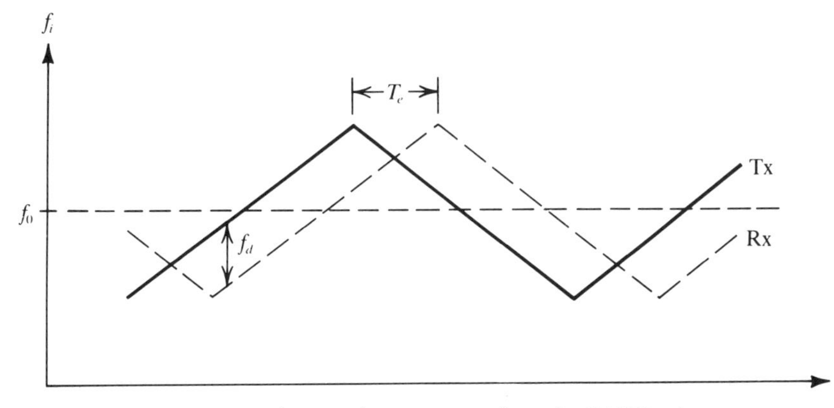

FIGURE 12.10. Transmit and receive frequency waveforms for FMCW radar.

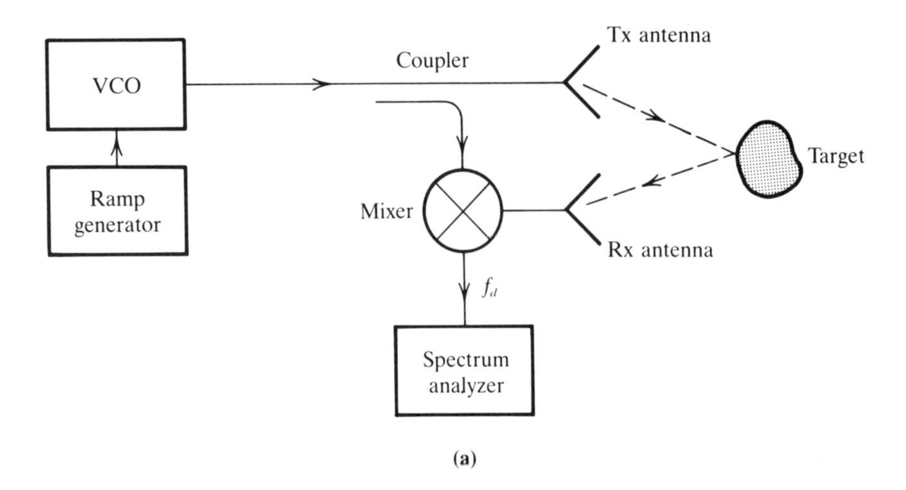

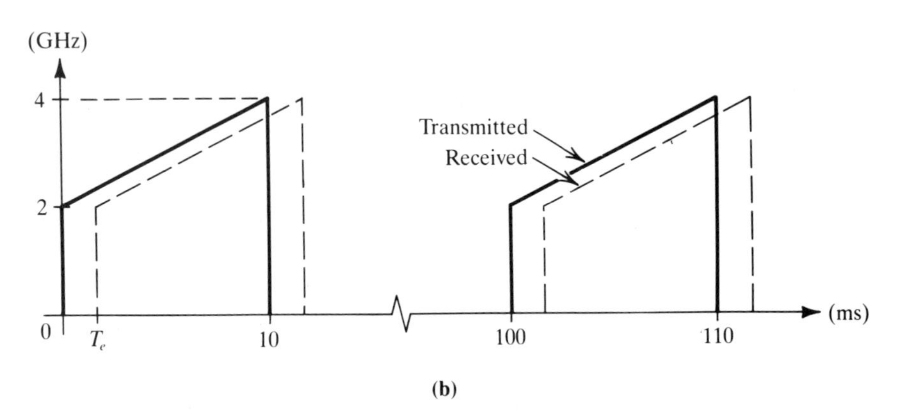

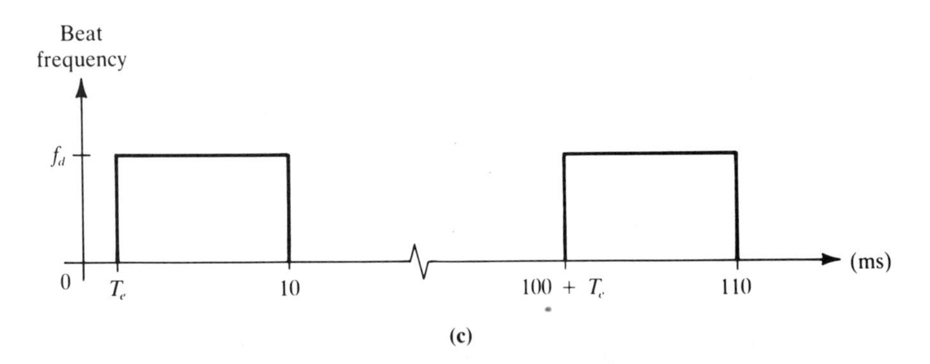

FIGURE 12.11. (a) Simplified FMCW configuration; (b) transmit and receive frequency variations with time; (c) resultant beat frequency. (*From A. E. Carr, L. G. Cuthbert, and A. D. Olver, "Digital signal processing for target detection in FMCW radar," IEE Proc., Vol. 128, Pt. F, No. 5, 1981.*)

Here f_0 is the carrier frequency and c is the free-space velocity of the wave. The wavelength λ is given by c/f_0.

If the target is moving away from the radar receiver, the received carrier frequency appears to be decreased by this amount. By measuring the Doppler frequency shift and noting whether this is positive or negative with respect to the transmitted frequency, the range rate for the target can be determined. The Doppler effect occurs whether or not the carrier is modulated, and although the description above refers to a moving target, in fact relative movement between source and target is all that is required.

Continuous-wave (CW) or unmodulated radar utilizes the Doppler effect in determination of velocities. FM Doppler radar is used where both range and range rate (velocity) must be determined. In coherent pulsed radar systems, the phase shift associated with the Doppler effect is used to provide a moving-target indicator (MTI). The three applications are described briefly in the following sections.

12.7 CW DOPPLER RADAR

The availability of microwave integrated circuits, together with small semiconductor sources and detectors, makes possible inexpensive low-power radars which can be used in a variety of applications, such as intruder alarms, speed measurements, and traffic signal control. The Doppler frequency shift provides the most convenient means of detection of relative movement between source and target. Figure 12.12 shows in block schematic form a simple CW radar system. This shows a single antenna and the use of a circulator to separate transmit and receive signals. Sufficient leakage is generally present from the transmit signal to provide the local oscillator input to the mixer. Isolation between transmit and receive functions may also be achieved by using separate waveguides or by suitably positioning the mixer diode in a common waveguide.

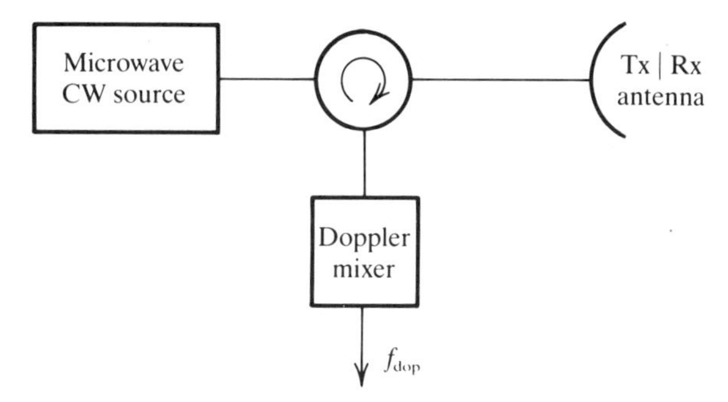

FIGURE 12.12. Simple continuous-wave Doppler radar system.

Figure 12.13 shows a number of arrangements for mounting source and detector. In Fig. 12.13(a), two side-by-side waveguides are used, one containing a Gunn diode source and the other a detector mixer diode. Sufficient leakage of the transmitted signal occurs around the front of the guides to provide the local oscillator input for the mixer diode. An adjustable shroud is mounted on the front to control the amount of LO injection. In the arrangement of Fig. 12.13(b), a single waveguide is used and the mixer diode is offset to one side to avoid being overloaded by the transmitted signal. This offset also reduces the sensitivity to the received signal. However, the single waveguide feed to the antenna enables a narrow beamwidth to be achieved, which is a requirement in some situations. The arrangement shown in Fig. 12.13(c) utilizes a single waveguide and a ferrite circulator. A magnetic field is applied along the axis of the ferrite post in such a direction to produce clockwise steering of the microwave signal. Thus the transmit signal approaching from the left is steered away from the mixer diode, which is displaced from the post as shown. The received signal coming from the right is also steered in a clockwise direction around the post, and thus it is directed to the mixer diode.

A Doppler speedometer scheme is shown in Fig. 12.14. Here the component of velocity along the main beam axis is

$$v_r = v \cos \gamma \tag{12.28}$$

This is the radial velocity and it replaces the range rate R in Eq. (12.27) so that the expression for Doppler frequency is

$$f_{\text{dop}} = \frac{2v \cos \gamma}{\lambda} \tag{12.29}$$

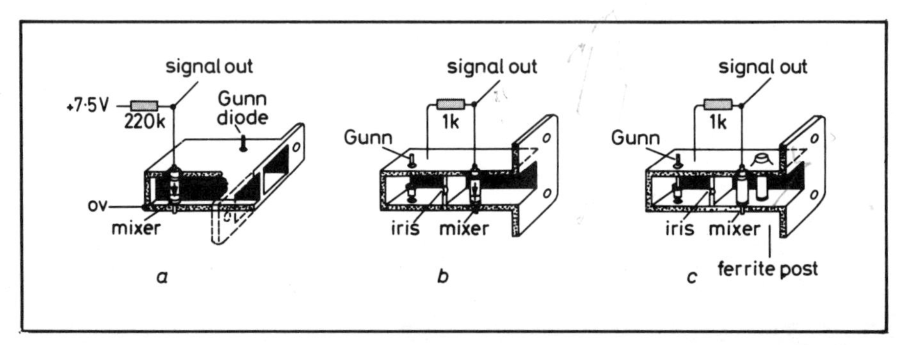

FIGURE 12.13. Microwave Doppler modules: (a) dual waveguide; (b) mixer offset; (c) ferrite post circulator. *(From K. Holford, "If it moves microwaves will detect it," Electr. Power, Vol. 27, No. 5, 1981; with permission.)*

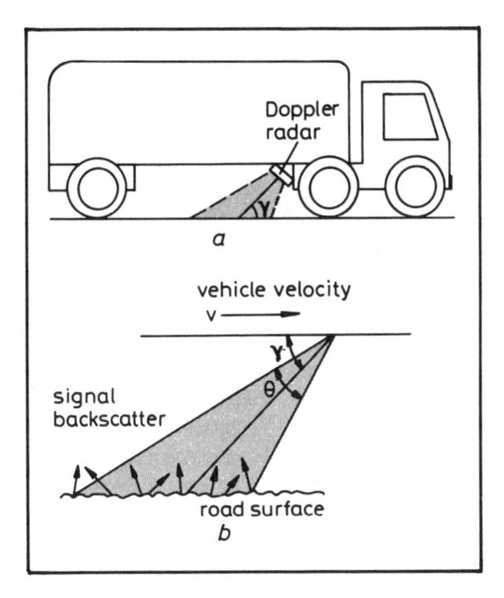

FIGURE 12.14. Doppler speedometer: (a) speed sensor mounted on vehicle; (b) back scattering from road surface. (*From K. Holford, "If it moves microwaves will detect it," Electr. Power, Vol. 27, No. 5, 1981; with permission.*)

This shows that the Doppler frequency is directly proportional to the true ground speed v. A phase-locked loop may be used to convert the Doppler frequency to a dc level which can be read out in digital form or used to drive a conventional analog speed indicator.

12.8 FM DOPPLER RADAR

With FM-CW radar it is shown in Fig. 12.10 that for a stationary target, the difference frequency between outgoing and received carriers is positive over the positive slope of the modulating triangular wave, and it is negative over the negative slope. With a moving target, this frequency difference will also contain the Doppler shift, as shown in Fig. 12.15. Assuming that the target is moving in toward the radar receiver, then over the positive slope the frequency difference is decreased by amount f_{dop}, since the received frequency is increased by this amount. Denoting the frequency difference over the positive slope by f_{d+}, then

$$f_{d+} = f_d - f_{\text{dop}} \tag{12.30}$$

Over the negative slope where the received frequency is above the transmit

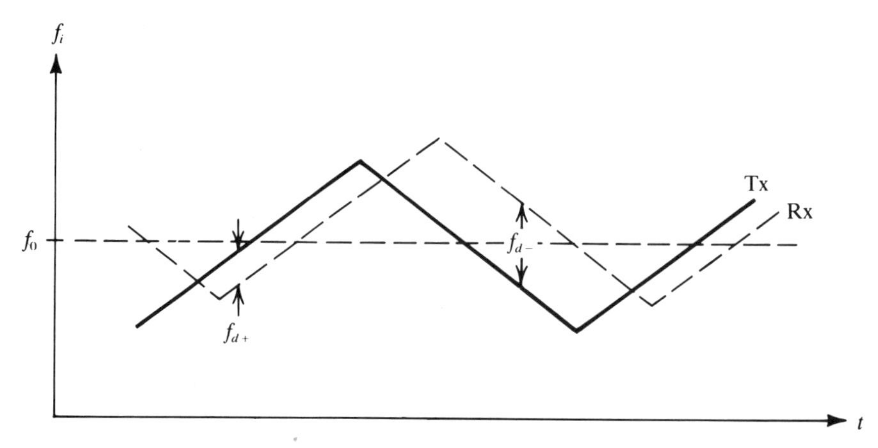

FIGURE 12.15. Transmit and receive frequency waveforms, including Doppler shift for FMCW radar.

frequency, as shown in Fig. 12.15, the frequency difference is given by

$$f_{d-} = f_d + f_{\text{dop}} \tag{12.31}$$

From the sum of these two equations f_d can be found as

$$2f_d = f_{d+} + f_{d-}$$

$$f_d = \frac{f_{d+} + f_{d-}}{2} \tag{12.32}$$

Substituting for f_d from Eq. (12.26) enables the range R to be determined

$$R = \frac{c(f_{d+} + f_{d-})}{4a} \tag{12.33}$$

From the difference of the two equations the Doppler frequency can be found as

$$f_{\text{dop}} = \frac{f_{d-} - f_{d+}}{2} \tag{12.34}$$

Substituting for f_{dop} from Eq. (12.27) enables the range rate to be determined

$$\dot{R} = \frac{c\,(f_{d-} - f_{d+})}{4f_0} \tag{12.35}$$

Note that since $c/f_0 = \lambda$, Eq. (12.35) can be written as

$$\dot{R} = \frac{\lambda(f_{d-} - f_{d+})}{4} \tag{12.36}$$

EXAMPLE 12.6

An FM-CW radar operates at a carrier frequency of 10 GHz. The modulation is triangular, as shown in Fig. 12.15, the magnitude of the slope of the modulation waveform being 1 GHz/s. The beat frequency from a target return is measured as 5 kHz over the negative slope and 4.8 kHz over the positive slope. Determine the target range, range rate, and whether it is moving toward or away from the radar.

SOLUTION

A frequency of 10 GHz has a free-space wavelength of 3 cm. From the data given, $f_{d-} = 5000$ Hz and $f_{d+} = 4800$ Hz. Since the frequency difference over the negative slope is greater than that over the positive slope, it can be concluded that the target is moving toward the radar. From Eq. (12.33),

$$\begin{aligned} R &= \frac{3 \times 10^8(4800 + 5000)}{4 \times 10^9} \\ &= \mathbf{735\ m} \end{aligned}$$

From Eq. (12.36),

$$\begin{aligned} \dot{R} &= \frac{3 \times 10^{-2}(5000 - 4800)}{4} \\ &= \mathbf{1.5\ m/s} \end{aligned}$$

FM-CW radar finds application in measurement of vehicle speeds and closing or separating distances. There is no restriction on minimum distance that can be measured, unlike pulsed radar systems, where the pulse width limits resolution. For these applications, the radar, including antennas, can be kept physically small, and power requirements are low, the transmit powers being of the order of a few watts. The aircraft altimeter is a major application of FM-CW radar. Another application is as an automotive warning radar, where it may be used to detect road vehicles ahead and give a warning if the closing speed or distance is potentially dangerous. Figure 12.16(a) shows the block diagram for one such radar, which operates at 0.9-cm wavelength (33.33 GHz). The microwave source

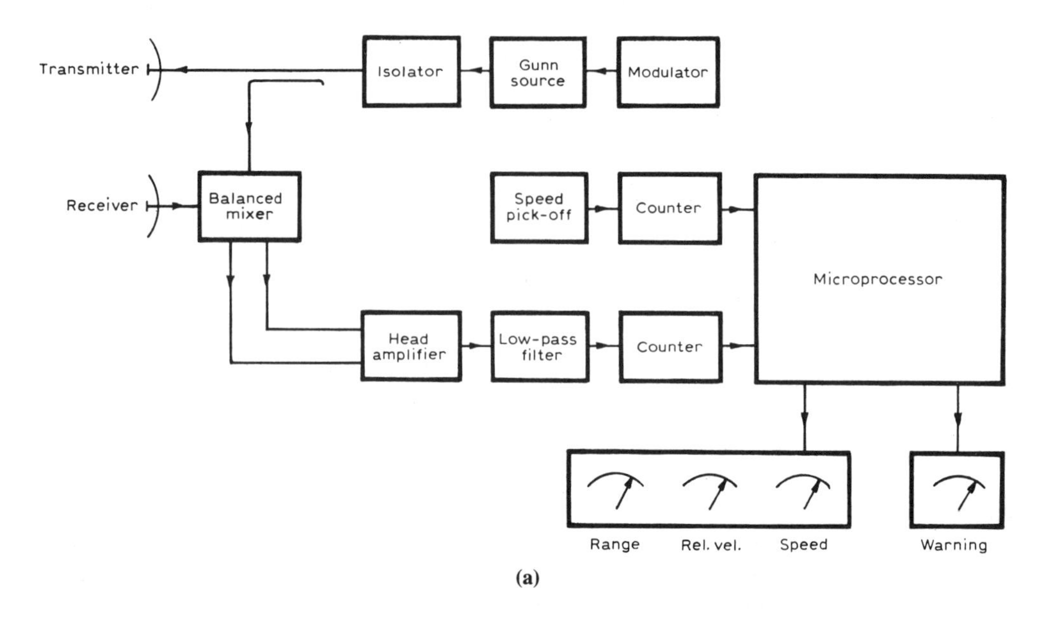

(a)

(b)

FIGURE 12.16. (a) FMCW radar headway warning system—block diagram; (b) 0.9-cm FMCW radar head unit mounted on vehicle (radome removed). (*From R. D. Codd, "Recent trends in automotive radar," Radio Electr. Eng., Vol. 47, No. 10, 1977; with permission.*)

is a Gunn oscillator (described in Section 7.13), and this is frequency modulated by a triangular wave similar to that shown in Fig. 12.15. An isolator is provided between the transmit antenna and the oscillator to prevent excessive loading on the oscillator, as too much loading would tend to create unwanted frequency changes. A directional coupler couples a small amount of the transmitted signal into a balanced mixer (described in Section 7.11). The receive antenna is coupled into this mixer also and the output from the mixer is at the difference frequency f_d, as noted previously. This output signal is amplified and filtered and passed to a frequency counter. At the same time, the vehicle's speed is monitored and both the vehicle's own speed and the range and speed of the vehicle ahead are fed into a microprocessor. As shown, the microprocessor can provide readouts of range, the relative velocity between vehicles, the speed of the radar vehicle, and a warning if conditions warrant this. As shown, separate transmit and receive antennas are used, and this obviates the need for special protection circuitry at the receiver input. The antennas are of the planar type described in Section 10.13, and these are shown in Fig. 12.16(b), mounted on front of a vehicle.

12.9 MOVING-TARGET INDICATOR

In a coherent radar, the transmitted signal is locked in phase to the reference signal, which is applied to a phase-sensitive detector in the receiver. The received echo maintains the phase relationship to the reference, and in addition has a phase shift determined by the path length, the nature of the reflecting surface, and whether or not the target is moving. The latter phase shift is directly related to the Doppler frequency shift. With coherent detection, successive radar scans are compared for phase shift. With stationary targets the phase does not change from scan to scan, so subtracting successive scans enables the stationary target returns to be canceled out. Moving targets produce a changing phase directly related to the Doppler frequency. This appears as a fluctuating output when successive scans are subtracted and is known as a moving-target indicator (MTI). Further details of the standard MTI system will be found in Kennedy (1970) and of digital signal processing for MTI in Hadjifotiou (1977).

12.10 SYNTHETIC APERTURE RADAR

Synthetic aperture radar (SAR) makes use of the Doppler effect to increase the resolution obtainable with mapping radar. With SAR, fine details such as terrain heights and contours can be resolved to a higher degree than with conventional radar. A result of diffraction theory [see, e.g., Glazier and Lamont (1958)] is that the minimum angle (in radians) in which radiation can be concentrated in a given plane is given to within an order of magnitude by

$$\theta = \frac{\lambda}{D} \tag{12.37}$$

D is the maximum dimension of the physical aperture of the antenna in the plane. Thus θ can be considered as the energy beamwidth of the antenna, in the chosen plane. At a range R from the ground, the linear distance covered by this beam is therefore

$$\rho = R\theta = \frac{R\lambda}{D} \tag{12.38}$$

Thus any element of the ground that is smaller than ρ will not be resolved, and ρ is defined as the resolution of the antenna. It is seen that the resolution is proportional to the range R. In the SAR system, the platform on which the radar is mounted moves at a known velocity with respect to the ground. The platform may be on an aircraft, or on a satellite. Normally, the boresight of the radar antenna is normal to the surface being mapped, and the radar is sometimes referred to as *sideways-looking radar* (SLR). As a target element enters the antenna beam at the beam edge and traverses across the beam, it produces a Doppler shift which changes in frequency because the range R changes from some maximum value at the beam edge to a minimum value R_0 at the center of the beam. In what is termed unfocused SAR [see Ellis (1983)], a bandpass filter is used to accept the frequency-modulated Doppler signal. By making the bandwidth f_B of the filter equal to the reciprocal of the traverse time of the target through the beam, the resolution can be shown to be given by [see Ellis (1983)]

$$\rho = \tfrac{1}{2}\sqrt{\lambda R} \tag{12.39}$$

Thus the resolution in this case is proportional to the square root of the range R, compared to being directly proportional in the standard radar situation given by Eq. (12.38).

With focused SAR, a matched filter is used, one method being to correlate the Doppler data against a predicted reference pattern. The optimum resolution in this case can be shown to be [see Ellis (1983)]

$$\rho = \frac{D}{2} \tag{12.40}$$

Here resolution is independent of the range R. Some figures are given in Table 12.3.

TABLE 12.3
Typical SAR Parameters

Range	10 km	100 km
Resolution	3 m	3 m
Platform velocity	100 m/s	100 m/s
Wavelength	3 cm	3 cm

Source: Data from Ellis (1983).

EXAMPLE 12.7

The angle θ for the SAR system above is approximately 0.58°. Determine the resolution at 10 km range for a conventional radar using this beamwidth.

SOLUTION Using Eq. (12.38), we obtain

$$P = 10^4 \times \frac{0.58}{57.3}$$

$$= \mathbf{101\ m}$$

12.11 MILLIMETER AND SUBMILLIMETER RADAR

The millimeter waveband extends from 10 to 1 mm, or a frequency range of 30 to 300 GHz, and the submillimeter band from 1 mm down to 100 μm or 300 to 3000 GHz. Atmospheric attenuation tends to increase with decrease in wavelength over these bands, but various "windows" exist where a decrease in attenuation occurs. These windows occur around frequencies of 35, 95, 140, 250, and 890 GHz, all but the latter being shown in Fig. 12.17.

The main advantages of these short-millimeter waves for radar compared to microwaves stem from the smaller antennas that can be used. Smaller antennas are obviously advantageous where portability and mobile radars are required. Also, as shown by Eq. (12.37), the angular resolution improves with shorter wavelength, and the narrower beamwidth gives increased immunity to interference. Increased bandwidth is also available at these higher frequencies, which increases the amount of information that can be received, and these radars are more sensitive to Doppler shift. Possible applications for which these short-millilmeter radars are suited are listed in Table 12.4.

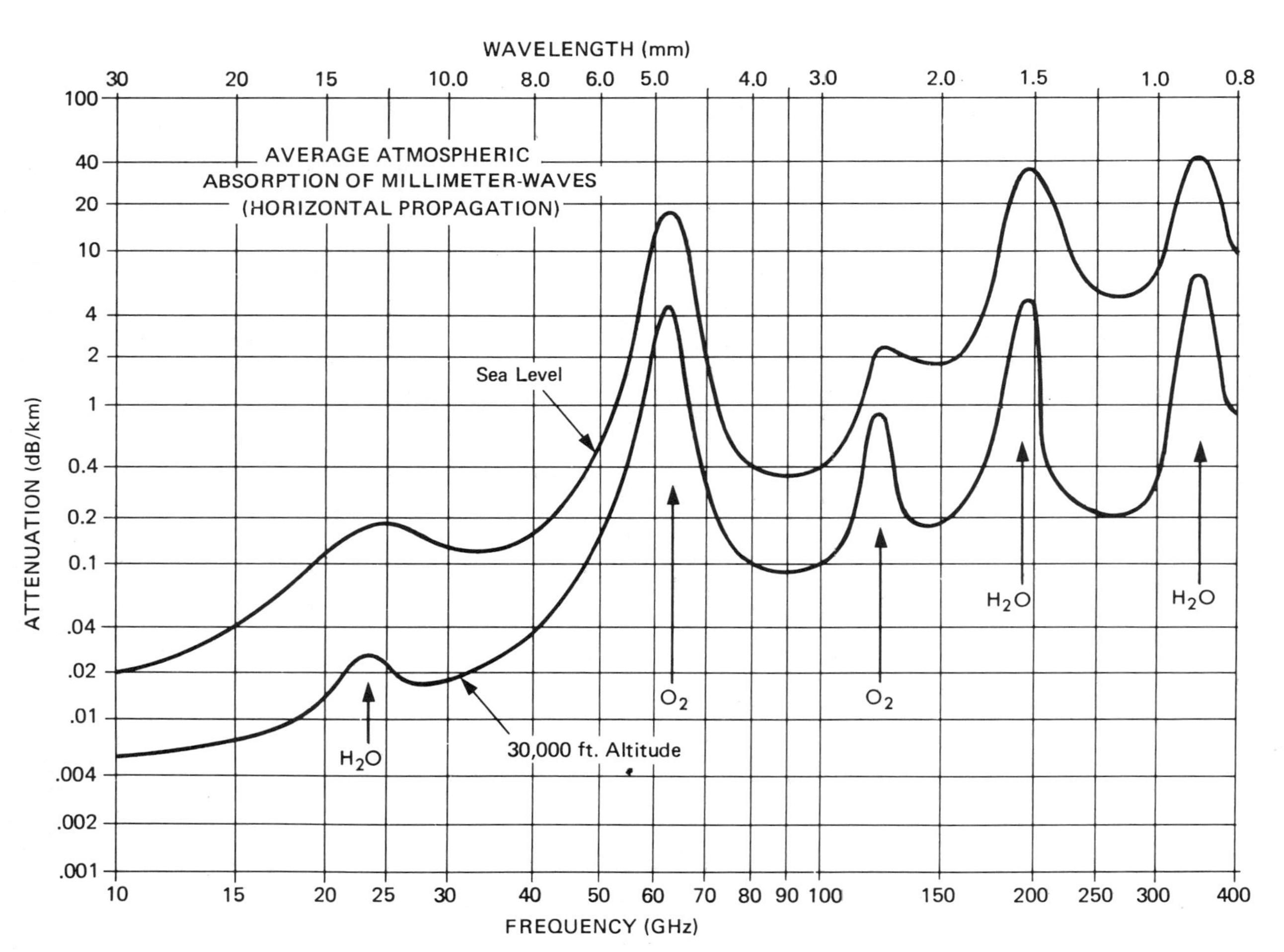

FIGURE 12.17. Average atmosphere absorption of millimeter waves. (*From Microwave Datamate 1982, courtesy of Marconi Instruments Ltd., Stevenage, Hertfordshire, England.*)

TABLE 12.4
Suggested Radar Applications

Low-angle tracking[a]	Remote sensing of the environment
"Secure" military radar[a]	Surveillance
Interference-free radar[a]	Target acquisition
Cloud-sensing radar[a]	Missile guidance
High-resolution radar[a]	Navigation
Imaging radar	Obstacle detection
Ground mapping	Clutter suppresion
Map matching	Fuzes
Space object identification	Harbor surveillance radar
Lunar radar astronomy	Airport surface detection radar
Target characteristics	Landing aids
Weather radar	Air traffic control beacons
Clear-air turbulence sensor	Jet engine exhaust and cannon blast

[a]Applications in which Skolnik believed that submillimeter waves offer more advantage than microwave frequencies.
Source: Johnson (1979).

One other use to which these radars can be put is scale modeling for microwave radars [Cram and Woolcock (1979)]. The determination of radar target data can be expensive and time consuming where full-scale tests have to be made. Scale modeling is achieved by scaling up the frequency by some factor S while scaling down all linear dimensions of the target by the same factor S; and also by scaling down the resistance of any lossy materials on the outer structure of the target by this factor. Such a scale model gives a correct representation of all the reflection phenomena that occur at the target.

Figure 12.18 shows the RF subsystem of a 95.6 GHz instrumentation radar used to obtain radar target data of tactical vehicles and naval vessels. A crystal-controlled master oscillator at 96.67 MHz supplies various reference signals and phase locks a Gunn oscillator at 92.7 GHz. This in turn is up-converted with a 2.9-GHz phase-coded carrier to provide a signal at 95.6 GHz which is used to injection lock a silicon Impatt diode oscillator at this frequency. The transmitted power is 17.8 dBm at this frequency.

The 92.7 GHz signal is also fed to the first receiver mixer to convert the received echo to a 2.9-GHz carrier. This is further reduced in frequency to an IF of 290 MHz, which is sent on to the signal-processing section. The complete radar consists of three subsystems, the RF section, the signal processing and analog recorder section, and the azimuth–elevation scanner section. Full details of the radar are given in Keicher (1982).

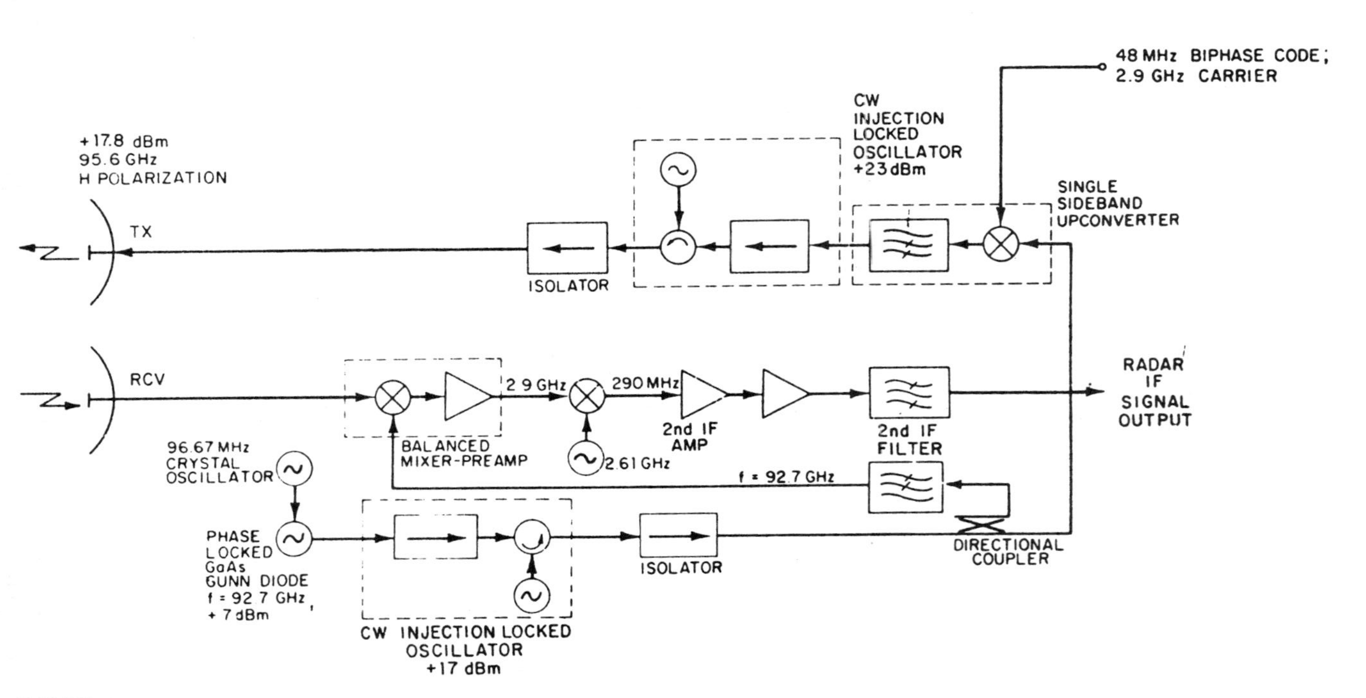

FIGURE 12.18. RF subsystem of a millimeter radar. (*From William E. Keicher and Henry E. Zieman, "Wideband phase coded millimeter wave instrumentation radar," Proc. SPIE, Vol. 337, May 1982.*)

12.12 PROBLEMS

1. A radar transmits an average power of 100 W, the transmit antenna gain being 40 dB. The range to the target is 5 km, and the target has an echoing area of 1 m^2. Calculate (a) the power density and (b) the reflected power at the target.
2. Repeat Problem 1 for a range of 5 miles and an echoing area of 10 ft^2.
3. Given that the same antenna is used to receive the reflected signal and that the wavelength is 3 cm, calculate the received power in Problems 1 and 2.
4. Convert to decilogs (a) a bandwidth of 10 MHz, (b) a distance of 3 km, and (c) the product kT_0, where k is Boltzmann's constant and T_0 is room temperature in kelvins.
5. The specifications for a radar are: frequency 2 GHz, transmit power 10 kW, antenna gain 35 dB, noise factor 3 dB, and noise bandwidth 3 kHz. Path losses of 10 dB and a target echoing area of 50 m^2 may be assumed. Calculate (a) the range for 0-dB signal-to-noise ratio and (b) the signal-to-noise ratio when the range is one-half the value calculated in part (a).
6. The range for a 0-dB S/N ratio for a radar is 100 km. Calculate the S/N when the range is (a) 30 km, (b) 150 km, and (c) 500 km.
7. For the radar of Problem 6, calculate the range (a) in decilogs, (b) in kilometers, and (c) in nautical miles when the signal-to-noise is 30 dB.
8. Explain what is meant by TR and anti-TR switches and why these are used in radar equipment. Describe the operation of one type of TR/anti-TR switch.
9. Explain how the pulse repetition time determines the maximum range of a pulse radar system. What parameter does the pulse width determine?
10. Distinguish between maximum power, peak power, and average power in a pulsed radar. Define the duty factor and show how it relates average power to peak power.
11. In a pulsed radar, the peak power is 3 MW and the duty factor is 0.01. Determine (a) the average power and (b) the power that occurs at a peak of the RF cycle.
12. Explain how echo signals may be resolved in a radar by use of frequency modulation. An FM-CW radar employs triangular-wave modulation as shown in Figure 12.10, the frequency modulation slope being 3 KHz/μs. The beat frequency produced by an echo signal is 5 kHz. Determine the range to the target.
13. Explain what is meant by Doppler shift applied to a radar signal. A

target is receding from a CW radar transmitter at the rate of 25 m/s. The radar wavelength is 3 cm. Calculate the Doppler frequency shift.

14. An FM-CW radar operates at a frequency of 9.25 GHz. A symmetrical triangular modulating waveform is used, the magnitude of the slope being 800 MHz/s. The return from a moving target produces a beat frequency of 3.85 kHz over the positive slope and 3.5 kHz over the negative slope of the FM. Determine (a) the target range, (b) the range rate, and (c) whether the target is moving toward or away from the radar.
15. The maximum physical dimension of a radar antenna is 3 m and the radar operates at 3-cm wavelength. Calculate (a) the approximate angular resolution and (b) the linear distance resolution for a range of 10 km.
16. Calculate the linear distance resolution for 3-cm unfocused synthetic aperture radar at a range of 10 km. Compare this with the result obtained in Problem 15.
17. Discuss the advantages and disadvantages of millimeter and submillimeter radar compared to microwave radar. Describe two applications of millimeter and submillimeter radar.

REFERENCES

BRADSELL, P., 1981. "Microwave radar," *Electronics & Power*, May.

CARR, A. E., L. G. CUTHBERT, and A. D. OLVER, 1981. "Digital signal processing for target detection in FMCW radar," *IEE Proc.*, Vol. 128, Pt. F, No. 5.

CODD, R. D., 1977. "Recent trends in automotive radar," *Radio Electron. Eng.*, Vol. 47, No. 10, pp. 472–476.

CRAM, L. A., and S. C. WOOLCOCK, 1979. "Development of model radar systems between 30 and 900 GHz," *Radio Electron. Eng.*, Vol. 49, No. 7/8.

ELLIS, A. B., 1983. "The processing of synthetic aperture radar signals," *Radio Electron. Eng.*, Vol. 53, No. 3.

GLAZIER, E. V. D., and H. R. L. LAMONT, 1958. *The Services' Textbook of Radio*, Vol. 5, *Transmission and Propagation*. London: Her Majesty's Stationery Office.

HADJIFOTIOU, ANAGNOSTIS, 1977. "Round-off error analysis in digital m.t.i. processors for radar," *Radio Electron. Eng.*, Vol. 47, No. 1/2.

HOLFORD, K., 1981. "If it moves, microwave will detect it," *Electron. Power*, Vol. 17, No. 5.

JOHNSTON, STEPHEN L., 1979. "Radar systems for operation at short millimetric wavelengths," *Radio Electron. Eng.*, Vol. 49, No. 7/8.

KEICHER, WILLIAM E., 1982. "Wideband, phase coded millimeter wave instrumentation radar," *Proc. SPIE*, Vol. 337.

KENNEDY, GEORGE, 1970. *Electronic Communication Systems*. New York: McGraw-Hill Book Company.

FURTHER READING

FIELDING, C. C., and J. B. G. ROBERTS, 1973. "The impact of solid-state electronics on the development of radar systems," *Radio Electron. Eng.*, Vol. 43, No. 1/2.

HOVANESSIAN, S. A., 1973. *Radar Detection and Tracking Systems*. Dedham, Mass.: Artech House, Inc.

WHEELER, GERSHON J., 1967. *Radar Fundamentals*. Englewood Cliffs, N.J.: Prentice-Hall, Inc.

thirteen
Industrial Applications of Microwaves

13.1 INTRODUCTION

Most industrial applications of microwaves utilize the heating effect of microwaves. Industrial microwave applications, categorized by Jolly (1976) into primary and secondary processes, are:

1. *Primary industrial processes:* cooking, baking, pasteurizing, puffing, blanching, thawing, tempering, drying, curing, evaporating, sterilizing, and boiling.
2. *Secondary industrial processes:* casting, extruding, calendering, molding, winding, forming, fusing, and sintering.

A second list, developed in a study by Freedman (1973) and quoted in Jolly's paper, shows the industries using microwave energy as: food, chemicals, rubber, textiles, plastics, graphic arts, foundry, building materials, paper, tobacco, pharmaceuticals, abrasives, lumber, ceramics, leather, coal, and cosmetics.

In addition to these industrial applications, microwaves have been proposed as a means of transmitting electrical energy, both for specialized ground-based systems, and for space-to-earth transmission, which would allow solar energy to be tapped on an almost continuous basis (Brown 1974).

Other applications are found in the areas of science and medicine, and the field is referred to generally as industrial, scientific, and medical (or ISM) applications. Frequencies presently assigned are 915, 2450, 3300,

5800, and 10,525 MHz (896 MHz in the United Kingdom). Some of the ISM applications are described in this chapter.

13.2 MICROWAVE HEATING

Materials that can be heated by microwaves are generally classed as lossy dielectrics. These exhibit conductivity as well as permittivity. Microwaves can penetrate into such materials, and the ohmic losses that occur in the equivalent conductance are dissipated as heat. The term *equivalent conductance* is used because it is made up of normal conductive effects, that is, free electron conduction, plus rotational and vibrational losses in molecules. The power density is given by

$$P_D = \sigma |E_i|^2 \qquad \text{W/m}^2 \tag{13.1}$$

where σ is the equivalent conductivity in S/m, and E_i is the internal electric field in V/m. The conductivity tends to increase with frequency for most lossy dielectrics, and a more useful parameter, termed the *loss tangent*, is used. The loss tangent is relatively independent of frequency for many materials and is defined as

$$\tan \delta = \frac{\sigma}{\omega \varepsilon_r' \varepsilon_0} \tag{13.2}$$

Here, ω is the angular frequency in rad/s of the microwave, ε_r' is the relative permittivity of the dielectric (the dielectric constant), and $\varepsilon_0 = 8.854$ pF/m is the permittivity of free space. Substituting Eq. (13.2) in Eq. (13.1) gives the basic heating as

$$P_D = \omega \varepsilon_r' \varepsilon_0 \tan \delta |E_i|^2 \qquad \text{W/m}^2 \tag{13.3}$$

It must be kept in mind that E_i is the internal electric field, and its relationship to the external field is complicated. The internal field will not necessarily be uniform, and local "hot spots" can be generated. This is discussed in more detail in Chapter 15.

Data on dielectric materials are usually presented in the form of complex permittivity. Thus the complex relative permittivity (i.e., dielectric constant) is written as

$$\varepsilon_r = \varepsilon_r' - j\varepsilon_r'' \tag{13.4}$$

Some authors use the form $\varepsilon_r' + j\varepsilon_r''$. In any case, ε_r' is the relative permittivity and ε_r'' is the loss factor. This in turn is given by

$$\varepsilon_r'' = \frac{\sigma}{\omega\varepsilon_0} \tag{13.5}$$

Thus the loss tangent can also be written as

$$\tan\delta = \frac{\varepsilon_r''}{\varepsilon_r'} \tag{13.6}$$

Values of complex permittivity for a wide range of biological substances are tabulated in Stuchly and Stuchly (1980).

EXAMPLE 13.1

The complex relative permittivity for human brain, in vitro, is given as $33 - j18$ at 2450 MHz. Calculate (a) the equivalent conductivity, (b) the loss tangent, and (c) the power density when the internal field is 100 mV/m.

SOLUTION (a) From Eq. (13.5),

$$\begin{aligned}\sigma &= \omega\varepsilon_0\varepsilon_r'' \\ &= 2\pi \times 2.45 \times 10^9 \times 8.854 \times 10^{-12} \times 18 \\ &= \mathbf{2.45\ S/m}\end{aligned}$$

(b) From Eq. (13.6),

$$\tan\delta = \frac{18}{33} = \mathbf{0.5454}$$

(c) From Eq. (13.1),

$$P_D = 2.45 \times (0.1)^2 = \mathbf{2.45 \times 10^{-2}\ W/m^2}$$

To the extent that the loss tangent is constant, the power density is seen to be proportional to frequency, which is one reason for using high frequencies. Unfortunately, high-power microwave sources in the higher-frequency bands are not readily available, and this is one limitation at present on the use of higher frequencies. For domestic microwave ovens, the cavity magnetron is the generally accepted microwave source, the standard power ratings being 600 W and 1000 W.

A measure of the penetration depth of microwaves is the depth at which the fields are reduced by a factor of $1/e$. This, termed the *penetration depth*, is given by (Osepchuk 1984):

$$D = \frac{0.225\lambda}{\sqrt{\varepsilon_r'(\sqrt{1 + \tan^2 \delta} - 1)}} \qquad (13.7)$$

This tends to suggest that greater penetration is obtained at lower frequencies, but to offset this, the magnitude of the internal field tends to decrease as the frequency is lowered. The penetration depth is discussed further in Section 15.3 and is plotted as a function of frequency for various materials in Fig. 15.2.

13.3 MICROWAVE OVENS

13.3.1 Domestic Ovens

The basic constructional features of a domestic microwave oven are shown in Fig. 13.1. Microwave energy, generated by the magnetron, is launched into the oven cavity through a waveguide aperture. A metallic stirrer, which is simply a rotating metallic blade, reflects the microwave energy and produces a multimode field in the cavity. This helps to ensure uniform heating throughout the cavity. The stirrer may be driven by convection heat currents. The food may also be rotated on a turntable to help ensure uniform cooking.

An electric heater is shown in Fig. 13.1, this being used to "brown" the food. It must be remembered that microwaves can cook from the "inside out." This may leave the outside appearing uncooked, and the browning element is used to give the food the conventional cooked appearance.

Because of the comparatively short cooking times required when using microwaves, accurate control of cooking time and temperature is essential. Domestic ovens often have some form of programmed control, possibly utilizing a microprocessor. Measurement of temperature may be made using a temperature probe in the food. This method has a number of drawbacks. The probe and leads may affect the cooking process by altering the field distribution. Also, microwave leakage may occur through the leads. An alternative, indicated in Fig. 13.1 consists of measuring the temperature of the exhaust air from the oven (Sato and Watanabe 1978). The exhaust temperature is sensed by a thermistor, which in turn controls the voltage to a comparator in the control circuit. The reference voltage for the comparator is proportional to the initial exhaust temperature, and therefore the circuit measures the change in temperature. The control

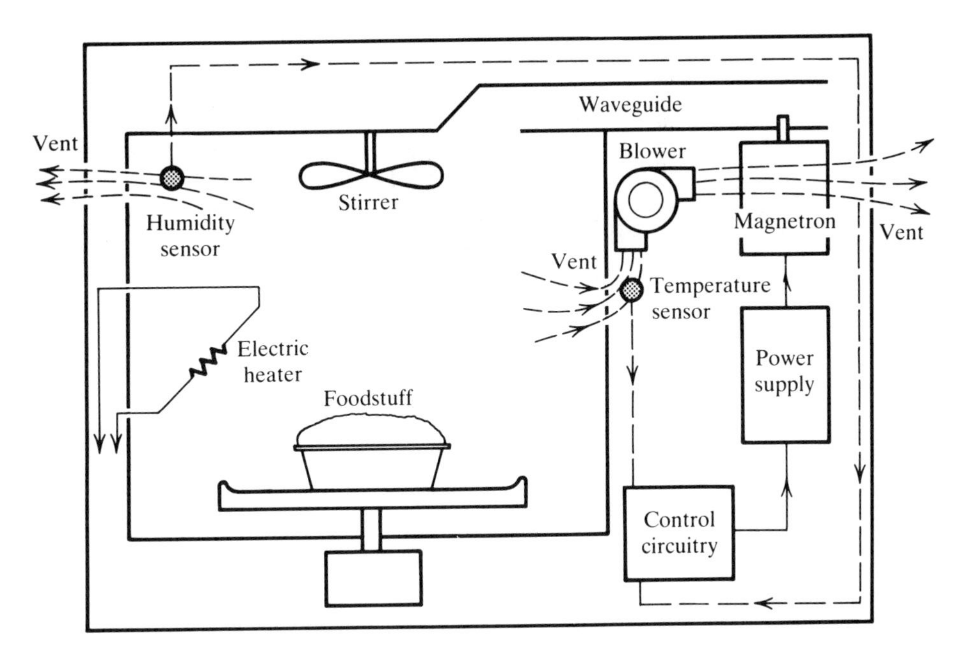

FIGURE 13.1. Basic construction of a microwave oven.

circuit has built-in compensation for variations in room temperature, differences in quantity and initial temperature of food, and changes in initial temperature resulting from quick, repeated use. It is interesting to note that analog circuitry is used in the arrangement described by Sato et al. to maintain low cost.

The relative humidity of the exhaust air may also be used for control purposes this being shown as an alternative to the temperature measurement in Fig. 13.1. The relative humidity decreases initially as the oven heats up, but for many foodstuffs, a sudden release of vapor occurs shortly before completion of cooking. The humidity sensor is a resistive element the resistance of which decreases with increase in humidity. A basic humidity-sensing type of oven is described by Nagamoto et al. (1978). The time to completion of cooking is predetermined for certain ranges of foods, and this information is stored in a microprocessor the control circuit being digital. Once the required change in humidity is detected, the power level of the magnetron may be reduced, and the time to completion of cooking is controlled automatically.

The door of the microwave oven must present a secure seal against leakage of microwave energy. Regulations are in force in most industrialized countries limiting the maximum permissible amount of radiation (see Chapter 15). The direct mechanical contact between door and oven cavity is not sufficient to stop leakage of microwave energy, and one commonly used method is to include a microwave choke in series with the door-to-

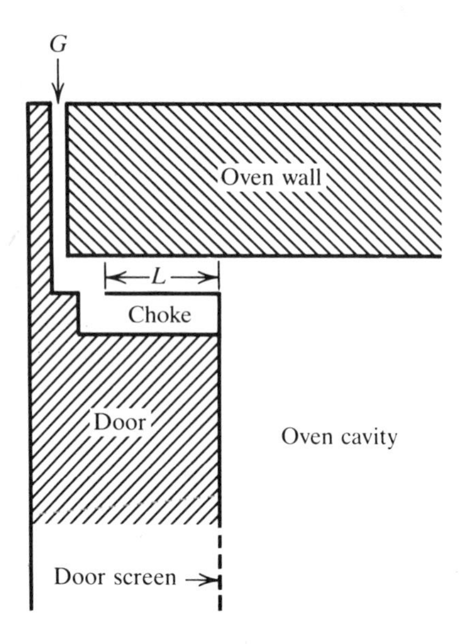

FIGURE 13.2. Basic choke arrangement for microwave oven door.

oven gap. This is illustrated in Fig. 13.2. A flange of depth L is fitted around the inside perimeter of the door. Length L is a quarter-wavelength, and therefore the flange forms a quarter-wave shorted section. From the theory of Section 4.6, this is seen to present an open circuit (in practice a very high impedance) at the gap G. The high impedance is in series with the leakage gap, and therefore effectively isolates this from the oven interior.

Figure 13.3 shows three variations of the basic choke arrangement. Tests conducted on these (Bucksbaum 1981) have shown that type C provides considerably better isolation than types A and B.

In practice the quarter-wave section is filled with dielectric material to prevent the entry of foreign matter, and the length must be cut accordingly. Even with the dielectric fill, the isolation provided by the choke tends to deteriorate with use. Various modifications have been made to the basic choke to maintain performance in normal use [see e.g., Osepchuk et al. (1973) and Ohkawa et al. (1978)].

13.3.2 Industrial Systems

Ovens and heaters for industrial systems generally have to be custom made for the particular application. Work may have to be conveyed through the microwave field on a conveyor belt, and exposure to the field may be in

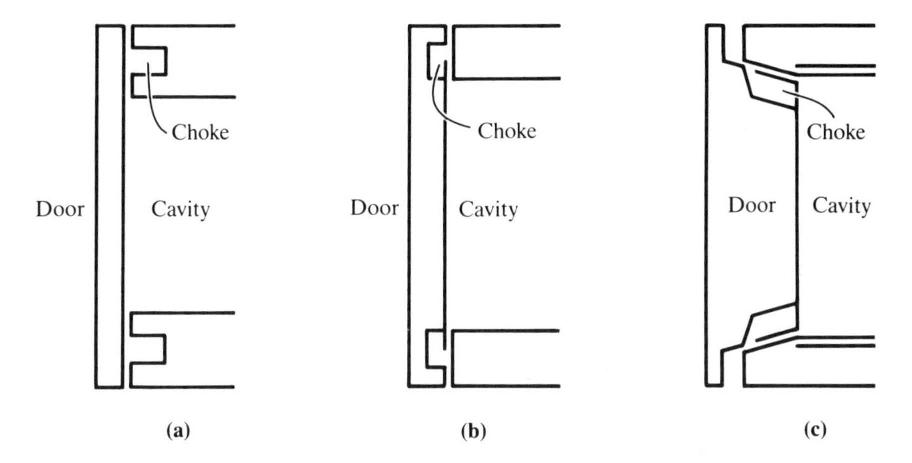

FIGURE 13.3. Variations on the basic choke arrangement.

an oven enclosure, in a waveguide, or possibly over radiating slots from a waveguide structure.

Microwave drying is gaining acceptance as an alternative to conventional drying in the chemical processing industry. In many instances it makes more efficient use of the total energy input, and provides faster drying at lower temperatures (Hubble (1982)). Operation may be at atmospheric pressure, or under vacuum where evaporation at lower temperatures is required. Fig. 13.4 shows the layout of a typical atmospheric-pressure microwave dryer. The material to be dried is fed into the tunnel-

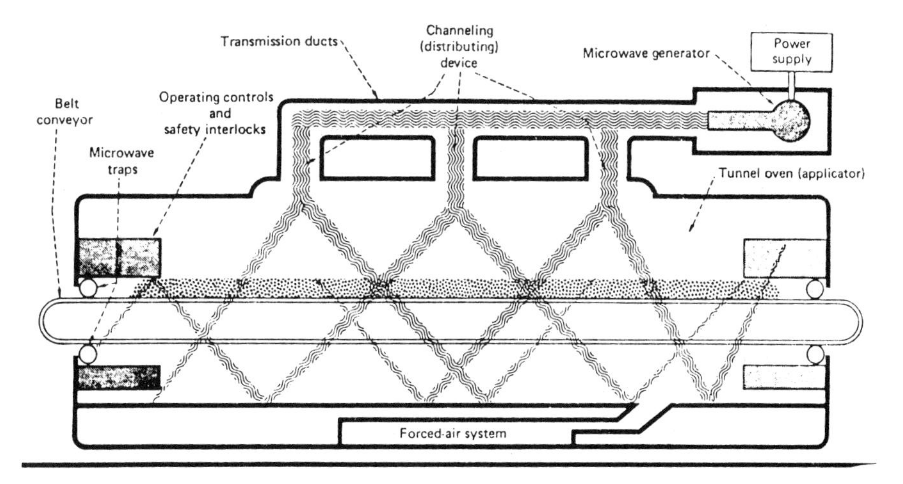

FIGURE 13.4. Schematic of a typical atmospheric-pressure microwave dryer. (*Excerpted by special permission from Chemical Engineering Oct. 4, 1982. Copyright © (1982) by McGraw-Hill: New York, N.Y. 10020*).

oven on a conveyor belt. Waveguide ducting distributes the microwave radiation throughout the oven as shown. The microwave traps shown in Fig. 13.4 consist of water flowing in plastic pipes. These traps prevent radiation of microwave energy from the end gaps of the conveyor system. The primary microwave source is a high power magnetron, a typical unit being the RCA 8684 magnetron as described in RCA data sheets and Application Note 4985. This tube can deliver up to 30 kW at 915 MHz, which is capable of evaporating about 75 lb of water per hour (Hubble (1982)).

Another illustration of an industrial process is a large-volume microwave plasma generator (LMP), described by Jullien and Bosisio (1983). This system has been used for the plasma deposition of polymer coatings on various surfaces. Use of microwaves enables a low-temperature plasma to be generated. In principle the plasma can be of any desired length, which makes the method of considerable interest to industrial processes requiring treatment of large areas. The microwave frequency is 2.45 GHz, and the power can be varied from zero to 2.5 kW.

A system presently under development for the vitrification of high-level radioactive wastes is described by Gayler and Hardwick (1978). Glass-fiber cylinders are continuously loaded into an oven cavity. Microwave energy is applied, which evaporates the radioactive waste and calcines it in the glass-fiber matrix.

13.4 POWER TRANSMISSION

Radio transmission involves the transfer of energy between points without the need for connecting wires. Although energy is transferred by this means in the traditional areas of communications and radar, the amounts are exceedingly small compared with the amounts required in the industrial and domestic power fields. Even so, the same principles of transmission can be applied to the power field, but with fundamentally different design constraints. The unique properties of microwaves for power transmission are listed by Brown (1974) as:

1. No mass either in the form of wire conductors or ferrying vehicles is required between the source of energy and the point of consumption.
2. Energy can be transferred at the velocity of light.
3. The direction of energy transfer can be rapidly changed by repointing the transmitting antenna.
4. There is no loss of energy in the transfer of energy through the vacuum of space. Further, over a relatively broad frequency range there is little loss of energy in the transfer of energy from space to the earth's surface.
5. The mass of the transducers between ordinary electrical power and microwave power at the transmitting and receiving points can be small.

Apart from property 2, which is hardly unique to radio transmission, since the velocity of electromagnetic propagation along wire conductors is very close to the velocity of light, these properties suggest that microwaves could be used profitably for certain applications. Some of these discussed by Brown (1974) are transmission over deep canyons or bodies of water, microwave powered (and control of) vehicles at high altitudes (e.g., 50,000 ft), and transmission of large amounts of power derived from solar energy. The latter application has received the most attention to date and is referred to as the *satellite solar power station* (SSPS) concept [see Brown (1973, 1974)].

Figure 13.5 shows the essential features of the SSPS for a 10-GW system. The space station, in synchronous orbit, contains the solar cell panels, the dc-to-microwave converters, and the transmitting antenna. The

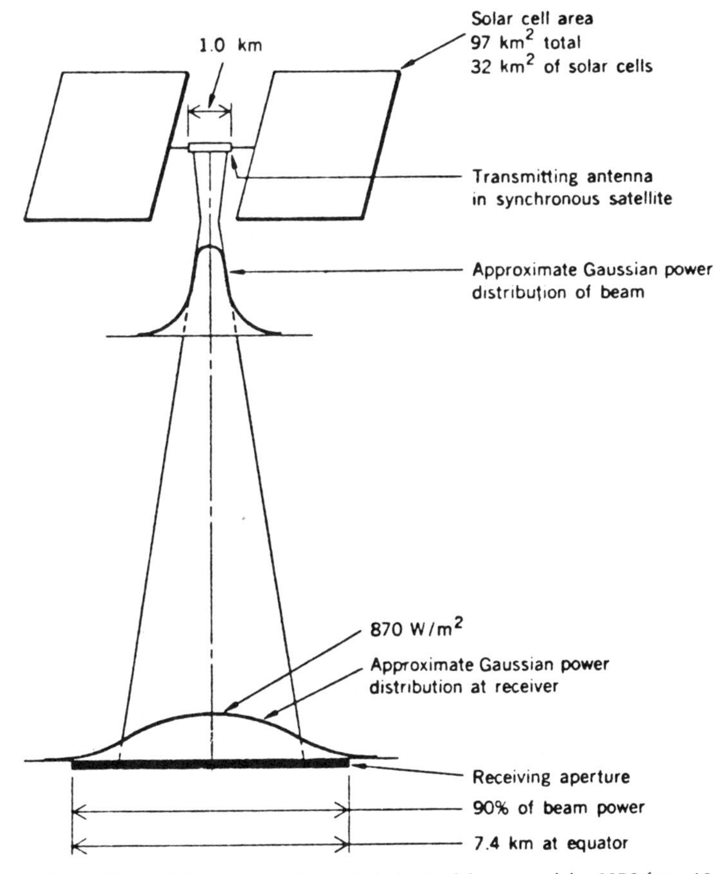

FIGURE 13.5. Dimensions of essential physical features of the SSPS for a 10-GW system. (*From W. C. Brown, "Satellite power stations: a new source of energy? IEEE Spectrum, Mar. 1973; © IEEE 1973; with permission.*)

solar cells convert sunlight to dc electrical energy. The proposal is to use microwave tubes of the crossed-field type (CFA, see Section 9.9) to convert the dc to microwaves. Two million 5-kW tubes would be needed to produce the 10 GW. A phased array antenna is required for transmission, and this must be divided into a large number of smaller arrays for phasing purposes. With correct phasing, a near-Gaussian distribution of power across the antenna surface is achieved, and the efficiency of transmission with such a beam can be on the order of 90%.

At the receiving station, the antenna array is 7.44 km in diameter, and this captures 90% of the transmitted energy. The array consists of a vast number of half-wave dipoles, each connected to its own rectifier circuit. The combination of dipole antenna and rectifier is known as a *rectenna* [see Brown (1969); rectennas are also described in Section 14.8]. Again, it is necessary to divide the receiving array into a large number of smaller arrays to allow for the fact that power density over the receive area is near-Gaussian, and to allow a series–parallel connection of outputs to be used. Figure 13.6 shows an interconnection arrangement of rectennas.

On strictly technological grounds, Brown (1973) has shown that wavelengths in the rather narrow range 7.5 to 15 cm should be used, and the design study for the 10 GW SSPS was carried out for a wavelength of 10 cm, or a frequency of 3 GHz. Figure 13.7 shows the projected power flow and losses, referenced to the solar cell output.

The advantages claimed for SSPS are that it is pollution free, and it

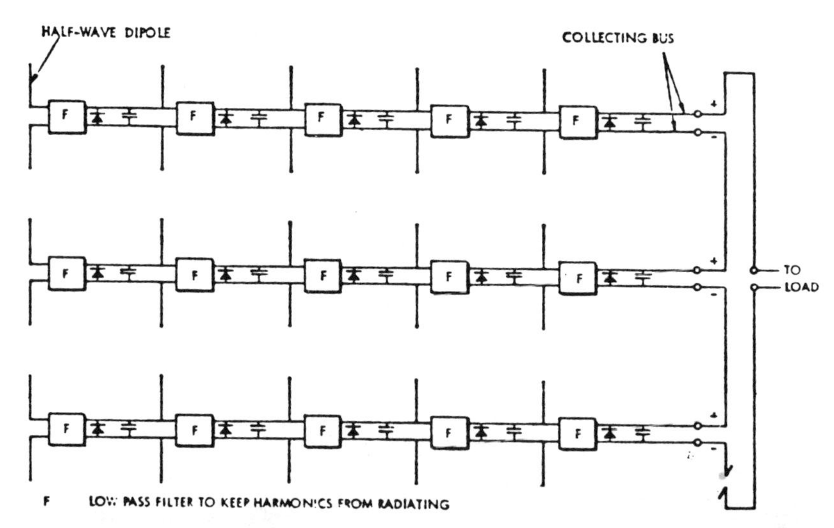

FIGURE 13.6. Proposed interconnection arrangement of half-wave dipoles, wave filters, rectifier circuits, and collecting buses. *(From W. C. Brown, "The technology and application of free space power transmission by microwave beam," Proc. IEEE Vol. 62, No. 1, 1974; copyright IEEE © 1974; with permission.)*

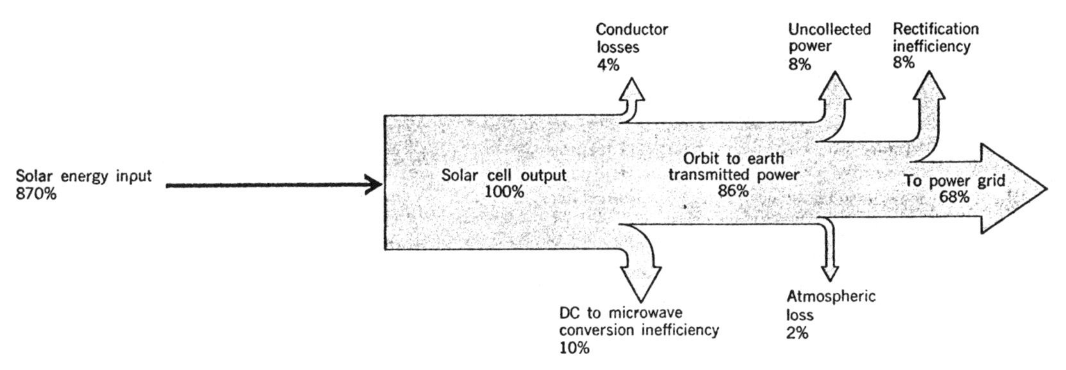

FIGURE 13.7. Projected power flow in the SSPS system indicating various losses. The power flow and losses are referenced to the solar cell output. *(From W. C. Brown, "Satellite Power Stations: a new source of energy?" IEEE Spectrum, Mar. 1973; pp. 38–47 © IEEE 1973)*

is based on an inexhaustible source of energy. At present it is not cost-effective compared with the more conventional methods of power generation. Experience has shown that even reasonable projections in the energy field, both for costs and utilization, are very difficult to make, and at present there is no clear indication that the SSPS concept will ever be feasible.

13.5 MICROWAVE AQUAMETRY

Microwave aquametry is the use of microwaves to measure water content in solids and liquids. The word *aquametry* is coined from the Latin *aqua*, water, and the Greek *metreo*, I measure (Kraszewski 1980). Applications are wide ranging, some examples being the determination of water content in wood, fabrics, grains, and seeds.

The basis for the method lies in the fact that the presence of water greatly changes the complex permittivity [see Eq. (13.4)] of a material. The complex permittivity controls the propagation coefficient, and more specifically, the attenuation coefficient and the phase-shift coefficient [see Eq. (1.9)]. The complex permittivity also controls the wave impedance of the medium and hence the voltage reflection coefficient at the air–wet material interface. In practice, therefore, a microwave aquametry instrument may measure the attenuation and phase shift through the material, these being known as *transmission-type measurements*, or it may measure the complex reflection coefficient (*reflection-type measurements*).

Figure 13.8 shows the block diagram for transmission type measurements. Both the attenuation and phase shift are measured. The relative moisture content is defined as

$$m_w = \frac{W_w}{W_w + W_0} \tag{13.8}$$

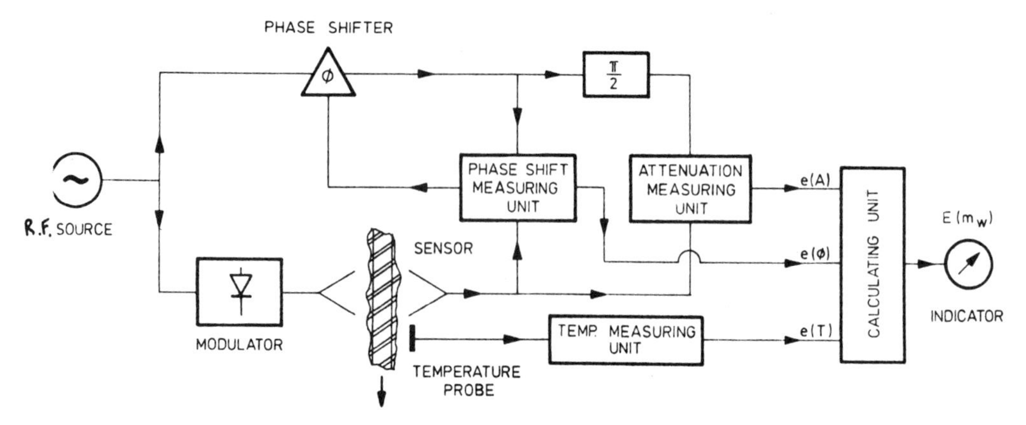

FIGURE 13.8. Block diagram of the two-parameter measuring set-up with a calculating unit and temperature compensating device. *(From A. Kraszewski and S. Kulinski, IEEE Trans. IECI-23, No. 4 © IEEE 1976 with permission.)*

Where W_w is the weight of water and W_0 the weight of dry material. Kraszewski (1976) has shown that the relative moisture content m_w is related to the total attenuation A and total phase shift ϕ through the sample by

$$m_w = \frac{Aa_4 - \phi a_2}{\phi(a_1 - a_2) - A(a_3 - a_4)} \tag{13.9}$$

where the a's are constants determined by a calibration procedure. The results of transmission measurements for sand for various degrees of moisture content are shown in Fig. 13.9. Good agreement is evident between calculated and measured results.

13.6 MEDICAL APPLICATIONS

Medical applications of microwaves can be broadly classified into three areas: therapeutic, diagnostic, and monitoring. Some of the advantages claimed for microwaves (Gupta 1981) are:

1. Less harmful than radioactive or X-ray exposure.
2. A better contrast is achieved in subcutaneous tissue.
3. Improved focusing of power and selective heating in tissue.
4. The body emits radiation in the microwave range the intensity of which is proportional to temperature of the emitting region.

Therapeutic applications. Therapeutic applications rely on the heating effect of microwaves. Because microwaves can penetrate the body fat layers to reach bones and areas inside muscles, localized heating can be achieved. In *diathermy*, the heating effect is used to relieve pain and promote healing of various inflammatory diseases and ailments. In *hyperthermia*, the heat treatment, when used in conjunction with radioactive and chemotherapy, assists in destroying cancerous cells.

Diagnostic applications. The body is known to produce radiations over a wide frequency range, which includes microwaves. The intensity of the microwave radiation is proportional to the temperature of the radiating region. Tissues with pathological conditions and malignant tumors have higher temperatures than their surroundings and a sensitive microwave receiver can be used to detect the difference in radiation levels.

Monitoring applications. Movement of heart and arterial walls can, in principle, be monitored by means of the Doppler shift. This is really a highly specialized application of the Doppler radar described in Chapter 12.

Changes in reflection coefficient have been used to monitor conditions in the lung. Accumulation of excessive fluid in the lungs (a condition referred to as *pulmonary edema*) produces a change in the complex permittivity, which changes the reflection coefficient of the tissue compared to healthy tissue. The amplitude of the reflection coefficient is found to

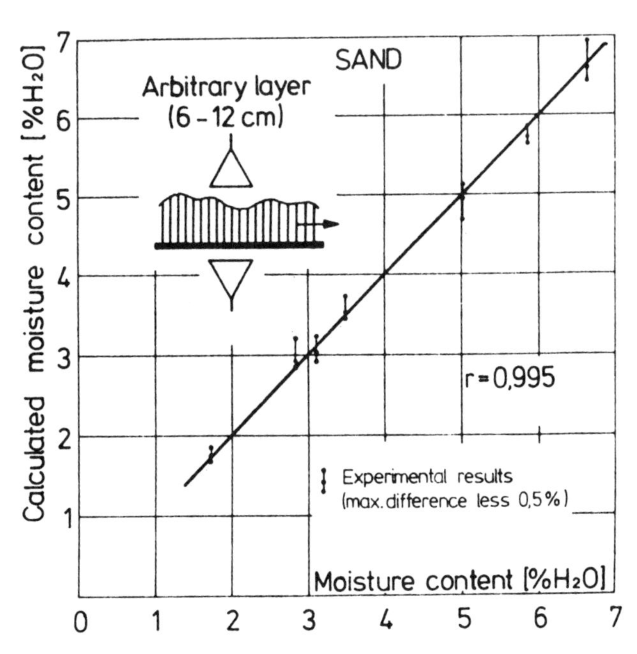

FIGURE 13.9. Relation between the real moisture content of an arbitrary layer of the sand and the moisture content calculated using Eq. 13.9. *(From Kraszewski & Kulinski* op. cit. © *IEEE 1976.)*

increase, and phase angle decrease, for the edematous condition compared to the normal lung.

Microwave energy is launched into a patient using special devices termed applicators. Applicators must meet a number of operational criteria. They must be safe, both for the patient and the operator, and they must be reasonably comfortable in use. They must provide good matching between microwave source and region of application, and they should provide the required focusing of energy. The design and optimization of multistepped waveguide applicators is described by Rebollar and Encinar (1984).

13.7 PROBLEMS

1. Define the term *loss tangent*. Explain its use in quantifying microwave heating effects.
2. Define the term *complex permittivity*. Explain its use in quantifying microwave heating effects. The complex relative permittivity of a certain lossy dielectric is $25 - j15$. Calculate the loss tangent.
3. The complex relative permittivity of human hemoglobin measured at a frequency of 3000 MHz under certain conditions is $59.9 - j19.9$. Calculate (a) the loss tangent, (b) the equivalent conductivity, and (c) the power density for an internal field of 90 mV/m.
4. Define the term *penetration depth*. Determine the penetration depth for the material in Problem 3.
5. Describe the main operational features of domestic microwave ovens. Explain how the temperature of the food in the oven may be accurately measured.
6. Explain the importance of having a tight microwave seal on the doors of microwave ovens. Describe one method of seal construction.
7. Discuss the possible use of microwaves in the utilization of solar energy.
8. Define and explain what is meant by microwave aquametry. Describe one area of application of microwave aquametry.
9. Explain how microwaves are utilized in medical applications for (a) therapeutic, (b) diagnostic, and (c) monitoring purposes.

REFERENCES

BROWN, W. C., 1969. "Progress in the design of rectennas," *J. Microwave Power*, Vol. 4, No. 3.

BROWN, W. C., 1973. "Satellite power stations: a new source of energy?" *IEEE Spectrum*, March, pp. 38–47.

BROWN, W. C., 1974. "The technology and application of free-space power transmission by microwave beam," *Proc. IEEE*, Vol. 62, No. 1, pp. 11–25.

FREEDMAN, G., 1973. "The future of microwave heating equipment in the food industries," *J. Microwave Power*, Vol. 7, No. 4.

GAYLER, R., and W. H. HARDWICK, 1978. "The vitrification of high level radioactive wastes using microwave power," Microwave Power Symp., Ottawa.

GUPTA, C., 1981. "Microwaves in medicine," *Electron. Power*, Vol. 27, No. 5, pp. 403–406.

JOLLY, JAMES A., 1976. "Economics and energy utilization aspects of the application of microwaves: a tutorial review," *J. Microwave Power*, Vol. 11, No. 3, pp. 233–245.

JULLIEN, HENRI, and RENATO G. BOSISIO, 1983. "Polymerization and surface treatment in a large volume microwave plasma generator," *J. Microwave Power*. Vol. 18, No. 4.

KRASZEWSKI, A. and STANISLAW KULINSKI, (1976). "An Improved Microwave Method of Moisture Content Measurement and Control." IEEE Trans. on Ind. Electron. and Control Inst. vol IECI-23 No. 4 Nov.

KRASZEWSKI, A., 1980. "Microwave aquametry—a review," *J. Microwave Power*, Vol. 15, No. 4.

NAGAMOTO, SHUN'ICHI, TSUNEHARU NITTA, and MICHIYO NAKANO, 1978. "Automatic microwave oven with humidity sensing," Microwave Power Symp., Ottawa.

OHKAWA, S., M. WATANABE, and K. KANEKO, 1978. "High performance door seal for microwave oven," *Microwave Power Symp. Digest.*

OSEPCHUK, JOHN M., 1984. "A history of microwave heating applications," *IEEE Trans. Microwave Theory Tech.*, Vol. MTT-32, No. 9, pp. 1200–1234.

OSEPCHUK, JOHN M., JAMES E. SIMPSON, and RICHARD A. FOERSTNER, 1973. "Advances in choke design for microwave oven door seals," *J. Microwave Power*, Vol. 8, Nos. 3/4.

REBOLLAR, J. M., and J. A. ENCINAR, 1984. "Design and optimization of multi-stepped waveguide applicators for medical applications," *J. Microwave Power*, Vol. 19, No. 4.

SATO, K., and M. WATANABE, 1978. "Automatic control system for domestic microwave ovens," Microwave Power Symp., Ottawa.

STUCHLY, M. A., and S. S. STUCHLY, 1980. "Dielectric properties of biological substances," *J. Microwave Power*, Vol. 15, No. 1.

FURTHER READING

BARBER, H., 1981. "Microwaves for cooking and heating," *Electronics & Power*, Vol. 27, No. 5, pp. 401–402.

BROWN, W. C., 1970. "High power microwave generators of the cross-field type," *J. Microwave Power*, Vol. 5, No. 4, pp. 245–260.

BUCKSBAUM, A. M., 1981. "Microwave oven door seal characteristics," *J. Microwave Power*, Vol. 16, No. 1.

fourteen
Microwave Measurements

14.1 INTRODUCTION

The microwave measurements described in this chapter cover those most frequently met with in laboratory or test and maintenance situations. They cover the measurement of frequency, power, attenuation, insertion loss, return loss, impedance, reflection coefficient, S-parameters, noise figure, and radiation levels. Specialized instruments are required for many of these measurements, and often these instruments incorporate microprocessors. The use of microprocessors adds very powerful signal-processing capabilities to instruments. For example, system nonlinearities can be automatically allowed for, and computation of ratios and conversion to decibels can be built-in functions of the instrument.

Many microwave quantities such as impedance, reflection coefficient, and S-parameters are complex; that is, they have both real and imaginary parts, or equivalently, magnitude and phase angle. The measurement of complex quantities at microwaves is difficult and requires expensive equipment. Where possible, therefore, devices and circuits are characterized by the measurement of the magnitude only of the quantities involved. These are referred to as *scalar measurements*, and are usually made on instruments known as *scalar network analyzers*.

The measurement of both magnitude and phase over a range of frequencies requires the use of a *complex network analyzer*. These complex measurements are sometimes referred to as *vector measurements*, even though the quantities being measured, such as impedance, are neither vectors nor phasors.

Microwave measurements differ in some important respects from low-frequency measurements. In particular, voltage and current are not easily measured at microwaves. Many of the properties of devices and circuits at microwaves are derived from measurements of reflection coefficient.

14.2 SWEPT FREQUENCY SCALAR MEASUREMENTS

The three main components of a swept frequency scalar system are shown in Fig. 14.1. These are the scalar analyzer, the sweep oscillator, and the reflectometer. A brief description of each is given in the following sections.

The modern scalar analyzer is a microprocessor-based instrument which enables power and power ratios to be measured. Thus it can be used to determine the insertion loss of a network, and the return loss of a device or network, as well as measure power. The *insertion loss* is a measure of the power lost in transmission through a device or network. If the network is an amplifier, power gain rather than loss is measured. Let P_1 represent the power received by the load when connected directly to source, and P_2 the power received by the load when the device or network is inserted between source and load, the incident power being held constant; then the insertion loss, in decibels, is

$$\mathrm{IL(dB)} = 10 \log \frac{P_2}{P_1} \tag{14.1}$$

The *return loss* is a measure of the power reflected by a device or network. In the case of a two-port device or network, it is the power reflected at the input port that is measured. Let P_{inc} represent the power in the incident wave and P_{refl} the power in the reflected wave at the input to the device under test. Then the return loss, in decibels, is

$$\mathrm{RL(dB)} = 10 \log \frac{P_{\mathrm{inc}}}{P_{\mathrm{refl}}} \tag{14.2}$$

It will be seen that only power, a scalar quantity, is involved in these equations and in principle, measurements could be made using a wattmeter. The scalar analyzer brings to these measurements a number of significant practical advantages. The instrument is used in conjunction with a sweep oscillator which enables the variation with frequency of measured quantities to be displayed. Use of the microprocessor enables the various decibel quantities to be computed and displayed directly. Data can be stored in the microprocessor memory and used to refresh the oscilloscope trace. The microprocessor can also be used to correct for nonlinearity in the measuring diode characteristic, resulting in accurate power measurements. Data obtained for variations with frequency of the measuring system characteristics can be stored and subtracted from the test measurements. In this way, the results for the device under test can be accurately determined.

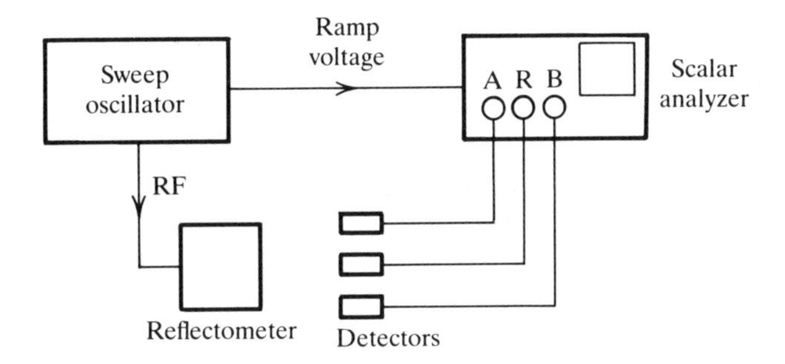

FIGURE 14.1. Block schematic of a swept-frequency scalar analyzer setup.

Figure 14.1 shows the basic arrangement for scalar measurements. The scalar analyzer itself usually has three input channels, a reference channel labeled R, and two other channels, A and B. Associated with each channel is a diode detector. This is used to rectify the microwave signal right at the point of measurement, and the resulting rectified voltage is fed to the channel input. Zero-biased Schottky barrier diodes are used, and as shown in Section 7.10, such diodes produce an output voltage proportional to input power when operated in the square-law region. In the Marconi 6500 scalar analyzer, for example, an EPROM holds a correction table to account for deviations from the square law with larger inputs, and a temperature sensor mounted in the detector head provides data for temperature correction. In any event, provision is made in most scalar analyzers to correct for the diode characteristics, and as a result an accurate power reading is obtained. Thus, in addition to being used for network reflection and transmission properties, the scalar analyzer can be used as a straight power meter.

The ramp voltage used to control the frequency sweep of the oscillator is also used to provide the horizontal sweep to the scalar analyzer. Again, the microprocessor is used to store the sweep data and to provide a frequency calibration for the horizontal scale. Of course, single-frequency measurements may also be made simply by disconnecting the sweep signal. The block diagram for the sweep oscillator is shown in Fig. 14.2. The frequency of the sweep oscillator is usually determined by a yttrium iron garnet (YIG) ferrite, and frequency control is exercised through control of the external magnetic field applied to the ferrite (see Section 6.15). The current to the magnetizing coil is a ramp function, and a corresponding ramp voltage is made available for the horizontal sweep of the scalar analyzer, as already explained.

A constant RF output voltage may be required while the frequency of the sweep oscillator is varied. This may be achieved through leveling circuits, and the voltage is then said to be leveled. The output may be leveled internally or externally, but the basic arrangement is the same in

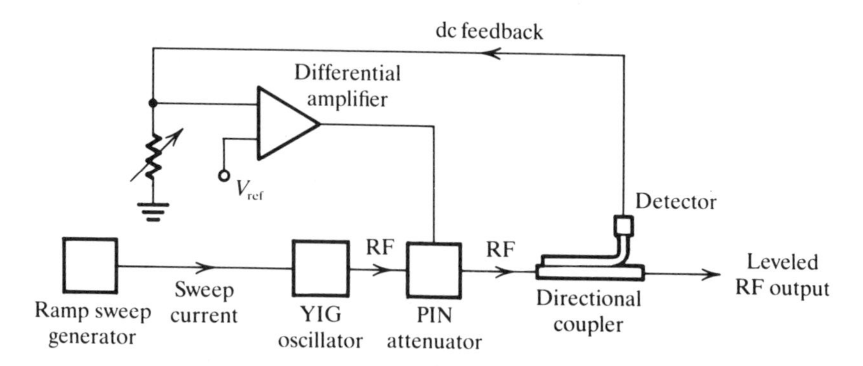

FIGURE 14.2. Block schematic of a leveled sweep oscillator.

each case and is shown in Fig. 14.2. A high-directivity directional coupler is used to monitor the incident wave. The monitored signal is rectified and the rectified voltage fed back to a leveling amplifier within the sweep oscillator. This is a differential amplifier which compares the feedback signal with a preset reference level. The output from the leveling amplifier is used to control a PIN-diode attenuator (see Sections 7.7 and 7.8). Note that only the incident wave is monitored by the directional coupler, not the reflected wave from the load. The situation can arise where a reflected wave from the load enters the oscillator and is again reflected by the internal impedance of the oscillator if this is not perfectly matched to the line. This secondary reflection will add to the incident wave (since they are both sine waves of the same frequency). This will be monitored by the leveling circuit and compensated for. Thus the leveling circuit virtually eliminates the effect of any source mismatch that might exist.

Where a scalar analyzer is used together with a sweep oscillator, leveling may not be absolutely essential, since the analyzer can be programmed to take into account variations in the output level. However, leveling may still be useful where the power input to the device under test must be kept below a specified limit. A *reflectometer* is a device that enables the incident wave and the reflected wave to be measured separately. A straightforward way of doing this is to use two directional couplers connected together, as shown in Fig. 14.3(a). For measurements with coaxial systems a dual directional coupler may be used, as shown in Fig. 14.3(b). As an alternative to the directional coupler arrangement, a *reflectometer bridge* can be used. The circuit is a basic Wheatstone bridge arrangement but with components and connections specially designed for microwave signals. Figure 14.4(a) shows the bridge arrangement for a 50-Ω system.

When the bridge is balanced, that is, $Z_x = 50\ \Omega$, the impedance seen looking into any of the ports 1, 2, or 3, is 50 Ω, and this fact enables 50-Ω lines to be used as connections to the ports. Also at balance, the detector voltage V_d is equal to zero. When Z_x is other than 50 Ω, an unbalance exists, which results in a finite value of detector voltage. Normalizing all impedances to 50 Ω, and letting $e = E/50$, the mesh equations for the bridge [Fig. 14.4(b)] are

$$e = 3I_1 + I_2 - I_3 \tag{14.3}$$

$$e = I_1 + I_2(2 + z_x) + I_3 \tag{14.4}$$

$$0 = -I_1 + I_2 + 3I_3 \tag{14.5}$$

Solving for I_3 yields

$$I_3 = \frac{\begin{vmatrix} 3 & 1 & e \\ 1 & 2 + z_x & e \\ -1 & 1 & 0 \end{vmatrix}}{\begin{vmatrix} 3 & 1 & -1 \\ 1 & 2 + z_x & 1 \\ -1 & 1 & 3 \end{vmatrix}}$$

$$= \frac{e(z_x - 1)}{8 + 8z_x} \tag{14.6}$$

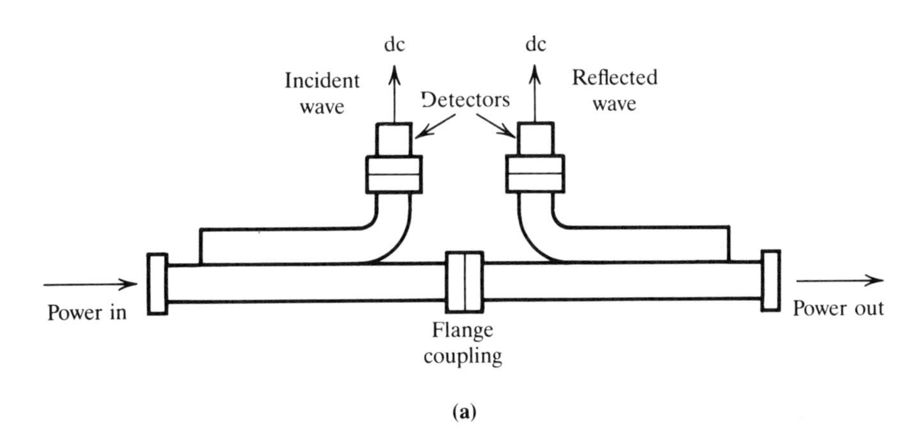

(a)

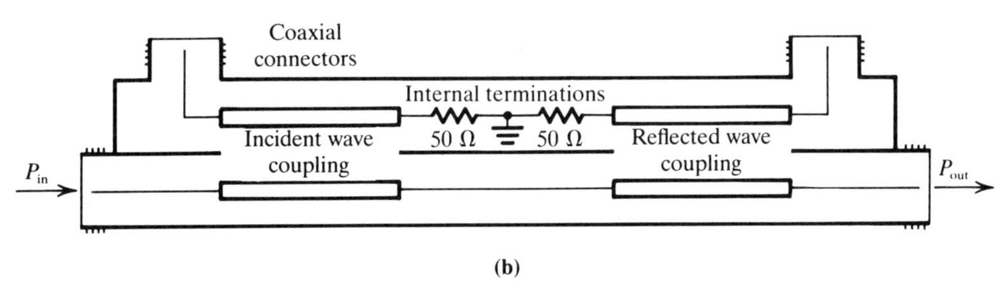

(b)

FIGURE 14.3. Directional couplers arranged as reflectometer: (a) waveguide reflectometer; (b) coaxial reflectometer.

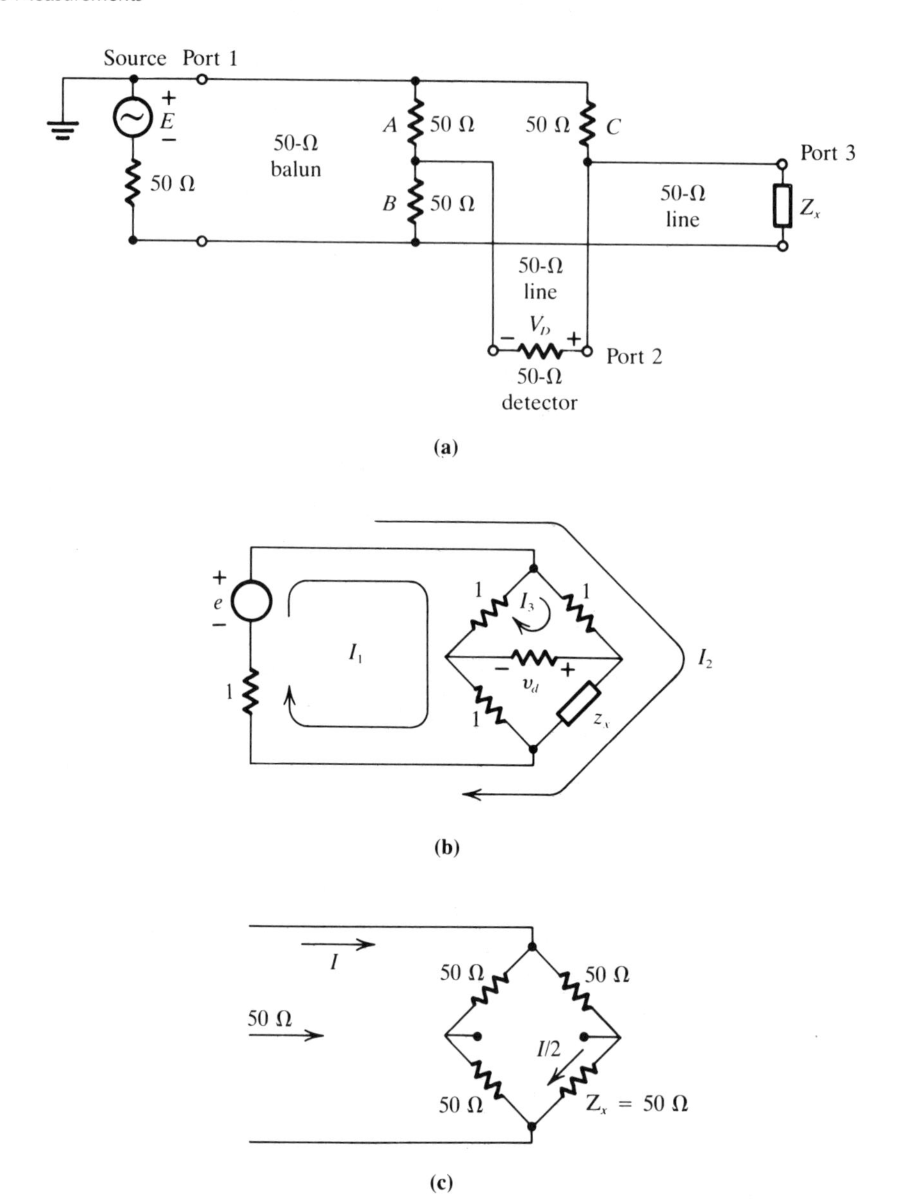

FIGURE 14.4. (a) Circuit for a 50 Ω Wheatstone bridge; (b) circuit normalized to 50 Ω; (c) input impedance is 50 Ω.

In terms of normalized values,

$$v_d = 1 \times I_3 = \frac{e(z_x - 1)}{8(z_x + 1)}$$

Therefore,

$$\frac{V_d}{E} = \frac{v_d}{e} = \frac{z_x - 1}{8(z_x + 1)}$$

$$= \frac{\Gamma}{8} \tag{14.7}$$

Thus, by holding E constant, or by using the scalar analyzer to correct for variations in E, the measured value of V_d is directly proportional to the reflection coefficient. Since the scalar analyzer indicates magnitudes only, only the magnitude of reflection coefficient will be obtained.

When $Z_x = 50\ \Omega$ (or $z_x = 1$), the bridge is balanced. Under these conditions, the detector carries no current and therefore the detector branch may be shown as an open circuit [Fig. 14.4(c)]. As a result, the input impedance as seen by the source is 50 Ω, and the load Z_x carries current $I/2$, as shown in Fig. 14.4(c), where I is the input current. Now, with the 50-Ω load connected directly to the source, the load current would also be I, and therefore the insertion loss of the bridge is 6 dB.

14.3 MEASUREMENT OF RETURN LOSS

Measurement of return loss requires a measurement of incident power, P_{inc} of Eq. (14.2) and also of P_{refl}. As shown in Fig. 14.3(a), the separation of incident and reflected waves may be achieved using two directional couplers, or alternatively, the reflected wave may be measured using a bridge. The equipment setup for measurement of return loss is shown in Fig. 14.5. In Fig. 14.5(a), the output from the first directional coupler is used for source leveling. Detector A measures the reflected wave. Calibration is carried out using a short circuit. This results in total reflection of incident power, or $P_{refl} = P_{inc}$ in Eq. (14.2) and hence in 0 dB return loss. Let A' represent the set of results for detected voltage over the swept frequency range with the short circuit in place, and let A represent the corresponding set with the device under test in place. Set A' is stored in the scalar analyzer memory, and the decibel values subtracted from the decibel values for A. Since the incident voltage is held constant by leveling, the computation $A'(\text{dB}) - A(\text{dB})$ gives the return loss. This is displayed as a function of frequency on the scalar analyzer CRT, or may be plotted

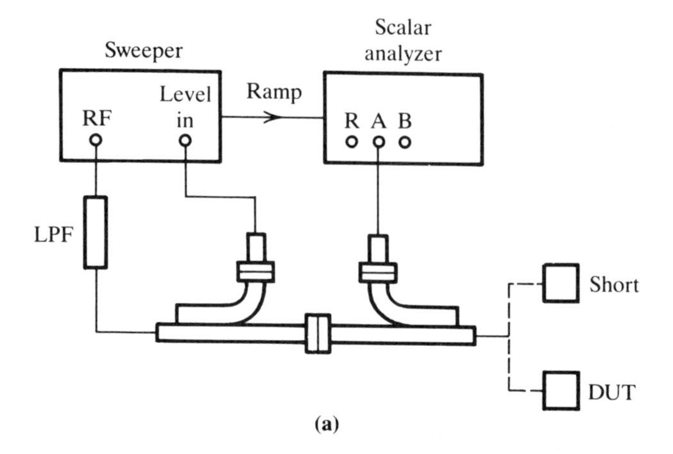

(a)

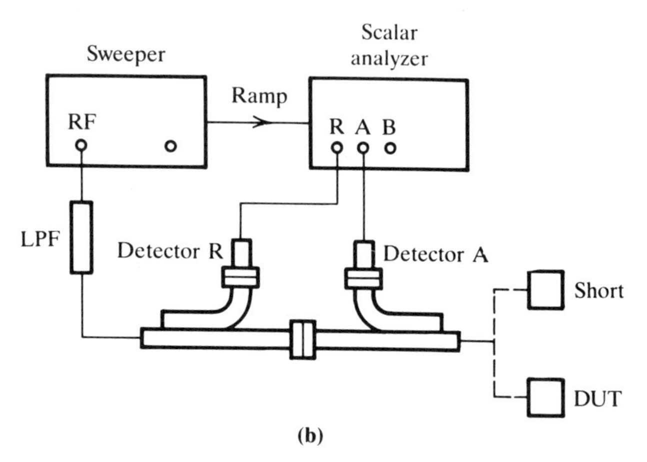

(b)

FIGURE 14.5. Return loss measurement: (a) leveled source arrangement; (b) ratio method.

out on an X-Y plotter. Instead of leveling, the input from the first directional coupler may be fed to the reference channel R, as shown in Fig. 14.5(b). A short circuit is used for calibration as before, but this time the set of values $T_1(\text{dB}) = A'(\text{dB}) - R'(\text{dB})$ is first stored, where R' is the input representing the incident wave as measured on detector R. Thus the system response is measured with the short circuit in place and variations of the incident wave are subtracted out. This is referred to as ratioing, since a difference of two quantities in decibels is the logarithmic equivalent of their ratio. The procedure is repeated with the device under test in place, yielding a new set of values $T_2(\text{dB}) = A(\text{dB}) - R(\text{dB})$, where R represents the set of incident wave values and A the reflected wave values. Finally, the scalar analyzer is programmed to calculate $T_1(\text{dB}) - T_2(\text{dB})$, which is the required return loss. An advantage of the ratioing method is that the source can deliver maximum power to the system, whereas with leveling, the power output is always leveled down to the minimum power available. Also, frequency shifts tend to occur with leveling and these are avoided in the ratioing method.

In both methods a low-pass filter is shown inserted between source and measurement system. This is to reduce the level of any harmonics that might be present in the source output, which could result in measurement errors. Note that in Fig. 14.5(a) the LPF is included within the leveling loop, and therefore any effect it has on source mismatch is leveled out.

14.4 MEASUREMENT OF INSERTION LOSS

As shown by Eq. (14.1), insertion loss requires a measurement of load power with and without the device under test inserted in circuit. As with return loss, either source leveling, or ratioing may be used to allow for the effects of variation in source output. Figure 14.6 shows the equipment setup for the ratioing method. The matched load with detector A is first connected to the output of the directional coupler and a calibration run made, which gives the ratio A'/R' or its decibel equivalent, $T_1(\text{dB}) = A'(\text{dB}) - R'(\text{dB})$. These data are stored in the scalar analyzer. The procedure is repeated with the device under test inserted between directional coupler and matched load, and the data $T_2(\text{dB}) = A(\text{dB}) - R(\text{dB})$ computed. A final computation of $T_2(\text{dB}) - T_1(\text{dB})$ gives the insertion loss of the device. Simultaneous measurements of return loss and insertion loss may be made using the equipment setup shown in Fig. 14.7. Return loss is measured using channel A and insertion loss using channel B. Calibration proceeds as for the individual measurements described previously. The return-loss calibration is made using a short circuit and the data, ratioed to the input,

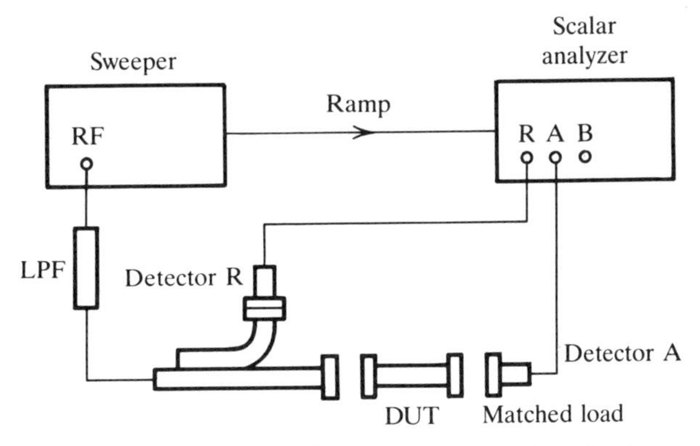

FIGURE 14.6. Ratio method for measurement of insertion loss.

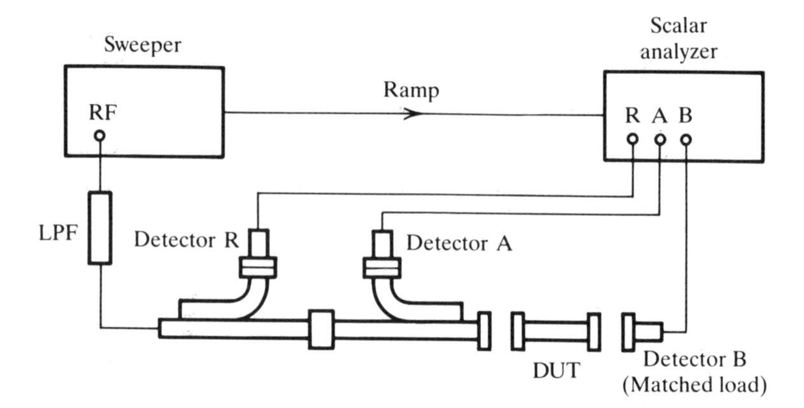

FIGURE 14.7. Arrangement for simultaneous measurement of insertion loss and return loss.

stored in the A memory. The insertion-loss calibration is made by removing the device under test and connecting the matched load with detector B directly to the directional coupler, the ratioed data being stored in the B memory. With the device under test inserted, the measured results can be ratioed with the input for the test run. With all data in decibels, the memory contents for each channel can be subtracted from the corresponding test results, yielding simultaneous values for return loss and insertion loss. One advantage of making the tests together is that often a device has to be adjusted to give a compromise between reflection and transmission, and with simultaneous measurements the adjustments are easier to make.

14.5 SLOTTED LINE MEASUREMENTS

Voltage reflection coefficient, and hence load impedance, may be measured using a slotted line. Equation (1.83) gives the relationship between the voltage standing-wave ratio (VSWR) and the magnitude of the reflection coefficient, and this equation is repeated here for convenience:

$$\rho_L = \frac{\text{VSWR} - 1}{\text{VSWR} + 1} \tag{14.8}$$

The slotted line allows the VSWR to be measured. As will be shown shortly, the phase angle may also be determined from the slotted line measurements. An example of commercially available slotted line equipment is shown in Fig. 14.8(a). The slotted section of the line allows a small probe to be inserted and moved along the line, the probe sampling the electric field. This field is proportional to the voltage between conductors, and

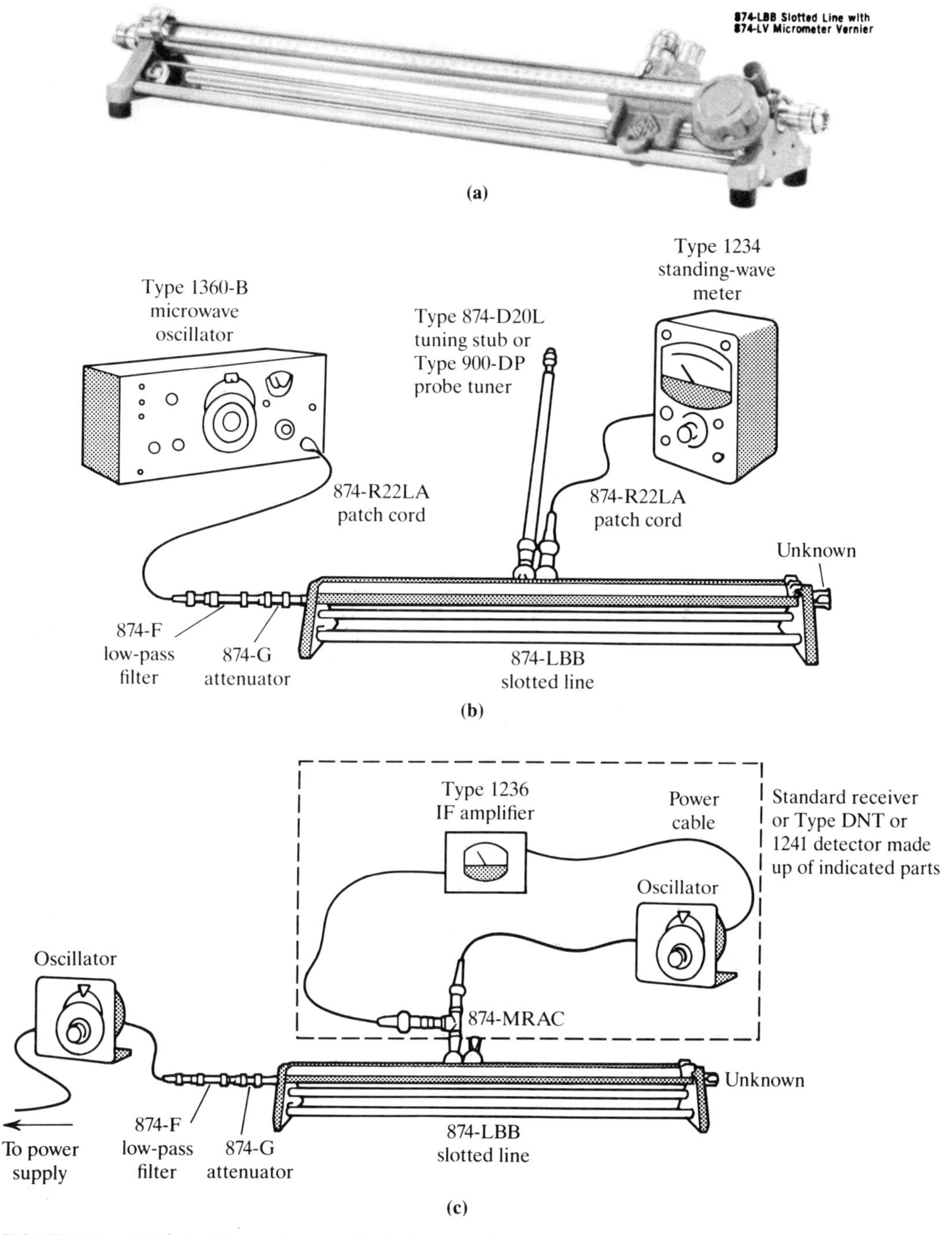

FIGURE 14.8. (a) Slotted line equipment; (b) simple tuned detector arrangement; (c) heterodyne detection. (*Courtesy of GenRad Inc., West Concord, Mass.*)

therefore the ratio of maximum field to minimum field is also the ratio V_{max}/V_{min}. Figure 14.8(b) shows how the internal detector in the carriage is tuned by means of a stub and the detected signal fed to a standing-wave meter, Fig. 14.8(c) shows a more sensitive arrangement where the diode mixer of Fig. 7.39 is used in a heterodyne system and the output IF at 30 MHz is amplified before detection. The line must be carefully constructed to minimize radiation from the slot and also the probe penetration must be carefully adjusted so as not to produce any significant reflections itself. The center conductor is supported at the ends only, and the line

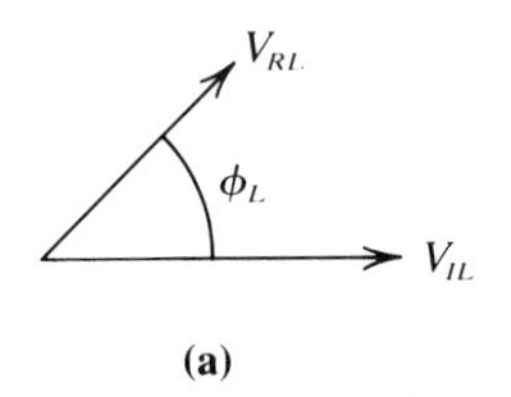

(a)

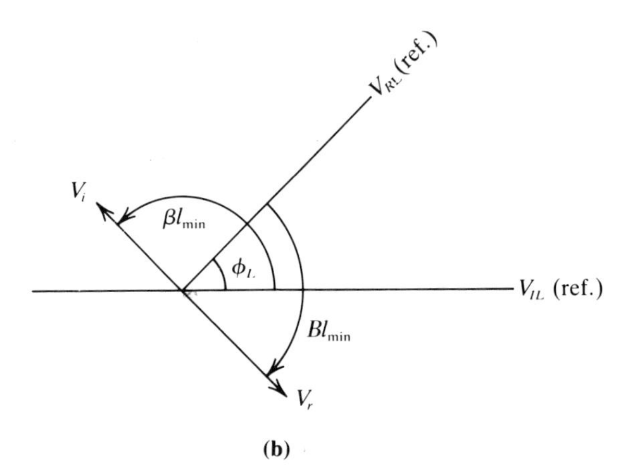

(b)

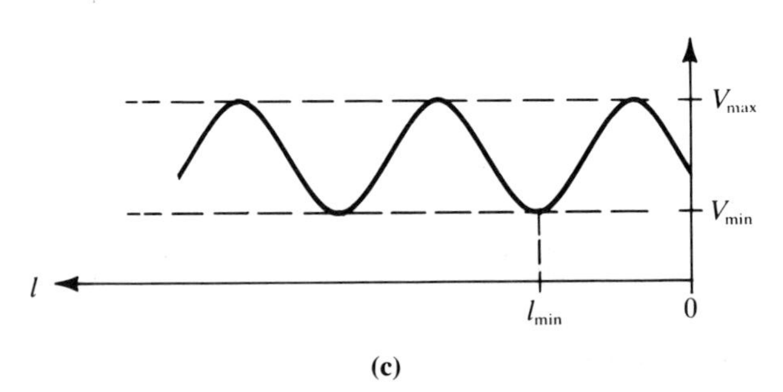

(c)

FIGURE 14.9. (a) Voltage phasors at the load; (b) voltage phasors at the first voltage minimum; (c) voltage standing wave.

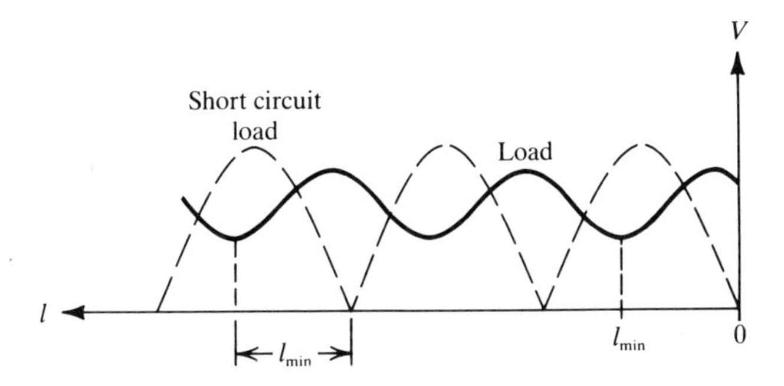

FIGURE 14.10. Technique for measuring l_{min}.

must be sufficiently rigid to prevent variations of line geometry introducing variations in the characteristic impedance. The line illustrated has a characteristic impedance of 50 Ω and can be used over the frequency range 300 to 2000 MHz.

In addition to allowing VSWR to be measured, the slotted line allows the distance between the first voltage minimum and load to be measured, and from this, the phase angle ϕ_L of the reflection coefficient can be determined as shown next. Figure 14.9(a) shows the phasor diagram for the voltage components at the load, and Fig. 14.9(b) for the voltage components at the first voltage minimum. Let the distance between load and first voltage minimum be represented by l_{min}, as shown in Fig. 14.9(c); then it will be seen from the phasor diagram of Fig. 14.9(b) that

$$2\beta l_{min} - \phi_L = \pi$$

Therefore, the desired result is

$$\phi_L = 2\beta l_{min} - \pi \tag{14.9}$$

In the theory, the first minimum is specified, but in fact any of the minima on the slotted section may be used (provided that losses can be ignored) because these are separated by $\lambda/2$, as shown previously. The addition of $\lambda/2$ to l_{min} in Eq. (14.9) adds an integer multiple of 2π radians, since $\beta = 2\pi/\lambda$, and this has no effect on ϕ_L since the principal value is all that is required. Often, it will not be possible to measure l_{min} directly, and a technique used is to establish an equivalent load reference on the slotted section by means of a short circuit load. This is illustrated in Fig. 14.10. The short circuit produces a series of minima, some of which will fall within the range of the slotted section, and the first of which must occur at the load point, by definition of a short circuit. Since both patterns repeat at intervals of $\lambda/2$, then l_{min} can be measured, as indicated on Fig. 14.10.

The method of using the slotted line to determine an unknown impedance may be summarized as follows:

1. Measure l_{min}.
2. Measure the distance between successive minima. This gives $\lambda/2$.
3. Measure the VSWR.
4. Use Eq. (14.9) to calculate ϕ_L.
5. Use Eq. (14.8) to calculate ρ_L.
6. Use Eq. (1.60) to calculate Z_L, remembering that $\Gamma_L = \rho_L e^{j\phi_L}$.

EXAMPLE 14.1

(a) Measurements on a lossless slotted line yielded the following values: $l_{min} = 5.6$ cm, distance betweeen successive minima = 15 cm, and VSWR = 3.2:1. Calculate the voltage reflection coefficient at the load.

(b) Given that the characteristic impedance of the slotted line is 50 Ω, calculate the load impedance.

SOLUTION

(a) $\lambda = 2 \times 15 = 30$ cm. From Eq. (14.9),

$$\phi_L = 2 \times \frac{2\pi}{30} \times 5.6 - \pi$$

$$= -0.7959 \text{ rad} \quad \text{or} \quad -45.6°$$

From Eq. (14.8),

$$\rho_L = \frac{3.2 - 1}{3.2 + 1} = 0.5238$$

Therefore,

$$\Gamma_L = \mathbf{0.5238} \; \underline{/-45.6°}$$

(b) From Eq. (1.60),

$$Z_L = Z_0 \frac{1 + \Gamma_L}{1 - \Gamma_L}$$

$$= \frac{50[1 + 0.5238(\cos(-45.6) + j\sin(-45.6)]}{1 - 0.5238[\cos(-45.6) + j\sin(-45.6))]}$$

$$= \frac{50(1.3665 - j0.3742)}{0.6335 + j0.3742}$$

$$= \mathbf{185.4 \; \underline{/-45.88°} \; \Omega}$$

Slotted line measurements are basically simple to make, involving only distance along line and voltage ratio. However, the calculations required to determine load impedance are lengthy and complex. One may, of course, use a programmable calculator to handle these, but the Smith chart provides a rapid and easy means of determining load impedance and reflection coefficient from slotted line measurements. This widely used method is described in Section 3.4.

The measurements described so far are single-frequency measurements, and these would have to be repeated at various frequencies throughout the band of interest in order to fully characterize the device under test. Apart from being very time consuming, such spot frequency measurements may in fact miss important resonant points in the response. For these reasons, swept frequency measurements are often made. These are limited to scalar measurements using a scalar analyzer and sweep oscillator similar to those used for return-loss and insertion-loss measurements described in the previous sections.

In making swept frequency measurements, the sweep frequency must always be much less than the signal frequency at any point. In effect, the frequency modulation aspect of sweeping can be ignored, and the signal treated as if it were sinusoidal at each frequency. The variation of voltage on a transmission line as a function of frequency can then be deduced from the CW theory given in Chapter 1. In particular, the phasor angle βl, which gives rise to voltage maxima and minima as a function of distance l, as shown in Fig. 14.9, can be written as

$$\beta l = \frac{2\pi l}{\lambda} = \frac{2\pi}{v_p} fl \tag{14.10}$$

This shows that either frequency or distance may be varied while the other is held constant to obtain the voltage standing wave shown in Fig. 1.14. There the frequency was held constant and distance varied, and of course with the sweep oscillator, just the opposite occurs. Figure 14.11(a) shows the slotted line with a sweep oscillator connected to the input, and a detector at position X1 which measures the line voltage at that point. The resulting voltage–frequency curve is shown in Fig. 14.11(b). By shifting the probe to some new position X2, the voltage–frequency curve will shift along the frequency axis resulting in a new curve, as shown in Fig. 14.11(b). This is because it is the product fl that governs the shape and position of the standing-wave curve. By slowly traversing the probe along the slotted line as the frequency is being swept, the individual curves will merge to form a single illuminated area on the CRT of the scalar analyzer as illustrated in Fig. 14.11(c). The V_{max}–frequency curve forms the upper boundary of this trace, and the V_{min}–frequency curve forms the lower boundary. In order for at least one maximum point and one minimum point to be included, the length of the slotted line must be at least half a wavelength

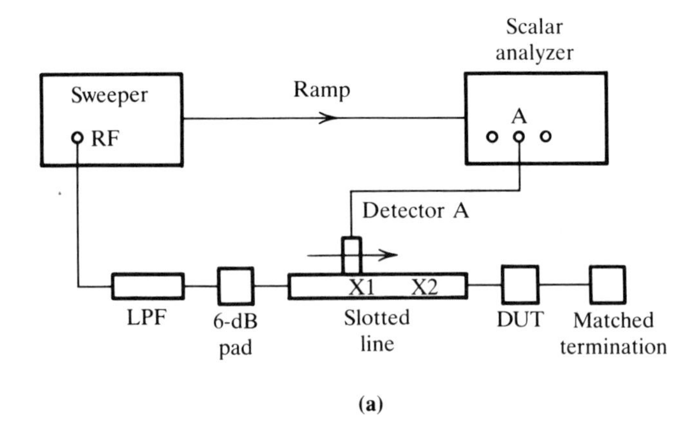

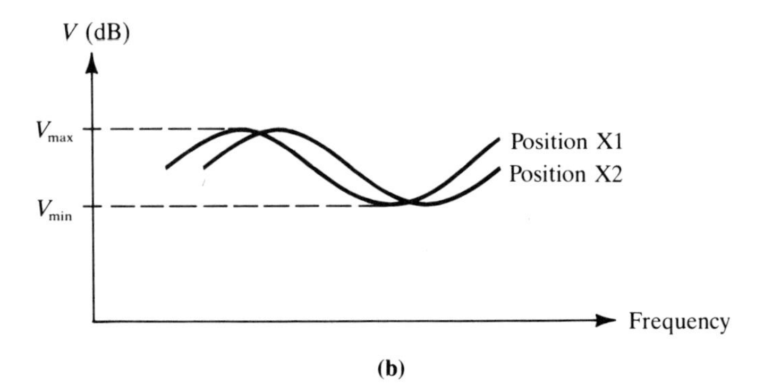

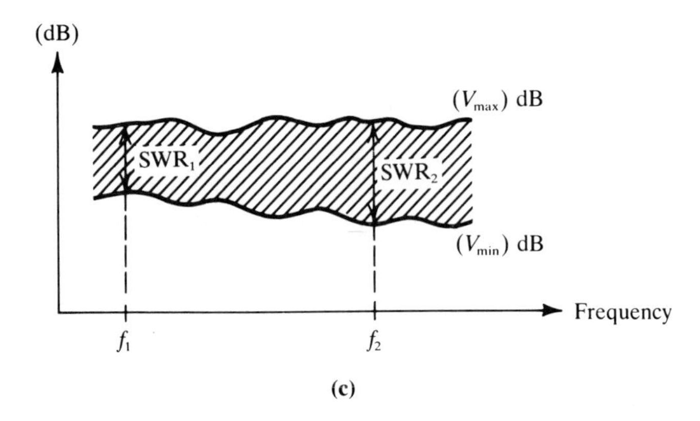

FIGURE 14.11. Sweep frequency measurement of VSWR: (a) equipment setup; (b) voltage–frequency curves at two different positions X1 and X2; (c) oscilloscope trace when probe is scanned along the line.

at the lowest frequency. A reference or calibration line may be obtained by terminating the line in a matched load, and storing the voltage–frequency data, for some fixed position of the probe. The scalar analyzer can be set to subtract the reference line amplitude, in decibels, from the decibel values obtained for the device under test.

By calibrating the vertical axis in decibels, the vertical distance between top and bottom of the envelope gives the standing-wave ratio in decibels since

$$\begin{aligned} \mathrm{SWR(dB)} &= 20 \log \mathrm{VSWR} \\ &= 20 \log V_{\max} - 20 \log V_{\min} \end{aligned} \tag{14.11}$$

The SWR in decibels is therefore read directly from the trace, as shown in Fig. 14.11(c), where, for example, the value is SWR_1 at frequency f_1, and SWR_2 at frequency f_2.

From the SWR the magnitude of the voltage reflection can be determined since

$$\mathrm{VSWR} = 10^{\mathrm{SWR}/20} \tag{14.12}$$

This may be substituted in Eq. (1.83) to give the magnitude of the reflection coefficient. Often, a knowledge of the variation of SWR with frequency is all that is required, and this is provided with the swept frequency method. For example, it may be desired to minimize the standing wave, and adjustments can be made to the device under test while observing the effect on the output trace. The slotted line is useful for measuring VSWRs close to unity, and the lower limit on accuracy is set by the residual VSWR of the line itself. For example, the Hewlett-Packard 817B slotted line has a residual VSWR of 1.04 at a frequency of 18 GHz. Similar measurements can be carried out using slotted waveguides, and Fig. 14.12 illustrates a slotted waveguide section with probe carriage.

14.6 COMPLEX NETWORK ANALYZER

Measurement of complex quantities requires a measurement of phase as well as amplitude. As described in the previous sections, scalar analyzers are used for amplitude measurements alone, while the slotted line may in fact be used for the determination of complex quantities. The disadvantage of the slotted line is that complex measurements are limited to single frequencies so that broadband testing is expensive and time consuming.

In order to determine phase, an accurate reference signal must be provided. Figure 14.13(a) shows one arrangement for the measurement of

FIGURE 14.12. Slotted waveguide sections. (*Courtesy of Marconi Instruments Ltd., Stevenage, Hertfordshire, England.*)

complex transmission parameters. The RF signal from the sweep oscillator is first divided by means of a power splitter into a test signal and a reference signal. The test signal is transmitted through the device to be tested, while the reference signal passes through an equalizing length of line. Comparison of phase at microwave frequencies is not practical, and therefore both the test and reference signals are reduced to a fixed intermediate frequency at which phase comparisons can be performed. Figure 14.13(b) shows the circuit schematic for the Hewlett-Packard 8411A harmonic frequency converter and the 8410A network analyzer. The frequency converter incorporates a phase-locked loop which enables the local oscillator to track the reference channel frequency, thus allowing for swept frequency measurements. The outputs of the first two mixers are at 20 MHz. These signals both pass through automatic gain control amplifiers. The AGC amplifier in the reference channel keeps the reference level constant and avoids the need for leveling in the sweep source. An error signal from this AGC amplifier is fed to the test channel AGC amplifier, which therefore compensates for common-mode variations of signal levels within the measurement system. A second mixing stage reduces both signals to 278 kHz. Ratio comparison of amplitudes may be made as described for the scalar analyzer, and phase measurement is made by passing both 278-kHz signals to a phase detector, not shown in Fig. 14.13(b).

A waveguide reflection–transmission test unit is shown in Fig. 14.14(a) and the schematic in Fig. 14.14(b). This is seen to be a balanced arrangement in which the reference line length can be balanced for transmission tests, and the device under test is compared to the sliding short for reflection tests. Accurately matched directional couplers are used to ensure balance. For example, with the sliding short adjusted to match a fixed short at the

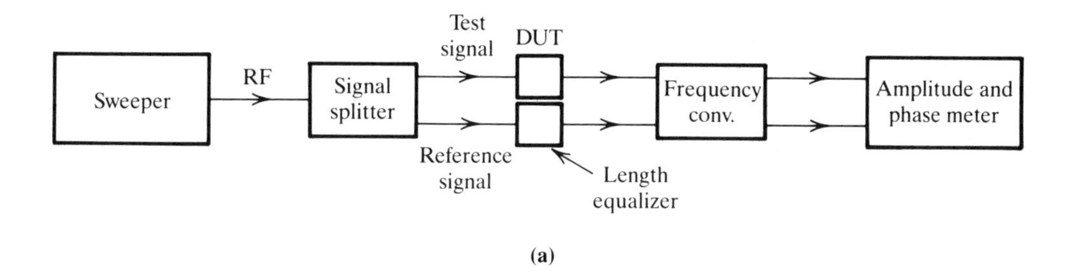

(a)

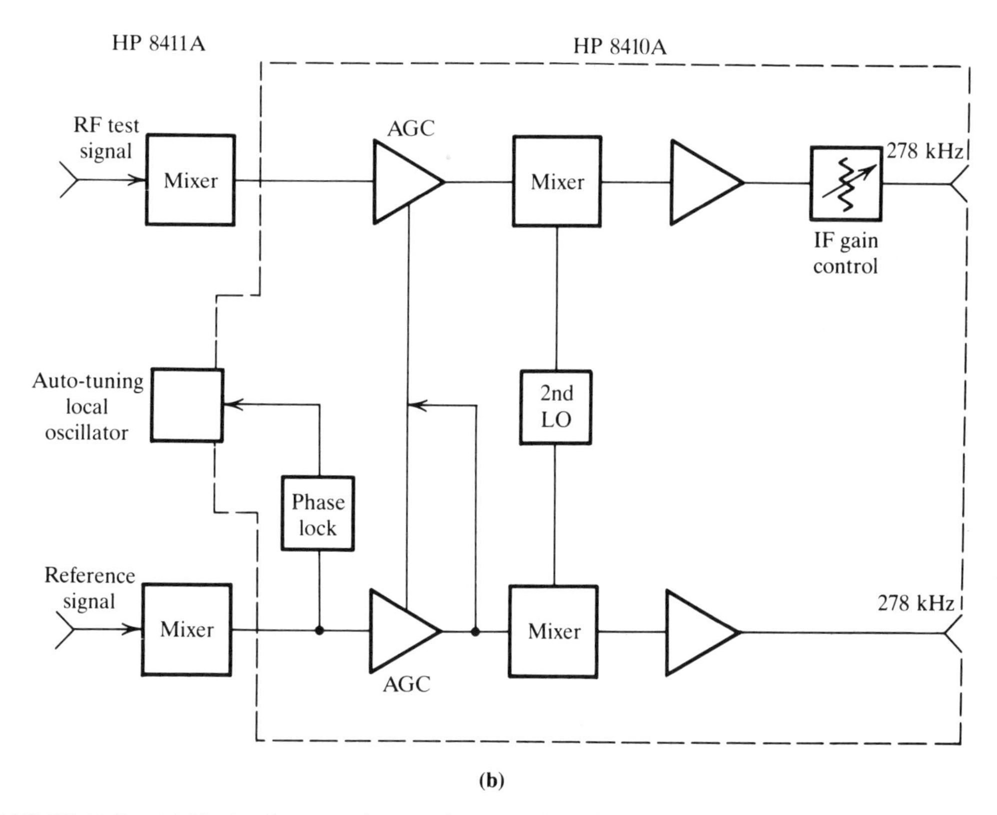

(b)

FIGURE 14.13. (a) Block schematic of a complex network analyzer; (b) simplified block diagram of the harmonic frequency converter (8411A) and network analyzer (8410A). [(b) courtesy of Hewlett-Packard, AN-117-1.]

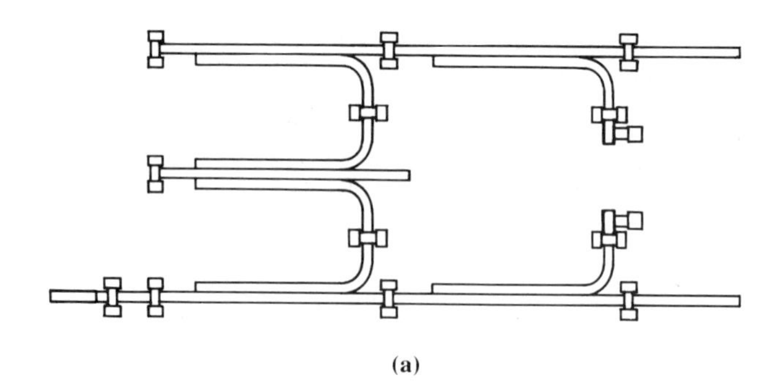

(a)

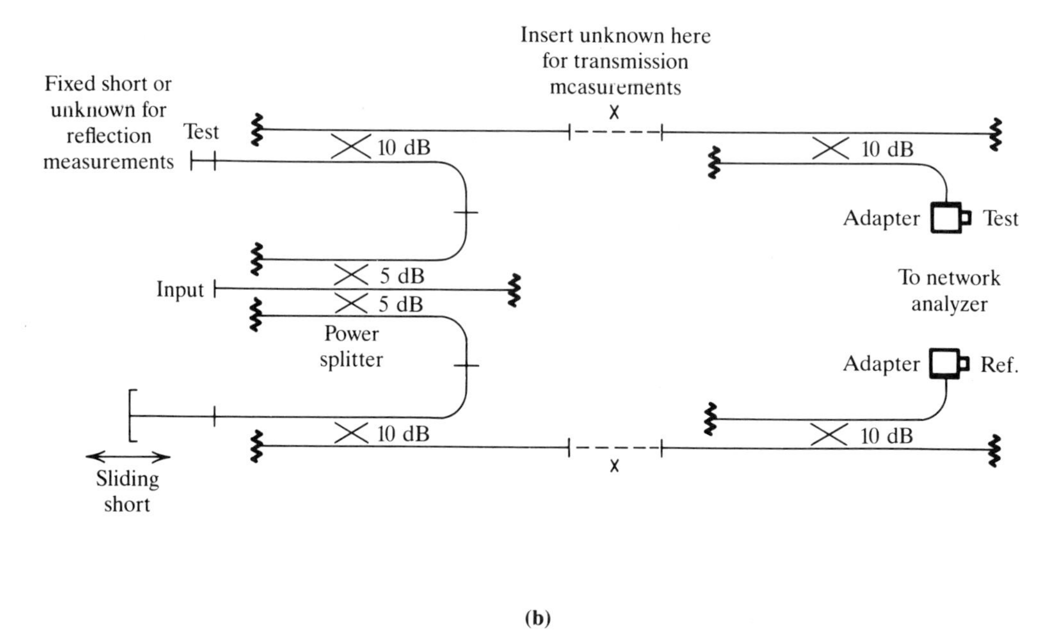

(b)

FIGURE 14.14. (a) HP-8747A waveguide reflection/transmission test unit; (b) HP-8747A schematic. (*Courtesy of Hewlett-Packard, AN-117-1.*)

reflection test port and identical connecting sections in the transmission paths, nearly identical signals should be observed at the test and reference output ports.

Figure 14.15(a) shows the test setup for measurement of S_{11} and S_{21} parameters, and Fig. 14.15(b) the setup for measurement of S_{22} and S_{12}. It will be seen that bias tees are included, which allow the correct operating bias to be applied when transistor parameters are being measured. The actual transistor jig must be made to accommodate the transistor type, and

Fig. 14.16(a) shows the transistor fixture used for stripline-type transistors. Various options are available with this to suit the various packages available, and Fig. 14.16(b) shows a transistor in a K-disk package inserted for common-emitter or common-source measurements.

14.7 POWER MEASUREMENTS

Microwave power measurements are usually divided into three groups depending on power level. These are low powers, which are less than 10 mW; medium powers, which range from 10 mW to 1 W; and high pow-

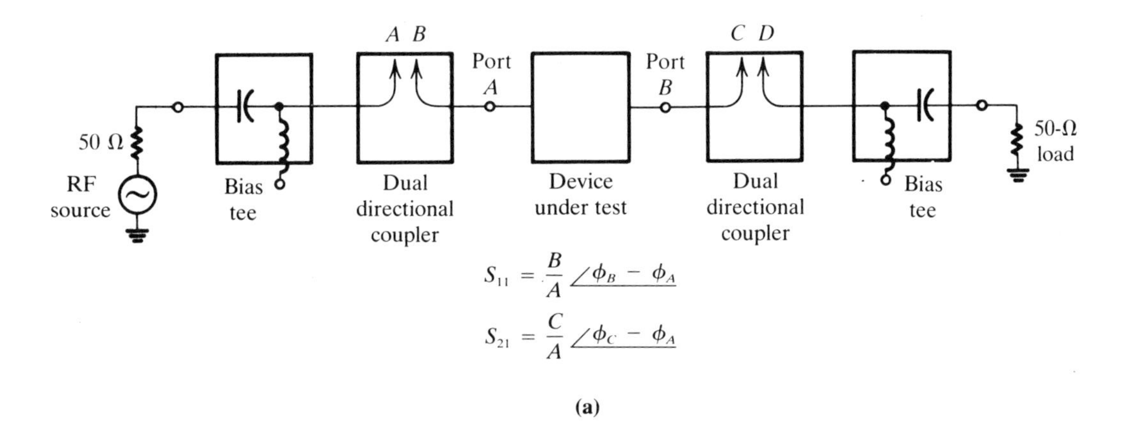

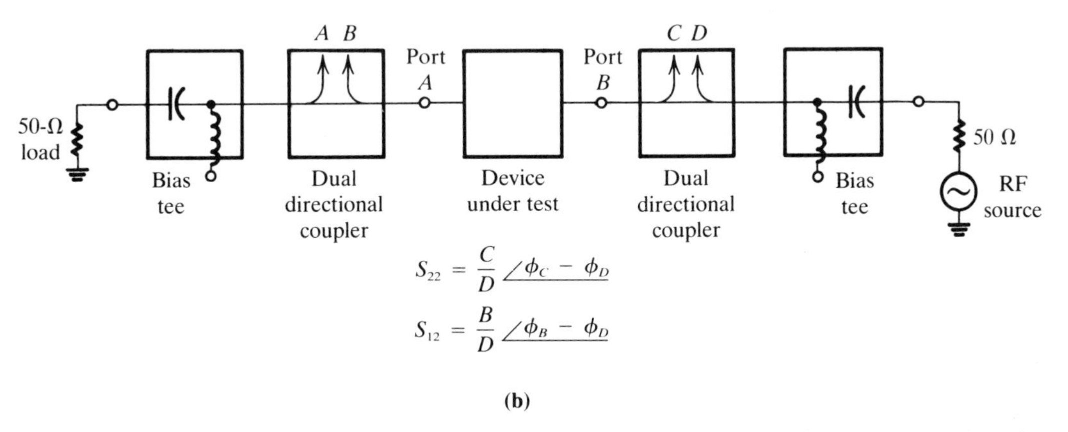

FIGURE 14.15. (a) Setup for measuring S_{11} and S_{12}; (b) setup for measuring S_{22} and S_{21}. *(Courtesy of Hewlett-Packard, AN-117-1.)*

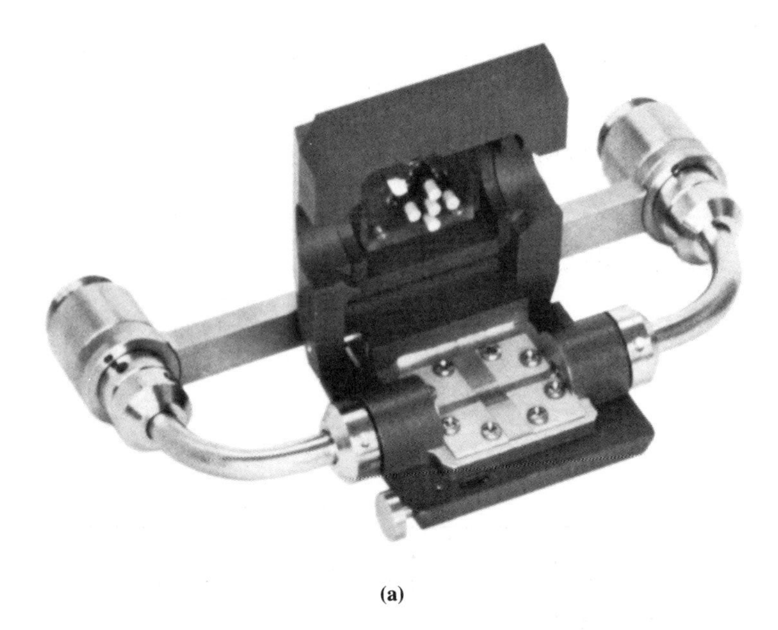

(a)

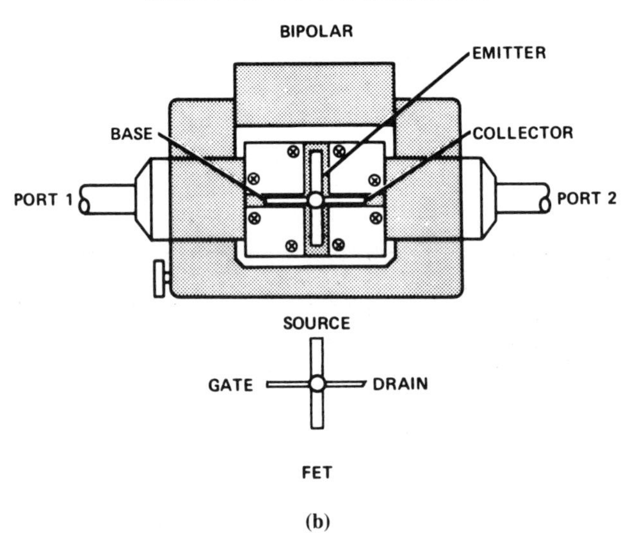

(b)

FIGURE 14.16. (a) 11608A transistor fixture; (b) transistor connections. (*Courtesy of Hewlett-Packard, AN-117-2.*)

ers, which are greater than 1 W. These groups are not sharply defined, and there is considerable overlap in measurement techniques. Most wattmeters are designed to measure average power (for power definitions, see Fig. 12.6) and the power sensors described next are average responding types.

Generally speaking, a microwave wattmeter consists of two parts, the power sensor, which converts the microwave power to some low-frequency change of state, and the low-frequency circuitry, which provides an output reading corresponding to the change measured by the sensor. The three most popular sensors used for low-power measurements are the Schottky barrier diode, the thermistor, and the thermocouple, the latter being the most widely used in modern instruments. Medium-power measurements are usually carried out using a low-power sensor and an accurate attenuator. Similarly, high-power measurements can be made using directional couplers and/or attenuators together with low-power sensors. In addition, microwave calorimeters are used for direct measurements of high powers. In a calorimeter, the microwave energy is converted into heat in a matched load and the corresponding temperature rise provides a direct measure of the power absorbed by the load. Calorimetric methods are seldom used for routine power measurements, being used mainly for standards and calibration purposes.

The design of the sensor mount is an important factor in determining the accuracy of the power measurement, irrespective of the type of sensor it contains. Power will be absorbed in the walls and supporting structures in the mount. Since this power does not reach the actual sensing element, it is not included in the measured result. The efficiency of the mount, denoted by the symbol η, is the ratio of power measured to total power dissipated in mount plus sensor. Efficiency does not take into account any mismatch, this being accounted for separately. Let ρ_L represent the magnitude of reflection coefficient of the sensor, including the effects of the sensing element and mount. Then, as shown by Eq. (10.2), the matching efficiency is given by

$$\eta_\Gamma = 1 - \rho_L^2 \tag{14.13}$$

The calibration factor for the sensor is then defined as

$$K = \eta\eta_\Gamma \tag{14.14}$$

A calibration factor should be provided with any particular sensor. For example, a calibration factor of 0.91 means that the meter indication is 91% of the input power. The calibration factor is determined by the manufacturer on the assumption that the source feeding the sensor is perfectly

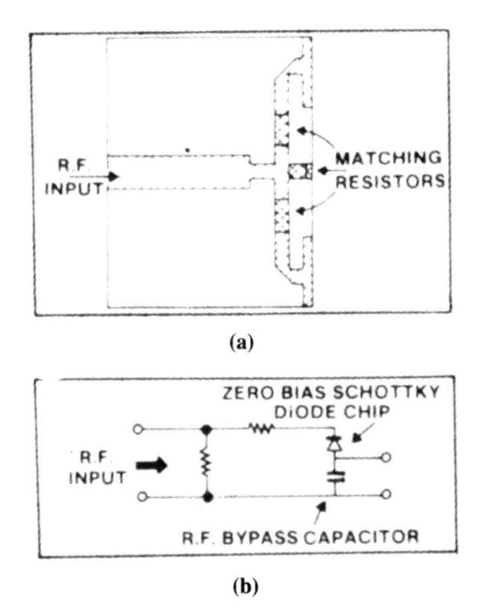

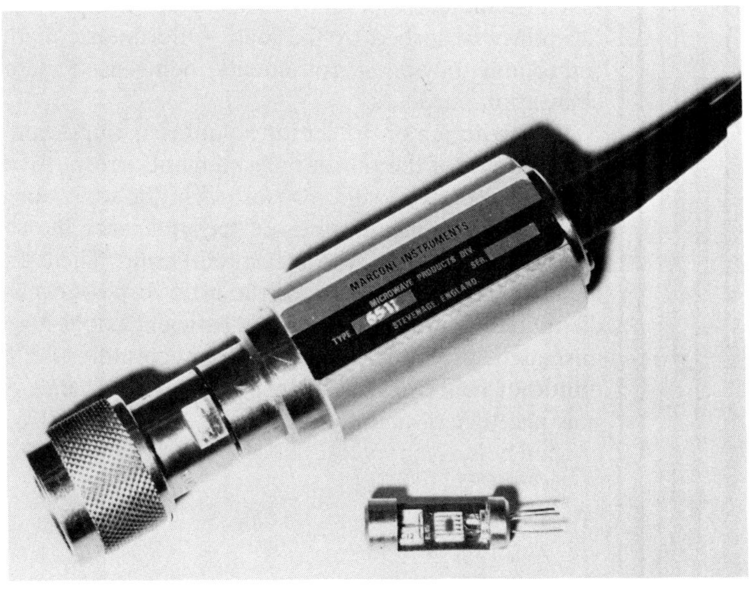

(c)

FIGURE 14.17. (a) Aluminum substrate and matching elements; (b) circuit of detector and matching elements; (c) detector head and field-replaceable RF module. *(Courtesy of Marconi Instruments Ltd., Stevenage, Hertfordshire, England.)*

matched. This will seldom be the case, and source mismatch will result in secondary reflections which contribute to the power input to the sensor. Since this mismatch factor will vary from source to source, it introduces a degree of uncertainty into the measurement which cannot be corrected for by the calibration factor.

As described in Section 14.2, a scalar analyzer can be used for power measurement, the sensor in this case being a zero-biased Schottky barrier diode. Because the diode resistance is a strong function of temperature, the detector circuit is arranged so that passive resistors are used to provide input matching, the effect of the diode resistance on the overall matching being negligible. One arrangement utilizing a thin-film network is shown in Fig. 14.17(a), and the detector circuit is shown in Fig. 14.17(b). The sensor head is shown in Fig. 14.17(c). This incorporates a temperature sensor which provides data to the scalar analyzer, used for temperature correction.

A bolometer is a power sensor in which the microwave energy is converted to heat, which in turn produces a change in resistance of the sensing element. Two types of bolometer are in use, the barretter, a thin metallic wire sensor, which has a positive temperature coefficient of resistance, and the thermistor, a semiconductor sensor which has a negative temperature coefficient of resistance. The thermistor is the more robust of the two. It is also smaller and more compact and is more easily fitted into mounts which are compatible with common types of transmission lines and waveguides.

Microwave power absorbed by a thermistor is converted to heat, and the resulting temperature rise increases the conduction carrier density, which in turn produces the decrease in resistance. The change of resistance could be monitored in a bridge circuit and used as a measure of the power dissipated. The two main disadvantages with this simple arrangement are that the change of resistance results in a mismatch at the microwave input, and the thermistor is also sensitive to changes in ambient temperature.

The circuit shown in Fig. 14.18 is designed to overcome both these problems. The upper bridge circuit measures the RF, and the lower bridge circuit is for temperature compensation. Considering first the mismatch problem, the solution is to decrease the dc power carried by the RF sensing thermistor until bridge balance is restored. The decrease in dc power exactly compensates for the microwave power added, and therefore the net change in thermistor resistance is zero. The RF bridge is self-balancing, so that the decrease in dc power is achieved automatically. Referring to Fig. 14.18, V_{rf} is the dc applied to the RF bridge. This V_{rf} is obtained from the bridge amplifier. When an RF input produces an initial unbalance, the unbalance voltage is amplified by the bridge amplifer and used as negative dc feedback. This decreases the level of V_{rf} and hence decreases the power

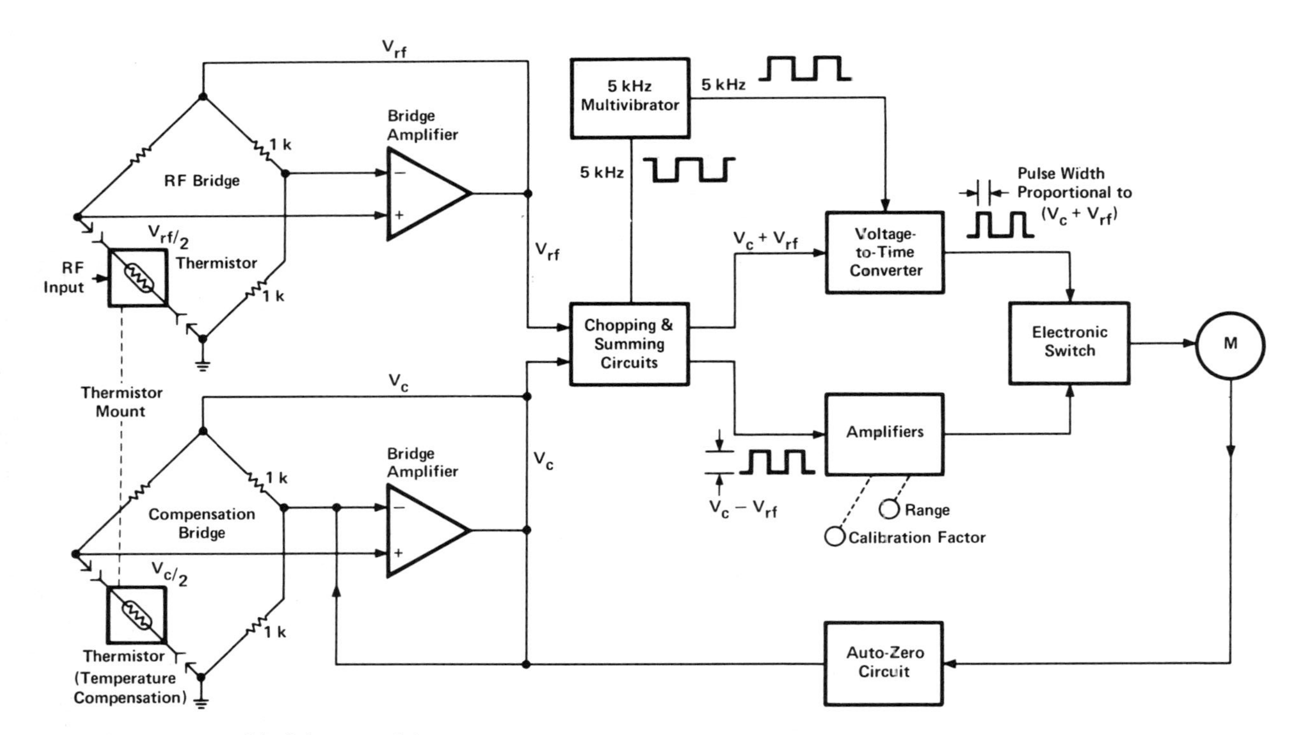

FIGURE 14.18. Simplified diagram of the HP-432A power meter. (*Courtesy of Hewlett-Packard, AN-64-1.*)

dissipation in the thermistor until balance is automatically restored with the thermistor resistance at its original value.

Ambient temperature compensation is achieved by using the compensation bridge to provide a dc reference voltage V_c as shown in Fig. 14.18. The bridge circuits are matched as closely as possible with regard to both electrical characteristics and temperature tracking. Ambient temperature changes will therefore result in both V_c and V_{rf} changing by the same amount. As shown next, these changes almost cancel each other out, so that negligible change occurs in the output reading.

The initial zero setting of the bridge consists of adjusting V_c to equal V_{rf} with no RF input signal applied. Denoting this value as V_{rf0}, then

$$V_c = V_{rf0} \tag{14.15}$$

From the circuit of Fig. 14.18, the dc voltage across the RF thermistor at balance is $0.5V_{rf0}$ initially, and $0.5V_{rf}$ with RF present and the bridge rebalanced. Let R represent the thermistor resistance at balance. Then the change in dc power, which equals the average RF input power P_{rf}, is

$$\begin{aligned} P_{rf} &= \frac{V_{rf0}^2}{4R} - \frac{V_{rf}^2}{4R} \\ &= \frac{(V_{rf0} - V_{rf})(V_{rf0} + V_{rf})}{4R} \\ &= \frac{(V_c - V_{rf})(V_c + V_{rf})}{4R} \end{aligned} \tag{14.16}$$

Suppose now that a change in ambient temperature results in a change ΔV in both V_c and V_{rf}. No net change occurs in the difference term. The sum term changes by amount $2\,\Delta V$ and becomes $(V_c + V_{rf} + 2\Delta V)$. In all practical situations $(V_c + V_{rf}) >> 2\,\Delta V$, and therefore this change is negligible. The meter logic circuitry shown in Fig. 14.18 responds to the product of the sum and difference terms given in Eq. (14.16). A full description of this power meter will be found in the *Hewlett-Packard Application Note 64-1*.

The third and most popular method of measuring low and medium powers is by means of a thermocouple. A thermocouple is a junction of two dissimilar metals or semiconductors, which when heated, generates a direct EMF. The microwave power to be measured is absorbed in a resistive load which heats up. The resistive load also forms one electrode of a thermocouple and the EMF generated as a result is proportional to the microwave power absorbed. Figure 14.19(a) shows one form of thermocouple element. The resistive load is a tantalum–nitride resistor, deposited

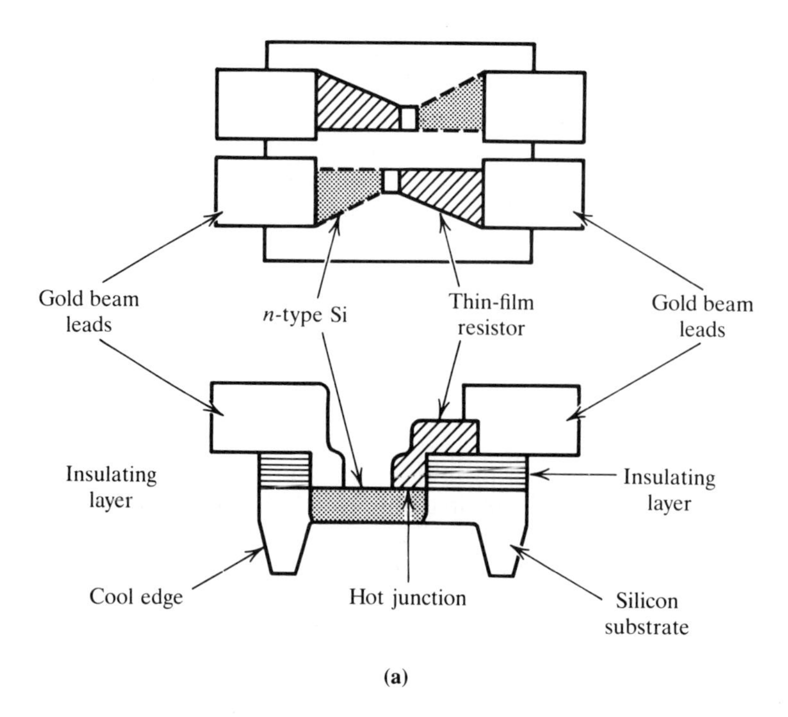

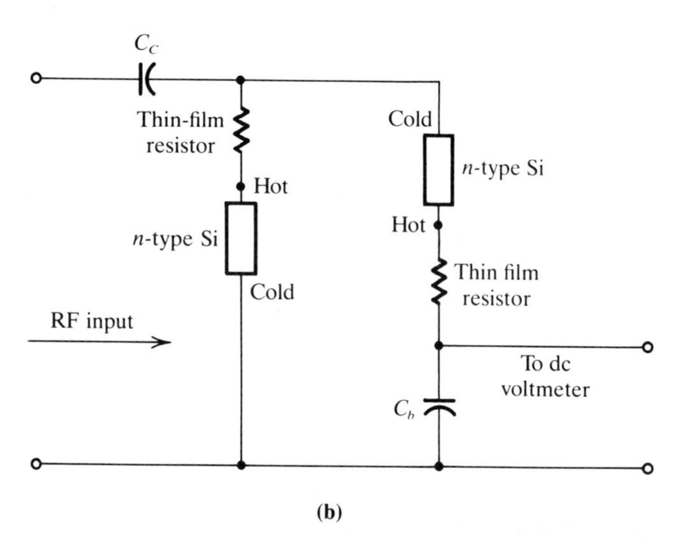

FIGURE 14.19. (a) Structure of the HP-8481A thermocouple chip; (b) schematic diagram of the HP-8481A thermocouple power sensor; (c) sketch of the thermocouple assembly for the HP-8481A. (*Courtesy of Hewlett-Packard, AN-64-1.*)

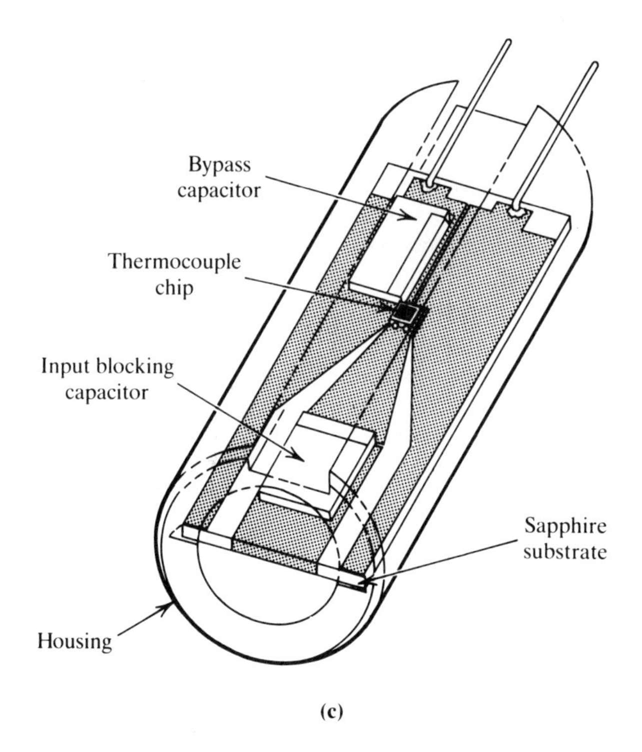

(c)

FIGURE 14.19. (continued)

as a thin film on a silicon substrate. The hot junction is at the center of the chip, as shown in Fig. 14.19. The free ends of the two materials form the cold junction, the temperature of which must be lower than that of the hot junction for the thermal EMF to be generated. At the same time, the cold ends must be separate electrically so that the EMF can be measured. With the arrangement shown in Fig. 14.19, the outer edge of the silicon chip, which forms part of the cold junction, is more massive than the center section, and it makes good thermal contact with the gold-beam connecting leads which act as heat sinks. As a result, the necessary temperature gradient is maintained between the hot and cold junctions. Two separate thermocouples are actually built onto the chip, and the equivalent circuit is shown in Fig. 14.19(b). Capacitor C_b is an RF bypass capacitor, and the two thermocouples therefore appear in parallel to the RF input, the effective input resistance being 50 Ω. Capacitor C_c is the input coupling capacitor, which prevents d.c. from the thermocouple reaching the microwave source.

The direct EMFs generated in the thermocouples add in series and the output is taken from across C_b. The advantage of this arrangement is that both leads going to the dc voltmeter are at ground potential for RF signals, which avoids the need for filtering these. A second advantage is that the series connection increases the available EMF. However, it will be seen that the cold junction of the second thermocouple is connected to the hot junction of the first. This reduces the temperature gradient across the second thermocouple and hence its output EMF. The individual thermocouples have a rated sensitivity of 100 μV/mW, while the overall sensitivity of the arrangement shown is 160 μV/mW. This means for example that the dc output voltage is 160 μV when the RF input is 1 mW.

Figure 14.19(c) shows a sketch of the thermocouple assembly. A coplanar transmission line on a sapphire substrate forms the interconnection between the input coupling capacitor C_c and the thermocouple chip. This line is tapered to accommodate the change in dimensions, while still maintaining a 50-Ω characteristic impedance. As a result, the sensor has a very low reflection coefficient.

Most thermocouple-type wattmeters contain a reference power source for periodic calibration of the sensor. The source is usually a 50-MHz power oscillator the output of which is accurately controlled at 1 mW. The sensor is connected to the reference source in exactly the same manner as to the RF source under test, and in this way the complete sensor system is included in the calibration. With the sensor connected to the reference source, the gain of the dc amplifier driving the output meter is adjusted until the meter indicates 1 mW.

It was mentioned at the start of this section that these wattmeters measured average power. Specialized instruments are available for measurement of peak power, which is the quantity usually of interest in radar systems. Where the microwave signal is amplitude modulated in an ON/OFF manner by a square wave, as shown in connection with Fig. 12.6, the peak power P is related to the average power P_{av} as

$$P = \frac{P_{av}}{D} \tag{14.17}$$

where $D = \tau/T_r$ is the duty factor.

14.8 MICROWAVE RADIATION MONITORS

In Chapter 15 some of the biological interactions with internally induced microwave fields are discussed. In practice it is not possible to measure the internal field; the best that can be done is to establish safety standards based on measurements of the field external to the subject. Some of the complexities of the external field are discussed in Section 15.2. Measurement of hazardous electromagnetic fields is discussed by Bowman (1978), and by Larsen and Ries (1981). Some of the desirable characteristics of EM survey meters listed are as follows. The instrument should be capable of measuring the electric or magnetic field strength or the energy density in either of these fields. The probe response should be independent of orientation, and it should not result in excessive scattering of the microwave energy. The probe should be able to resolve the fine structure of the field, down to dimensions of the order of 1 cm. The instrument should have a dynamic range of at least 20 dB without a change of probes, and the calibration factor should be essentially constant over a wide frequency range. The instrument should be direct reading; that is, it should not require calibration charts, nulling, or frequent rezeroing. The instrument must be well shielded against EM interference, and in addition it should be stable, rugged, and easily portable. Instruments available at present are capable of measuring average field levels, but it would be desirable to be able to measure peak field levels as well.

Most commercially available instruments measure the E field, and two types of sensors are in common use, the dipole–diode (or rectenna) type and the thermocouple type. In both of these, the leads to the detector element (diode or thermocouple) form the dipole antenna, the length of which is shorter than $\frac{1}{2}\lambda$ for the shortest wavelength involved.

For the dipole probe to be independent of orientation, a three-dimensional antenna structure is required. One such arrangement, the subject of U.S. patent No. 3,750,017 (Bowman et al. 1973) is illustrated in Fig. 14.20(a) . The dipoles (10), (12) and (14), are mounted at right angles to one another. The voltages developed across the center gaps of the dipoles will be: $V_x = l_{\text{eff}}E_x$; $V_y = l_{\text{eff}}E_y$; and $V_z = l_{\text{eff}}E_z$. Here, l_{eff} is the effective length (see Sec. 10.7) of each dipole (assumed equal) and E_x, E_y, and E_z are the respective field strengths. The dipole arms labelled (24) and (26) in Fig. 14.20(b), are made of tubular metal, and a rectifying diode (28) (or other sensing element) can be connected as shown in Fig. 14.20(b). High resistance output leads (30) and (32) are also inserted in the dipole tubes, these then being filled with conductive glue (38) to ensure good electrical connection.

High-resistance leads are required between the detectors and their respective preamplifiers to reduce to negligible levels the currents induced in these by the electromagnetic field. Keeping in mind that these leads will be of the same order of size as the antennas, they can couple directly from the field, and also to the antennas, giving rise to signals of arbitrary magnitude at the detectors. The high-resistance leads (1 to 100 kΩ) along with their distributed capacitances, shown as (41) in Fig. 14.20(c), provide sufficient filtering to reduce RF currents to negligible levels. The non-linear summing amplifier can be adjusted to compensate for diode non-linearities,

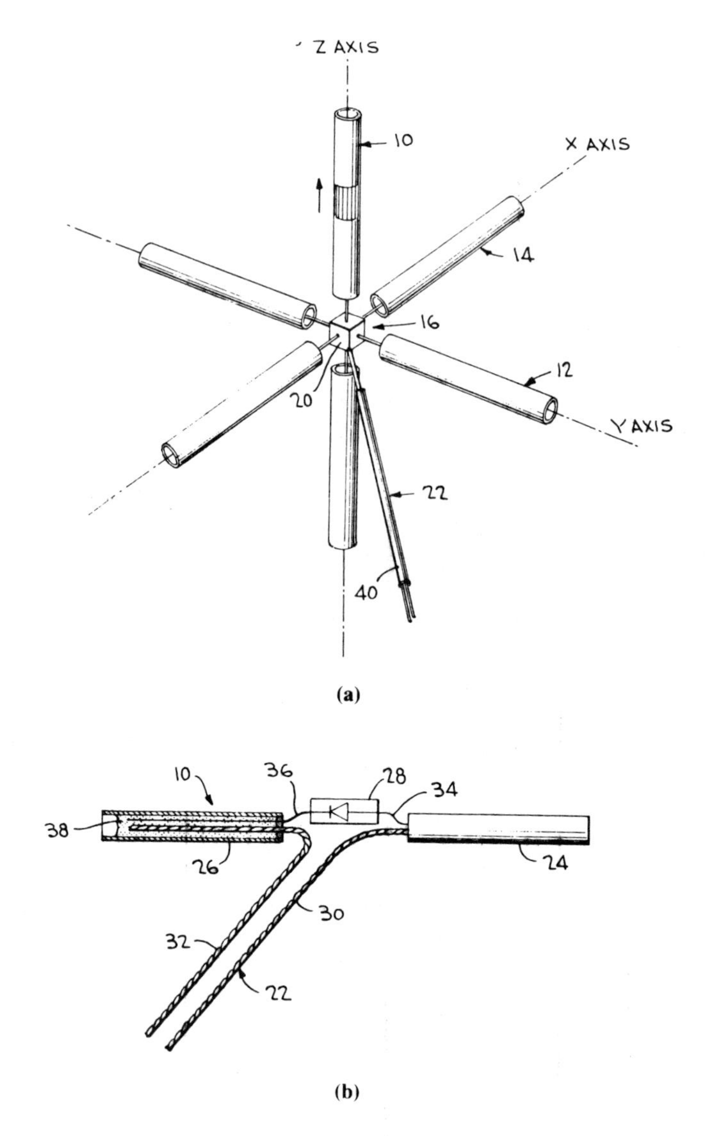

FIGURE 14.20. (a) Three mutually orthogonal dipoles; (b) schematic of the instrument showing the signal processing. (*From Bowman, 1978.*)

and arranged to give an output proportional to either $V_x^2 + V_y^2 + V_z^2$ or $(V_x^2 + V_y^2 + V_z^2)^{1/2}$ as desired. The effective field strength is E where

$$\begin{aligned} E &= (E_x^2 + E_y^2 + E_z^2)^{1/2} \\ &= (V_x^2 + V_y^2 + V_z^2)^{1/2}/l_{\text{eff}} \end{aligned} \tag{14.18}$$

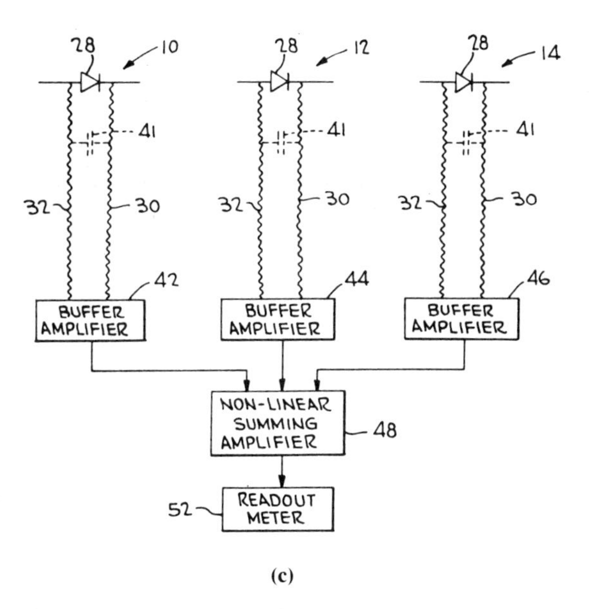

(c)

FIGURE 14.20. (continued)

It should be kept in mind that l_{eff} will vary with frequency, and in all cases should be less than the value given for the $\lambda/2$ dipole in Sec. 10.7. In practice the instrument is calibrated using standard test cells, which provide known electromagnetic field strengths, Larsen et al. (1981).

Figure 14.21(a) shows a thin-film dipole with a built-in thermocouple sensor. The thermocouple materials are antimony and bismuth, and these are deposited on a substrate of plastic or mica, which is then fixed on a more rigid dielectric material for support. The length of the antenna is a small fraction of a wavelength and is a compromise between minimizing the disturbance of the field being measured while delivering adequate EMF to permit detection of the field. Figure 14.21(b) shows a simplified schematic of the three-dipole arrangement, and Fig. 14.21(c) shows a commercial monitoring instrument with probes enclosed in radomes. The two probes between them cover the frequency range 10 MHz to 26 GHz. Various probes are also available to cover different power density ranges, and typically a dynamic range of 0.01 to 200 mW/cm^2 can be monitored with various combinations of probes and instruments.

14.9 FREQUENCY MEASUREMENT

Frequency counters are available which allow direct measurement of frequencies up to about 500 MHz. This limit is set by the maximum speed of the logic circuitry. For higher frequencies, some form of frequency down-

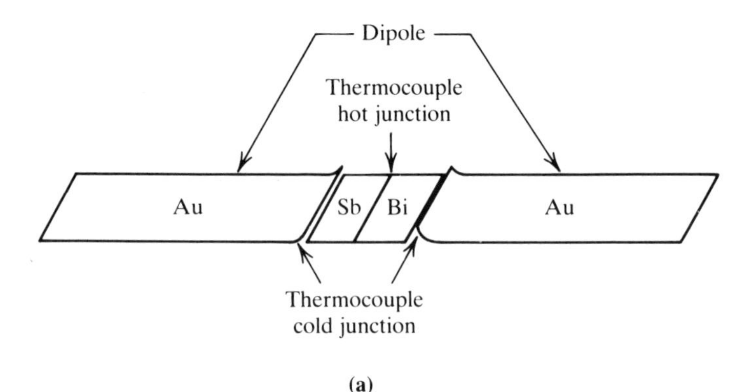

(a)

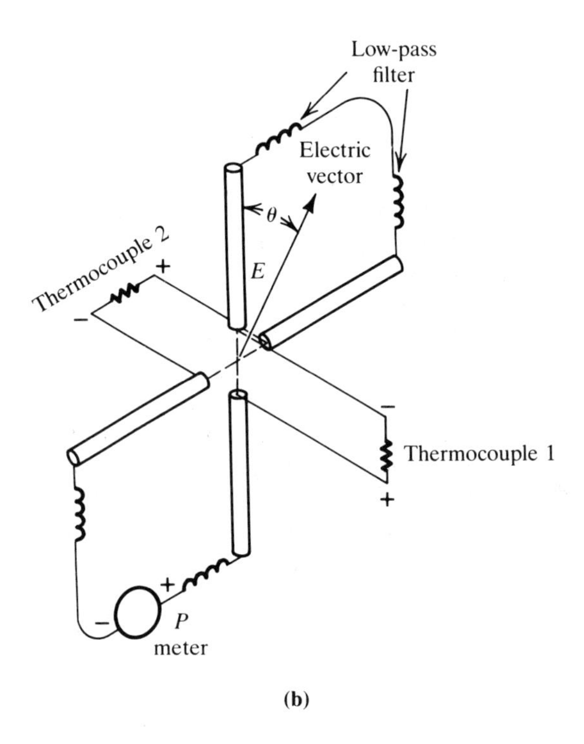

(b)

FIGURE 14.21. (a) Thin evaporated film dipole-detector element that comprises one of the two orthogonal elements in the probe; (b) simplified diagram of the radiation survey meter and sensor; (c) radiation monitor with probes covering the range 10 MHz to 26 GHz. (*Courtesy of Narda, Hauppauge, N.Y.*)

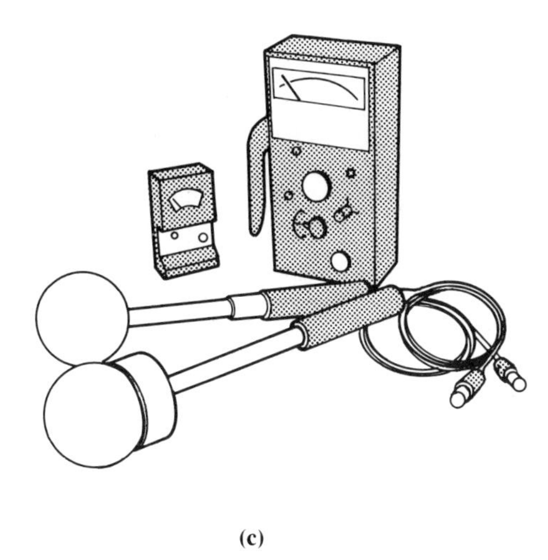

(c)

FIGURE 14.21. (continued)

conversion must be employed to bring the frequency within the range of the conventional counter. The simplest arrangement is to use a prescaler to divide the unknown input frequency by some factor N. A conventional frequency counter is used to determine the output of the prescaler, which is below the 500-MHz limit. The counter's gate time can then be extended by the known factor N so that the output displays the microwave input frequency. With this technique, frequency measurements can be extended up to a few gigahertz.

A much higher frequency range can be achieved using heterodyne down-conversion of frequency, this method being useful for frequencies up to about 20 GHz. In this method, the incoming frequency is mixed with a highly stable local oscillator, the beat frequency being within the 500-MHz range of a conventional counter. Figure 14.22 shows the block diagram for the heterodyne down-conversion technique, the section within the dashed line being the down-converter. The time-base signal of the instrument is applied to a frequency multiplier to produce a highly stable reference frequency in the range 100 to 500 MHz. This is shown as f_{in} in Fig. 14.22. The f_{in} signal is applied to a harmonic generator which produces a "comb line" of harmonics spaced at f_{in}. One harmonic at a time is selected by the YIG/PIN switched filter under microprocessor control and is mixed with the unknown f_x to produce a beat frequency $f_x - Kf_{in}$, where K is the harmonic number. The microprocessor steps the filter through its range starting at $K = 1$, until a mixer output is obtained. The output from the mixer is amplified by the video amplifier and passed to the signal detector,

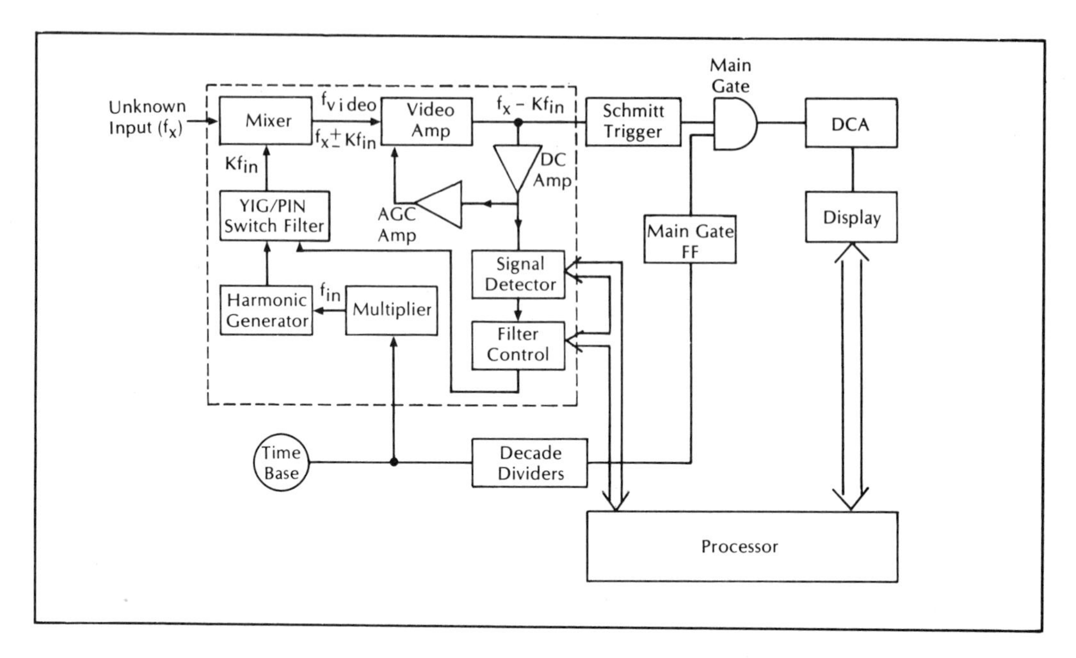

FIGURE 14.22. Block diagram of the heterodyne down-conversion technique. (*Courtesy of Hewlett-Packard, AN-144 and AN-200-1.*)

which signals the processor to stop and store the final value of K. The acquisition time for the counter, which is the time required for the counter to reach the correct value of K, is about 150 ms. At this stage both K and f_{in} are known, and since the conventional counter measures $f_x - Kf_{in}$, f_x can be determined by having the processor add Kf_{in} to the counter frequency, and displaying the sum, which is f_x.

A phase-locked loop may also be used to determine frequency. In this method, the incoming signal is phase locked to some harmonic of the VCO in the phase-locked loop. The fundamental VCO frequency is typically within the range 100 to 200 MHz and can be measured on a conventional counter. Additional circuitry is required to determine the harmonic number, and the unknown frequency can then be calculated. This method is referred to as a "transfer oscillator" method, since the measurement problem is transferred to the VCO frequency. The acquisition time is also about 150 ms.

By combining the heterodyne and transfer oscillator techniques, the frequency range can be extended to about 40 GHz, but at the expense of a longer acquisition time, typically 350 ms. The combined method is known as the *harmonic heterodyne converter*. Details of the latter two methods will be found in *Hewlett-Packard Application Note 200-1*.

Microwave counters of the types mentioned above have accuracies limited mainly by the time-base accuracy. Typically, such time bases operate at 10 MHz, and have a short-term accuracy of 10^{-9} averaged over 1 s. The aging rate is typically 10^{-7} per month. In many cases such extreme accuracy is not required, and simple frequency measurements can be made with the use of resonant cavities and slotted lines. In the theory of the slotted line it is seen that successive minima are spaced by half a wavelength, and thus wavelength may be measured to an accuracy determined by the carriage mechanism, which typically is of the order of 0.05%.

With resonant cavity wavemeters, the resonant cavity is weakly coupled into the main line and tuned to resonance by adjusting the length of the cavity as discussed in Section 4.17. Figure 14.23 illustrates an absorption wavemeter suitable for the 8.2- to 12.4-GHz band. The resonator operates in the TE_{011} mode and the micrometer is calibrated directly in guide wavelength. The through section of waveguide forms part of the waveguide run, or it may be connected directly between a source and an output indicator. When the cavity is tuned to resonance, the output indication dips by about 2 dB as a result of the energy absorbed by the cavity. This dip in output is quite sharp, and for example the resolution of the wavemeter shown is specified as 1 MHz at 8.2 GHz and 2.1 MHz at 12.4 GHz.

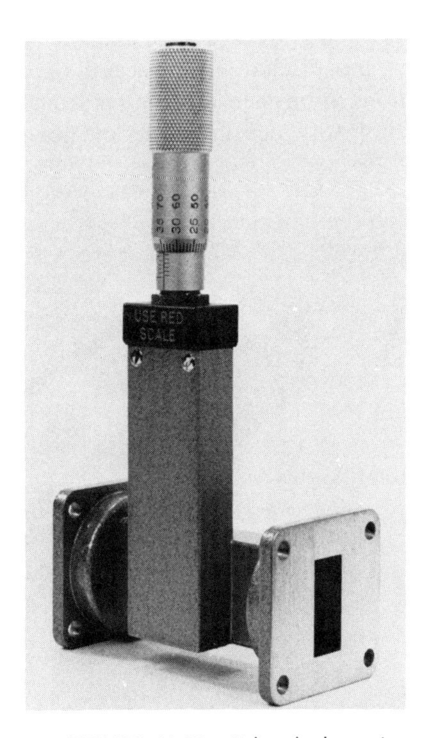

FIGURE 14.23. X-band absorption wavemeter. *(Courtesy of Marconi Instruments Ltd., Stevenage, Hertfordshire, England.)*

14.10 NOISE MEASUREMENT

In Chapter 8 it is seen that noise factor F and noise temperature T_e are two ways of specifying the noise performance of an amplifier or other two-port network. Noise factor and noise temperature are simply related as

$$T_e = (F - 1)T_0 \tag{14.19}$$

where T_0 is room temperature, 290 K. The choice of F or T_e to characterize a device or network is, to a certain extent, arbitrary. However, for terrestrial microwave systems, noise factor, or noise figure, which is simply 10 log F or F in decibels, proves to be the more convenient quantity, whereas for satellite systems, noise temperature is the more convenient. Noise temperatures for terrestrial systems range from about 600 to 3000 K, the equivalent noise figures being 4.87 to 10.55 dB. For satellite systems, noise temperatures may range from about 35 to 120 K, the corresponding noise figures being from about 0.5 to 1.5 dB.

An excess noise source may be used to measure noise factor. Excess noise is noise over and above the room-temperature thermal noise of the source. Such sources include hot and cold resistive terminations, thermionic diodes, gas discharge tubes, and silicon avalanche diodes. Resistive terminations at physically different temperatures are not convenient for routine laboratory measurements and are used mainly for calibration purposes.

FIGURE 14.24. Hewlett-Packard broadband noise source, 10 MHz to 18 GHz. (*Courtesy of Hewlett-Packard.*)

Thermionic diodes have limited frequency response and cannot be used above a few hundred megahertz. Until the advent of the silicon avalanche diode, the gas discharge tube was the main source for excess noise at microwave frequencies. However, the tube is physically large and requires a turn-on voltage of several thousand volts, all of which make it awkward to use. Where possible, gas discharge tubes are being replaced with avalanche diode sources. When a silicon diode is operated in the reverse-bias avalanche region, the spontaneous generation of charge results in wideband noise, typically covering the spectrum from 10 MHz to about 18 GHz. The diode is physically small, so it fits readily into standard microwave enclosures, and it operates from a comparatively low voltage, low power supply.

Figure 14.24 shows the Hewlett-Packard noise source, type 346B. This operates from a 28-V supply, and each source comes with a calibration table showing excess noise ratio (ENR) at 20 selected frequencies over the band 10 MHz to 18 GHz. Because the source delivers noise over and above thermal noise, it may be considered to be at some high temperature T_h. For the 346B source, for example, T_h is about 10,000 K. The excess noise ratio is then defined as

$$\text{ENR(dB)} = 10 \log \frac{T_h - T_0}{T_0} \tag{14.20}$$

where T_0 is room temperature, 290 K.

When the diode is operated at its rated dc input to produce avalanching, its effective temperature is T_h, and when the dc is switched off, its effective temperature is T_0 (sometimes this is referred to as the cold temperature T_c). The noise source is connected to the input port of the device under test, and an output power meter is connected to the output port to read the noise output. Let N_2 represent the noise power output obtained when the noise source is switched on, the equivalent noise temperature of the source being T_h. Let N_1 represent the noise power output when the source is switched off and therefore at equivalent noise temperature T_0. Also, let T_e represent the noise temperature of the device under test, this being the quantity that is being measured. Then, from the definition of T_e given in Chapter 8, the following equations can be written:

$$N_2 = k(T_h + T_e)B \tag{14.21}$$

$$N_1 = k(T_0 + T_e)B \tag{14.22}$$

The ratio N_2/N_1 is termed the Y-factor:

$$Y = \frac{N_2}{N_1} \tag{14.23}$$

From these three equations it is readily shown that

$$T_e = \frac{T_h - YT_0}{Y - 1} \tag{14.24}$$

From the calibration table of ENR versus frequency supplied with the noise source, T_h can be determined from Eq. (10.20). Using this together with the measured value of Y, T_e can be calculated from Eq. (10.24). In modern noise meters, the calibration values are stored in a random access memory and a microprocessor is used to perform the necessary calculations, so that T_e and F can be displayed as output.

14.11 PROBLEMS

1. Definc and explain what is meant by (a) scalar and (b) vector measurements in relation to microwaves. Describe how swept-frequency scalar measurements may be made.
2. Describe the principles of operation of (a) a reflectometer and (b) a reflectometer bridge, and give one measurement application of each. Show that the insertion loss of a reflectometer bridge is 6 dB.
3. Define and explain the terms (a) insertion loss and (b) return loss. Explain how these quantities may be measured, and the function of leveling in such measurements.
4. What is the value of the return loss of a matched load?
5. The measured value of return loss for a device is 4 dB. Calculate the magnitude of the reflection coefficient.
6. In a measurement of return loss using the circuit of Fig. 14.5(a), the reading at A with the short circuit in place was 0 dB. With the device under test (DUT) in place, the reading dropped to -5 dB. The incident signal is held constant by leveling. Determine the return loss in decibels.
7. Describe the ratioing method of measuring return loss. What are the advantages of this method over the leveling method?
8. Describe the ratioing method for measurement of insertion loss. The power delivered directly to the matched load in the arrangement of Fig. 14.6 produces a reading of 3 dB at A relative to the reference level. With the device under test inserted, the reading at A drops to -7 dB, the reference level remaining constant. Determine the insertion loss of the device under test.
9. A slotted line has its distance scale calibrated in centimeters starting from the input end. With a short-circuit termination on the slotted line, two successive minima were measured at 7.5 cm and 23.8 cm.

With the load under test connected, the VSWR was 3.1:1, the minimum occurring at the 15-cm position. Determine (a) the voltage reflection coefficient at the load and (b) the load impedance given that $Z_0 = 50\ \Omega$.

10. Describe the swept-frequency method of measuring the standing-wave ratio using a slotted line. A width of 5 dB on the oscilloscope trace was measured using this method, at a given frequency. Determine the magnitude of the voltage reflection coefficient.

11. Assuming that the impedance measured in Problem 10 is complex and that the characteristic impedance of the line is known, state, with reasons, whether or not the magnitude of the load impedance can be determined from the measurement.

12. Describe how transistor scattering parameters would be measured using a complex network analyzer.

13. Define and explain what is meant by the calibration factor in microwave power measurements. A microwave power sensor has an efficiency of 95% and results in a reflection coefficient of magnitude 0.1 when used for a certain power measurement. Calculate the calibration factor.

14. Describe how a thermistor may be used in microwave power measurements. Show how mismatch resulting from change in thermistor resistance can be avoided. Explain also how temperature compensation may be achieved.

15. Describe with the aid of suitable diagrams how a thermocouple may be used in microwave power measurements.

16. Describe with the aid of suitable diagrams how microwave radiation may be monitored. List the desirable characteristics of a radiation monitor.

17. Describe the heterodyne method of frequency measurement for microwave frequencies.

18. Describe how a resonant cavity wavemeter works and state the typical accuracy expected from such an instrument.

19. Describe how an excess noise source may be used to measure noise factor. Derive Eq. (14.24).

20. Show that noise factor F can be expressed as

$$F = \frac{10^{\mathrm{ENR}/10}}{Y - 1}$$

where ENR is the excess noise figure in decibels and Y is given by Eq. (14.23).

21. Measurement of the noise figure using an excess noise source resulted in $Y = 4.91$ dB, with an ENR of 5.83 dB. Assuming that $T_0 = 290$ K, determine (a) the noise figure in decibels, and (b) the equivalent noise temperature.

22. The noise temperature of an amplifier is known to lie in the range 70 to 120 K. Determine the range expected for Y in decibels when an excess noise source with an ENR of 6.02 dB is used to measure the noise temperature. Assume that $T_0 = 290$ K.

REFERENCES

BOWMAN, R. R., E. B. LARSEN and D. R. BELSHER, (1973). "Electromagnetic Field Measuring Device." U.S. Patent No. 3,750,017.

BOWMAN, R. R., 1978. "Quantifying hazardous microwave fields," *Trans. Int. Microwave Power Inst.*, Vol. 8.

Hewlett-Packard Application Note 64-1, Fundamentals of RF and Microwave Power Measurement.

Hewlett-Packard Application Note 117-1, Microwave Network Analyzer Applications.

Hewlett-Packard Application Note 117-2, Stripline Component Measurements with the 8410A Network Analyzer.

Hewlett-Packard Application Note 200-1, Fundamentals of Microwave Frequency Counters.

LARSEN, E. B. and F. X. RIES, (1981). "Design and Calibration of the NBS Isotropic Electric-Field Monitor (EFM-5) 0.2 to 1000 MHz. NBS Technical Note 1033.

FURTHER READING

Hewlett-Packard Application Note 57-1, Fundamentals of RF and Microwave Noise Figure Measurements.

Hewlett-Packard Application Note 183, High Frequency Swept Frequency Measurements.

HOOK, A. P., 1983. "Microwave scalar analysis: practical points for accurate measurement," *Electron. Power*, February.

Pacific Measurements Application Note AN20.

fifteen
Microwave Radiation Hazards

15.1 INTRODUCTION

The fact that microwaves can be used for cooking purposes and in heating applications suggests that they have the potential for causing biological damage. A great deal of research has gone into studying "microwave bioeffects," with the result that a lot is known about what happens, but unfortunately comparatively little about why.

It is true in general that any material which has the properties of a poor dielectric—some conductivity and a relative permittivity usually greater than unity—will absorb energy from a microwave field and dissipate this as heat; blood, brain matter, bone, muscle, and fat are no exceptions, as all of these biological substances behave as conductive dielectrics. Some of the microwave bioeffects are described in this chapter, together with the precautions taken to safeguard against excessive radiation exposure, but before going into this, some general points are summarized here:

1. Microwave radiation must be carefully distinguished from ionizing radiation (e.g., radioactivity). Both types of radiation can be modeled in terms of electromagnetic waves, but this is not a very useful model for the ionizing type of radiation. Ionizing radiation causes cellular damage (e.g., disruption of atomic bonds) by what is best modeled as particle interaction. Microwaves produce molecular vibration, which in turn produces heat. Evidence to date is that ionizing radiation is *much* more damaging than microwaves. It should be noted that in many cases, microwaves are generated by high-voltage devices (see, e.g., Chapter 9), and these present a possible X-ray hazard, but this is a separate problem from that of microwave hazards.

2. Microwave wavelengths are of the same order of magnitude as dimensions of the human body. This leads to "close coupling" in many cases between the body and the microwave field, with consequent large absorption of energy.
3. Biological damage seems to be related directly to the electric field of the microwave, direct interaction with the magnetic field being negligible. However, this is not necessarily true for all radio frequencies, there being some evidence to suggest that the magnetic fields of very low frequency electromagnetic waves do interact with the human system.
4. Microwave effects do not appear to be cumulative, except of course where permanent damage has been caused by previous exposures.
5. Microwave effects are nearly always reversible (i.e., when the microwave radiation is removed, the system returns to normal). However, see Section 15.4 for a counterexample.

15.2 RADIATION LEVELS

Power density is the most widely used measure of microwave radiation level. In some situations, power density is *not* the most suitable measure, but before discussing these in detail, some typical power density levels are given below in Table 15.1 to indicate the order of magnitude involved.

These are average power density levels. Note that whereas sunshine may produce a surface heat sensation (and maybe a tan), the microwave heating may be experienced under the skin—and no tan.

For a plane, sinusoidal, TEM wave in free space, the average power density is given by any of the equivalent expressions:

$$
\begin{aligned}
P_D &= \frac{E^2}{377} \qquad \text{W/m}^2 \\
&= H^2 \times 377 \\
&= EH
\end{aligned}
\tag{15.1}
$$

Here, E is the rms electric field strength (V/m), H is the rms magnetic field strength (A/m), and 377 Ω is the numerical value of wave impedance in free space. E (and less frequently, H) are the quantities measured in practice, and P_D is calculated from the expressions above. Unfortunately, most hazardous fields in practice are more complicated than the single, plane, sinusoidal TEM wave, so the relationships given by Eq. (15.1) may not hold true. For example, a measuring instrument (see Section 14.8) may measure E of a complicated field but be calibrated for P_D from (Eq. 15.1). This may be termed "the equivalent plane wave power density" and it is assumed that maximum permissible levels can be specified in this way even when the exact nature of the field is not known.

TABLE 15.1

Typical Power Density Levels	**(mW/cm²)**
Sunshine on a warm day	100
At the antenna of a trailer unit for 12-GHz satellite communications (3-m-diameter antenna) (Day 1980)	32
Microwave oven, maximum permissible level, at 5 cm	1
Warmth sensation threshold, 1-s exposure at 3000 MHz (Michaelson 1978)	58.6
Threshold of pain, 20 s at 3000 MHz (Michaelson 1978)	3.1

Some of the factors that can give rise to a complicated field are:

1. *Standing waves:* Reflected waves from nearby objects and surfaces can combine with the incident wave to produce standing waves in the manner described in Section 1.12. Such reflections can occur from the person in the field, and from the measuring probe, thus upsetting the actual quantity that one is attempting to measure. With standing waves the electric field strength can be high (at one of the maxima) while in the same region the average power density can be low (or zero). It would be important in this case that the instrument respond to electric field strength.
2. *Near fields:* The theory leading to Eq. (15.1) considers the *radiated* component of field only. Near fields exist which alternately store and return energy to the source (i.e., these are reactive fields similar to the electric field of a capacitor and the magnetic field around an inductance). As with standing waves, it is possible for the electric field component to be high even where the radiated power density is low or zero. Near fields are probably more of a hazard at low radio frequencies, where they extend further in space around the antenna, hand-held CB radio transmitters being an example of where the operator is in close proximity to the antenna.
3. *Polarization* (see Section 10.6): In many cases the waves making up the total field will have differing polarizations, and the resultant polarization may not therefore be fixed. For example, leakage from vertical and horizontal cracks in a microwave oven can combine to produce complex polarization. It is important, therefore, that the measuring probe be designed to be independent of orientation, as discussed in Section 14.8.
4. *Modulation:* Amplitude modulation may occur in various ways. With pulsed radar, very high peak fields of short duration can occur, too short for the measuring instrument to respond to. Rotating antennas will also result in effective amplitude modulation at a fixed location, as will the stirrers in microwave ovens. The frequency response of the measuring probe should cover the spectrum of the modulated wave, as discussed in Section 14.8.

As a result of these complications, maximum permissible levels in protection standards are more involved than simply specifying average power density. Different authorities have different ideas on how to go about specifying safe levels. Tables 15.2 and 15.3 show the exposure stand-

TABLE 15.2
USSR, Polish and Czechoslovakian Exposure Standards

Standard	*Type*	*Frequency*	*Exposure limit*	*Exposure duration*	*CW/ pulsed*	*Antenna Stationary/ Rotating*	*Remarks*
USSR Government 1977	Occupa-tional	10–30 MHz	20 V/m	working day	both	both	Military units and establishments of the Ministry of Defence excluded
		30–50 MHz	10 V/m	working day	both	both	
			0.3 A/m	working day	both	both	
		50–300 MHz	5 V/m	working day	both	both	
		0.3–300 GHz	10 $\mu W/cm^2$	working day	both	stationary	
			100 $\mu W/cm^2$	working day	both	rotating	
			100 $\mu W/cm^2$	2 hours	both	stationary	
			1 mW/cm^2	2 hours	both	rotating	
			1 mW/cm^2	20 min.	both	stationary	
USSR Government 1970	General public	0.3–300 GHz	1 $\mu W/cm^2$	24 hours	both	both	

Czechoslovakia Government 1970	Occupational	10–30 MHz	50 V/m	working day	both	both	
		30–300 MHz	10 V/m	working day	both	both	
		0.3–300 GMz	25 μW/cm²	working day	CW	both	
			10 μW/cm²	working day	pulsed	both	max peak 1 kW/cm²
			1.6 mW/cm²	1 hour	CW	both	
			0.64 mW/cm²	1 hour	pulsed	both	
	General public	30–300 MHz	1 V/m	24 hours	both	both	
		0.3–300 GMz	2.5 μW/cm²	24 hours	CW	both	
			1 μW/cm²	24 hours	pulsed	both	
		30–300 MHz	1 V/m	24 hours	both	both	
		10–30 MHz	2.5 V/m	24 hours	both	both	
Poland Government 1972	Occupational	0.3–300 GHz	0.2 mW/cm²	10 hours	both	stationary	
			0.2–10 mW/cm²	32/P² (hours)	both	stationary	P-power density in W/m²
			1 mW/cm²	10 hours	both	rotating	
			1–10 mW/cm²	800/P² (hours)	both	rotating	P-power density in W/m²
	General public	0.3–300 GHz	10 μW/cm²	24 hours	both	stationary	
			0.1 mW/cm²	24 hours	both	stationary	
Poland Government 1975 proposed	Occupational	10–300 MHz	20 V/m	working day	both	both	
			20–300 V/m	3200/E² (hours)	both	both	E-electric field intensity in V/m
	General public	10–300 MHz	7 V/m	24 hours	both	both	

Source: Health and Welfare, Canada, 1978.

TABLE 15.3
United States, Canadian, West European Exposure Standards

Standard	*Type*	*Frequency*	*Exposure limit*	*Exposure duration*	*CW/pulsed*	*Antenna Stationary/ Rotating*	*Remarks*
U.S. ANSI 1974	Occupational	10 MHz–100 GHz	10 mW/cm^2 200 V/m 0.5 A/m	no limit	CW	both	
			1 mWhr/cm^2	0.1 hour	pulsed	both	
U.S. Army and Air Force 1965	Occupational	10 MHz–300 GHz	10 mW/cm^2	no limit	both	both	
			10–100 mW/cm^2	6000/X^2 (min.)	both	both	X-power density in mW/cm^2
U.S. Indust. Hygienist 1971	Occupational	100 MHz–100 GHz	10 mW/cm^2	8 hours	both	both	
			25 mW/cm^2	10 min.	both	both	
Canada Can. Standards Assoc. 1966	Occupational	10 MHz–100 GHz	10 mW/cm^2	no limit	CW	both	
			1 mWhr/cm^2	0.1 hour	pulsed	both	
Canada H&W proposed	Occupational	10 MHz–1 GHz	1 mW/cm^2	no limit	both	both	
		1–300 GHz	5 mW/cm^2	no limit	both	both	
		10 MHz–300 GHz	25 mW/cm^2	1 min.	both	both	
	General public	10 MHz–300 GHz	1 mW/cm^2	no limit	both	both	
Sweden Worker Prot. Authority 1976	Occupational	0.3–300 GHz	1 mW/cm^2	8 hours	both	both	
		10–300 MHz	5 mW/cm^2	8 hours	both	both	
		0.3–300 GHz	1–25 mW/cm^2	60/X (min).	both	both	X-power density in mW/m^2
		10 MHz–300 GHz	25 mW/cm^2	any	CW, pulsed averaged over 1 sec.	both	

Source: Health and Welfare, Canada, 1978.

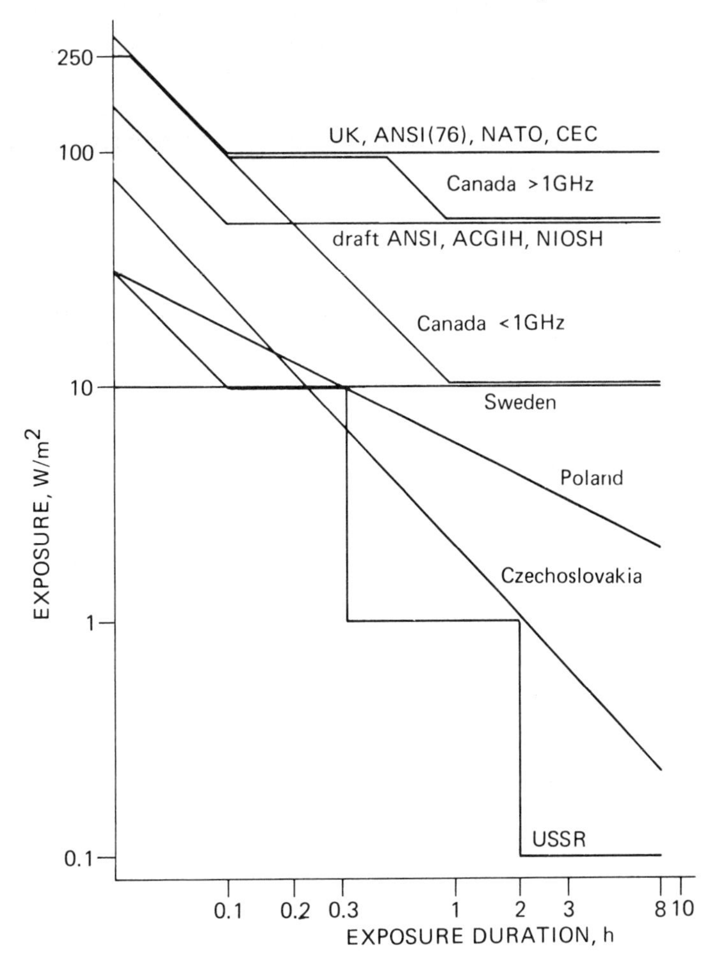

FIGURE 15.1. Comparison of whole-body exposure standards for nonionizing radiation. *(From Microwave Datamate, 1982, courtesy of Marconi Instruments Ltd., Stevenage, Hertfordshire, England.)*

ards in force in various countries, (Health and Welfare Canada, 1978). These standards are discussed extensively in the Health and Welfare Canada reference cited. Some of the standards are summarized in graphical form in Figure 15.1 (Marconi 1982).

15.3 MICROWAVE–BIOLOGICAL INTERACTIONS

Computer modeling, phantom modeling, and laboratory tests are all ways in which microwave bioeffects have been extensively studied.

Computer modeling. Here, the wave equations for quite complicated situations are solved to yield such parameters as *penetration depth* and *specific absorption rate* (SAR). Penetration depth (also known as skin depth) is the depth at which the electric field has dropped to e^{-1} of its boundary value (on the assumption of a plane TEM wave incident at the boundary). The electric field in these circumstances decays exponentially with distance, being given by

$$E_x = E_0 e^{-x/D} \tag{15.2}$$

where E_x is the field strength at position x within the medium, E_0 the field strength at the boundary ($x = 0$), and D, the penetration depth. From Eq. (15.2), when $x = D$ (the penetration depth), $E_x = E_0 e^{-1}$.

Penetration depth is a function of frequency and Fig. 15.2 shows the results for some biological materials (after Lin 1978). It will bc seen that the penetration for fat is much greater than for either muscle or blood over most of the microwave frequency range. This is because both the dielectric constant and conductivity for fat are less than those for the other substances. In particular, the higher conductivities of muscle and blood lead to higher power dissipation in these substances, and the wave is more

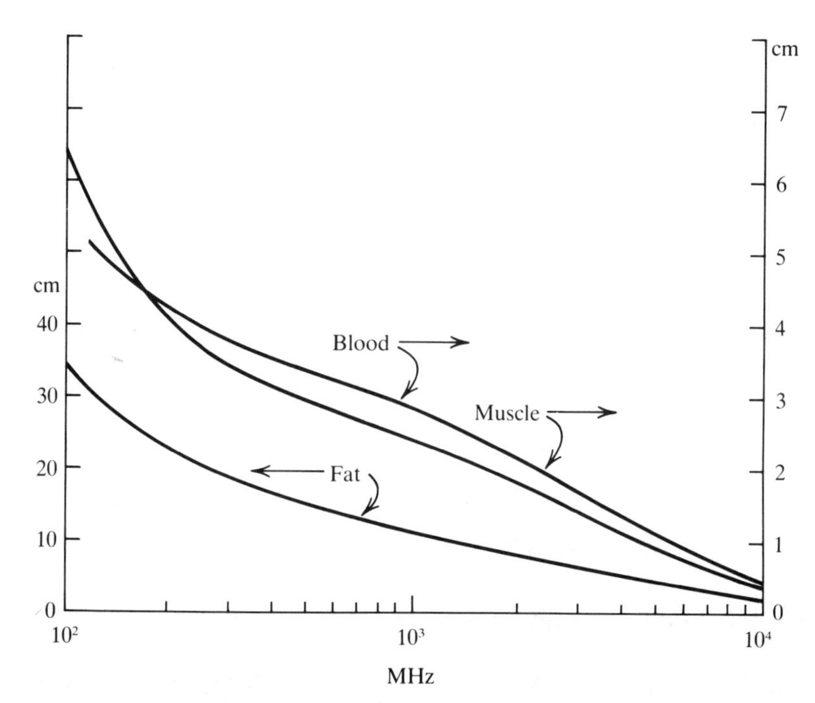

FIGURE 15.2. Depth of penetration in blood, muscle, and fatty tissue as a function of frequency. (*After Lin, 1978.*)

rapidly attenuated with increasing depth of penetration. It also follows that the rate of heating will be greater in these substances.

Specific absorption rate (SAR) is the power absorbed per unit mass of substance. In terms of the electric field within the biological material, the SAR is given by

$$\text{SAR} = \frac{\sigma E^2}{d} \quad \text{W/kg} \tag{15.3}$$

Here, E(V/m) is the rms electric field within the material; σ(S/m) is the conductivity of the material; and d(kg/m^3) is the mass density of the material.

Computation of SAR shows up two important effects. The first of these is the formation of "hot spots," which result from the focusing action of high-dielectric-constant materials (see, e.g., Section 10.11.1), and also from the shape of the cavities involved. Computer modeling predicts that hot spots should occur within the human skull at frequencies in the region of 918 to 2450 MHz, and within the eyeball at frequencies around 1500 MHz.

The second important effect is the significance of size scaling. It is found that variations in body size and orientation are highly significant, in relation to frequency, in determining SAR. Great care must therefore be taken in trying to relate to human beings experimental results obtained with small laboratory animals. For example, computed results show that at 700 MHz, the average SAR for rats peaks at 0.8 W/kg, while the value for human beings is about 0.03 W/kg, for an incident power density of 1 mW/cm^2. At 70 MHz the average SAR for human beings peaks at 0.25 W/kg, while for rats it is 0.0125 W/kg at this frequency (Lin 1978). It is interesting to note that a height of 1.75 m was used for the human model, which corresponds to about 0.4λ, where λ is the free-space wavelength at 70 MHz (i.e., this approximates a $\frac{1}{2}\lambda$ receptor).

Phantom modeling consists of making a physical model out of synthetic materials which have electrical and heat properties similar to the biological material being studied (Lin 1978). For example, for brain matter, a composition is used which consists of saline solution, powdered polyethylene, and a jelling agent known as "super stuff." The phantom model shell, made out of polyfoam to the appropriate shape, is filled with the synthetic material and exposed to the desired radiation for a short known period of time. The model is then opened (it is made in two half-sections) and the temperature pattern over the internal surfaces obtained by means of a thermographic camera. Now knowing the temperature rise ΔT(°C) for exposure time Δt(s), the SAR can be calculated from

$$\text{SAR} = 4185c \frac{\Delta T}{\Delta t} \quad \text{(W/kg)} \tag{15.4}$$

Here, the constant 4185 J/kcal is the mechanical equivalent of heat, and c is the specific heat of the substance in kcal/kg·°C. Measurements based on the phantom model method show very good agreement with results obtained from computer modeling.

Small laboratory animals such as mice, rats, dogs, and cats have been experimented on in an attempt to determine the effects and the mechanisms of microwave–biological interactions. The literature is extensive and discussed comprehensively in Lin (1978), Michaelson (1978), Health and Welfare Canada (1978). Some of these results are given in the next section.

15.4 INTERACTIVE MECHANISMS

These are generally classified as thermal, microthermal, and uncertain.

Thermal. The energy from the microwave field is absorbed by the biological material and converted to heat. The differential equation relating rate of change of temperature per unit mass to net heat is

$$\frac{dT}{dt} = \frac{Q}{c} \quad (°\text{C/s}) \tag{15.5}$$

where $Q = \text{SAR} + W_M - W_L$. W_M is the metabolic rate of heat production per unit mass, and W_L the rate of heat loss per unit mass, both in W/kg. SAR and c are as given in Eq. (15.4)

In a living animal, Q will be a function of time even though SAR is constant. For high SAR [when SAR is the dominant term in Eq. (15.5)] the temperature will initially increase rapidly over a period of minutes. The thermoregulatory system of the body will then function so as to stabilize the temperature, and possibly reduce it through vasodilation. However, if the system cannot remove the excess heat as fast as it is produced, the body temperature will start to increase once again, leading to collapse through hyperthermia. Figure 15.3 shows the temperature variation in a dog subjected to steady microwave radiation, which clearly illustrates these points (Michaelson et al. 1967).

Low-level radiation [SAR not the dominant term in Eq. (15.5)], can also upset bodily behavior through thermal effects. The metabolic rate of heat production may automatically reduce to compensate for the additional SAR component, with the result that the subject appears to suffer from general lethargy and loss of appetite.

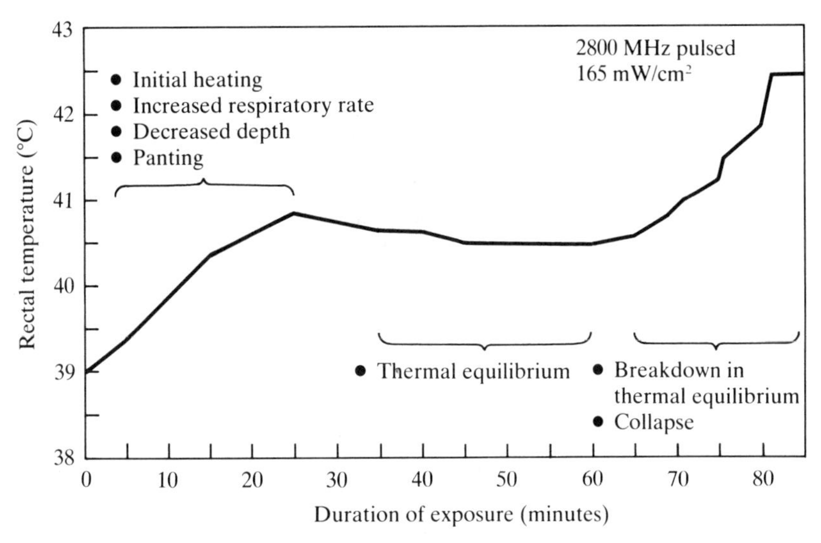

FIGURE 15.3. The triphasic response of the rectal temperature of dogs exposed to 2800 MHz pulsed microwaves at 165 mW/cm². Initial heating is followed by a period of thermal equilibrium and finally, if the exposure is continued, by thermoregulatory collapse.

Microthermal mechanism. Here the temperature rise is very small, being of the order of a few microdegrees Celsius, but taking place over a short period of time (a few microseconds). One peculiar effect which is thought to be attributable to this cause is that of "microwave hearing." Some people claim they can hear "clicks and buzzes" when exposed to pulsed microwaves. Of course, the auditory system responds to sound pressure waves, not to electromagnetic waves, and the upper-frequency limit of hearing is of the order of a few tens of kilohertz for normal individuals. Evidence suggests that a microthermal mechanism occurs in the brain matter, the resulting thermal stress sets up a pressure wave which reaches the cochlea through bone conduction.

One of the major problems encountered in studying microwave–biological interactions is that biological behavior in many cases is extremely temperature sensitive and it becomes difficult, if not impossible, to separate out thermal effects from others (possibly direct electromagnetic wave interaction with the biological material). The range of effects studied, including those known to be thermal or microthermal, is summarized in Table 15.4 and these are discussed in considerable detail in Michaelson (1978). The effects of microwaves on the blood, which comes under the heading of *Hematopoiesis*, deserves special mention. Small animals exposed to relatively low level radiation showed certain blood changes, and the blood had not returned to normal some months after the tests had been stopped. Blood changes were also observed in large groups of workers occupationally exposed to low- and medium-level radiation.

TABLE 15.4
Exposure Effects of Microwaves

Cellular effects
Neuroendocrine responses
Effects on nervous system
Cardiovascular effects
Hematopoiesis
Auditory response
Cataract
Implanted electronic cardiac pacemaker interference

In a similar connection, Repacholi (1978) reported that microwave blood warmers (used to warm the blood prior to transfusion) spoiled the blood (hemolysis) through overheating or heating too fast.

Regarding the cardiac pacemakers (the last item in Table 15.4), the *asynchronous* type of pacemaker, which is clocked from a fixed-frequency oscillator, is relatively insensitive to electromagnetic interference. Noncompetitive types (meaning that they do not compete with the human system), which are triggered by some well-defined aspect of the circulatory system, are very susceptible to all kinds of electromagnetic interference, not just microwaves, if not properly shielded. Units now manufactured are well shielded and as an additional precaution they are designed to transfer to asynchronous mode if the triggered mode is interrupted by electromagnetic interference.

15.5 HAZARDOUS SITUATIONS

Microwaves are being increasingly used in situations where persons other than technical personnel engaged in using microwaves are being exposed to such radiation. Two examples which best illustrate these new developments are (1) the use of microwave ovens in many households, and (2) the use of transportable ground stations for two-way communications via satellites.

Regulations governing microwave ovens are issued by responsible government authorities. In Canada this is the Radiation Protection Bureau of the Department of Health and Welfare. The Canadian regulations governing all types of microwave ovens—domestic, commercial, and industrial—state that the radiation leakage at a distance of 5 cm from the outer surface must not exceed 1 mW/cm^2 with a minimum operating load, and 5 mW/cm^2 unloaded. Two independent interlocks are required which will switch off the microwave power when the oven door is opened, and an interlock monitor is also required which will disable the oven should the door interlocks fail.

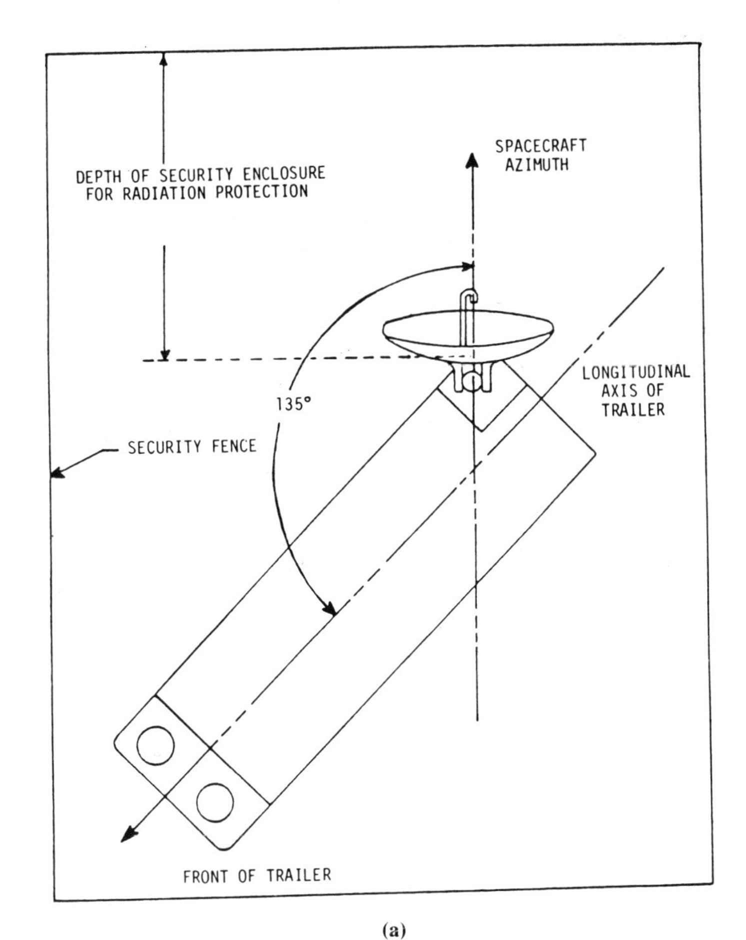

(a)

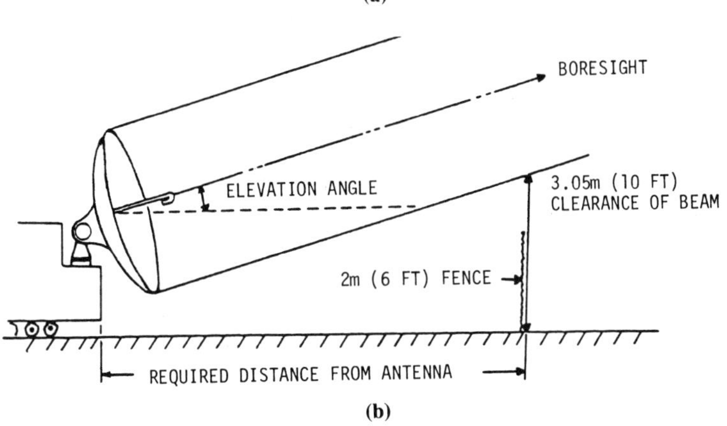

(b)

FIGURE 15.4. TV terminal orientation on site locations: (a) plan view; (b) side elevation. (*From J. W. B. Day, Anik-B DOC Communications System and Ground Terminals, Communications Research Centre, Department of Communications, Ottawa, Canada, 1980; with permission.*)

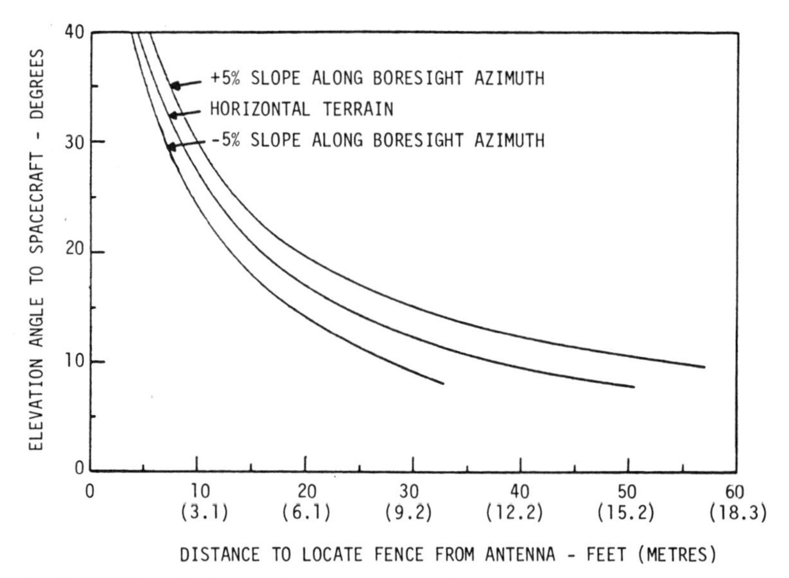

FIGURE 15.5. Depth of safety enclosure as a function of elevation angle in site location shown in Fig. 15.4 (*From Day* op. cit.)

In a survey [Stuchly, Repacholi, and Lecuyer (1978)], it was found that since the issuance of regulations, oven safety standards had improved noticeably. Two relatively weak areas observed were (1) lack of microwave radiation monitors in repair shops, and (2) user neglect leading to faulty door hinges and dirty door seals, leading to increased radiation levels. A good illustration of the radiation intensity from a typical microwave oven will be found in Thomsit (1979).

With transportable microwave transmit/receive stations, it is of course necessary to prevent people from entering into the beam of the transmit antenna, and generally they should be kept well clear of the antenna. The problem is that these particular units are intended mainly for use in remote communities, where proper supervision of security arrangements may not be possible. Figure 15.4 shows a typical TV terminal layout and Fig. 15.5 the dimensions of the enclosure in relation to angle of elevation of antenna, required by certain of the TVT trailer units in use with the Anik-B satellite communications system (Day 1980).

15.6 PROBLEMS

1. Distinguish between ionizing and nonionizing radiation. Into which category does microwave radiation fall?

2. Which field component, electric or magnetic, of microwave radiation interacts with biological materials?
3. In Table 15.2, the USSR exposure standard for the frequency range 50 to 300 MHz is 5 V/m. Calculate the corresponding power density. How does this compare with the power density specified for frequencies greater than 300 MHz in the same table?
4. Explain what is meant by penetration depth. Using Eq. (15.2) and Fig. 15.2, determine the electric field strength, normalized to the surface value, at a depth of 2 cm for fat, blood, and muscle.
5. Explain what is meant by the specific absorption rate. A certain biological substance has a conductivity of 150 S/m and a mass density of 1.3×10^3 kg/m^3. Calculate the specific absorption rate for an electric field strength of 5 V/m.
6. The specific heat of the substance in Problem 5 is 1.19 kcal/kg·C. Calculate the rate of temperature rise in °C per second.
7. In a selected region in an experimental animal under exposure to microwave radiation, the rate of metabolic heat production is 0.2 W/kg, the rate of heat loss is 0.18 W/kg, the SAR is 0.08 W/kg, and the specific heat is 0.8 kcal/kg·C. Calculate the rate of change of temperature in the region in °C per second.

REFERENCES

BOWMAN, R. R., 1978. "Quantifying hazardous microwave fields," *Trans. Int. Microwave Power Inst.*, Vol. 8.

DAY, J. W. B., 1980. Department of Communications, Canada, Anik-B DOC Communications Systems and Ground Terminals.

HEALTH AND WELFARE, CANADA, 1978. *Health Aspects of Radio Frequency and Microwave Exposure.*

LIN, JAMES C., 1978. "Microwave biophysics," *Trans. Int. Microwave Power Inst.*, Vol. 8.

MARCONI INSTRUMENTS LTD., 1982. *Microwave Datamate.*

MICHAELSON, S. M., R. A. E. THOMSON and J. W. HOWLAND, 1967. "Biologic effects of microwave exposure," RADC-TR-67-461, 138 pp.

MICHAELSON, S. M., 1978. "Biologic and pathophysiologic effects of exposure to microwaves," *Trans. Int. Microwave Power Inst.*, Vol. 8.

REPACHOLI, M. H., 1978. "Control of microwave exposure in Canada," *Trans. Int. Microwave Power Inst.*, Vol. 8.

STUCHLY, M. A., and M. H. REPACHOLI, 1978. "Microwave and radio frequency protection standards," *Trans. Int. Microwave Power Inst.*, Vol. 8.

STUCHLY, M. A., M. H. REPACHOLI, and D. LECUYER, 1978. "Regulatory control of microwave ovens in Canada," *IMPI Digest*, Microwave Power Symposium, Ottawa.

THOMSIT, D. I., 1979. "Colour display of microwave fields," *Electronics & Power*, June.

appendix A
Answers to Odd Numbered Problems

Chapter 1

1. $LC = \mu\varepsilon$
3. 60 cm
5. (a) 0.051 N/m, (b) 1.711
7. (a) 1.836, (b) 28.38 rad/m, (c) 0.0661 N/m
9. (a) 1.566, (b) 262 rad/m, (c) 0.0831 N/m, (d) 49.86 W/m
11. (a) 0.571 V, (b) 0.429 V, (c) −0.75
13. $0.732 \angle -60.4^\circ$
15. $115.3 + j133.7\ \Omega$
17. $41.44 \angle 35.5^\circ\ \Omega$
19. (a) $j36.33\ \Omega$, (b) 11.56 nH;
21. $-j68.82\ \Omega$
23. $j0.01453$ S

Chapter 2

3. *V volts*	*I amps*	*W watts.*
(a) $29.54 + j5.21$	$1.879 - j0.684$	51.96
(b) $5 + j8.66$	$2.598 - j1.5$	0
(c) $100 + j0$	$1.813 + j0.845$	181.26

5. $V = 49.5 + j14.14$ V ($51.48 \angle 15.95^\circ$V); $I = 0.424 - j1.13$ A ($1.208 \angle -69.44^\circ$A)

7. (a) 12.98W, (b) 2450 W

9. (a) $a_n = 16.5 - j8$ $(18.34\angle -25.87°)$, $b_n = 13.5 - j12$ $(18.06\angle -41.63°$; (b) $a_n = 44.17 + j27.35$ $(51.94\angle 31.77°)$, $b_n = 42.45 + j22.65$ $(48.11\angle 28.09°)$

11. (a) $-0.2 - j0.4$ or $0.447\angle 63.43°$
(b) $0.6896 - j1.724$ or $1.857\angle -68.2°$
(c) $0.459 - j0.547$ or $0.714\angle -50°$
(d) $0 + j0.066$ or $0.066\angle 90°$

13. (a) $0.31 - j0.276$ or $0.415\angle -41.63°$
(b) $0.2 - j0.4$ or $0.447\angle -63.43°$
(c) $0.372 + j0.3188$ or $0.49\angle 40.6°$
(d) $-0.333 + j0$ or $0.333\angle 0$
(e) $-0.9802 + j0.198$ or $1\angle -11.42°$

15. $\Gamma_s = 0.4 + j0.2$ or $0.447\angle 26.57°$; $b_s = (4.24 - j1.414)10^{-4}$ or 4.47×10^{-4} $\angle -18.44°$

19. (a) $a_1 = 2.12 \times 10^{-4}$ $a_2 = 0$ $b_1 = 10^{-4}(-0.426 + j1.39)$ $b_2 = 10^{-4} \times (1.88 - j3.127)$
(b) $a_1 = 2.12 \times 10^{-4}$ $a_2 = 10^{-4}(0.6096 - j0.905)$ $b_1 = 10^{-4}(-0.517 + j1.31)$ $b_2 = 10^{-4}(1.83 - j2.71)$

21. For 19(a) $A_v = -0.2436 - j1.645$ or $1.66\angle 81.58°$
For 19(b) $A_v = -0.1927 - j2.099$ or $2.108\angle 84.75°$
For 19(c) $A_v = -0.124 - j2.426$ or $2.43\angle 87.07°$
For 20(a) $A_v = -2.605 + j0.714$ or $2.7\angle -15.33°$
For 20(b) $A_v = -3.586 - j2.265$ or $4.241\angle 32.28°$
For 20(c) $A_v = -2.118 + j0.7817$ or $2.258\angle -20.27°$

23. GT values for problems 19 and 20 are:
19(a) 2.958, (b) 2.359, (c) 1.567
20(a) 21.02, (b) 18.89, (c) 12.14

25. For the conditions set in problem 19, $\Gamma_{22} = s_{22} = 0.381\angle 153°$
For the conditions set in problem 20:
(a) $0.302 + j0.0171$ or $0.303\angle 3.23°$
(b) $0.518 - j0.209$ or $0.559\angle -22°$
(c) $0.0987 - j0.801$ or $0.807\angle -82.98°$

Chapter 3

1. (a) $0.34\angle 0$, (b) $0.45\angle -64°$, (c) $1\angle 67.5°$

3. (a) $0.01 + j0.03$ S, (b) $7.07 - j7.07$ mS, (c) $j20$ mS

5. (a) $55 - j45$ Ω, (b) $11 + j9$ mS, (c) 90.9 Ω, $-j111.1$ Ω.

7. $155 + j69$ Ω

9. (a) $5.25 - j9$ Ω, (b) $0.046 + j0.084$ S, (c) $-j11.9$ Ω.

11. 0.098 λ; 0.145 λ; a short circuit.

13. 0.313 λ; 0.089 λ; a short circuit.
15. 0.169 λ; 0.367 λ.

Chapter 4

1. 3.16 cm; 3 cm; 9.56 cm; 956 Mm/s; 94.1 Mm/s
5. 1.36
9. $P_1 = P_3 = -2.5$ mW (negative sign shows power flow is out of the ports).
15. $P_2 = 4.5$ mW; $P_3 = 0.5$ μW; $P_4 = 0.5$ mW.
17. $P_2 = 0.912$ mW; $P_3 = 0.1$ μW; $P_4 = 10$ μW; −10.6 dBm.
19. 625.5 MHz; 2 GHz.
21. $f < 6.309$ GHz; TE_{10}.

Chapter 5

1. 119.2 Ω
3. 0.121 in.
5. (a) 23.86 Ω; (b) 6.89; (c) 3.81 cm.
7. (a) 51.42 Ω; (b) 6.05; (c) 4.07 cm.
9. 0.405; 5.707
11. 69.37 Ω; 36.04 Ω.

Chapter 6

1. 28.43 kA/m
3. 3.938 GHz.
5. (a) 0.3517 GHz; (b) 10.
7. (a) 0.3 T; (b) 5.57 kA/m; (c) 2500 G; (d) 4272.6 Oe.
11. 0.7741; 1.191; 0.938
13. 1.45 cm.
17. 9.3 GHz.

Chapter 7

7. (a) 2.515; (b) 6.326 or 8.01 dB.
9. Idler, 6 GHz; Output, 9 GHz.
11. The break frequency is 7.96 MHz. Allowing a factor of 10:1, $f > 80$ MHz.
13. 2 + j8860 μS.
15. (a) 0.26 dB; (b) 4 nH; (c) 0 dB.

23. 0.4717 mA.

25. 12 mV/μW

27. The −40 dBm diode.

33. 6 dB.

41. 11.6 dB.

Chapter 8

3. 7.34 dB.

5. 1.18 dB

7. $0.67\angle -143°$; $0.66\angle 142°$; 10.94 − j16 Ω; 11.4 + j16.4 Ω.

9. 16.65 − j31.09 Ω; 9.5 + j1.68 Ω; 8.05 dB.

11. For the input network, Z_{01} = 30.99 Ω, Z_{02} = 108.8 Ω; For the output network, Z'_{01} = 84.9 Ω, Z'_{02} − 41.25 Ω.

13. For the output network, Z'_{01} = 64.53 Ω, Z'_{02} = 54.82 Ω.

15. Input network, Z_{01} = 26.48 Ω, Z_{02} = 19.45 Ω; Output network, Z'_{01} = 12.48 Ω, Z'_{02} = 26.16 Ω.

21. $F = 2$.

23. 0.04 pW.

25. 4 dB.

31. 6.76 dB.

37. (a) 2610 K; (b) 1450 K; (c) 290 K.

43. 22.5 dBm.

Chapter 9

5. $1.04u_o$; $0.96u_o$

9. 1.76×10^{16}m/s^2

Chapter 10

1. (a) 0.87; (b) 0.625 kW.

3. (a) $\lambda^2/4\mu$; (b) 4.8 sq.cm

5. 7.5 dB.

7. 0.9063

11. (a) 468 sq.cm; (b) 19.1 cm.

13. 73.46 + j41.28 Ω; 85.26 + j74.45 Ω; 63.03 + j8.61 Ω.

15. 4.86 sq.cm

17. 4.85 dB

19. 39.4 dB

21. 50.3 dB

25. (a) 563.7 Ω; (b) 375.9 − *j*204 Ω; 426.7 + *j*311.7 Ω.

Chapter 11

1. (a) Carrier frequencies in MHz are:
4037.5, 4066.5, 4095.5, 4124.5, 4153.4, 4182.5, 3969.5, 3940.5, 3911.5, 3882.5, 3853.5, 3842.5
(b) 29 mHz; (c) 19.5 MHz

5. 36.88 km

Chapter 12

1. (a) 3.18 mW/sq.m; (b) 3.18 mW

3. 7.26 pW; 1.0 pW

5. (a) 700 km; (b) 12 dB

7. (a) 42.5 dB; (b) 17.783 km; (c) 9.6 n.mi

11. (a) 0.03 MW; (b) 6 MW

13. 1666.7 Hz

15. (a) 0.573 deg; (b) 100 m

Chapter 13

3. (a) 0.3322; (b) 3.32 S/m; (c) 26.9 mW/sq.m

Chapter 14

5. 0.631

9. $0.512\angle 14.36^\circ$; (b) $144\angle 19^\circ$.

13. 0.94

21. (a) 2.61 dB; (b) 239 K

Chapter 15

3. 6.63 μW/sq.cm

5. 2.88 W/kg

7. 0.055 °C/s

INDEX